Undergraduate Texts in Mathematics

Undergraduate Texts in Mathematics

Undergraduate Texts in Mathematics are generally aimed at third- and fourth-year undergraduate mathematics students at North American universities. These texts strive to provide students and teachers with new perspectives and novel approaches. The books include motivation that guides the reader to an appreciation of interrelations among different aspects of the subject. They feature examples that illustrate key concepts as well as exercises that strengthen understanding.

More information about this series at http://www.springer.com/series/666

Miklós Laczkovich · Vera T. Sós

Real Analysis

Series, Functions of Several Variables, and Applications

Springer

Miklós Laczkovich
Department of Analysis
Eötvös Loránd University—ELTE
Budapest
Hungary

Vera T. Sós
Alfréd Rényi Institute of Mathematics
Hungarian Academy of Sciences
Budapest
Hungary

Translated by Gergely Bálint

ISSN 0172-6056 ISSN 2197-5604 (electronic)
Undergraduate Texts in Mathematics
ISBN 978-1-4939-8464-0 ISBN 978-1-4939-7369-9 (eBook)
https://doi.org/10.1007/978-1-4939-7369-9

Mathematics Subject Classification (2010): MSC2601, MSC26BXX, MSC2801, MSC28AXX

Softcover reprint of the hardcover 1st edition 2017
Translation from the second Hungarian language edition: *Valós Analízis II* by Miklós Laczkovich and Vera T. Sós, © Laczkovich Miklós, T. Sós Vera, Typotex, 2013
1st edition: *Analízis II* by Laczkovich Miklós and T. Sós Vera, © Laczkovich Miklós, T. Sós Vera, Nemzeti Tankönykiadó Zrt., Budapest, 2007

Printed on acid-free paper

This Springer imprint is published by Springer Nature
The registered company is Springer Science+Business Media LLC
The registered company address is: 233 Spring Street, New York, NY 10013, U.S.A.

Preface

Analysis forms an essential basis of both mathematics and statistics, as well as most of the natural sciences. Moreover, and to an ever increasing extent, mathematics has been used to underpin our understanding of the social sciences. It was Galileo's insight that "Nature's great book is written in the language of mathematics." And it is the theory of analysis (specifically, differentiation and integration) that was created for the express purpose of describing the universe in the language of mathematics. Working out the precise mathematical theory took almost 300 years, with a large portion of this time devoted to creating definitions that encapsulate the essence of limit and continuity. This task was neither easy nor self-evident.

In postsecondary education, analysis is a foundational requirement whenever mathematics is an integral component of a degree program. Mastering the concepts of analysis can be a difficult process. This is one of the reasons why introductory analysis courses and textbooks introduce the material at many different levels and employ various methods of presenting the main ideas. This book is not meant to be a first course in analysis, for we assume that the reader already knows the fundamental definitions and basic results of one-variable analysis, as is discussed, for example, in [7]. In most of the cases we present the necessary definitions and theorems of one-variable analysis, and refer to the volume [7], where a detailed discussion of the relevant material can be found.

In this volume we discuss the differentiation and integration of functions of several variables, infinite numerical series, and sequences and series of functions. We place strong emphasis on presenting applications and interpretations of the results, both in mathematics itself, like the notion and computation of arc length, area, and volume, and in physics, like the flow of fluids. In several cases, the applications or interpretations serve as motivation for formulating relevant mathematical definitions and insights. In Chapter 8 we present applications of analysis in apparently distant fields of mathematics.

It is important to see that although the classical theory of analysis is now more than 100 years old, the results discussed here still inspire active research in a broad spectrum of scientific areas. Due to the nature of the book we cannot delve into such matters with any depth; we shall mention only a small handful of unsolved problems.

Many of the definitions, statements, and arguments of single-variable analysis can be generalized to functions of several variables in a straightforward manner, and we occasionally omit the proof of a theorem that can be obtained by repeating the analogous one-variable proof. In general, however, the study of functions of several variables is considerably richer than simple generalizations of one-variable theorems. In the realm of functions of several variables, new phenomena and new problems arise, and the investigations often lead to other branches of mathematics, such as differential geometry, topology, and measure theory. Our intent is to present the relevant definitions, theorems, and their proofs in full detail. However, in some cases the seemingly intuitively obvious facts about higher-dimensional geometry and functions of several variables prove remarkably difficult to prove in full generality. When this occurs (for example, in Chapter 5, during the discussion of the so-called integral theorems) with results that are too important for either the theory or its applications, we present the facts, but not the full proofs.

Our explicit intent is to present the material gradually, and to develop precision based on intuition with the help of well-designed examples. Mastering this material demands full student involvement, and to this end we have included about 600 exercises. Some of these are routine, but several of them are problems that call for an increasingly deep understanding of the methods and results discussed in the text. The most difficult exercises require going beyond the text to develop new ideas; these are marked by (∗). Hints and/or complete solutions are provided for many exercises, and these are indicated by (H) and (S), respectively.

Budapest, Hungary — Miklós Laczkovich
February 2017 — Vera T. Sós

Contents

Functions of Several Variables

Functions of several variables are needed in order to describe complex processes. A detailed meteorological relief map indicating the temperature as it changes during the day needs four variables: three coordinates of the place (longitude, latitude, altitude) and one coordinate of the time. The mathematical description of complex systems, e.g., the motion of gases or fluids, may need millions of variables.

If a system depends on p parameters, then we can describe a quantity determined by the system using a function that assigns the value of the quantity to the sequences of length p that characterize the state of the system.

We say that f is a **function of p variables** if every element of the domain of f is a sequence of length p. For example, if we assign to every date (year, month, day) the corresponding day of the week, then we obtain a function of three variables, for which f (2016, July, 18) = Monday.

In the sequel we will mainly consider functions that depend on sequences of *real* parameters.

Chapter 1
$\mathbb{R}^p \longrightarrow \mathbb{R}$ functions

1.1 Euclidean Spaces

In mathematical analysis, points of the plane are associated with ordered pairs of real numbers, and the plane itself is associated with the set $\mathbb{R} \times \mathbb{R} = \mathbb{R}^2$. We will proceed analogously in representing three-dimensional space. The coordinate system in three-dimensional space can be described as follows. We consider three lines in space intersecting at a point that are mutually perpendicular, which we call the x-, y-, and z-**axes.** We call the plane spanned by the x- and y-axes the xy-plane, and we have similar definitions for the xz- and yz-planes. We assign an ordered triple (a, b, c) to every point P in space, in which a, b, and c denote the distance (with positive or negative sign) of the point from the yz-, xz-, and xy-planes, respectively. We call the numbers a, b, and c the **coordinates** of P. The geometric properties of space imply that the map $P \mapsto (a, b, c)$ that we obtain in this way is a bijection. This justifies our representation of three-dimensional space by ordered triples of real numbers.

Thus if we want to deal with questions both in the plane and in space, we need to deal with sets that consist of ordered p-tuples of real numbers, where $p = 2$ or $p = 3$. We will see that the specific value of p does not usually play a role in the definitions and proofs that arise. Therefore, for every positive integer p we can define **p-dimensional Euclidean space**, by which we simply mean the set of all sequences of real numbers of length p, with the appropriately defined addition, multiplication by a constant, absolute value, and distance. If $p = 1$, then this Euclidean space is just the real line; if $p = 2$, then it is the plane; and if $p = 3$, then it is 3-dimensional space. For $p > 3$, p-dimensional space does not have an observable meaning, but it is very important for both theory and applications.

Definition 1.1. $\mathbb{R}^p$ denotes the set of ordered p-tuples of real numbers, that is,

$$\mathbb{R}^p = \{(x_1, \ldots, x_p) \colon x_1, \ldots, x_p \in \mathbb{R}\}.$$

M. Laczkovich and V.T. Sós, *Real Analysis*, Undergraduate Texts in Mathematics, https://doi.org/10.1007/978-1-4939-7369-9_1

The points of the set $\mathbb{R}^p$ are sometimes called *p-dimensional vectors*. The *sum of the vectors* $x = (x_1, \ldots, x_p)$ and $y = (y_1, \ldots, y_p)$ is the vector

$$x + y = (x_1 + y_1, \ldots, x_p + y_p),$$

and the *product of the vector x and a real number c* is the vector

$$c \cdot x = (cx_1, \ldots, cx_p).$$

The *absolute value* of the vector x is the nonnegative real number

$$|x| = \sqrt{x_1^2 + \cdots + x_p^2}.$$

(The absolute value of the vector x is also called the **norm** of the vector x. In order to be consistent with the usage of [7], we will use the term absolute value.)

It is clear that for all $x \in \mathbb{R}^p$ and $c \in \mathbb{R}$ we have $|cx| = |c| \cdot |x|$. It is also easy to see that if $x = (x_1, \ldots, x_p)$, then

$$|x| \le |x_1| + \cdots + |x_p|. \tag{1.1}$$

The **triangle inequality** also holds:

$$|x + y| \le |x| + |y| \qquad (x, y \in \mathbb{R}^p). \tag{1.2}$$

To prove this it suffices to show that $|x + y|^2 \le (|x| + |y|)^2$, since both sides are nonnegative. By the definition of the absolute value this is exactly

$$(x_1 + y_1)^2 + \cdots + (x_p + y_p)^2 \le$$
$$(x_1^2 + \cdots + x_n^2) + 2 \cdot \sqrt{x_1^2 + \cdots + x_p^2} \cdot \sqrt{y_1^2 + \cdots + y_p^2} + y_1^2 + \cdots + y_p^2,$$

that is,

$$x_1y_1 + \cdots + x_py_p \le \sqrt{x_1^2 + \cdots + x_p^2} \cdot \sqrt{y_1^2 + \cdots + y_p^2},$$

which is the Cauchy[1]–Schwarz[2]–Bunyakovsky[3] inequality (see [7, Theorem 11.19]).

The **distance** between the vectors x and y is the number $|x - y|$. By (1.2) it is clear that

$$||x| - |y|| \le |x - y| \qquad \text{and} \qquad |x - y| \le |x - z| + |z - y|$$

---

[1] Augustin Cauchy (1789–1857), French mathematician.

[2] Hermann Amandus Schwarz (1843–1921), German mathematician.

[3] Viktor Yakovlevich Bunyakovsky (1804–1889), Russian mathematician.

for all $x, y, z \in \mathbb{R}^p$. We can consider these to be variants of the triangle inequality.

If we apply (1.1) to the difference of the vectors $x = (x_1, \dots, x_p)$ and $y = (y_1, \dots, y_p)$, then we get that

$$||x| - |y|| \le |x - y| \le |x_1 - y_1| + \cdots + |x_p - y_p|. \tag{1.3}$$

The **scalar product** of the vectors $x = (x_1, \dots, x_p)$ and $y = (y_1, \dots, y_p)$ is the real number $\sum_{i=1}^{p} x_i y_i$, which we denote by $\langle x, y \rangle$. One can prove that if $x \neq 0$ and $y \neq 0$, then $\langle x, y \rangle = |x| \cdot |y| \cdot \cos \alpha$, where α denotes the angle enclosed by the two vectors. (For $p = 2$ see [7, Remark 14.57].) We say that the vectors $x, y \in \mathbb{R}^p$ are **orthogonal** if $\langle x, y \rangle = 0$.

1.2 Real Functions of Several Variables and Their Graphs

We say that f is a **p-variable real function** if $D(f) \subset \mathbb{R}^p$ and $R(f) \subset \mathbb{R}$. (Recall that $D(f)$ denotes the domain and $R(f)$ denotes the range of the function f.)

Similarly to the case of single-variable functions, multivariable functions are best illustrated by their graphs. The **graph** of a function $f \colon H \to \mathbb{R}$ is the set of pairs $(u, f(u))$, where $u \in H$. If $H \subset \mathbb{R}^p$, then graph $f \subset \mathbb{R}^p \times \mathbb{R}$; in other words, graph f is the set of pairs $((x_1, \dots, x_p), x_{p+1})$, where $(x_1, \dots, x_p) \in H$ and $x_{p+1} = f(x_1, \dots, x_p)$. In this case it is useful to "identify" $\mathbb{R}^p \times \mathbb{R}$ as the set $\mathbb{R}^{p+1}$ in the sense that instead of the pair $((x_1, \dots, x_p), x_{p+1})$, we consider the vector $(x_1, \dots, x_p, x_{p+1}) \in \mathbb{R}^{p+1}$. From now on, if $f \colon H \to \mathbb{R}$, where $H \subset \mathbb{R}^p$, then by the graph of f we mean the set

$$\text{graph } f = \{(x_1, \dots, x_p, x_{p+1}) \colon (x_1, \dots, x_p) \in H \text{ and } x_{p+1} = f(x_1, \dots, x_p)\}.$$

For example, if $f \colon H \to \mathbb{R}$, where $H \subset \mathbb{R}^2$, then graph $f \subset \mathbb{R}^3$. Just as we can visualize the graph of a function as a curve in the plane in the $p = 1$ case, we can also visualize the graph of a function as a surface in three-dimensional space in the $p = 2$ case.

Aside from using the usual coordinate notation (x_1, x_2) and (x_1, x_2, x_3), we will also use the traditional notation (x, y) and (x, y, z) in the $p = 2$ and $p = 3$ cases, respectively.

Example 1.2. **1.** The graph of the constant function $f(x, y) = c$ is a horizontal plane (in other words, it is parallel to the xy-plane). (See Figure 1.1.)
2. The graph of the function $f(x, y) = x^2$ is an infinite trough-shaped surface, whose intersections with the planes orthogonal to the y-axis are parabolas. (See Figure 1.2.)

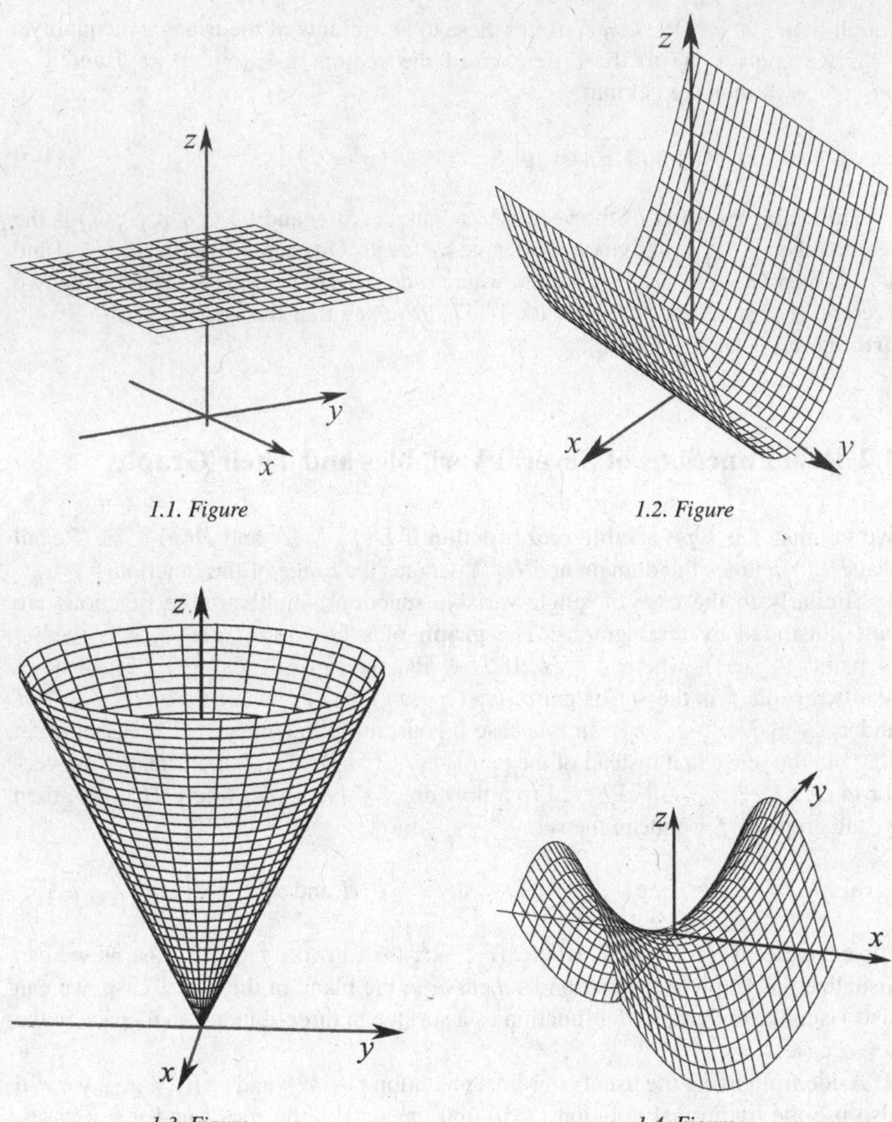

1.1. Figure *1.2. Figure*

1.3. Figure *1.4. Figure*

3. The graph of the function $f(x,y) = |(x,y)| = \sqrt{x^2 + y^2}$ is a cone. (See Figure 1.3.)
4. The graph of the function $f(x,y) = xy$ is called a **saddle surface**. (See Figure 1.4.)

We may ask whether multivariable analysis is "more difficult" or more complicated than its single-variable counterpart. The answer is twofold. On the one hand, the answer is that it is not harder at all, since it makes no difference whether we define our mappings[4] on subsets of $\mathbb{R}$ or on subsets of $\mathbb{R}^p$. On the other hand, the answer is "to a great extent," since we have "much more room" in a multidimensional space; that is, the relative positions of points in space can be much more complicated than their relative positions on a line. On the real line, a point can be to the left or to the right to another point, and there is no other option.

There is truth to both answers. While it is true that the relative positions of points can be much more complicated in a multidimensional space, this complication mostly falls in the topics of geometry and topology. For a good portion of our studies of multivariable analysis we can follow the guideline that more variables only complicate the notation but not the ideas themselves. We will warn the reader when this guideline is no longer applicable.

1.3 Convergence of Point Sequences

Definition 1.3. We say that a sequence (x_n) of the points $x_n \in \mathbb{R}^p$ *converges to* a point $a \in \mathbb{R}^p$ if for every $\varepsilon > 0$ there exists n_0 such that $|x_n - a| < \varepsilon$ holds for every $n > n_0$. We denote this fact by $\lim_{n\to\infty} x_n = a$ or simply by $x_n \to a$. We say that the sequence of points (x_n) is *convergent* if there exists an $a \in \mathbb{R}^p$ to which it converges. In this case we say that a is the limit of the sequence (x_n). If a sequence of points is not convergent, then it is *divergent.*

We denote by $B(a,r)$ the **open ball centered at a with radius r:** $B(a,r) = \{x \in \mathbb{R}^p : |x - a| < r\}$. Note that if $p = 1$, then $B(a,r)$ is the open interval $(a - r, a + r)$, and if $p = 2$, then $B(a,r)$ is the open disk with center a and radius r.

Theorem 1.4. *The following statements are equivalent:*

(i) $x_n \to a$.

(ii) *For every $\varepsilon > 0$ there are only finitely many points of the sequence (x_n) that fall outside of the open ball $B(a,\varepsilon)$.*

(iii) $|x_n - a| \to 0$.

Proof. The implication (i)⇒(ii) is clear from the definition of $x_n \to a$.

Suppose (ii), and let $\varepsilon > 0$ be given. Then there is an n_0 such that $|x_n - a| < \varepsilon$ holds for every $n > n_0$. By the definition of the convergence of sequences of real numbers, this means that $|x_n - a| \to 0$; that is, (iii) holds.

Now suppose (iii), and let $\varepsilon > 0$ be given. Then there is an n_0 such that $|x_n - a| < \varepsilon$ holds for every $n > n_0$. By the definition of the convergence of sequences of points of $\mathbb{R}^p$, this means that $x_n \to a$; that is, (i) holds. □

The following theorem states that the convergence of a sequence of points is equivalent to the convergence of the sequences of their coordinates.

---

[4] We use the terms *function* and *mapping* interchangeably.

Theorem 1.5. *Let $x_n = (x_{n,1}, \ldots, x_{n,p}) \in \mathbb{R}^p$ for every $n = 1, 2, \ldots$, and let $a = (a_1, \ldots, a_p)$. The sequence (x_n) converges to a if and only if $\lim_{n\to\infty} x_{n,i} = a_i$ for every $i = 1, \ldots, p$.*

Proof. Suppose $x_n \to a$. Since $0 \le |x_{n,i} - a_i| \le |x_n - a|$ for every $i = 1, \ldots, p$ and $|x_n - a| \to 0$, we have that $|x_{n,i} - a_i| \to 0$ follows from the squeeze theorem (see [7, Theorem 5.7]).

On the other hand, if $|x_{n,i} - a_i| \to 0$ for every $i = 1, \ldots, p$, then the inequality

$$|x_n - a| \le \sum_{i=1}^{p} |x_{n,i} - a_i|$$

and the repeated use of the squeeze theorem give us $x_n \to a$. □

We can generalize several theorems for sequences of real numbers to sequences of points of $\mathbb{R}^p$ with the help of the above theorem. The proofs of the next two theorems (which are left to the reader) are just applications of the respective theorems for sequences of real numbers to sequences of coordinates of a point-sequence.

Theorem 1.6.

(i) *If a sequence of points is convergent, then the deletion of finitely many of its terms, addition of finitely many new terms, or the reordering of its terms affect neither the convergence of the sequence nor the value of its limit.*

(ii) *If a sequence of points is convergent, then its limit is unique.*

(iii) *If a sequence of points converges to a, then each of its subsequences also converges to a.*

□

Theorem 1.7. *If $x_n \to a$ and $y_n \to b$, then $x_n + y_n \to a + b$ and $c \cdot x_n \to c \cdot a$, for every $c \in \mathbb{R}$.* □

Theorem 1.8. (Cauchy's criterion) *A sequence of points (x_n) is convergent if and only if for every $\varepsilon > 0$ there exists an index N such that $|x_n - x_m| < \varepsilon$ for every $n, m \ge N$.*

Proof. If $|x_n - a| < \varepsilon$ for every $n > N$, then $|x_n - x_m| < 2\varepsilon$ for every $n, m \ge N$. This proves the "only if" direction of our statement.

Let $\varepsilon > 0$ be given, and suppose that $|x_n - x_m| < \varepsilon$ for every $n, m \ge N$. If $x_n = (x_{n,1}, \ldots, x_{n,p})$ $(n = 1, 2, \ldots)$, then for every $i = 1, \ldots, p$ and $n, m > N$ we have

$$|x_{n,i} - x_{m,i}| \le |x_n - x_m| < \varepsilon.$$

This means that for every fixed $i = 1, \ldots, p$ the sequence $(x_{n,i})$ satisfies Cauchy's criterion (for real sequences), and thus it is convergent. Therefore, (x_n) is convergent by Theorem 1.5. □

We say that a set $A \subset \mathbb{R}^p$ is **bounded** if there exists a box $[a_1, b_1] \times \ldots \times [a_p, b_p]$ that covers (contains) it. It is obvious that a set A is bounded if and only if the set of the ith coordinates of its points is bounded in $\mathbb{R}$, for every $i = 1, \ldots, p$ (see Exercise 1.1).

A sequence of points (x_n) is **bounded** if the set of its terms is bounded.

Theorem 1.9. (Bolzano[5]–Weierstrass[6] theorem) *Every bounded sequence of points has a convergent subsequence.*

Proof. Let us assume that the sequence of points (x_n) is bounded, and let $x_n = (x_{n,1}, \ldots, x_{n,p})$ $(n = 1, 2, \ldots)$. The sequence of the ith coordinates $(x_{n,i})$ is bounded for every $i = 1, \ldots, p$. Based on the Bolzano–Weierstrass theorem for real sequences (see [7, Theorem 6.9]), we can choose a convergent subsequence $(x_{n_k,1})$ from $(x_{n,1})$. The sequence $(x_{n_k,2})$ is bounded, since it is a subsequence of the bounded sequence $(x_{n,2})$. Thus, we can choose a convergent subsequence $(x_{n_{k_l},2})$ of $(x_{n_k,2})$. If $p \geq 3$, then $(x_{n_{k_l},3})$ is bounded, since it is a subsequence of the sequence $(x_{n,3})$. Therefore, we can choose another convergent subsequence again. Repeating the process p times yields a subsequence (m_j) of the indices for which the ith coordinate sequence of (x_{m_j}) is convergent for every $i = 1, \ldots, p$. Thus, by Theorem 1.5, the subsequence (x_{m_j}) is convergent. □

Exercises

1.1. Prove that for every set $A \subset \mathbb{R}^p$, the following statements are equivalent.

(a) The set A is bounded.
(b) There exists an $r > 0$ such that $A \subset B(0, r)$.
(c) For all $i = 1, \ldots, p$ the ith coordinates of the points of A form a bounded set in $\mathbb{R}$.

1.2. Show that

(a) if $x_n \to a$, then $|x_n| \to |a|$;
(b) if $x_n \to a$ and $y_n \to b$, then $\langle x_n, y_n \rangle \to \langle a, b \rangle$.

(Here $x_n, y_n \in \mathbb{R}^p$ and $\langle x_n, y_n \rangle$ is the scalar product of x_n and y_n.)

1.3. Show that $x_n \in \mathbb{R}^p$ does not have a convergent subsequence if and only if $|x_n| \to \infty$.

1.4. Show that if every subsequence of (x_n) has a convergent subsequence converging to a, then $x_n \to a$.

1.5. Show that if $x_n \in \mathbb{R}^p$ and $|x_{n+1} - x_n| \leq 2^{-n}$ for every n, then (x_n) is convergent.

---

[5] Bernhard Bolzano (1781–1848), Italian-German mathematician, and

[6] Karl Weierstrass (1815–1897), German mathematician.

1.6. Let $x_0 = (0,0)$, $x_{n+1} = x_n + (2^{-n}, 0)$ if n is even, and $x_{n+1} = x_n + (0, 2^{-n})$ if n is odd. Show that (x_n) is convergent. What is its limit?

1.7. Construct a sequence $x_n \in \mathbb{R}^2$ having a subsequence that converges to $x \in \mathbb{R}^2$ for every x.

1.4 Basics of Point Set Theory

In order to describe the basic properties of subsets of the space $\mathbb{R}^p$, we need to introduce a few notions. We define some of these by generalizing the corresponding notions from the case $p = 1$ to an arbitrary p. Since we do not exclude the $p = 1$ case from our definitions, everything we say below holds for the real line as well.

First, we generalize the notion of neighborhoods of points. The **neighborhoods** of a point $a \in \mathbb{R}^p$ are the open balls $B(a, r)$, where r is an arbitrary positive real number.

By fixing an arbitrary set $A \subset \mathbb{R}^p$, we can divide the points of $\mathbb{R}^p$ into three classes.

The first class consists of the points that have a neighborhood that is a subset of A. We call these points the **interior points** of A, and denote the set of all interior points of A by int A. That is,

$$\text{int } A = \{x \in \mathbb{R}^p : \exists\, r > 0,\ B(x, r) \subset A\}.$$

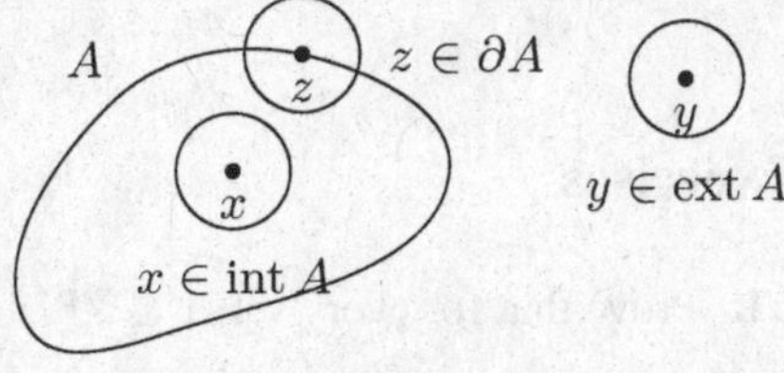

1.5. Figure

The second class consists of those points that have a neighborhood that is disjoint from A. We call these points the **exterior points** of A, and denote the set of all exterior points of A by ext A. That is,

$$\text{ext } A = \{x \in \mathbb{R}^p : \exists\, r > 0,\ B(x, r) \cap A = \emptyset\}.$$

The third class consists of the points that do not belong to any of the first two classes. We call these points the **boundary points** of A. In other words, a point x is a boundary point of A if every neighborhood of x has a nonempty intersection with both A and the complement of A. We denote the set of all boundary points of A by ∂A. That is,

$$\partial A = \{x \in \mathbb{R}^p : \forall\, r > 0,\ B(x, r) \cap A \neq \emptyset \text{ and } B(x, r) \setminus A \neq \emptyset\}.$$

It is easy to see that ext $A = \text{int }(\mathbb{R}^p \setminus A)$, int $A = \text{ext }(\mathbb{R}^p \setminus A)$, and $\partial A = \partial(\mathbb{R}^p \setminus A)$ hold for every set $A \subset \mathbb{R}^p$.

Example 1.10. **1.a.** Every point of the open ball $B(a, r)$ is an interior point. Indeed, if $x \in B(a, r)$, then $|x - a| < r$. Let $\delta = r - |x - a|$. Now $\delta > 0$ and

$B(x,\delta) \subset B(a,r)$, since $y \in B(x,\delta)$ implies $|y-x| < \delta$, and thus

$$|y-a| \le |y-x| + |x-a| < \delta + |x-a| = r,$$

i.e., $y \in B(a,r)$.
1.b. If $|x-a| > r$, then x is an exterior point of the open ball $B(a,r)$. Indeed, $\eta = |x-a| - r > 0$ and $B(x,\eta) \cap B(a,r) = \emptyset$, since if $y \in B(x,\eta)$, then $|y-x| < \eta$ and

$$|y-a| \ge |x-a| - |y-x| > |x-a| - \eta = r.$$

1.c. We now prove that the boundary of $B(a,r)$ is the set $S(a,r) = \{x \in \mathbb{R}^p : |x-a| = r\}$ (Figure 1.6). (In the case $p = 1$, the set $S(a,r)$ consists of the points $a-r$ and $a+r$, while in the case $p = 2$ the set $S(a,r)$ consists of the boundary of the circle with center a and radius r. In the case $p = 3$, $S(a,r)$ contains the surface of the ball with center a and radius r.)

Indeed, if $x \in S(a,r)$, then $x \notin B(a,r)$; therefore, every neighborhood of x has nonempty intersection with the complement of $B(a,r)$. We show that every neighborhood of x also has nonempty intersection with $B(a,r)$. Intuitively, it is clear that for every $\varepsilon > 0$, the open sphere $B(x,\varepsilon)$ contains those points of the segment connecting a and x that are close enough to x.

To formalize this idea, it is enough to show that for a well-chosen $\eta \in (0,1)$ we have $x - t(x-a) \in B(a,r) \cap B(x,\varepsilon)$ if $t \in (0,\eta)$. Since

$$|(x - t(x-a)) - a| = (1-t) \cdot |x-a| = (1-t) \cdot r < r,$$

it follows that $x - t(x-a) \in B(a,r)$. On the other hand,

$$|(x - t(x-a)) - x| = t \cdot |x-a| < \eta \cdot r < \varepsilon$$

for $\eta < \varepsilon/r$, and then $x - t(x-a) \in B(x,\varepsilon)$ also holds for every $t \in (0,\eta)$.

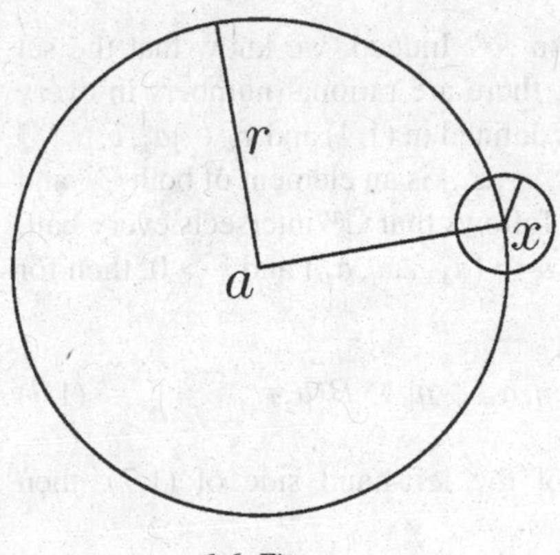

1.6. Figure

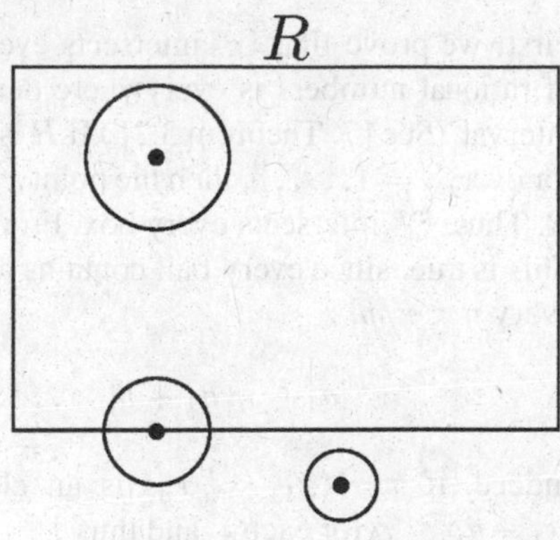

1.7. Figure

2. By an **axis-parallel rectangle** in $\mathbb{R}^p$, or just a **rectangle** or a **box** for short, we will mean a set of the form

$$[a_1, b_1] \times \cdots \times [a_p, b_p],$$

where $a_i < b_i$ for every $i = 1, \ldots, p$. The boxes in the Euclidean spaces $\mathbb{R}$, $\mathbb{R}^2$, and $\mathbb{R}^3$ are the nondegenerate and bounded closed intervals, the axis-parallel rectangles, and the rectangular boxes, respectively.

The interior of the box

$$R = [a_1, b_1] \times \ldots \times [a_p, b_p] \tag{1.4}$$

is the **open box**

$$(a_1, b_1) \times \ldots \times (a_p, b_p). \tag{1.5}$$

For every point $x = (x_1, \ldots, x_p)$ of this open box, we have $a_i < x_i < b_i$ for every $i = 1, \ldots, p$. If $\delta > 0$ is small enough, then

$$a_i < x_i - \delta < x_i < x_i + \delta < b_i \tag{1.6}$$

for every $i = 1, \ldots, p$. Then $B(x, \delta) \subset R$, since $y = (y_1, \ldots, y_p) \in B(x, \delta)$ implies $|y - x| < \delta$, which gives $|y_i - x_i| < \delta$ for every i, and thus, by (1.6), $a_i < y_i < b_i$ for every i.

If the point $x = (x_1, \ldots, x_p)$ is not in the open box defined in (1.5), then x is not an interior point of R. Indeed, if there exists an i such that $x_i < a_i$ or $x_i > b_i$, then we can find an appropriate neighborhood of x that is disjoint from R. Therefore, in this case x is an exterior point. On the other hand, if $x \in R$ and there exists i such that $x_i = a_i$ or $x_i = b_i$, then every neighborhood of x intersects both R and its complement, and thus x is a boundary point of R (Figure 1.7).

3. Let $\mathbb{Q}^p$ be the set of those points $x \in \mathbb{R}^p$ for which every coordinate of x is rational. We show that

$$\operatorname{int} \mathbb{Q}^p = \operatorname{ext} \mathbb{Q}^p = \emptyset.$$

First, we prove that $\mathbb{Q}^p$ intersects every box in $\mathbb{R}^p$. Indeed, we know that the set of rational numbers is everywhere dense; i.e., there are rational numbers in every interval. (See [7, Theorem 3.2].) If R is the box defined in (1.4) and $x_i \in [a_i, b_i] \cap \mathbb{Q}$ for every $i = 1, \ldots, p$, then the point $x = (x_1, \ldots, x_p)$ is an element of both $\mathbb{Q}^p$ and R. Thus, $\mathbb{Q}^p$ intersects every box. From this it follows that $\mathbb{Q}^p$ intersects every ball. This is true, since every ball contains a box: if $a = (a_1, \ldots, a_p)$ and $r > 0$, then for every $\eta < r/p$,

$$[a_1 - \eta, a_1 + \eta] \times \ldots \times [a_p - \eta, a_p + \eta] \subset B(a, r). \tag{1.7}$$

Indeed, if $x = (x_1, \ldots, x_p)$ is an element of the left-hand side of (1.7), then $|x_i - a_i| \le \eta$ for each i, and thus

$$|x-a| \le \sum_{i=1}^{p} |x_i - a_i| \le p\eta < r,$$

and $x \in B(a,r)$. We have proved that $\mathbb{Q}^p$ intersects every ball, and thus ext $\mathbb{Q}^p = \emptyset$.

Now we prove that each ball $B(a,r)$ has a point that is not an element of $\mathbb{Q}^p$. We need to find a point in $B(a,r)$ that has at least one irrational coordinate. We can, however, go further and find a point that has only irrational coordinates. We know that the set of irrational numbers is also dense everywhere (see [7, Theorem 3.12]). Thus we can repeat the same steps as above, and then int $\mathbb{Q}^p = \emptyset$ follows.

In the end we get that $\mathbb{Q}^p$ has neither interior nor exterior points, i.e., every point $x \in \mathbb{R}^p$ is a boundary point of $\mathbb{Q}^p$.

Definition 1.11. We say that a point $a \in \mathbb{R}^p$ is a *limit point* of the set $A \subset \mathbb{R}^p$ if every neighborhood of the point a contains infinitely many points of A. We call the set of all limit points of the set A the *derived set* of A, and denote it by A'.

We say that a point $a \in \mathbb{R}^p$ is an *isolated point* of A if there exists $r > 0$ such that $B(a,r) \cap A = \{a\}$.

Remark 1.12. **1.** The limit points of A are not necessarily elements of the set A. For example, every point y that satisfies $|y-x| = r$ is a limit point of the ball $B(x,r)$ (see Example 1.10.1.c). Thus $S(a,r) \subset B(a,r)'$. However, $S(a,r) \cap B(a,r) = \emptyset$.
2. By our definitions, the isolated points of A need to be elements of A. It is easy to see that the set of all isolated points of A is nothing other than the set $A \setminus A'$. It follows that every point of A is either an isolated point or a limit point of A.
3. It is also easy to see that a point a is a limit point of the set A if and only if there exists a sequence $x_n \in A \setminus \{a\}$ that converges to a.

We say that the set $A \subset \mathbb{R}^p$ is **open** if every point of A is an interior point of A, i.e., if $A = \text{int}\, A$. The open balls and open boxes are indeed open sets by Example 1.10.1a and Example 1.10.2. The empty set and $\mathbb{R}^p$ are also open.

Obviously, the set A is open if and only if $A \cap \partial A = \emptyset$.

Theorem 1.13. *The following hold for an arbitrary set $A \subset \mathbb{R}^p$:*

(i) int A *and* ext A *are open sets;*

(ii) int A *is the largest open set contained by A.*

Proof. Part (i) follows from the definition and from the fact that every ball is an open set.

If $G \subset A$ is open and $x \in G$, then there exists $r > 0$ such that $B(x,r) \subset G$. In this case, $B(x,r) \subset A$ also holds, and thus $x \in \text{int}\, A$. We have proved that int A contains every open set contained by A. Since int A is also open by part (i), it follows that (ii) holds. □

Theorem 1.14. *The intersection of finitely many open sets and the union of arbitrarily many open sets are also open.*

Proof. If A and B are open sets and $x \in A \cap B$, then $x \in \text{int } A$, and $x \in \text{int } B$ means that there exist positive numbers r and s such that $B(x,r) \subset A$ and $B(x,s) \subset B$. In this case, $B(x, \min(r,s)) \subset A \cap B$, and thus $x \in \text{int } (A \cap B)$. We have proved that every point of $A \cap B$ is an interior point of $A \cap B$, and thus the set $A \cap B$ is open. By induction we have that the intersection of n open sets is open, for every $n \in \mathbb{N}^+$.

Let G_i be an open set for each $i \in I$, where I is an arbitrary (finite or infinite) index set, and let $G = \bigcup_{i \in I} G_i$. If $x \in G$, then x is in one of the sets G_{i_0}. Since G_{i_0} is open, it follows that $x \in \text{int } G_{i_0}$, i.e., $B(x,r) \subset G_{i_0}$ for some $r > 0$. Now $B(x,r) \subset G$ holds, and thus $x \in \text{int } G$. This is true for every $x \in G$, which implies that the set G is open. □

Remark 1.15. The intersection of infinitely many open sets is not necessarily open. For example, the intersection of the sets $B(x, 1/n)$ is the singleton $\{x\}$. This set is not open, since its interior is empty.

We say that a ball $B(x,r)$ is a **rational ball** if each of the coordinates of its center x, along with its radius, is a rational number.

Lemma 1.16. *Every open set is the union of rational balls.*

Proof. Let G be an open set and $x \in G$. Then $B(x,r) \subset G$ holds for some $r > 0$. As shown in Example 1.10.3, every ball contains a point with rational coordinates. Let $y \in B(x, r/2)$ be such a point. If $s \in \mathbb{Q}$ and $|x - y| < s < r/2$, then $B(y,s)$ is a rational ball that contains x, since $|x - y| < s$. On the other hand, $B(y,s) \subset B(x,r)$, since $z \in B(y,s)$ implies

$$|z - x| \le |z - y| + |y - x| < s + (r/2) < r.$$

We have proved that every point in G is in a rational ball contained by G. Therefore, G is equal to the union of all the rational balls it contains. □

We say that a set $A \subset \mathbb{R}^p$ is **closed** if it contains each of its boundary points, i.e., $\partial A \subset A$. Thus every box is closed. The set $\overline{B}(a,r) = \{x \in \mathbb{R}^p : |x - a| \le r\}$ is also closed. We call this set the **closed ball** with center a and radius r.

Theorem 1.17. *For every set $A \subset \mathbb{R}^p$ the following are equivalent:*

(i) *A is a closed set.*

(ii) *$\mathbb{R}^p \setminus A$ is an open set.*

(iii) *If $x_n \in A$ for every n and $x_n \to a$, then $a \in A$.*

Proof. (i)⇒(ii): If A is closed and $x \notin A$, then $x \notin \text{int } A$ and $x \notin \partial A$, and thus $x \in \text{ext } A$. Thus $B(x,r) \cap A = \emptyset$ holds for some $r > 0$, i.e., $B(x,r) \subset \mathbb{R}^p \setminus A$. We have shown that every point of $\mathbb{R}^p \setminus A$ is an interior point of $\mathbb{R}^p \setminus A$; that is, $\mathbb{R}^p \setminus A$ is open.

(ii)⇒(iii): We prove by contradiction. Assume that $x_n \to a$, where $x_n \in A$ for every n, but $a \notin A$, i.e., $a \in \mathbb{R}^p \setminus A$. Since $\mathbb{R}^p \setminus A$ is open, $B(a,r) \subset \mathbb{R}^p \setminus A$ for some $r > 0$. On the other hand, as $x_n \to a$, we have $x_n \in B(a,r) \subset \mathbb{R}^p \setminus A$ for every n large enough. This is a contradiction, since $x_n \in A$ for every n.
(iii)⇒(i): Let $a \in \partial A$. Then for every $n \in \mathbb{N}^+$ we have $B(a,1/n) \cap A \neq \emptyset$. Choose a point $x_n \in B(a,1/n) \cap A$ for each n. Then $x_n \to a$, and thus $a \in A$ by (iii). We have proved that $\partial A \subset A$, i.e., A is closed. □

It follows from our previous theorem that *the boundary of every set is a closed set.* Indeed, $\partial A = \mathbb{R}^p \setminus (\operatorname{int} A \cup \operatorname{ext} A)$, and by Theorems 1.13 and 1.14, $\operatorname{int} A \cup \operatorname{ext} A$ is open. It is also easy to see that *the set of limit points of an arbitrary set is closed* (see Exercise 1.22).

Theorem 1.18. *The union of finitely many closed sets and the intersection of arbitrarily many closed sets is also a closed set.*

Proof. This follows from Theorems 1.14 and 1.17. □

Obviously, there are sets that are neither open nor closed (for example, the set $\mathbb{Q}$ as a subset of $\mathbb{R}$). On the other hand, the empty set and $\mathbb{R}^p$ are both open and closed at the same time. We will show that there is no other set in $\mathbb{R}^p$ that is both open and closed.

For every $a, b \in \mathbb{R}^p$ we denote by $[a,b]$ the set $\{t \in [0,1] : a + t(b-a)\}$. It is clear that $[a,b]$ is the segment connecting the points a and b.

Theorem 1.19. *If $A \subset \mathbb{R}^p$, $a \in A$, and $b \in \mathbb{R}^p \setminus A$, then the segment $[a,b]$ intersects the boundary of A, i.e., $[a,b] \cap \partial A \neq \emptyset$.*

Proof. Let $T = \{t \in [0,1] : a + t(b-a) \in A\}$. The set T is nonempty (since $0 \in T$) and bounded; thus it has a least upper bound. Let $t_0 = \sup T$. We show that the point $x_0 = a + t_0(b-a)$ is in the boundary set of A. Obviously, for every $\varepsilon > 0$, the interval $(t_0 - \varepsilon, t_0 + \varepsilon)$ intersects both T and $[0,1] \setminus T$. (This is also true in the case $t_0 = 1$, since $1 \notin T$.) If $t \in (t_0 - \varepsilon, t_0 + \varepsilon) \cap T$, then the point $x = a + t(b-a)$ is an element of A, and $|x - x_0| < \varepsilon \cdot |b-a|$. However, if $t \in (t_0 - \varepsilon, t_0 + \varepsilon) \setminus T$, then the point $y = a + t(b-a)$ is not an element of A, and $|y - x_0| < \varepsilon \cdot |b-a|$. We have proved that every neighborhood of x_0 intersects both A and the complement of A, i.e., $x_0 \in \partial A$. □

Corollary 1.20. *If a set $A \subset \mathbb{R}^p$ is both open and closed, then $A = \emptyset$ or $A = \mathbb{R}^p$.*

Proof. If A is an open set, then $A \cap \partial A = \emptyset$. If, however, A is a closed set, then $\partial A \subset A$. Only if $\partial A = \emptyset$ can these conditions both hold. Now Theorem 1.19 states that if $\emptyset \neq A \neq \mathbb{R}^p$, then $\partial A \neq \emptyset$. □

The connected open sets play an important role in multivariable analysis.

Definition 1.21. We say that an open set $G \subset \mathbb{R}^p$ is *connected* if G cannot be represented as the union of two disjoint nonempty open sets.

Theorem 1.22.

(i) *An open set G is connected if and only if every pair of its points can be connected with a polygonal line*[7] *contained entirely in G.*

(ii) *Every open set can be written as the union of pairwise disjoint connected open sets (the number of which can be finite or infinite).*

Proof.

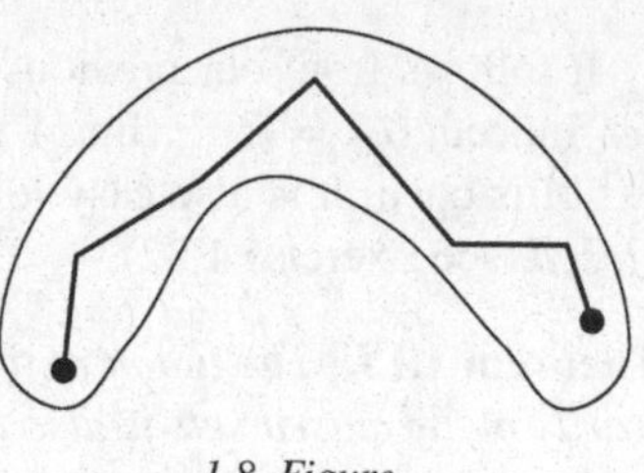

1.8. Figure

Let $G \subset \mathbb{R}^p$ be an open set. We call the points $x, y \in G$ equivalent if they can be connected by a polygonal line that lies entirely in G. We will denote this fact by $x \sim y$. Obviously, this is an equivalence relation in G. If $x \in G$, then $B(x,r) \subset G$ for some $r > 0$. The point x is equivalent to every point y of $B(x,r)$, since $[x,y] \subset B(x,r) \subset G$. It follows that every equivalence class (the set of points equivalent to an arbitrary fixed point) is an open set. Since the different equivalence classes are disjoint, we have a system of pairwise disjoint open sets whose union is G.

If G is connected, then there is only one equivalence class, for otherwise, we could write G as the union of two disjoint nonempty open sets (e.g., take a single class and the union of the rest). Thus we have proved that if G is connected, then every pair of its points are equivalent to each other.

To prove the converse, let us assume that every pair of points in G are equivalent to each other, but G is not connected. Let $G = A \cup B$, where A and B are nonempty disjoint open sets. Let $x \in A$, $y \in B$, and let T be a polygonal line connecting the two points. Let T be the union of the segments $[x_{i-1}, x_i]$ $(i = 1, \dots, n)$, where $x_0 = x$ and $x_n = y$. Since $x_0 \in A$ and $x_n \notin A$, there exists i such that $x_{i-1} \in A$ and $x_i \notin A$. The segment $[x_{i-1}, x_i]$ contains a boundary point of A by Theorem 1.19. This is impossible, since every point of $[x_{i-1}, x_i]$ is either an exterior or an interior point of A, as implied by $[x_{i-1}, x_i] \subset G = A \cup B$. This contradiction proves (i).

We showed that an arbitrary open set G can be written as the union of pairwise disjoint open sets G_i, where each G_i contains every point from the same equivalence class. We also proved that each G_i is also a connected set, which proves (ii). □

We call the connected open sets **domains.**

The proof of Theorem 1.22 also shows that the decomposition in part (ii) of the theorem is unique: the open sets of the composition are just the equivalence classes of the $x \sim y$ equivalence relation. We call the domains of this decomposition of the set G the **components** of G.

Definition 1.23. We call the set $A \cup \partial A$ the *closure* of the set A, and use the notation $\operatorname{cl} A$.

---

[7] By a **polygonal line** we mean a set of the form $[a_0, a_1] \cup [a_1, a_2] \cup \ldots \cup [a_{n-1}, a_n]$, where $a_0, \dots, a_n$ are arbitrary points in $\mathbb{R}^n$.

Theorem 1.24. *For an arbitrary set $A \subset \mathbb{R}^p$, the following hold.*

(i) *the point x is in* cl A *if and only if every neighborhood of x intersects A;*

(ii) cl $A = A \cup A'$;

(iii) cl $A = \mathbb{R}^p \setminus \operatorname{ext} A = \mathbb{R}^p \setminus \operatorname{int}(\mathbb{R}^p \setminus A)$;

(iv) cl A *is the smallest closed set containing A.*

Proof. We leave the proof of (i)–(iii) to the reader, while (iv) follows from (iii) and Theorem 1.13. □

Our next theorem is a generalization of Cantor's axiom[8] (see [7, p. 33]). Note that Cantor's axiom states only that if the sets $A_1 \supset A_2 \supset \ldots$ are closed intervals in $\mathbb{R}$, then their intersection is nonempty. As the following theorem shows, it follows from Cantor's axiom and from the other axioms of the real numbers that the statement is also true in $\mathbb{R}^p$ (for every p) and for much more general sets. From now on, we consider only subsets of $\mathbb{R}^p$.

Theorem 1.25. (Cantor's Theorem) *If the sets $A_1 \supset A_2 \supset \ldots$ are bounded, closed, and nonempty, then the set $\bigcap_{n=1}^{\infty} A_n$ is also nonempty.*

Proof. Choose a point x_n from each set A_n. The sequence (x_n) is bounded, since it is contained in the bounded set A_1. The Bolzano–Weierstrass theorem (Theorem 1.9) states that (x_n) has a convergent subsequence. Let (x_{n_k}) be one such subsequence, and let its limit be a. We show that $a \in \bigcap_{n=1}^{\infty} A_n$.

Let n be fixed. For k large enough, we have $n_k > n$, and thus $x_{n_k} \in A_{n_k} \subset A_n$. Therefore, the sequence (x_{n_k}) is contained in A_n, except for at most finitely many of its points. Since A_n is closed, we have $a \in A_n$ (Theorem 1.17). Also, since n was arbitrary, it follows that $a \in \bigcap_{n=1}^{\infty} A_n$. □

Theorem 1.26. (Lindelöf's[9] Theorem) *If the set A is covered by the union of some open sets, then we can choose countably many of those open sets whose union also covers A.*

Lemma 1.27. *The set of rational balls is countable.*

Proof. Let $(r_n)_{n=1}^{\infty}$ be an enumeration of the rational numbers. If $x = (r_{n_1}, \ldots, r_{n_p})$ and $r = r_m$, then we call $n_1 + \ldots + n_p + m$ the weight of $B(x, r)$. Obviously, there are only finitely many balls with a given weight w for every $w \geq p + 1$. It follows that there exists a sequence that contains every rational ball. Indeed, first we enumerate the rational balls with weight $p + 1$ (there is at most one such ball). Then we list the rational balls with weight $p + 2$, and so on. In this way we list every rational ball in a single infinite sequence, which proves that the set of rational balls is countable. □

---

[8] Georg Cantor (1845–1918), German mathematician.

[9] Ernst Lindelöf (1870–1946), Finnish mathematician.

Remark 1.28. The proof above also shows that the set $\mathbb{Q}^p$ (the set of points with rational coordinates) is countable. Combining this result with Example 1.10.3, we get that *there exists a set in $\mathbb{R}^p$ that is countable and everywhere dense.*

Proof of Theorem 1.26. Let $(B_n)_{n=1}^{\infty}$ be an enumeration of the rational balls. (By Lemma 1.27, there is such a sequence.)

Let $\mathcal{G}$ be a system of open sets whose union covers A. For every ball B_n that is contained by at least one of the open sets $G \in \mathcal{G}$ we choose an open set $G_n \in \mathcal{G}$ such that $B_n \subset G_n$. In this way we have chosen the countable subsystem $\{G_n\}$ of $\mathcal{G}$. The union of the sets of this subsystem is the same as the union of all sets in $\mathcal{G}$. Indeed, if $x \in \bigcup \mathcal{G}$, then there is a set $G \in \mathcal{G}$ containing x. By Lemma 1.16, there is a ball B_n such that $x \in B_n \subset G$. Since $B_n \subset G_n$ holds, it follows that $x \in \bigcup_{n=1} G_n$.

Therefore, if the union of $\mathcal{G}$ covers A, then the union of the sets G_n also covers A. □

Example 1.29. **1.** The balls $B(0,r)$ cover the whole space $\mathbb{R}^p$. Lindelöf's theorem claims that countably many of these also cover $\mathbb{R}^p$, e.g., $\bigcup_{n=1}^{\infty} B(0,n) = \mathbb{R}^p$. On the other hand, it is obvious that finitely many of the balls $B(0,r)$ cannot cover the whole of $\mathbb{R}^p$.
2. The open sets $G_r = \mathbb{R}^p \setminus \overline{B}(0,r) = \{x \in \mathbb{R}^p \colon |x| > r\}$ cover the set $A = \mathbb{R}^p \setminus \{0\}$. Lindelöf's theorem claims that countably many of these also cover A, e.g., $\bigcup_{n=1}^{\infty} G_{1/n} = A$. On the other hand, it is obvious that finitely many of the sets G_r do not cover A.

The examples above show that we cannot replace the word "countable" by "finite" in Lindelöf's theorem. That is, we cannot always choose a finite subcovering system from a covering system of open sets. The sets that satisfy this stronger condition form another important class of sets.

Definition 1.30. We call a set $A \subset \mathbb{R}^p$ *compact* if we can choose a finite covering system from each of its covering systems of open sets.

Theorem 1.31. (Borel's[10] Theorem) *A set $A \subset \mathbb{R}^p$ is compact if and only if it is bounded and closed.*

Proof. Let A be compact. Since $A \subset \mathbb{R}^p = \bigcup_{n=1}^{\infty} B(0,n)$, there exists N such that $A \subset \bigcup_{n=1}^{N} B(0,n) = B(0,N)$ (this follows from the compactness of A). Thus A is bounded.

Now we prove that A is closed. We shall do so by showing that $\mathbb{R}^p \setminus A$ is open. Let $a \in \mathbb{R}^p \setminus A$. Then

$$A \subset \mathbb{R}^p \setminus \{a\} = \bigcup_{k=1}^{\infty} \left(\mathbb{R}^p \setminus \overline{B}(a,1/k)\right)$$

[10] Émile Borel (1871–1956), French mathematician.

is an open cover of A, and then, by the compactness of A, there exists an integer K such that

$$A \subset \bigcup_{k=1}^{K} (\mathbb{R}^p \setminus \overline{B}(a, 1/k)) = \mathbb{R}^p \setminus \overline{B}(a, 1/K).$$

Thus $B(a, 1/K) \cap A = \emptyset$ and $B(a, 1/K) \subset \mathbb{R}^p \setminus A$. Since $a \in \mathbb{R}^p \setminus A$ was arbitrary, this proves that $\mathbb{R}^p \setminus A$ is open.

Now suppose that A is bounded and closed; we shall show that A is compact. Let $\mathcal{G}$ be a system of open sets covering A. By Lindelöf's theorem there exists a countable subsystem $\{G_1, G_2, \ldots\}$ of $\mathcal{G}$ that also covers A. Let

$$F_n = A \setminus \bigcup_{i=1}^{n} G_i = A \cap \left(\mathbb{R}^p \setminus \bigcup_{i=1}^{n} G_i \right)$$

for each n. The sets F_n are closed (since $\bigcup_{i=1}^{n} G_i$ is open, $A_n = \mathbb{R}^p \setminus \bigcup_{i=1}^{n} G_i$ is closed, and thus $F_n = A \cap A_n$ is also closed), and they are bounded (since they are contained in A), and $F_1 \supset F_2 \supset \ldots$ holds. If the sets F_n are all nonempty, then by Cantor's theorem, their intersection $A \setminus \bigcup_{i=1}^{\infty} G_i$ is also nonempty. However, this is impossible, since $A \subset \bigcup_{i=1}^{\infty} G_i$. Thus, there exists n such that $F_n = A \setminus \bigcup_{i=1}^{n} G_i = \emptyset$; that is, $A \subset \bigcup_{i=1}^{n} G_i$. This shows that finitely many of the sets G_i cover A. □

If A and B are nonempty sets in $\mathbb{R}^p$, then the **distance** between A and B is

$$\operatorname{dist}(A, B) = \inf\{|x - y| : x \in A,\ y \in B\}.$$

The distance between two disjoint closed sets can be zero (see Exercise 1.36). Our next theorem shows that this is possible only if neither A nor B is bounded.

Theorem 1.32. *Let A and B be disjoint nonempty closed sets, and suppose that at least one of them is bounded. Then*

(i) *there exist points $a \in A$ and $b \in B$ such that* $\operatorname{dist}(A, B) = |a - b|$, *and*

(ii) $\operatorname{dist}(A, B) > 0$.

Proof. Let $\operatorname{dist}(A, B) = d$, and let the points $a_n \in A$ and $b_n \in B$ be chosen such that $|a_n - b_n| < d + (1/n)$ $(n = 1, 2, \ldots)$. Since at least one of the sets A and B is bounded, it follows that both of the sequences (a_n) and (b_n) are bounded.

By the Bolzano–Weierstrass theorem (Theorem 1.9) we can select a convergent subsequence of (a_n). Replacing (a_n) by this subsequence, we may assume that (a_n) itself is convergent. Then we select a convergent subsequence of (b_n). Turning to this subsequence, we may assume that (a_n) and (b_n) are both convergent.

If $a_n \to a$ and $b_n \to b$, then $a \in A$ and $b \in B$, since A and B are both closed. Now $|a - b| = \lim_{n\to\infty} |a_n - b_n| \le d$. Using the definition of the distance between sets, we get $|a - b| \ge d$, and thus $|a - b| = d$. This proves (i), while (ii) follows immediately from (i). □

Exercises

1.8. Let $p = 2$. Find int A, ext A, and ∂A for each of the sets below.

(a) $\{(x, y) \in \mathbb{R}^2 : x, y > 0,\ x + y < 1\}$;
(b) $\{(x, 0) \in \mathbb{R}^2 : 0 < x < 1\}$;
(c) $\{(x, y) \in \mathbb{R}^2 : x = 1/n\ (n = 1, 2, \ldots), 0 < y < 1\}$.

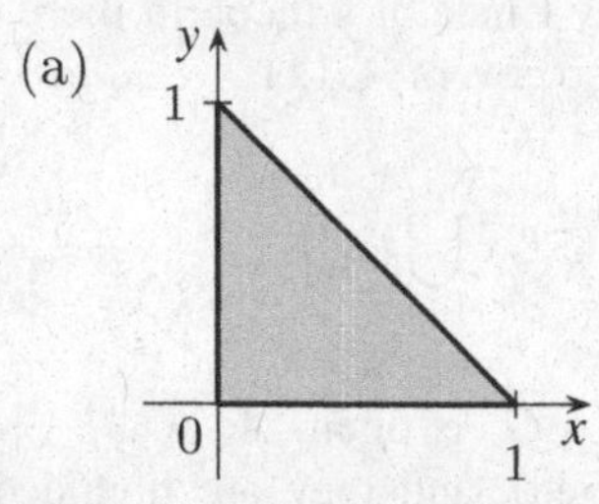

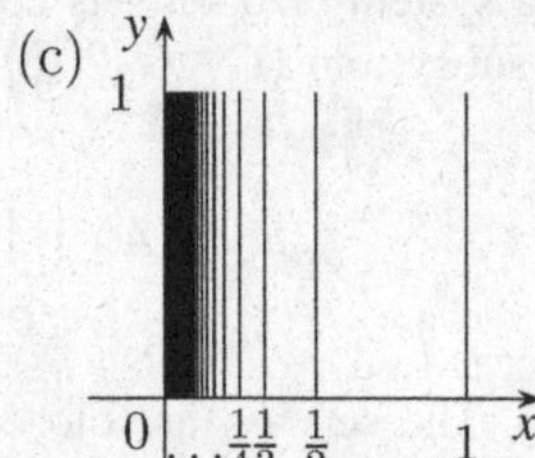

1.9. Figure

1.9. Find every set $A \subset \mathbb{R}^p$ such that int A has exactly three elements. (S)

1.10. Show that $\partial(A \cup B) \subset \partial A \cup \partial B$ and $\partial(A \cap B) \subset \partial A \cup \partial B$ hold for every $A, B \subset \mathbb{R}^p$-re. (S)

1.11. Is there a set $A \subset \mathbb{R}^2$ such that $\partial A = \{(1/n, 0) : n = 1, 2, \ldots\}$?

1.12. Let $A \subset \mathbb{R}^2$ be a closed set. Show that $A = \partial H$ for a suitable set $H \subset \mathbb{R}^2$.

1.13. Show that $\partial\,\partial A \subset \partial A$ for every set $A \subset \mathbb{R}^p$. Also show that $\partial\,\partial A = \partial A$ is not always true.

1.14. Show that if the set $A \subset \mathbb{R}^p$ is open or closed, then $\partial\,\partial A = \partial A$ and int $\partial A = \emptyset$.

1.15. Show that the union of infinitely many closed sets is not necessarily closed.

1.16. Show that every open set of $\mathbb{R}^p$ can be written as the union of countable many boxes.

1.17. What are the sets whose boundary consists of exactly three points?

1.18. Show that if $A \subset \mathbb{R}^p$, where $p > 1$, and if ∂A is countable, then one of A and $\mathbb{R}^p \setminus A$ is countable.

1.19. Which are the sets satisfying

(a) $\operatorname{int} A = \partial A$?
(b) $\operatorname{int} A = \operatorname{cl} A$?
(c) $\operatorname{ext} A = \operatorname{cl} A$?

1.20. Show that every infinite bounded set has a limit point.

1.21. What are the sets with no limit points? What are the sets with exactly three limit points?

1.22. Show that for every set $A \subset \mathbb{R}^p$, the set A' is closed. (S)

1.23. Find every set $A \subset \mathbb{R}^2$ that satisfies $A' = A$ and $(\mathbb{R}^2 \setminus A)' = \mathbb{R}^2 \setminus A$.

1.24. Let $A \subset \mathbb{R}^2$ be bounded, $G \subset \mathbb{R}^2$ open, and let $A' \subset G$. Show that $A \setminus G$ is finite.

1.25. Construct a set A such that the sets $A,\ A',\ A''$, etc. are distinct.

1.26. Is there a bounded infinite set every point of which is an isolated point?

1.27. Show that the number of isolated points of an arbitrary set is countable. (H)

1.28. A set $A \subset \mathbb{R}^p$ is called **everywhere dense** if it has a point in every ball. Construct an everywhere dense set $A \subset \mathbb{R}^2$ that does not contain three collinear points.

1.29. Decompose $\mathbb{R}^2$ into infinitely many pairwise disjoint everywhere dense sets.

1.30. Construct a function $f : \mathbb{R} \to \mathbb{R}$ whose graph is everywhere dense in $\mathbb{R}^2$.

1.31. We call a set $A \subset \mathbb{R}^2$ a **star** if it is the union of three segments that have a common endpoint but are otherwise disjoint. Show that every system of pairwise disjoint stars is countable. (∗ H)

1.32. Show that a system of pairwise disjoint stars in $\mathbb{R}^2$ cannot cover a line. (∗)

1.33. Construct a sequence of sets $A_1 \supset A_2 \supset \ldots$ that satisfy $\bigcap_{n=1}^{\infty} A_n = \emptyset$ and are

(a) bounded and nonempty;
(b) closed and nonempty.

1.34. Show that a set $A \subset \mathbb{R}^p$ is bounded and closed if and only if every sequence $x_n \in A$ has a subsequence converging to a point of A.

1.35. Is there a sequence $x_n \in \mathbb{R}$ such that $[0,1] \subset \bigcup_{n=1}^{\infty}(x_n - 2^{-n}, x_n + 2^{-n})$? How about a sequence with $[0,1] \subset \bigcup_{n=1}^{\infty}(x_n - 2^{-n-1}, x_n + 2^{-n-1})$? (H)

1.36. Give examples of two disjoint nonempty closed sets with distance zero (a) in $\mathbb{R}^2$, and (b) in $\mathbb{R}$. (S)

1.37. A set $G \subset \mathbb{R}^p$ is called a **regular open set** if $G = \text{int cl } G$. Show that for every $G \subset \mathbb{R}^p$ the following statements are equivalent.

(i) The set G is regular open.
(ii) There is a set A with $G = \text{int cl } A$.
(iii) There is a set A with $G = \text{ext int } A$.
(iv) $G = \text{ext ext } G$.

1.38. Which of the following sets in $\mathbb{R}^2$ are regular open?

(i) $\{(x, y) \colon x^2 + y^2 < 1\}$.
(ii) $\{(x, y) \colon 0 < x^2 + y^2 < 1\}$.
(iii) $\{(x, y) \colon x^2 + y^2 < 1,\ y \neq 0\}$.
(iv) $\{(x, y) \colon x^2 + y^2 \in [0, 1) \setminus \{1/2\}\}$.

1.39. Show that for every set $A \subset \mathbb{R}^p$ the following are true:

$$\begin{aligned} &\text{ext ext ext ext } A = \text{ext ext } A, && \text{ext ext ext int } A = \text{ext int } A, \\ &\text{ext ext int } \partial A = \text{int } \partial A, && \text{ext ext } \partial A = \text{int } \partial A, \\ &\partial \text{ ext ext int } A = \partial \text{ ext int } A, && \partial \text{ ext ext ext } A = \partial \text{ ext ext } A, \\ &\partial \text{ ext int } \partial A = \partial \text{ int } \partial A. \end{aligned} \tag{1.8}$$

1.40. Show that applying the operations int , ext , ∂ to an arbitrary set $A \subset \mathbb{R}^p$ (repeated an arbitrary number of times and in an arbitrarily chosen order) cannot result in more than 25 different sets. (* H)

1.41. Show that the estimate in the previous exercise is sharp; i.e., give an example of a set $A \subset \mathbb{R}^p$ such that we get 25 different sets by applying the operations int, ext, ∂ an arbitrary number of times and in an arbitrarily chosen order.

1.42. Show that applying the operations int , ext , ∂ together with the closure operation and the complement operation on an arbitrary set $A \subset \mathbb{R}^p$ (repeated an arbitrary number of times and in an arbitrarily chosen order) cannot result in more than 34 different sets.

1.5 Limits

At the core of multivariable analysis—as in the case of one-variable analysis—lies the investigation and application of the limit, continuity, differentiation, and integration of functions.

The concept of limit of a multivariable function—similarly to the single-variable case—is the idea that if x is close to a point a, then the value of the function at x is close to the limit.

Definition 1.33. Let the real-valued function f be defined on the set $A \subset \mathbb{R}^p$, and let a be a limit point of A. We say that *the limit of the function f at the point a restricted to the set A is $b \in \mathbb{R}$* if the following condition is satisfied. For every $\varepsilon > 0$ there exists $\delta > 0$ such that whenever $x \in A$ and $0 < |x - a| < \delta$, then $|f(x) - b| < \varepsilon$. **Notation:** $\lim_{x \to a,\, x \in A} f(x) = b$.

If the domain of f is A (i.e., if $D(f)$ is not larger than A), then we can omit the part "restricted to the set A" from the definition and instead we can say that *the limit of the function f at the point a is b*. In this case, the notation is $\lim_{x \to a} f(x) = b$ or $f(x) \to b$ as $x \to a$.

Example 1.34. **1.** Let $p = 2$. We show that $\lim_{(x,y)\to(0,0)} \frac{x^2 y}{x^2+y^2} = 0$. For $\varepsilon > 0$ fixed, $0 < |(x,y)| = \sqrt{x^2+y^2} < \varepsilon$ implies $|y| < \varepsilon$; thus

$$\left| \frac{x^2 y}{x^2+y^2} \right| \le |y| < \varepsilon.$$

2. We show that the limit $\lim_{(x,y)\to(0,0)} \frac{xy}{x^2+y^2}$ does not exist.

Since the function is zero on the axes, there exists a point in every neighborhood of $(0,0)$ where the function is zero. On the other hand, the function is $1/2$ at the points of the line $y = x$, whence there exists a point in every neighborhood of $(0,0)$ where the function is $1/2$. This implies that the limit does not exist: we cannot find an appropriate δ for $\varepsilon = 1/4$, regardless of the value of b. (See Figure 1.10.)

Note, however, that the function $xy/(x^2+y^2)$ has a limit at the origin when restricted to a line that passes through it, since the function is constant on every such line (aside from the origin itself).

Definition 1.35. Let the function f be defined on the set $A \subset \mathbb{R}^p$, and let a be a limit point of A. We say that *the limit of the function f at the point a restricted to the set A is infinity (negative infinity)* if for every K there exists $\delta > 0$ such that $f(x) > K$ ($f(x) < K$) for every $x \in A$ satisfying $0 < |x - a| < \delta$. **Notation:** $\lim_{x \to a,\, x \in A} f(x) = \infty \; (-\infty)$.

If the domain of f is A (i.e., if it is not larger than A), then we can omit the part "restricted to the set A" of the definition and instead we can say that *the limit of the function f at the point a is infinity (negative infinity)*. In this case, the notation is $\lim_{x \to a} f(x) = \infty \; (-\infty)$.

Example 1.36. Let A be the $\{(x,y) : y > x\}$ half-plane.
Then $\lim_{\substack{(x,y)\to(0,0) \\ (x,y)\in A}} \frac{1}{y-x} = \infty$. Indeed, if $K > 0$ is fixed and $0 < |(x,y)| = \sqrt{x^2+y^2} < 1/K$, then $|x|, |y| < 1/K$, thus $|y - x| < 2/K$. On the other hand, if $(x,y) \in A$ also holds, then $x < y$ and $0 < y - x < 2/K$, and thus $1/(y-x) > K/2$.

By the same argument, $\lim_{\substack{(x,y)\to(0,0) \\ (x,y)\in B}} \frac{1}{y-x} = -\infty$, where $B = \{(x,y) : y < x\}$.
It also follows that the limit $\lim_{(x,y)\to(0,0)} \frac{1}{y-x}$ does not exist.

These three kinds of limits can be described by a single definition with the help of punctured neighborhoods (sometimes called deleted neighborhoods). The **punctured neighborhoods** of a point $a \in \mathbb{R}^p$ are the sets $B(a, r) \setminus \{a\}$, where r is an arbitrary positive number.

Recall that the neighborhoods of ∞ and $-\infty$ are defined as the half-lines (a, ∞) and $(-\infty, a)$, respectively.

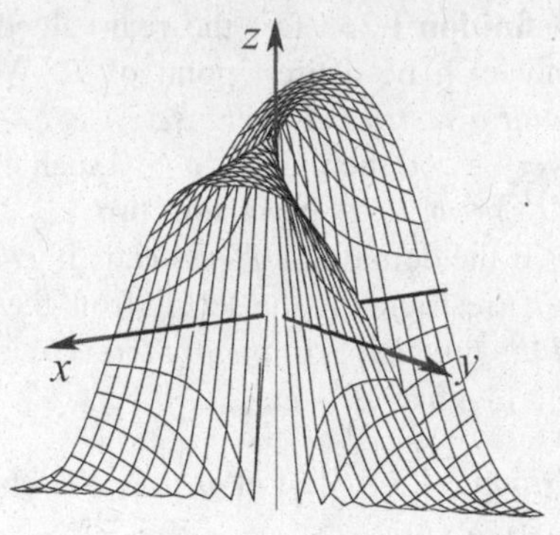

1.10. Figure

Theorem 1.37. *Let the function f be defined on the set $A \subset \mathbb{R}^p$, and let a be a limit point of A. Let β be a real number b or one of $\pm\infty$. Then $\lim_{x \to a,\ x \in A} f(x) = \beta$ holds if and only if for every neighborhood V of β, there exists a punctured neighborhood $\dot{U}$ of a such that $f(x) \in V$ for every $x \in A \cap \dot{U}$.* □

The proof of the following theorem is exactly the same as the proof of its single-variable counterpart (see [7, Theorem 10.19]).

Theorem 1.38. (Transference principle) *Let the function f be defined on the set $A \subset \mathbb{R}^p$, and let a be a limit point of A. Let β be a real number b or one of $\pm\infty$. Then $\lim_{x \to a,\ x \in A} f(x) = \beta$ holds if and only if for every sequence (x_n) with $x_n \to a$ and $x_n \in A \setminus \{a\}$ for every n, we have that $f(x_n) \to \beta$.* □

The following three statements follow easily from the definitions and from the theorems above, combined with their single-variable counterparts. (See [7, Theorems 10.29-10.31].)

Theorem 1.39.

(i) **(Squeeze theorem)** *If $f(x) \le g(x) \le h(x)$ for every $x \in A \setminus \{a\}$ and*

$$\lim_{\substack{x \to a \\ x \in A}} f(x) = \lim_{\substack{x \to a \\ x \in A}} h(x) = \beta,$$

then $\lim_{x \to a,\ x \in A} g(x) = \beta$.

(ii) *If*

$$\lim_{\substack{x \to a \\ x \in A}} f(x) = b < c = \lim_{\substack{x \to a \\ x \in A}} g(x),$$

then there exists a punctured neighborhood $\dot{U}$ of a such that $f(x) < g(x)$ holds for every $x \in \dot{U} \cap A$.

(iii) *If the limits $\lim_{x \to a,\ x \in A} f(x) = b$ and $\lim_{x \to a,\ x \in A} g(x) = c$ exist, and furthermore, if $f(x) \leqq g(x)$ holds at the points of the set $A \setminus \{a\}$, then $b \le c$.* □

From the squeeze theorem and from the corresponding theorems on real sequences we obtain the following.

Theorem 1.40. *Let the limits* $\lim_{x\to a,\, x\in A} f(x) = b$ *and* $\lim_{x\to a,\, x\in A} g(x) = c$ *exist and be finite. Then we have* $\lim_{x\to a,\, x\in A}(f(x) + g(x)) = b + c$, $\lim_{x\to a,\, x\in A} (f(x) \cdot g(x)) = b \cdot c$, *and, assuming also* $c \neq 0$, $\lim_{x\to a,\, x\in A}(f(x)/g(x)) = b/c$. □

Remark 1.41. In the case of one-variable functions, one can define 15 kinds of limits, considering five different options for the location of the limit (a finite point, left- or right-sided limit at a finite point, ∞, and $-\infty$), and three options for the value of the limit (finite, ∞, and $-\infty$).

In the case of multivariable functions the notion of left- and right-sided limits and limits at ∞ and $-\infty$ are meaningless. The reason is clear; for $p > 1$ we have infinitely many directions in $\mathbb{R}^p$, instead of merely two. Obviously, it would be pointless to define limits for every direction; if we really need to talk about limits in a given direction, we can simply take the limit of the function restricted to the corresponding line.

The limit at infinity in a given direction can be viewed as the limit at ∞ of an appropriate single-variable function. For example, if v is a unit vector in the plane, then a half-line starting from the origin in the direction of v is the set of vectors tv $(t > 0)$. Thus the limit of a function at infinity in the direction of v can be viewed as the limit of the single-variable function $t \mapsto f(tv)$ at infinity.

Exercises

1.43. Evaluate the following limits or prove that the limits do not exist for the following two-variable functions at the given points. If the limit exists, find a suitable δ for every $\varepsilon > 0$ (based on the definition of the limit).

(a) $\dfrac{x-2}{y-3}$, $(2,3)$; (b) $\dfrac{x^2y}{x^2+y}$, $(0,0)$;

(c) $x \cdot \sin\dfrac{1}{y}$, $(0,0)$; (d) $\dfrac{x^2-y^2}{x^2+y^2}$, $(0,0)$;

(e) $x + \dfrac{1}{y}$, $(3,2)$; (f) $\dfrac{\sin xy}{y}$, $(0,0)$;

(g) x^y $(x > 0,\ y \in \mathbb{R})$, $(0,0)$; (h) $(1+x)^y$, $(0,0)$;

(i) $\dfrac{x^2y^2}{x+y}$, $(0,0)$; (j) $\dfrac{xy-1}{x-1}$, $(1,1)$;

(k) $\dfrac{\log x}{x-1}$, $(1,1)$; (l) $\dfrac{\sqrt[3]{x^2y^5}}{x^2+y^2}$, $(0,0)$;

(m) $\dfrac{\sin x - \sin y}{x-y}$, $(0,0)$.

1.44. Show that if $A \subset \mathbb{R}^p$ is countable, then there exists a function $f\colon A \to \mathbb{R}$ such that $\lim_{x\to a} f(x) = \infty$ for every point $a \in A'$.

1.45. Show that if $A \subset \mathbb{R}^p$, $f\colon A \to \mathbb{R}$, and $\lim_{x\to a} f(x) = \infty$ for every point $a \in A'$, then A is countable. (H)

1.6 Continuity

Definition 1.42. Let the function f be defined on the set $A \subset \mathbb{R}^p$, and let $a \in A$. We say that f is *continuous at the point a restricted to the set A* if for every $\varepsilon > 0$, there exists $\delta > 0$ such that $x \in A$, $|x-a| < \delta$ imply $|f(x) - f(a)| < \varepsilon$.

If the domain of f is equal to A, we can omit the part "restricted to the set A" in the above definitions, and instead we can say that f is continuous at a.

If the function f is continuous at every point $a \in A$, we say that f *is continuous on the set A*.

Intuitively, the continuity of a function f at a point a means that the graph of f at the point $(a, f(a))$ "does not break."

Remark 1.43. It is obvious from the definition that f is continuous at a point a restricted to the set A if and only if one of the following statements holds:

(i) the point a is an isolated point of A;
(ii) $a \in A \cap A'$ and $\lim_{x\to a,\ x\in A} f(x) = f(a)$.

We can easily prove the following theorem, called the **transference principle for continuity**, with the help of Theorem 1.38.

Theorem 1.44. *The function f is continuous at the point a restricted to the set A if and only if for every sequence (x_n) with $x_n \to a$ and $x_n \in A$ we have $f(x_n) \to f(a)$.* □

While investigating multivariable functions, fixing certain variables at a given value and considering our original function as a function of the remaining variables can make the investigation considerably easier. The functions we get this in way are the **sections** of the original function. For example, the sections of the two-variable function $f(x,y)$ are the single-variable functions $y \mapsto f_a(y) = f(a,y)$ and

$x \mapsto f^b(x) = f(x, b)$, for every $a, b \in \mathbb{R}$. The section f_a is defined at those points y for which the point (a, y) is in the domain $D(f)$ of the function f. Similarly, the section f^b is defined at those points x for which $(x, b) \in D(f)$.

Remark 1.45. It is easy to see that if a function is continuous at the point $(a_1, \ldots, a_p)$, then fixing a subset of the coordinates at the appropriate numbers a_i, we obtain a section that is continuous at $(a_{i_1}, \ldots, a_{i_s})$, where the $i_1, \ldots, i_s$ denote the indices of the nonfixed coordinates. For example, if a two-variable function f is continuous at the point (a, b), then the section f_a is continuous at b, and the section f^b is continuous at a. The converse of the statement is not true. *The continuity of the sections does not imply the continuity of the original function.*

Consider the function $f\colon \mathbb{R}^2 \to \mathbb{R}$, where $f(x, y) = xy/(x^2 + y^2)$ if $(x, y) \neq (0, 0)$, and $f(0, 0) = 0$. (See Figure 1.10.) Every section of f is continuous. Indeed, if $a \neq 0$, then the function $f_a(y) = ay/(a^2 + y^2)$ is continuous everywhere, since it can be written as a rational function whose denominator is never zero (see Theorem 1.48 below). However, for $a = 0$ the function f_a is constant, with the value zero, and thus it is continuous as well. Similarly, the section f^b is continuous for every b.

On the other hand, the function f is not continuous at the point $(0, 0)$, since by Example 1.34.2, it does not even have a limit at $(0, 0)$.

Theorem 1.40 implies the following theorem.

Theorem 1.46. *If the functions f and g are continuous at the point a restricted to the set A, then the same is true for the functions $f + g$ and $f \cdot g$. Furthermore, if $g(a) \neq 0$, then the function f/g is also continuous at the point a.* □

Definition 1.47. We call the function $x = (x_1, \ldots, x_p) \mapsto x_i$, defined on $\mathbb{R}^p$, *the ith coordinate function.*

We call the function $f\colon \mathbb{R}^p \to \mathbb{R}$ a *p-variable polynomial function* (polynomial for short) if we can obtain f from the coordinate functions $x_1, \ldots, x_p$ and constants using only addition and multiplication. Clearly, the polynomials are finite sums of terms of the form $c \cdot x_1^{n_1} \cdots x_p^{n_p}$, where the c coefficients are real numbers and the exponents $n_1, \ldots, n_p$ are nonnegative integers.

We call the quotients of two p-variable polynomials p-variable *rational functions.*

Theorem 1.48. *The polynomials are continuous everywhere. The rational functions are continuous at every point of their domain.*

Proof. First we show that the coordinate functions are continuous everywhere. This follows from the fact that if $|x - a| < \varepsilon$, where $x = (x_1, \ldots, x_p)$ and $a = (a_1, \ldots, a_p)$, then $|x_i - a_i| < \varepsilon$ for every $i = 1, \ldots, p$. From this it is clear, by Theorem 1.46, that the polynomials are continuous everywhere.

If p and q are polynomials, then the domain of the rational function p/q consists of the points where q is not zero. Again, Theorem 1.46 gives that p/q is continuous at those points. □

The following theorem concerns the limits of composite functions.

Theorem 1.49. *Suppose that*

(i) $A \subset \mathbb{R}^p$, $g\colon A \to \mathbb{R}$ *and* $\lim_{x\to a} g(x) = \gamma$, *where* γ *is a real number or one of* $\pm\infty$;

(ii) $g(A) \subset H \subset \mathbb{R}$, $f\colon H \to \mathbb{R}$, *and* $\lim_{y\to\gamma} f(y) = \beta$, *where* β *is a real number or one of* $\pm\infty$;

(iii) $g(x) \neq \gamma$ *in a punctured neighborhood of* a, *or* $\gamma \in H$ *and* f *is continuous at* γ *restricted to* H.

Then

$$\lim_{x\to a} f(g(x)) = \beta. \tag{1.9}$$

Proof. By the transference principle, we have to show that if $x_n \to a$ is a sequence with $x_n \in A \setminus \{a\}$ for each n, then $f(g(x_n)) \to \beta$.

It follows from Theorem 1.38 that $g(x_n) \to \gamma$. If $g(x) \neq \gamma$ in a punctured neighborhood of a, then $g(x_n) \neq \gamma$ for every n large enough. Then, applying Theorem 1.38 again, we find that $f(g(x_n)) \to \beta$. Also, if f is continuous at γ, then Theorem 1.44 gives $f(g(x_n)) \to f(\gamma) = \beta$. Therefore, applying Theorem 1.38, we obtain (1.9). □

Corollary 1.50. *If* g *is continuous at a point* $a \in \mathbb{R}^p$ *restricted to the set* $A \subset \mathbb{R}^p$ *and if the single-variable function* f *is continuous at* $g(a)$ *restricted to* $g(A)$, *then* $f \circ g$ *is also continuous at* a *restricted to* A. □

This corollary implies that all functions obtained from the coordinate functions using elementary functions[11] are continuous on their domain. For example, the three-variable function

$$(x, y, z) \mapsto \frac{e^{\cos(x^2+y)} - z}{1 - xyz}$$

is continuous at every point (x, y, z) such that $xyz \neq 1$.

The familiar theorems concerning continuous functions on bounded and closed intervals (see [7, Theorems 10.52 and 10.55]) can be generalized as follows.

Theorem 1.51. (Weierstrass's theorem) *Let* $A \subset \mathbb{R}^p$ *be nonempty, bounded, and closed, and let* $f\colon A \to \mathbb{R}$ *be continuous. Then* f *is bounded on the set* A, *and the range of* f *has a greatest as well as a least element.*

Proof. Let $M = \sup f(A)$. If f is not bounded from above, then $M = \infty$, and for every n there exists a point $x_n \in A$ such that $f(x_n) > n$. On the other hand, if f is bounded from above, then M is finite, and for every positive integer n there

---

[11] By the elementary functions we mean the polynomial, rational, exponential, power, logarithmic, hyperbolic, and trigonometric functions and their inverses, and all functions that can be obtained from these using basic operations and composition.

exists a point $x_n \in A$ such that $f(x_n) > M - (1/n)$. In both cases we have found a sequence $x_n \in A$ with the property $f(x_n) \to M$.

The sequence (x_n) is bounded (since its terms are in A). Then, by the Bolzano–Weierstrass theorem, it has a convergent subsequence (x_{n_k}). Let $\lim_{k\to\infty} x_{n_k} = a$. Since A is closed, it follows that $a \in A$ by Theorem 1.17. Now, f is continuous at a, and thus the transference principle implies $f(x_{n_k}) \to f(a)$. Thus $M = f(a)$. We obtain that M is finite, whence f is bounded from above, and that $M \in f(A)$; that is, $M = \max f(A)$.

The proof of the existence of $\min f(A)$ is similar. □

Definition 1.52. We say that a function f is *uniformly continuous* on the set $A \subset \mathbb{R}^p$ if for every $\varepsilon > 0$ there exists a uniform δ, i.e., a $\delta > 0$ independent of the location in A such that $x, y \in A$ and $|x - y| < \delta$ imply $|f(x) - f(y)| < \varepsilon$.

Theorem 1.53. (Heine's[12] theorem) *Let $A \subset \mathbb{R}^p$ be bounded and closed, and let $f\colon A \to \mathbb{R}$ be continuous. Then f is uniformly continuous on A.*

Proof. We prove the statement by contradiction. Suppose that f is not uniformly continuous in A. Then there exists $\varepsilon_0 > 0$ for which there does not exist a "good" $\delta > 0$; that is, there is no δ satisfying the requirement formulated in the definition of uniform continuity. Then in particular, $\delta = 1/n$ is not "good" either, that is, for every n there exist $\alpha_n, \beta_n \in A$ for which $|\alpha_n - \beta_n| < 1/n$ but $|f(\alpha_n) - f(\beta_n)| \geq \varepsilon_0$.

Since $\{\alpha_n\} \subset A$ and A is bounded, there exists a convergent subsequence (α_{n_k}) whose limit, α, is also in A, since A is closed. Now we have

$$\beta_{n_k} = (\beta_{n_k} - \alpha_{n_k}) + \alpha_{n_k} \to 0 + \alpha = \alpha.$$

Since f is continuous on A, it is continuous at α (restricted to A). Thus, by the transference principle, $f(\alpha_{n_k}) \to f(\alpha)$ and $f(\beta_{n_k}) \to f(\alpha)$, so

$$\lim_{k\to\infty} \left(f(\alpha_{n_k}) - f(\beta_{n_k})\right) = 0.$$

This, however, contradicts $|f(\alpha_n) - f(\beta_n)| \geq \varepsilon_0$. □

In many different applications of analysis we need to replace the functions involved by simpler functions that approximate the original one and are much easier to handle. An important example is the Weierstrass approximation theorem, which in the one-variable case states that if $f\colon [a, b] \to \mathbb{R}$ is continuous, then for every $\varepsilon > 0$ there exists a polynomial g such that $|f(x) - g(x)| < \varepsilon$ for every $x \in [a, b]$. (See [7, Theorem 13.19].) Our next theorem is the generalization of this theorem to continuous functions of several variables.

Theorem 1.54. (Weierstrass's approximation theorem) *Let the real-valued function f be continuous on the box $R \subset \mathbb{R}^p$. Then for every $\varepsilon > 0$ there exists a p-variable polynomial g such that $|f(x) - g(x)| < \varepsilon$ for every $x \in R$.*

---

[12] Heinrich Eduard Heine (1821–1881), German mathematician.

Proof. We prove the theorem by induction on p. The case $p = 1$ is covered by [7, Theorem 13.19]. (See also Remark 7.85 of this volume, where we give an independent proof.) We now consider the $p = 2$ case.

Let $R = [a,b] \times [c,d]$, and let $0 < \varepsilon < 1$ be fixed. If f is continuous on R, then by Heine's theorem (Theorem 1.53), f is uniformly continuous on R. Choose a positive δ such that $|f(x_1,y_1) - f(x_2,y_2)| < \varepsilon$ holds for every $(x_1,y_1), (x_2,y_2) \in R$ satisfying $|(x_1,y_1) - (x_2,y_2)| < \delta$. We fix an integer $n > 2(b-a)/\delta$ and divide the interval $[a,b]$ into n equal subintervals. Let $a = t_0 < t_1 < \ldots < t_n = b$ be the endpoints of these subintervals.

For every $i = 0, \ldots, n$, let u_i denote the continuous one-variable function that is zero everywhere outside of (t_{i-1}, t_{i+1}), equals 1 at the point t_i, and is linear on the intervals $[t_{i-1}, t_i]$ and $[t_i, t_{i+1}]$. (The numbers $t_{-1} < a$ and $t_{n+1} > b$ can be arbitrarily chosen for the functions u_0 and u_n.) The functions $u_0, \ldots, u_n$ are continuous, and $\sum_{i=0}^{n} u_i(x) = 1$ for every $x \in [a,b]$. Consider the function

$$f_1(x,y) = \sum_{i=0}^{n} f(t_i, y) \cdot u_i(x). \tag{1.10}$$

We show that $|f(x,y) - f_1(x,y)| < \varepsilon$ for every $(x,y) \in R$. For a fixed $(x,y) \in R$, $u_i(x)$ is nonzero only if $|t_i - x| < 2(b-a)/n < \delta$. For every such factor $u_i(x)$ we have $|(t_i, y) - (x,y)| < \delta$, and thus $|f(t_i,y) - f(x,y)| < \varepsilon$ by the choice of δ. Since the sum of the numbers $u_i(x)$ is 1, it follows that

$$\begin{aligned} |f_1(x,y) - f(x,y)| &= \left| \sum_{i=0}^{n} (f(t_i,y) - f(x,y)) \cdot u_i(x) \right| \le \\ &\le \sum_{u_i(x) \ne 0} \varepsilon \cdot u_i(x) = \varepsilon \cdot \sum_{i=0}^{n} u_i(x) = \varepsilon. \end{aligned}$$

By the single-variable version of Weierstrass's approximation theorem, we can choose the polynomials g_i and h_i such that $|f(t_i,y) - g_i(y)| < \varepsilon/(n+1)$ for every $y \in [c,d]$, and $|u_i(x) - h_i(x)| < \varepsilon/(n+1)$ for every $x \in [a,b]$ $(i = 1, \ldots, n)$. Consider the two-variable polynomial $g(x,y) = \sum_{i=0}^{n} g_i(y) \cdot h_i(x)$. We show that g approximates f_1 well on R. Indeed,

$$\begin{aligned} &|f(t_i,y) \cdot u_i(x) - g_i(y) \cdot h_i(x)| \le \\ &\quad \le |f(t_i,y) - g_i(y)| \cdot u_i(x) + |g_i(y)| \cdot |u_i(x) - h_i(x)| \\ &\quad \le (\varepsilon/(n+1)) \cdot 1 + (K+\varepsilon) \cdot (\varepsilon/(n+1)) \le (K+2)\varepsilon/(n+1), \end{aligned}$$

where K denotes an upper bound of $|f|$ on R. Thus

$$|f_1(x,y) - g(x,y)| \le \sum_{i=0}^{n} |f(t_i,y) \cdot u_i(x) - g_i(y) \cdot h_i(x)| \le (K+2)\varepsilon$$

for every $(x,y) \in R$. We get $|f-g| \le |f-f_1| + |f_1-g| < (K+3)\varepsilon$ for each point in the box R. Since ε was arbitrary, we have proved the theorem for $p=2$.

In the general case of the induction step a similar argument works. We leave the details to the reader. □

Remark 1.55. In the previous theorem one can replace the box R by an arbitrary bounded and closed set. More precisely, the following is true: *if the set $A \subset \mathbb{R}^p$ is bounded and closed, and the function $f\colon A \to \mathbb{R}$ is continuous, then for every $\varepsilon > 0$ there exists a p-variable polynomial g such that $|f(x)-g(x)| < \varepsilon$ holds for every $x \in A$.* See Exercises 1.59–1.63.

Exercises

1.46. Let $f(x,y) = xy/(x^2+y^2)^\alpha$ if $(x,y) \neq 0$, and $f(0,0)=0$. For what values of α will f be continuous at the origin?

1.47. Let $f(x,y) = |x|^\alpha |y|^\beta$ if $x \neq 0$ and $y \neq 0$, and let $f(x,y)=0$ otherwise. For what values of α, β will f be continuous at the origin?

1.48. Let $A \subset \mathbb{R}^p$ and $f\colon A \to \mathbb{R}$. Show that if the limit $g(x) = \lim_{y\to x} f(y)$ exists and is finite for every $x \in A$, then g is continuous on A.

1.49. Construct a function $f\colon \mathbb{R}^2 \to \mathbb{R}$ such that f is continuous when restricted to any line, but f is not continuous everywhere. (H)

1.50. Let the function $f\colon \mathbb{R}^2 \to \mathbb{R}$ be such that the section f_a is continuous for every a, and the section f^b is monotone and continuous for every b. Show that f is continuous everywhere.

1.51. Let the set $A \subset \mathbb{R}^p$ be such that every continuous function $f\colon A \to \mathbb{R}$ is bounded. Show that A is bounded and closed.

1.52. Is there a two-variable polynomial with range $(0,\infty)$? (H S)

1.53. Show that if $A \subset \mathbb{R}^p$ is closed and $f\colon A \to \mathbb{R}$ is continuous, then the graph of f is a closed set in $\mathbb{R}^{p+1}$.

1.54. True or false? If the graph of $f\colon [a,b] \to \mathbb{R}$ is a closed set in $\mathbb{R}^2$, then f is continuous on $[a,b]$. (H)

1.55. Let $A \subset \mathbb{R}^p$ and $f\colon A \to \mathbb{R}$. Show that the graph of f is bounded and closed in $\mathbb{R}^{p+1}$ if and only if A is bounded and closed, and f is continuous on A.

1.56. Let $A \subset \mathbb{R}^p$. Which of the following statements is true?

(a) If every function $f\colon A \to \mathbb{R}$ is continuous, then A is closed.
(b) If every function $f\colon A \to \mathbb{R}$ is continuous, then A is bounded.
(c) If every function $f\colon A \to \mathbb{R}$ is uniformly continuous, then A is closed.
(d) If every function $f\colon A \to \mathbb{R}$ is uniformly continuous, then A is bounded.

1.57. Let $A \subset \mathbb{R}^p$. Show that the function $f\colon A \to \mathbb{R}$ is continuous on A if and only for every open interval $I \subset \mathbb{R}$ there exists an open set $G \subset \mathbb{R}^p$ such that $f^{-1}(I) = A \cap G$.

1.58. Show that if $f\colon \mathbb{R}^p \to \mathbb{R}$ is continuous and $g_1, \ldots, g_p\colon [a, b] \to \mathbb{R}$ are integrable on $[a, b]$, then the function $x \mapsto f(g_1(x), \ldots, g_p(x))$ is also integrable on $[a, b]$.

In the next five exercises $A \subset \mathbb{R}^p$ is a fixed bounded and closed set, and $f\colon A \to \mathbb{R}$ is a fixed continuous function.

1.59. Show that for every polynomial h and $\varepsilon > 0$, there exists a polynomial g such that $||h(x)| - g(x)| < \varepsilon$ for every $x \in A$. (S)

1.60. Let $h_1, \ldots, h_n$ be polynomials. Show that for every $\varepsilon > 0$, there exist polynomials g_1, g_2 such that $|\max(h_1(x), \ldots, h_n(x)) - g_1(x)| < \varepsilon$ and $|\min(h_1(x), \ldots, h_n(x)) - g_2(x)| < \varepsilon$ for every $x \in A$. (S)

1.61. Show that for every $a, b \in A$ there exists a polynomial $g_{a,b}$ such that $g_{a,b}(a) = f(a)$ and $g_{a,b}(b) = f(b)$. (S)

1.62. Let $\varepsilon > 0$ be fixed. Show that for every $a \in A$, there exists a polynomial g_a such that $g_a(x) > f(x) - \varepsilon$ for every $x \in A$, and $g_a(a) < f(a) + \varepsilon$. (S)

1.63. Show that if $A \subset \mathbb{R}^p$ is a bounded and closed set and $f\colon A \to \mathbb{R}$ is a continuous function, then for every $\varepsilon > 0$ there exists a p-variable polynomial g such that $|f(x) - g(x)| < \varepsilon$, for every $x \in A$. (S)

1.7 Partial Derivatives

Differentiation of multivariable functions shows more diversity than limits or continuity. Although some of the equivalent definitions of differentiability of one-variable functions have a straightforward generalization to functions of several variables, the notion of derivative is more complicated than that for functions of one variable. For this reason we postpone the discussion of differentiability and the derivative of functions of several variables to the next section and begin with those derivatives that we get by fixing all but one variable and differentiating the resulting single-variable function.

Definition 1.56. Let the function f be defined in a neighborhood of the point $a = (a_1, \ldots, a_p) \in \mathbb{R}^p$. Let us fix each of the coordinates $a = (a_1, \ldots, a_p)$, except for the ith one, and consider the corresponding section of the function:

$$t \mapsto f_i(t) = f(a_1, \ldots, a_{i-1}, t, a_{i+1}, \ldots, a_p). \tag{1.11}$$

We call the derivative of the single-variable function f_i at the point a_i (when it exists) the ith *partial derivative* of the function f at a, and use any of the following notation:[13]

$$\frac{\partial f}{\partial x_i}(a),\ f'_{x_i}(a),\ f_{x_i}(a),\ D_{x_i} f(a),\ D_i f(a).$$

So, for example,

$$D_i f(a) = \lim_{t \to a_i} \frac{f(a_1, \ldots, a_{i-1}, t, a_{i+1}, \ldots, a_p) - f(a)}{t - a_i}, \tag{1.12}$$

assuming that the (finite or infinite) limit exists.

Let the function f be defined on a subset of $\mathbb{R}^p$. By the ith *partial derivative function* of f we mean the function $D_i f$, where $D_i f$ is defined at every point a, where the ith partial derivative of f exists and is finite, and its value is $D_i f(a)$ at these points.

Example 1.57. We get the partial derivatives by fixing all but one of the variables and differentiating the resulting function as a single-variable function. For example, if $f(x, y) = xy(x^2 + y^2 - 1)$, then

$$\frac{\partial f}{\partial x} = D_1 f(x, y) = y(x^2 + y^2 - 1) + xy \cdot 2x = y \cdot (3x^2 + y^2 - 1)$$

and

$$\frac{\partial f}{\partial y} = D_2 f(x, y) = x(x^2 + y^2 - 1) + xy \cdot 2y = x \cdot (x^2 + 3y^2 - 1)$$

at every point (x, y).

Remark 1.58. Continuity does not follow from the existence of finite partial derivatives. Let $f(x, y) = xy/(x^2 + y^2)$ if $(x, y) \neq (0, 0)$, and let $f(0, 0) = 0$. Both partial derivatives of f exist at the origin, and they are both zero, since the sections f_0 and f^0 are both constant with value zero. (It is also clear that the partial derivatives of f exist and are finite at every other point $(x, y) \neq (0, 0)$.)

However, by Example 1.34.2, f is not continuous at the origin.

---

[13] Each of these symbols appears in practice. The symbol $\partial f / \partial x_i$ is used mostly by engineers and physicists and in older books on mathematics; the symbol f_{x_i} appears in the field of partial differential equations. The symbol D_i is used in contemporary pure mathematics; most of the time (though not exclusively) we will also write D_i for the ith partial derivative.

According to one of the most important applications of differentiation of one-variable functions, if a is a local extremum point of the function f and if f is differentiable at a, then $f'(a) = 0$. (See [7, Theorem 12.44, part (v)].) This theorem can easily be generalized to multivariable functions.

Definition 1.59. We say that a function f has a *local maximum* (or *local minimum*) at the point a if a has a neighborhood U such that f is defined on U and for every $x \in U$ we have $f(x) \leq f(a)$ (or $f(x) \geq f(a)$). In this case we say that the point a is a *local maximum point* (or *local minimum point*) of the function f.

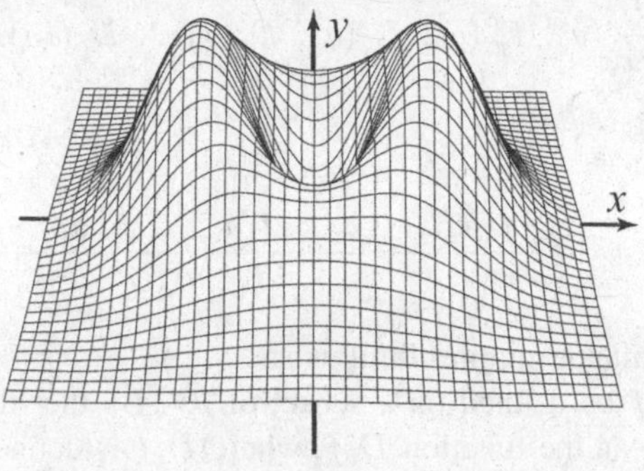

1.11. Figure

If for every point $x \in U \setminus \{a\}$ we have $f(x) < f(a)$ (or $f(x) > f(a)$), then we say that a is a *strict local maximum point (or strict local minimum point)*.

We call the local maximum and local minimum the *local extrema*, while we call the local maximum points and local minimum points *local extremal points*, collectively.

Let f have a local maximum at the point $a = (a_1, \ldots, a_p)$. Obviously, for every $i = 1, \ldots, p$, the function f_i defined by (1.11) also has a local maximum at a_i. If f_i is differentiable at a_i, then $f_i'(a_i) = 0$. It is easy to see that $f_i'(a_i) = \pm\infty$ cannot happen, and thus we have proved the following theorem.

Theorem 1.60. *If the function f has a local extremum at the point $a \in \mathbb{R}^p$, and if the partial derivatives of f exist at a, then $D_i f(a) = 0$ for each $i = 1, \ldots, p$.* □

Applying Theorems 1.51 and 1.60, we can determine the extrema of functions that are continuous on a bounded and closed set A and have partial derivatives in the interior of A. This method, described in the next theorem, corresponds to the technique that finds the extrema of functions of one variable that are continuous on an interval $[a, b]$ and differentiable in (a, b). (See Example 12.46, Remark 12.47, and Example 12.48 in [7].)

Theorem 1.61. *Let $A \subset \mathbb{R}^p$ be bounded and closed, let $f\colon A \to \mathbb{R}$ be continuous, and let the partial derivatives of f exist at every internal point of A. Every point where f takes its greatest (least) value is either a boundary point of A, or else an internal point of A where the partial derivatives $D_i f$ are zero for every $i = 1, \ldots, p$.*

Proof. By Weierstrass's theorem (Theorem 1.51), f has a maximal value on A. Let $a \in A$ be a point where f is the largest. If $a \in \partial A$, then we are done. If, on the other hand, $a \in \operatorname{int} A$, then f has a local maximum at a. By the assumptions of the theorem, the partial derivatives of f exist at the point a; thus $D_i f(a) = 0$ for every $i = 1, \ldots, p$ by Theorem 1.60. □

Example 1.62. **1.** Find the maximum value of the function $f(x, y) = xy(x^2 + y^2 - 1)$ on the disk $K = \{(x, y) \colon x^2 + y^2 \leq 1\}$. In Example 1.10.1.c we saw that the boundary of K is the circle $S = \{(x, y) \colon x^2 + y^2 = 1\}$. Since $S \subset K$, it follows that K is closed. The function f is a polynomial; thus it is continuous (see Theorem 1.48), and then, by Weierstrass's theorem, f has a maximal value on K. The value of f is zero on the whole set S. Since the function f is positive for every $(x, y) \in \operatorname{int} K$, $x > 0$, $y < 0$, it follows that f takes its largest value somewhere in the interior of K.

Let $(a, b) \in \operatorname{int} K$ be a point where the value of f is the largest. Now, $0 = D_1 f(a, b) = b \cdot (3a^2 + b^2 - 1)$ and $0 = D_2 f(a, b) = a \cdot (a^2 + 3b^2 - 1)$.

If $a = 0$, then $b = 0$ (since $|b| < 1$), which is impossible, since the value of the function at the origin is zero, even though its maximal value is positive. Similarly, we can exclude the $b = 0$ case. So, $a \neq 0 \neq b$, whence $a^2 + 3b^2 - 1 = 3a^2 + b^2 - 1 = 0$, and we get $a = \pm 1/2$ and $b = \pm 1/2$. Of these cases, f takes the value $1/8$ at the points $(\pm 1/2, \mp 1/2)$, while it takes the value $-1/8$ at the points $(\pm 1/2, \pm 1/2)$. Thus, the largest value of f is $1/8$, and f takes this value at two points, namely at $(\pm 1/2, \mp 1/2)$.

2. Find the triangle with largest area that can be inscribed in a circle with a fixed radius.

Consider a triangle H defined by its three vertices that lie on the circle $S = \{(u, v) \colon u^2 + v^2 = 1\}$. If the angles between the segments connecting the origin and the vertices are x, y, z, then we can compute the area of H with the help of the formula $(\sin x + \sin y + \sin z)/2$. (This follows from the fact that if the two equal sides of an isosceles triangle are of unit length, and the angle between these two sides is α, then the area of the triangle is $\frac{1}{2} \cdot \sin \alpha$.) This is true even if one of the angles x, y, z is larger than π. Since $z = 2\pi - x - y$, we need to find the maximum value of the function $f(x, y) = \sin x + \sin y - \sin(x + y)$ on the set $A = \{(x, y) \colon x \geq 0,\ y \geq 0,\ x + y \leq 2\pi\}$.

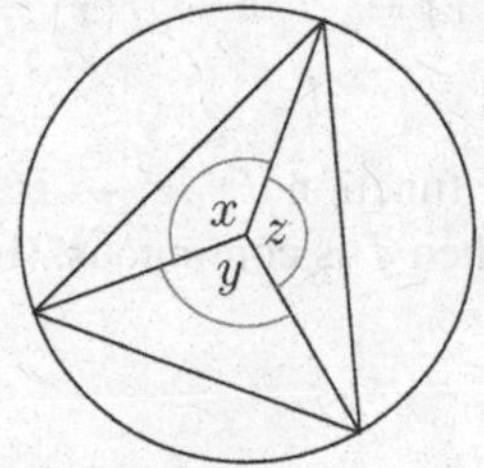

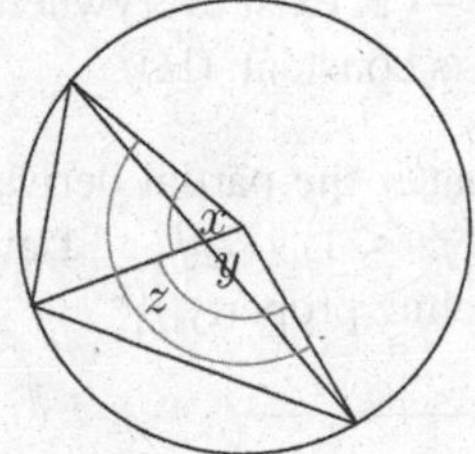

1.12. Figure

The set A is nothing other than the triangle defined by the points $(0,0)$, $(2\pi,0)$, and $(0,2\pi)$. Obviously, this is a bounded and closed set, and thus Theorem 1.61 can be applied.

It is easy to see that the function f is zero on the boundary of the set A. Since f takes a positive value (e.g., at the point $(\pi/2,\pi/2)$), it follows that f takes its maximum at an internal point (x,y), for which $D_1f(x,y) = \cos x - \cos(x+y) = 0$ and $D_2f(x,y) = \cos y - \cos(x+y) = 0$. We get $\cos x = \cos y$, so either $y = 2\pi - x$ or $y = x$. In the first case, (x,y) lies on the boundary of A, which is impossible. Thus $y = x$ and $\cos x = \cos 2x$. Since $x = 2x$ is not possible (it would imply that $x = 0$, whence (x,y) would be on the boundary of A again), we must have $2x = 2\pi - x$, and $x = y = 2\pi/3$. We have proved that *the triangle with the largest area that can be inscribed in a circle with fixed radius is an equilateral triangle.* □

Exercises

1.64. Find the points where the partial derivatives of the following two-variable functions exist.

(a) $|x+y|$;
(b) $\sqrt[3]{x^3+y^3}$;
(c) $f(x,y) = x$ if $x \in \mathbb{Q}$, $f(x,y) = y$ if $x \notin \mathbb{Q}$.

1.65. Show that the partial derivatives of the function $f(x,y) = xy/\sqrt{x^2+y^2}$, $f(0,0) = 0$ exist and are bounded everywhere in the plane.

1.66. Construct a two-variable function whose partial derivatives exist everywhere, but the function is unbounded in every neighborhood of the origin.

1.67. Let $f\colon \mathbb{R}^2 \to \mathbb{R}$. Show that if $D_1f \equiv 0$, then f depends only on the variable y. If $D_2f \equiv 0$, then f depends only on the variable x.

1.68. Show that if $f\colon \mathbb{R}^2 \to \mathbb{R}$, $D_1f \equiv 0$, and $D_2f \equiv 0$, then f is constant.

1.69. Show that if $G \subset \mathbb{R}^p$ is a connected open set, the partial derivatives of the function $f\colon G \to \mathbb{R}$ exist everywhere, and $D_1f(x) = \ldots = D_pf(x) = 0$ for every $x \in G$, then f is constant. (H)

1.70. Show that if the partial derivatives of the function $f\colon \mathbb{R}^2 \to \mathbb{R}$ exist everywhere and $|D_1f| \le 1$, $|D_2f| \le 1$ everywhere, then f is continuous. (Furthermore, f has the Lipschitz property.)[14]

---

[14] Rudolph Otto Sigismund Lipschitz (1832–1903), German mathematician. A function f is said to have the *Lipschitz property* (is Lipschitz, for short) on a set A if there exists a constant $K \ge 0$ such that $|f(x_1) - f(x_0)| \le K \cdot |x_1 - x_0|$ for all $x_0, x_1 \in A$.

1.71. Construct a two-variable polynomial that has two local maximum points but no local minimum points. (H S)

1.72. Find the local extremal points of the function $x^2 + xy + y^2 - 4\log x - 10\log y$.

1.73. Find the maximum of $x^3 + y^2 - xy$ on the square $[0,1] \times [0,1]$.

1.74. Find the minimum of $x + \frac{y^2}{4x} + \frac{z^2}{y} + \frac{2}{z}$ in the octant $x, y, z > 0$. (First prove that the function can be restricted to a bounded and closed set.)

1.75. Find the minimum of $(x^3 + y^3 + z^3)/(xyz)$ on the set $\{(x, y, z \in \mathbb{R}^3 : x, y, z > 0\}$.

1.76. Find the maximum and minimum values of $xy \cdot \log(x^2 + y^2)$ on the disk $x^2 + y^2 \le R^2$.

1.77. Find the maximum of $x^{\sqrt{2}} \cdot y^e \cdot z^\pi$ restricted to $x, y, z \ge 0$ and $x + y + z = 1$.

1.78. Find the minimum value of the function $2x^4 + y^4 - x^2 - 2y^2$.

1.79. What is the minimum value of $xy + \frac{50}{x} + \frac{20}{y}$ on the set $x, y > 0$?

1.8 Differentiability

Weierstrass's approximation theorem states that if f is a continuous function defined on a box (or, more generally, on a closed and bounded set), then f can be approximated by polynomials (see Theorem 1.54 and Exercises 1.59–1.63). However, we cannot control the degree of the approximating polynomials: in general, it may happen that we need polynomials of arbitrarily high degrees for the approximation. The situation is different in the case of *local* approximation, when we want to approximate a function in a neighborhood of a given point. For an important class of functions, good local approximation is possible using polynomials of first degree.

In the case of single-variable analysis, differentiability is equivalent to local approximability by first-degree polynomials (see [7, Theorem 12.9]). For multivariable functions, differential quotients do not have an immediate equivalent (since we cannot divide by vectors), and therefore, we cannot define differentiability via the limits of differential quotients. Approximability by first-degree polynomials, however, can be generalized verbatim to multivariable functions.

We call the function $\ell \colon \mathbb{R}^p \to \mathbb{R}$ a **linear function** if there exist real numbers $\alpha_1, \ldots, \alpha_p$ such that $\ell(x) = \alpha_1 x_1 + \ldots + \alpha_p x_p$ for every $x = (x_1, \ldots, x_p) \in \mathbb{R}^p$.

Definition 1.63. Let the function f be defined in a neighborhood of the point $a \in \mathbb{R}^p$. We say that f is *differentiable at the point a if there exists a linear function $\ell(x)$ such that*

$$f(x) = f(a) + \ell(x - a) + \varepsilon(x) \cdot |x - a| \tag{1.13}$$

for every $x \in D(f)$, where $\varepsilon(x) \to 0$ as $x \to a$.

Remark 1.64. **1.** It is clear that the function f is differentiable at the point a if and only if it is defined in a neighborhood of $a \in \mathbb{R}^p$ and there exists a linear function $\ell(x)$ such that

$$\lim_{x\to a} \frac{f(x) - f(a) - \ell(x-a)}{|x-a|} = 0.$$

2. If $p = 1$, then the notion of differentiability defined above is equivalent to the "usual" definition, that is, to the existence of a finite limit of the differential quotient $(f(x) - f(a))/(x-a)$ as $x \to a$.
3. If a function depends only on one of its variables, then the differentiability of the function is equivalent to the differentiability of the corresponding single-variable function. More precisely, let $a_1 \in \mathbb{R}$, and let a single-variable function f be defined in a neighborhood of a_1. Let $g(x_1,\ldots,x_p) = f(x_1)$ for every $x_1 \in D(f)$ and $x_2,\ldots,x_p \in \mathbb{R}$. For arbitrary $a_2,\ldots,a_p$, the function g is differentiable at the point $a = (a_1,\ldots,a_p)$ if and only if f is differentiable at a_1 (see Exercise 1.82).

Example 1.65. **1.** It follows from the definition that every polynomial of degree at most one is differentiable everywhere.
2. Let $f(x,y) = \frac{x^2y^2}{x^2+y^2}$ if $(x,y) \neq (0,0)$, and let $f(0,0) = 0$. We prove that f is differentiable at the origin. Indeed, if ℓ is the constant zero function and $(x,y) \neq (0,0)$, then we have

$$\left|\frac{f(x,y) - \ell(x,y)}{|(x,y)|}\right| = \frac{x^2y^2}{(x^2+y^2)\cdot\sqrt{x^2+y^2}} = \frac{x^2y^2}{(x^2+y^2)^{3/2}} \le$$
$$\le \frac{\max(x^2,y^2)^2}{\max(x^2,y^2)^{3/2}} = \max(x^2,y^2)^{1/2},$$

and (1.13) holds.

We know that every single-variable, differentiable function is continuous (see [7, Theorem 12.4]). The following theorem generalizes this fact for functions with an arbitrary number of variables.

Theorem 1.66. *If the function f is differentiable at a point a, then f is continuous at a.*

Proof. Since the right-hand side of (1.13) converges to $f(a)$ as $x \to a$, it follows that

$$\lim_{x\to a} f(x) = f(a). \square$$

The following fundamental theorem represents the linear functions of the definition of differentiability with the help of partial derivatives.

Theorem 1.67. *If a function f is differentiable at a point $a = (a_1, \ldots, a_p) \in \mathbb{R}^p$, then*

(i) *every partial derivative of f exists and is finite at a, and furthermore,*
(ii) *there is only one function ℓ satisfying Definition 1.63, namely the function*

$$\ell(x) = D_1 f(a)x_1 + \ldots + D_p f(a)x_p.$$

Proof. Suppose that (1.13) holds for the linear function $\ell = \alpha_1 x_1 + \ldots + \alpha_p x_p$. Let i be fixed, and apply (1.13) with the point $x = (a_1, \ldots, a_{i-1}, t, a_{i+1}, \ldots, a_p)$. We get that

$$f_i(t) = f(a) + \alpha_i(t - a_i) + \varepsilon(x) \cdot |t - a_i|,$$

where f_i is the function defined at (1.11). Since $f_i(a_i) = f(a)$, we have

$$\frac{f_i(t) - f_i(a_i)}{t - a_i} = \alpha_i \pm \varepsilon(x),$$

and thus by $\lim_{x \to a} \varepsilon(x) = 0$, we obtain that f_i is differentiable at the point a_i, and $f_i'(a_i) = \alpha_i$. Therefore, by the definition of the partial derivatives, $D_i f(a) = \alpha_i$. This is true for every $i = 1, \ldots, p$, and thus (i) and (ii) are proved. □

Corollary 1.68. *Let f be defined in a neighborhood of $a \in \mathbb{R}^p$. The function f is differentiable at the point $a \in \mathbb{R}^p$ if and only if all partial derivatives of f exist at a, they are finite, and*

$$f(x) = f(a) + D_1 f(a)(x_1 - a_1) + \ldots + D_p f(a)(x_p - a_p) + \varepsilon(x) \cdot |x - a| \tag{1.14}$$

for every $x \in D(f)$, where $\lim_{x \to a} \varepsilon(x) = 0$. □

Example 1.69. **1.** We show that the function $f(x, y) = xy$ is differentiable at $(1, 2)$. Since $D_1 f(1, 2) = 2$ and $D_2 f(1, 2) = 1$, we need to prove

$$\lim_{(x,y) \to (1,2)} \frac{xy - 2 - 2(x - 1) - (y - 2)}{\sqrt{(x - 1)^2 + (y - 2)^2}} = 0.$$

Considering that the numerator is $(x - 1)(y - 2)$ and

$$\left| \frac{(x - 1)(y - 2)}{\sqrt{(x - 1)^2 + (y - 2)^2}} \right| \le |y - 2| \to 0$$

as $(x, y) \to (1, 2)$, we obtain that indeed, f is differentiable at $(1, 2)$.
2. The function $|x|$ is continuous but not differentiable at 0. This is true in the multivariable case as well. Indeed, the partial derivatives of $|x| = \sqrt{x_1^2 + \ldots + x_p^2}$ do not exist at the origin. Since $|x| = |t|$ at the point $x = (0, \ldots, 0, t, 0, \ldots, 0)$, the

fraction of the right-hand side on (1.12) is $\frac{|t|-|0|}{t-0}$, which does not have a limit as $t \to 0$. Therefore, by Theorem 1.67, $|x|$ is not differentiable at the origin.
3. Consider the function $f(x, y) = \sqrt{|xy|}$ on $\mathbb{R}^2$. By Corollary 1.50, f is continuous everywhere. We prove that f is not differentiable at the origin. In contrast to our previous example, the partial derivatives do exist at the origin. Indeed, the sections f_0 and f^0 are both zero, and hence their derivatives are also constant and equal to zero, i.e., $D_1 f(0, 0) = D_2 f(0, 0) = 0$.

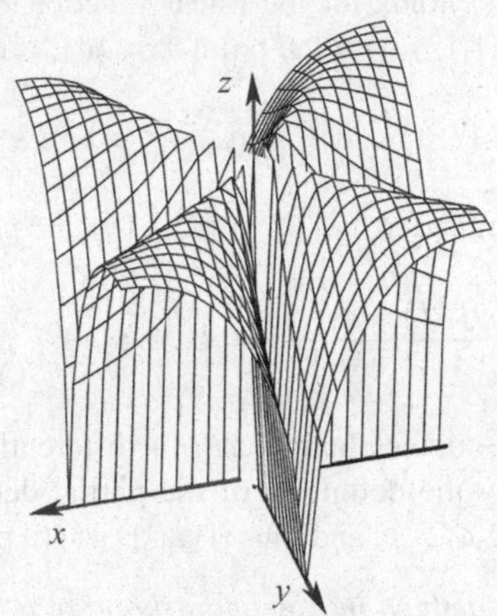

1.13. Figure The graph of the function $\sqrt{|xy|}/\sqrt{x^2 + y^2}$

Now, f is differentiable at the origin if and only if

$$\lim_{(x,y)\to(0,0)} \frac{\sqrt{|xy|}}{\sqrt{x^2 + y^2}} = 0 \tag{1.15}$$

holds (see Corollary 1.68). However, the value of the fraction on the line $y = x$ is $1/\sqrt{2}$, and consequently, there exists a point in every neighborhood of $(0, 0)$ where the fraction is $1/\sqrt{2}$. Thus (1.15) does not hold, and f is not differentiable at the point $(0, 0)$.

The right-hand side of the equality (1.14) can be simplified if we notice that $D_1 f(a)(x_1 - a_1) + \ldots + D_p f(a)(x_p - a_p)$ is nothing other than the scalar product of the vectors $(D_1 f(a), \ldots, D_p f(a))$ and $x - a$. This motivates the following definition.

Definition 1.70. If f is differentiable at the point $a \in \mathbb{R}^p$, then the vector

$$(D_1 f(a), \ldots, D_p f(a))$$

is said to be the *derivative vector* of f at a and is denoted by $f'(a)$.

Using the notation above, (1.14) becomes $f(x) = f(a) + \langle f'(a), x - a\rangle + \varepsilon(x) \cdot |x - a|$. In the single-variable case this is the well-known formula $f(x) = f'(a) \cdot (x - a) + f(a) + \varepsilon(x) \cdot |x - a|$.

The following theorem gives a useful sufficient condition for differentiability.

Theorem 1.71. *Let f be defined in a neighborhood of $a \in \mathbb{R}^p$. If the partial derivatives of f exist in a neighborhood of a and they are continuous at a, then f is differentiable at a.*

Proof. We prove the result for $p = 3$. It is straightforward to generalize the proof for an arbitrary p.

Let $\varepsilon > 0$ be fixed. Since the partial derivatives of f exist in a neighborhood of a and they are continuous at a, there exists $\delta > 0$ such that $|D_i f(x) - D_i f(a)| < \varepsilon$ for every $x \in B(a, \delta)$ and $i = 1, 2, 3$.

Let us fix $x = (x_1, x_2, x_3) \in B(a, \delta)$ and connect the points $a = (a_1, a_2, a_3)$ and x with a polygonal line consisting of at most three segments, each parallel to one of the axes. Let $u = (x_1, a_2, a_3)$ and $v = (x_1, x_2, a_3)$. The segment $[a, u]$ is parallel to the x-axis (including the possibility that it is reduced to a point), the segment $[u, v]$ is parallel to the y-axis, and the segment $[v, x]$ is parallel to the z-axis.

The partial derivative $D_1 f$ exists and is finite at each point of the segment $[a, u]$, and thus the section $t \mapsto f(t, a_2, a_3)$ is differentiable on the interval $[a_1, x_1]$, and its derivative is $D_1 f(t, a_2, a_3)$ there. By the mean value theorem,[15] there is a point $c_1 \in [a_1, x_1]$ such that

$$f(u) - f(a) = f(x_1, a_2, a_3) - f(a_1, a_2, a_3) = D_1 f(c_1, a_2, a_3) \cdot (x_1 - a_1).$$

Since $(c_1, a_2, a_3) \in B(a, \delta)$, we have $|D_1 f(c_1, a_2, a_3) - D_1 f(a)| < \varepsilon$, and thus

$$|f(u) - f(a) - D_1 f(a)(x_1 - a_1)| \le \varepsilon \cdot |x_1 - a_1| \le \varepsilon \cdot |x - a| \tag{1.16}$$

follows. Similarly, the partial derivative $D_2 f$ exists and is finite everywhere on the segment $[u, v]$; thus the section $t \mapsto f(x_1, t, a_3)$ is differentiable on the interval $[a_2, x_2]$, and its derivative is $D_2 f(x_1, t, a_3)$ there. By the mean value theorem, there is a point $c_2 \in [a_2, x_2]$ such that

$$f(v) - f(u) = f(x_1, x_2, a_3) - f(x_1, a_2, a_3) = D_2 f(x_1, c_2, a_3) \cdot (x_2 - a_2).$$

Since $(x_1, c_2, a_3) \in B(a, \delta)$, it follows that $|D_2 f(x_1, c_2, a_3) - D_2 f(a)| < \varepsilon$, and

$$|f(v) - f(u) - D_2 f(a)(x_2 - a_2)| \le \varepsilon \cdot |x_2 - a_2| \le \varepsilon \cdot |x - a|. \tag{1.17}$$

---

[15] The mean value theorem states that if $g \colon [a, b] \to \mathbb{R}$ is continuous on $[a, b]$ and differentiable on (a, b), then there is a point $c \in (a, b)$ such that $g'(c) = (g(b) - g(a))/(b - a)$. See [7, Theorem 12.50].

By the same argument we obtain

$$|f(x) - f(v) - D_3 f(a)(x_3 - a_3)| \le \varepsilon \cdot |x - a|. \tag{1.18}$$

Applying the triangle inequality yields

$$\begin{aligned}\big|f(x) - (D_1 f(a)(x_1 - a_1) + &D_2 f(a)(x_2 - a_2) + D_3 f(a)(x_3 - a_3) + f(a))\big| \le \\ &\le |f(u) - f(a) - D_1 f(a)(x_1 - a_1)| + \\ &+ |f(v) - f(u) - D_2 f(a)(x_2 - a_2)| + \\ &+ |f(x) - f(v) - D_3 f(a)(x_3 - a_3)|,\end{aligned}$$

whence the approximations (1.16), (1.17), and (1.18) give

$$\big|f(x) - (D_1 f(a)(x_1 - a_1) + D_2 f(a)(x_2 - a_2) + D_3 f(a)(x_3 - a_3) + f(a))\big| \le 3\varepsilon \cdot |x - a|.$$

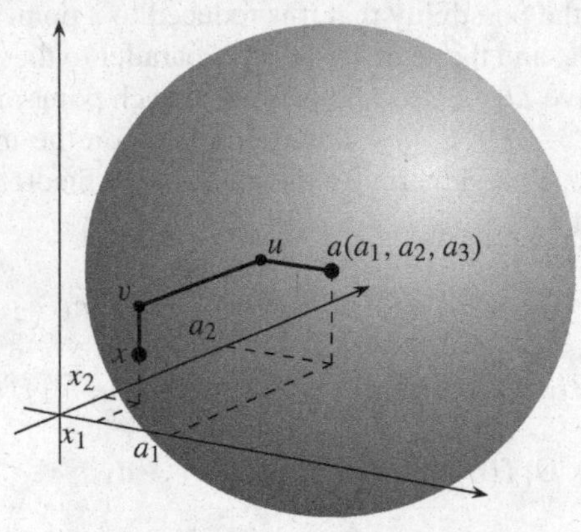

1.14. Figure

Since ε was arbitrary, we have

$$\lim_{x \to a} \frac{f(x) - (D_1 f(a)(x_1 - a_1) + D_2 f(a)(x_2 - a_2) + D_3 f(a)(x_3 - a_3) + f(a))}{|x - a|} = 0,$$

and f is differentiable at the point a. □

Corollary 1.72. *The polynomial functions are differentiable everywhere. The rational functions are differentiable at every point of their domain.*

Proof. The partial derivative functions of a polynomial p are also polynomials, and by Theorem 1.48, they are continuous everywhere. Hence, by Theorem 1.71, p is differentiable everywhere.

The partial derivative functions of a rational function r are also rational functions, and they have the same domain as r. These partial derivatives are continuous in the domain of r by Theorem 1.48, and thus Theorem 1.71 gives that r is differentiable on its whole domain. □

Remark 1.73. By Theorems 1.66, 1.67, and 1.71 we have the following:

(i) *if f is differentiable at a point a, then f is continuous at a, and its partial derivatives exist and are finite at a;* furthermore,

(ii) *if the partial derivatives of f exist in a neighborhood of a and are continuous at a, then f is differentiable at a.*

We prove that the converses of these implications are not true.

Let $f(x,y) = \frac{x^2 y}{x^2+y^2}$ if $(x,y) \neq (0,0)$, and let $f(0,0) = 0$. In Example 1.34.1 we proved that the limit of f at $(0,0)$ is zero, and thus f is continuous at the origin. (Furthermore, f is continuous everywhere by Theorem 1.48.) The partial derivatives of f exist everywhere. If $a \neq 0$, then the section $f_a(y) = a^2 y/(a^2+y^2)$ is differentiable everywhere, and if $a = 0$, then f_a is zero everywhere; thus it is also differentiable everywhere. The same is true for the sections f^b. Thus the partial derivatives $D_1 f$, $D_2 f$ exist everywhere and $D_1 f(0,0) = D_2 f(0,0) = 0$.

By Theorem 1.67, f is differentiable at the origin if and only if

$$\lim_{(x,y)\to(0,0)} \frac{x^2 y}{(x^2+y^2)\sqrt{x^2+y^2}} = 0. \tag{1.19}$$

However, the value of the fraction is $\pm 1/2\sqrt{2}$ at every point of the line $y = x$, and hence there exists a point in every neighborhood of $(0,0)$ where the fraction takes the value $\pm 1/2\sqrt{2}$. Therefore, (1.19) does not hold, and f is not differentiable at $(0,0)$. We have shown that the converse of statement (i) is not true.

One can check that the function $f(x) = x^2 \cdot \sin(1/x)$, $f(0) = 0$, is differentiable everywhere on $\mathbb{R}$, but its derivative is not continuous at zero (see [7, Example 13.43])). This function shows that the converse of statement (ii) is not true for single-variable functions. By Remark 1.64.3, $g(x_1,\ldots,x_p) = f(x_1)$ is differentiable everywhere on $\mathbb{R}^p$, and since $D_1 g(x_1,\ldots,x_p) = f'(x_1)$ for every $x \in \mathbb{R}^p$, the partial derivative $D_1 g$ is not continuous at the origin. We have therefore shown that the converse of (ii) is also not true for p-variable functions.

If f is a differentiable function of one variable, then the graph of the first-degree polynomial approximating f in a neighborhood of a is nothing but the tangent of the graph of f at the point $(a, f(a))$. We want to find an analogous statement in the multivariable case.

In three dimensions, planes are given by equations of the form $a_1x_1 + a_2x_2 + a_3x_3 = b$, where at least one of the coefficients a_1, a_2, a_3 is nonzero. This can be shown as follows. Let S be a plane and let c be a point in S. Let a be a nonzero vector perpendicular to the plane S. We know that a point x is a point of the plane S if and only if the vector $x - c$ is perpendicular to a, i.e., if $\langle x - c, a\rangle = 0$. Thus, $x \in S$

if and only if $\langle a, x\rangle = \langle a, c\rangle$. Using the notation $x = (x_1, x_2, x_3)$, $a = (a_1, a_2, a_3)$, and $c = (c_1, c_2, c_3)$ we have that $x \in S$ if and only if $a_1x_1 + a_2x_2 + a_3x_3 = b$, where $b = \langle a, c\rangle$.

Now suppose that $a_1, a_2, a_3, b \in \mathbb{R}$, and at least one of a_1, a_2, a_3 is nonzero. Let $a = (a_1, a_2, a_3)$. Choose a vector c such that $\langle a, c\rangle = b$. Obviously, the equality $a_1x_1 + a_2x_2 + a_3x_3 = b$ holds if and only if $\langle x - c, a\rangle = 0$, i.e., if the vector $x - c$ is perpendicular to a. We get that $a_1x_1 + a_2x_2 + a_3x_3 = b$ is the equation of the plane containing the point c and perpendicular to the vector a.

Let $g(x_1, x_2) = a_1x_1 + a_2x_2 + b$ be a first-degree polynomial. Then the graph of g, i.e., the set $\{(x_1, x_2, x_3) : x_3 = a_1x_1 + a_2x_2 + b\}$, is a plane. Conversely, if $a_1x_1 + a_2x_2 + a_3x_3 = b$ is the equation of a plane S that satisfies $a_3 \neq 0$, then S is the graph of the first-degree polynomial $g(x_1, x_2) = -(a_1/a_3)x_1 - (a_2/a_3)x_2 + (b/a_3)$.

We can now generalize the definition of the tangent to the case of two-variable functions. Let us switch from the coordinate notation (x_1, x_2, x_3) to the notation (x, y, z).

Definition 1.74. Let $(a, b) \in \mathbb{R}^2$ be fixed, and let f be defined in a neighborhood of the point (a, b). We say that the plane S is the **tangent plane** of graph f at the point $(a, b, f(a, b))$ if S contains the point $(a, b, f(a, b))$, and S is the graph of a first-degree polynomial g that satisfies

$$\lim_{(x,y)\to(a,b)} \frac{f(x, y) - g(x, y)}{|(x, y) - (a, b)|} = 0.$$

It is clear from Remark 1.64.1 that the graph of f has a tangent plane at the point $(a, b, f(a, b))$ if and only if f is differentiable at (a, b). Using the definition above and Corollary 1.68, it is also obvious that the equation of the tangent plane is

$$z = D_1 f(a, b)(x - a) + D_2 f(a, b)(y - b) + f(a, b).$$

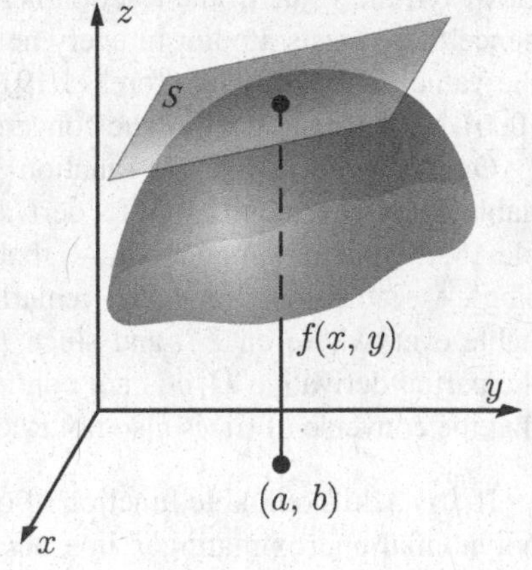

1.15. Figure

These concepts can be generalized to functions with an arbitrary number of variables. We call the set of points of the space $\mathbb{R}^{p+1}$ that satisfy the equality $a_1x_1 + \ldots + a_{p+1}x_{p+1} = b$ a **hyperplane** of $\mathbb{R}^{p+1}$, where at least one of the coefficients $a_1, \ldots, a_{p+1}$ is nonzero.

Definition 1.75. Let f be defined in a neighborhood of the point $u = (u_1, \ldots, u_p) \in \mathbb{R}^p$. We say that the hyperplane $H \subset \mathbb{R}^{p+1}$ is the *tangent hyperplane* of the graph graph f at the point $v = (u_1, \ldots, u_p, f(u_1, \ldots, u_p))$ if S contains the point v, and H is the graph of a first-degree polynomial g that satisfies

$$\lim_{x \to u} (f(x) - g(x))/|x - u| = 0.$$

It is easy to see that the graph of f has a tangent hyperplane at the point v if and only if f is differentiable at u. In this case, the equation of the tangent hyperplane is $x_{p+1} = \langle f'(a), x - a\rangle + f(a)$.

Note that the concept of the tangent and the tangent plane can be defined for every subset of $\mathbb{R}^p$. The tangent and the tangent plane of the graph of a function are just special cases of the general definition. The reader can find more on this in the appendix of this chapter.

Let f be defined in a neighborhood of $a \in \mathbb{R}^p$, and let $v \in \mathbb{R}^p$ be a unit vector. The function $t \mapsto f(a + tv)$ $(t \in \mathbb{R})$ is defined in a neighborhood of 0. The value of $f(a + tv)$ is the height of the graph of the function f at the point $a + tv$. (If $p = 2$, then the graph of the function $t \mapsto f(a + tv)$ can be illustrated by intersecting the graph of f by the vertical plane containing the line $a + tv$ $(t \in \mathbb{R})$ and the point $(a, f(a))$ of the graph.) In this way, $t \mapsto f(a + tv)$ describes the "climbing" we do as we start from the point $(a, f(a))$ on the graph of f and walk in the direction of v. Intuitively it is clear that the derivative of the function $t \mapsto f(a + tv)$ at the point 0 (if it exists) tells us how steep a slope we need to climb at the point $(a, f(a))$. We arc descending when the derivative is negative, and ascending when the derivative is positive.

Definition 1.76. Let $v \in \mathbb{R}^p$ be a unit vector. We call the derivative of the function $t \mapsto f(a + tv)$ at the point 0 (if it exists) the *directional derivative* of the function f at the point a and in the direction v. Notation: $\frac{\partial f}{\partial v}(a)$ or $D_v f(a)$. In other words,

$$D_v f(a) = \lim_{t \to 0} \frac{f(a + tv) - f(a)}{t},$$

assuming that the limit exists.

Theorem 1.77. *If the function f is differentiable at $a \in \mathbb{R}^p$, then the single-variable function $t \mapsto f(a + tv)$ is differentiable at 0 for every vector $v \in \mathbb{R}^p$, and its derivative is $\langle f'(a), v\rangle$. In particular, if $|v| = 1$, then the directional derivative $D_v f(a)$ exists and its value is $D_v f(a) = \langle f'(a), v\rangle$.*

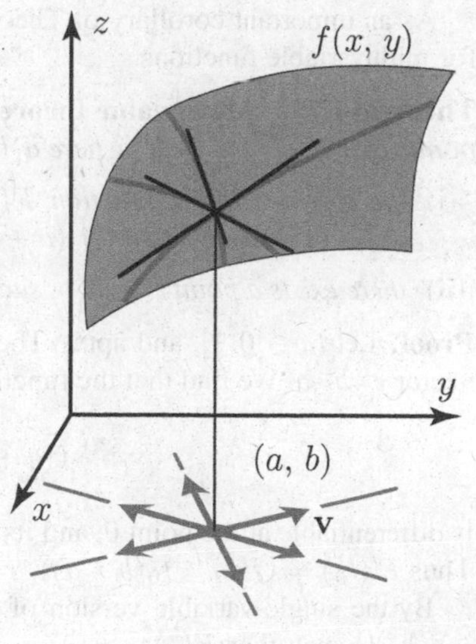

1.16. Figure

Proof. By Corollary 1.68 we have

$$f(a+tv) = f(a) + \langle f'(a), tv\rangle + \varepsilon(a+tv)\cdot |tv|,$$

i.e.,

$$\frac{f(a+tv)-f(a)}{t} = \langle f'(a), v\rangle \pm \varepsilon(a+tv)\cdot |v|$$

for every $t \neq 0$ satisfying $a+tv \in D(f)$. Since $\lim_{x\to a}\varepsilon(x) = 0$ implies $\lim_{t\to 0}$ $\varepsilon(a+tv) = 0$, we have $(f(a+tv)-f(a))/t \to \langle f'(a), v\rangle$ as $t \to 0$. Thus we have proved the first statement of the theorem. The second statement is obvious from the first one. □

Remark 1.78. **1.** The partial derivative $D_i f(a)$ is the same as the directional derivative in the direction v_i, where v_i is the vector whose coordinates are all zero except for its ith coordinate, which is 1. This follows directly from the definitions. Furthermore, if f is differentiable at a, this also follows from the formula $D_v f(a) = \langle f'(a), v\rangle$.
2. Suppose that at least one of the partial derivatives $D_i f(a)$ is nonzero, i.e., the derivative vector $f'(a)$ is not the zero vector. If $|v| = 1$, then $\langle f'(a), v\rangle = |f'(a)|\cdot\cos\alpha$, where α is the angle between vectors $f'(a)$ and v (see page 3). Therefore, $\langle f'(a), v\rangle \le |f'(a)|$, and equality holds only if the directions of the vectors v and $f'(a)$ are the same. In other words, the "climbing" of the graph of f is the steepest in the direction of the vector $f'(a)$. Because of this, we also call the derivative vector $f'(a)$ the **gradient**.
3. It is possible that the directional derivative $D_v f(a)$ exists for every $|v| = 1$ yet f is not differentiable at a (see Exercise 1.89).

As an important corollary of Theorem 1.77, we obtain the mean value theorem for multivariable functions.

Theorem 1.79. (Mean value theorem) *Let the function f be differentiable at the points of the segment $[a, b]$, where $a, b \in \mathbb{R}^p$. Then*

(i) *the single-variable function $F(t) = f(a+t(b-a))$ $(t \in [0,1])$ is differentiable in $[0,1]$, $F'(t) = \langle f'(a+t(b-a)), b-a\rangle$ for every $t \in [0,1]$, and*

(ii) *there exists a point $c \in [a,b]$ such that $f(b) - f(a) = \langle f'(c), b-a\rangle$.*

Proof. Let $t_0 \in [0,1]$, and apply Theorem 1.77 to the point $a + t_0(b-a)$ and the vector $v{=}b{-}a$. We find that the function

$$t \mapsto f(a + (t_0 + t)(b-a))$$

is differentiable at the point 0, and its derivative is $\langle f'(a+t_0(b-a)), b-a\rangle$ there. Thus $F'(t_0) = \langle f'(a+t_0(b-a)), b-a\rangle$, which proves (i).

By the single-variable version of the mean value theorem, there exists a point $u \in [0,1]$ such that $F(1) - F(0) = F'(u)$. Since $F(0) = f(a)$ and $F(1) = f(b)$, by applying (i) we have $f(b) - f(a) = \langle f'(c), b-a\rangle$, where $c = a + u(b-a)$. □

Exercises

1.80. Which of the following functions are differentiable at the origin?

(a) $\sqrt{x^2+y^2}$; (b) $\sqrt{|x^2-y^2|}$;

(c) $\sqrt{|x^3-y^3|}$; (d) $\sqrt{|x^3+y^3|}$;

(e) $\sqrt{|x^2y+xy^2|}$;

(f) $f(x,y)=xy/\sqrt{x^2+y^2}$, $f(0,0)=0$;

(g) $\sqrt[3]{x^3+y^3}$; (h) $\sqrt[3]{x^3+y^4}$ (H S);

(i) $x\cdot\sqrt{|y|}$;

(j) $f(x,y)=xy(x^2-y^2)/(x^2+y^2)$, $f(0,0)=0$;

(k) $f(x,y)=(x^3+y^5)/(x^2+y^4)$, $f(0,0)=0$;

(l) $f(x,y)=x^2\cdot\sin(x^2+y^2)^{-1}$, $f(0,0)=0$;

(m) $f(x,y)=\frac{x^3}{x^2+y^2}$, $f(0,0)=0$; (n) $\frac{\sqrt[3]{x^2y^5}}{\sqrt{x^2+y^2}}$, $f(0,0)=0$.

(o) $f(x,y)=x\cdot\sin\frac{1}{y}$, $f(x,0)=0$.

1.81. Let $f(x,y)=|x|^\alpha\cdot|y|^\beta$ if $xy\neq 0$, and let $f(x,y)=0$ if $xy=0$. For what values of α,β is f differentiable at the origin? For what values of α,β is f differentiable everywhere?

1.82. Show that if $f:\mathbb{R}\to\mathbb{R}$ is differentiable at a, then the function $g(x,y)=f(x)$ is differentiable at (a,b) for every b. (S)

1.83. For what functions $f:\mathbb{R}^2\to\mathbb{R}$ will the function $x\cdot f(x,y)$ be differentiable at the origin?

1.84. Show that if the function $f:\mathbb{R}^2\to\mathbb{R}$ is differentiable at the origin, then for every $c\in\mathbb{R}$ the single-variable function $g(x)=f(x,cx)$ is differentiable at 0.

1.85. Show that if the function $f:\mathbb{R}^p\to\mathbb{R}$ is differentiable at a and $f(a)=D_1f(a)=\ldots=D_pf(a)=0$, then $f\cdot g$ is also differentiable at a for every bounded function $g:\mathbb{R}\to\mathbb{R}$.

1.86. True or false? If f is differentiable at $a\in\mathbb{R}^2$ and f has a strict local minimum at a restricted to every line going through a, then f has a strict local minimum at a. (H)

1.87. Find the directional derivatives of $f(x,y)=\sqrt[3]{x^3+y^3}$ at the origin. Can we choose the vector a such that the directional derivative in the direction u equals $\langle a,u\rangle$ for every $|u|=1$? Prove that f is not differentiable at the origin.

1.88. Find the directional derivatives of $f(x,y)=\frac{x^3}{x^2+y^2}$, $f(0,0)=0$, at the origin. Can we choose a vector a such that the directional derivative in the direction u equals $\langle a,u\rangle$ for every $|u|=1$?

1.89. Construct two-variable functions f whose every directional derivative at the origin is 0, but

(a) f is not differentiable at the origin,
(b) f is not continuous at the origin,
(c) there does not exist a neighborhood of the origin on which f is bounded.

1.90. Let $G \subset \mathbb{R}^p$ be a connected open set, and let $f\colon \mathbb{R}^p \to \mathbb{R}$ be differentiable. Show that if $f'(x) = 0$ for every $x \in G$, then f is a constant function. (H)

1.91. Let $f\colon \mathbb{R}^2 \to \mathbb{R}$ be differentiable in the plane, and let $D_1 f(x,x) = D_2 f(x,x) = 0$, for every x. Show that $f(x,x)$ is a constant function.

1.92. Let the real functions f and g be differentiable at the point $a \in \mathbb{R}^p$. Find a formula for the partial derivatives of the functions $f \cdot g$ and (when $g(a) \neq 0$) of f/g at the point a in terms of the partial derivatives of f and g.

1.93. Verify that the gradient of $\sqrt{x^2+y^2}$ at $(a,b) \neq (0,0)$ is parallel to and points in the same direction as (a,b). Why is this obvious intuitively?

1.94. Verify that the gradient of $\sqrt{1-x^2-y^2}$ at the point (a,b) is parallel to and points in the opposite direction as (a,b) when $0 < a^2+b^2 < 1$. Why is this obvious intuitively?

1.95. Let $a, b > 0$, and let $T_{a,b}$ denote the tetrahedron bounded by the xy, xz, yz coordinate planes and by the tangent plane of the graph of the function $f(x,y) = 1/(xy)$ at the point (a,b). Show that the volume of $T_{a,b}$ is independent of a and b.

1.9 Higher-Order Derivatives

Definition 1.80. Let f be defined in a neighborhood of $a \in \mathbb{R}^p$. If the partial derivative $D_j f$ exists in a neighborhood of a and the ith partial derivative of $D_j f$ exists at a, then we call this the ijth *second-order partial derivative* of the function f at the point a, and we use any of the following notations:

$$\frac{\partial^2 f}{\partial x_i \partial x_j}(a),\ f''_{x_j x_i}(a),\ f_{x_j x_i}(a),\ D_i D_j f(a),\ D_{ij} f(a).$$

(The function f has at most p^2 different second-order partial derivatives at the point a.)

Example 1.81. **1.** The partial derivatives of the two-variable function $f(x, y) = \sin(x^2y)$ exist everywhere, with $D_1f(x, y) = \cos(x^2y) \cdot (2xy)$ and $D_2f(x, y) = \cos(x^2y) \cdot x^2$ for every (x, y). Since the partial derivatives of these functions exist everywhere, each of f's four second-order derivatives exist everywhere, with

$$\begin{aligned} D_{11}f(x, y) &= D_1D_1f(x, y) = -\sin(x^2y) \cdot 4x^2y^2 + \cos(x^2y) \cdot 2y, \\ D_{21}f(x, y) &= D_2D_1f(x, y) = -\sin(x^2y) \cdot 2x^3y + \cos(x^2y) \cdot 2x, \\ D_{12}f(x, y) &= D_1D_2f(x, y) = -\sin(x^2y) \cdot 2x^3y + \cos(x^2y) \cdot 2x, \\ D_{22}f(x, y) &= D_2D_2f(x, y) = -\sin(x^2y) \cdot x^4. \end{aligned}$$

Note that $D_{12}f(x, y) = D_{21}f(x, y)$ everywhere. This is surprising, since there is no obvious reason why the two calculations should lead to the same results. Our next example shows that $D_{12}f = D_{21}f$ is not always true.
2. Let $f(x, y) = xy \cdot (x^2 - y^2)/(x^2 + y^2)$ if $(x, y) \neq (0, 0)$, and let $f(0, 0) = 0$. First we prove that the partial derivative D_1f exists everywhere. The section f^0 is zero everywhere, and thus $D_1f(x, 0)$ exists for every x, and its value is zero everywhere. If $b \neq 0$, then the section f^b is differentiable everywhere; thus $D_1f(x, b)$ also exists for every x. If $b \neq 0$, then

$$D_1f(0, b) = \lim_{x \to 0} \frac{f(x, b) - f(0, b)}{x} = \lim_{x \to 0} \frac{xb \cdot (x^2 - b^2)}{(x^2 + b^2) \cdot x} = b \cdot \lim_{x \to 0} \frac{x^2 - b^2}{x^2 + b^2} = -b.$$

We have shown that $D_1f(x, y)$ exists everywhere, and $D_1f(0, y) = -y$ for every y. It follows that $D_{21}f(0, 0) = D_2D_1f(0, 0) = -1$.

Now let us consider the partial derivatives D_2f. The section f_0 is zero everywhere, and thus $D_2f(0, y)$ exists for all y, and its value is zero everywhere. If $a \neq 0$, then f_a is differentiable everywhere; thus $D_2f(a, y)$ also exists for every y. If $a \neq 0$, then

$$D_2f(a, 0) = \lim_{y \to 0} \frac{f(a, y) - f(a, 0)}{y} = \lim_{y \to 0} \frac{ay \cdot (a^2 - y^2)}{(a^2 + y^2) \cdot y} = a \cdot \lim_{y \to 0} \frac{a^2 - y^2}{a^2 + y^2} = a.$$

We have shown that $D_2f(x, y)$ exists everywhere, and $D_2f(x, 0) = x$ for every x. It follows that $D_{12}f(0, 0) = D_1D_2f(0, 0) = 1$, and thus $D_{12}f(0, 0) \neq D_{21}f(0, 0)$. □

The following theorem explains why $D_{12}f = D_{21}f$ was true for Example 1.81.1.

Theorem 1.82. (Young's[16] theorem) *Let $f(x, y)$ be a two-variable function. If the partial derivative functions $D_1f(x, y)$ and $D_2f(x, y)$ exist in a neighborhood of $(a, b) \in \mathbb{R}^2$ and they are differentiable at (a, b), then $D_{12}f(a, b) = D_{21}f(a, b)$.*

---

[16] William Henry Young (1863–1942), British mathematician.

Lemma 1.83.

(i) *If the partial derivative* $D_1 f(x,y)$ *exists in a neighborhood of* (a,b) *and it is differentiable at* (a,b), *then*

$$\lim_{t\to 0} \frac{f(a+t,b+t)-f(a+t,b)-f(a,b+t)+f(a,b)}{t^2} = D_{21}f(a,b). \tag{1.20}$$

(ii) *If the partial derivative* $D_2 f(x,y)$ *exists in a neighborhood of* (a,b) *and it is differentiable on* (a,b), *then*

$$\lim_{t\to 0} \frac{f(a+t,b+t)-f(a+t,b)-f(a,b+t)+f(a,b)}{t^2} = D_{12}f(a,b). \tag{1.21}$$

Proof. (i) Let us use the notation

$$H(t) = (f(a+t,b+t)-f(a+t,b)) - (f(a,b+t)-f(a,b))$$

and, for a fixed t, $F(u) = f(u,b+t) - f(u,b)$. Clearly, $H(t) = F(a+t) - F(a)$. The main idea of the proof is to use the mean value theorem for the latter formula, and then use the differentiability of $D_1 f$ at a to show that $H(t)$ is close to $D_{21}f(a)\cdot t^2$ when t is small.

Let $\varepsilon > 0$ be fixed. Since $D_1 f(x,y)$ is differentiable at (a,b), we can choose $\delta > 0$ such that

$$\begin{aligned} \left|D_1 f(x,y) - (D_{11}f(a,b)(x-a) + D_{21}f(a,b)(y-b) + D_1 f(a,b))\right| &\le \\ \le \varepsilon\cdot(|x-a|+|y-b|) \end{aligned} \tag{1.22}$$

holds for every point $(x,y) \in B((a,b),\delta)$.

Let $0 < |t| < \delta/2$ be fixed. The function F is differentiable in the interval $[a,a+t]$, since $u \in [a,a+t]$ implies

$$(u,b+t) \in B((a,b),\delta)$$

and

$$(u,b) \in B((a,b),\delta).$$

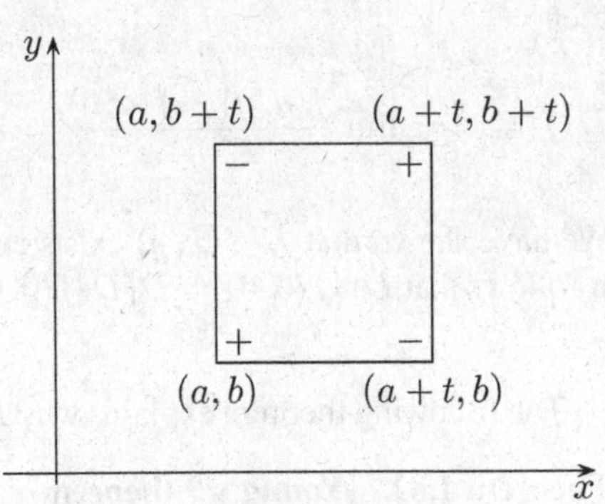

1.17. Figure

Furthermore, the sections f^{b+t} and f^b are differentiable at $[a,a+t]$, with derivatives $D_1 f(u,b+t)$ and $D_1 f(u,b)$, respectively. Thus $F'(u) = D_1 f(u,b+t) - D_1 f(u,b)$ for every $u \in [a,a+t]$. By the mean value theorem we have

$$F(a+t) - F(a) = (D_1 f(c,b+t) - D_1 f(c,b))\cdot t$$

for an appropriate choice of $c \in [a, a+t]$, and thus

$$H(t) = (D_1 f(c, b+t) - D_1 f(c, b)) \cdot t. \tag{1.23}$$

Plugging $(x, y) = (c, b+t)$ and $(x, y) = (c, b)$ into (1.22), we get

$$\left| D_1 f(c, b+t) - (D_{11} f(a, b)(c-a) + D_{21} f(a, b)t + D_1 f(a, b)) \right| \le$$
$$\le \varepsilon \cdot (|c-a| + |t|) \le 2\varepsilon \cdot t$$

and

$$\left| D_1 f(c, b) - (D_{11} f(a, b)(c-a) + D_1 f(a, b)) \right| \le$$
$$\le \varepsilon \cdot |c-a| \le \varepsilon \cdot |t|,$$

respectively. Applying the triangle inequality yields

$$|D_1 f(c, b+t) - D_1 f(c, b) - D_{21} f(a, b)t| \le 3\varepsilon \cdot |t|.$$

Comparing with (1.23), we get

$$\left| \frac{H(t)}{t^2} - D_{21} f(a, b) \right| \le 3\varepsilon.$$

Since ε was arbitrary, and this is true for every $0 < |t| < \delta/2$, (1.20) is proved.
(ii) Let $0 < |t| < \delta/2$ be fixed, and let $G(v) = f(a+t, v) - f(a, v)$ for every v for which f is defined at the points $(a+t, v)$ and (a, v). We have $H(t) = G(b+t) - F(b)$ for every t small enough. Repeating the steps of the proof of (i), we get (1.21). □

Proof of Theorem 1.82. By the assumptions of the theorem, the conditions of both statements of Lemma 1.83 are satisfied. Therefore, both of (1.20) and (1.21) hold, and thus $D_{12} f(a, b) = D_{21} f(a, b)$. □

Let us revisit Example 1.81.1. One can see that the second-order partial derivatives of f are continuous everywhere. By Theorem 1.71 this implies that the first-order partial derivatives of f are differentiable. Thus, by Young's theorem, $D_{12} f = D_{21} f$ everywhere.

Definition 1.84. Let f be differentiable in a neighborhood of $a \in \mathbb{R}^p$. If the partial derivative functions of f are differentiable at a, then we say that f is *twice differentiable* at the point a.

Lemma 1.85. *Let $p > 2$, let f be defined in a neighborhood of $a = (a_1, a_2, \ldots, a_p) \in \mathbb{R}^p$, and consider the section*

$$g(u, v) = f(u, v, a_3, \ldots, a_p).$$

If f is twice differentiable at a, then g is twice differentiable at $(a_1, a_2) \in \mathbb{R}^2$. Furthermore, $D_{21}g(a_1, a_2) = D_{21}f(a)$ and $D_{12}g(a_1, a_2) = D_{12}f(a)$.

Proof. From the definition of the partial derivative, we have $D_1g(u, v) = D_1 f(u, v, a_3, \dots, a_p)$ and $D_2g(u, v) = D_2f(u, v, a_3, \dots, a_p)$ whenever the right-hand sides exist. Thus, D_1g and D_2g are defined in a neighborhood of (a_1, a_2). By assumption, D_1f is differentiable at a, and thus

$$D_1f(x) = D_1f(a) + \sum_{i=1}^{p} D_{i1}f(a)(x_i - a_i) + \varepsilon(x) \cdot |x - a|,$$

where $\varepsilon(x) \to 0$ as $x \to a$. Applying this with $x = (u, v, a_3, \dots, a_p)$, we obtain

$$\begin{aligned} D_1g(u, v) =& D_1g(a_1, a_2) + D_{11}f(a)(u - a_1) + D_{21}f(a)(v - a_2) + \\ &+ \varepsilon(u, v, a_3, \dots, a_p) \cdot |(u, v) - (a_1, a_2)|. \end{aligned}$$

Since $\varepsilon(u, v, a_3, \dots, a_p) \to 0$ if $(u, v) \to (a_1, a_2)$, it follows that D_1g is differentiable at (a_1, a_2), and $D_{21}g(a_1, a_2) = D_{21}f(a)$. Similarly, D_2g is differentiable at (a_1, a_2), and $D_{12}g(a_1, a_2) = D_{12}f(a)$. □

Theorem 1.86. *If f is twice differentiable at $a \in \mathbb{R}^p$, then $D_{ij}f(a) = D_{ji}f(a)$ for every $i, j = 1, \dots, p$.*

Proof. We may assume $i \neq j$. Since the role of the coordinates is symmetric, we may also assume, without loss of generality, that $i = 1$ and $j = 2$. Consider the section

$$g(u, v) = f(u, v, a_3, \dots, a_p).$$

Combining Young's theorem and our previous lemma yields $D_{12}g(a_1, a_2) = D_{21} g(a_1, a_2)$, and thus $D_{12}f(a) = D_{21}f(a)$. □

Definition 1.87. We define the *kth-order partial derivatives* recursively on k. Assume that we have already defined the kth-order partial derivatives of the function f. Then we define the $(k+1)$st-order partial derivatives as follows.

Let $1 \le i_1, \dots, i_{k+1} \le p$ be arbitrary indices, and suppose that the kth-order partial derivative $D_{i_2 \dots i_{k+1}}f(x)$ exists and is finite in a neighborhood of $a \in \mathbb{R}^p$. If the i_1th partial derivative of the function $x \mapsto D_{i_2 \dots i_{k+1}}f(x)$ exists at a, then we call this the $(k+1)$st-order partial derivative of f at a, and use the notation $D_{i_1 \dots i_{k+1}}f(a)$. (Obviously, f has at most p^k different kth-order partial derivatives at a.)

Some other usual notation for $D_{i_1 \dots i_k}f(a)$:

$$\frac{\partial^k f}{\partial x_{i_1} \dots \partial x_{i_k}}(a),\ f^{(k)}_{x_{i_k} \dots x_{i_1}}(a),\ f_{x_{i_k} \dots x_{i_1}}(a),\ D_{i_1} \dots D_{i_k}f(a).$$

Definition 1.88. Suppose that we have already defined k-times differentiability. (We did so in the cases of $k = 1$ and $k = 2$.) We say that a function f is $(k+1)$ *times differentiable at* $a \in \mathbb{R}^p$ if f is k times differentiable on a neighborhood of a, furthermore, every kth-order partial derivative of f exists and is finite in a neighborhood of a, and these partial derivatives are differentiable at a.

Thus, we have defined k times differentiability for every k.

We say that a function f is *infinitely differentiable at* a if f is k times differentiable at a for every $k = 1, 2, \ldots$.

Remark 1.89. It follows from Theorem 1.67 that if f is k times differentiable at a, then every kth-order partial derivative of f exists and is finite at a.

Theorem 1.90. *The polynomials are infinitely differentiable everywhere. The rational functions are infinitely differentiable at every point of their domains.*

Proof. By Corollary 1.72, polynomials are differentiable everywhere. Suppose we have already proved that polynomials are k times differentiable. Since the kth-order partial derivatives of a polynomial are also polynomials, these are differentiable, showing that the polynomials are also $(k+1)$ times differentiable. Thus, the polynomials are infinitely differentiable.

The proof for rational functions is similar. □

Theorem 1.91. *Let the function f be k times differentiable at $a \in \mathbb{R}^p$. If the ordered k-tuples $(i_1, \ldots, i_k)$ and $(j_1, \ldots, j_k)$ are permutations of each other (i.e., each $i = 1, \ldots, p$ appears the same number of times in both k-tuples), then $D_{i_1 \ldots i_k} f(a) = D_{j_1 \ldots j_k} f(a)$.*

Proof. The statement is trivial for the $k = 1$ case, and the $k = 2$ case is covered by Theorem 1.86. For $k > 2$, the statement can be proved by induction by applying Theorem 1.86. □

Exercises

1.96. Find every function $f \colon \mathbb{R}^2 \to \mathbb{R}$ such that $D_2(D_1 f)$ is zero everywhere. (H)

1.97. Young's theorem implies that the function $f(x, y) = xy \cdot (x^2 - y^2)/(x^2 + y^2)$, $f(0, 0) = 0$, cannot be twice differentiable at the origin. Verify, without using the theorem, that $D_1 f$ is not differentiable at the origin.

1.98. For what values of $\alpha, \beta > 0$ is $|x|^\alpha \cdot |y|^\beta$ twice differentiable at the origin?

1.99. Show that if $D_{12} f$ and $D_{21} f$ exist in a neighborhood of (a, b) and are continuous at (a, b), then $D_{12} f(a, b) = D_{21} f(a, b)$.

1.100. Let the partial derivatives $D_1 f$, $D_2 f$, and $D_{12} f$ exist in a neighborhood of (a, b), and let $D_{12} f$ be continuous at (a, b). Show that $D_{21} f(a, b)$ exists and is equal to $D_{12} f(a, b)$ (Schwarz's theorem).

1.101. Let $f : \mathbb{R}^2 \to \mathbb{R}$ be twice differentiable everywhere. Show that if $D_{21} f$ is nonnegative everywhere, then $f(b, d) - f(a, d) - f(b, c) + f(a, c) \geq 0$ for every $a < b$ and $c < d$.

1.10 Applications of Differentiation

The most important applications of differentiation—in the cases of multi- and single-variable functions alike—is the analysis of functions, finding the greatest and the smallest values, and finding good approximations using simpler functions (e.g., polynomials).

Since each of the applications below is based on Taylor[17] polynomials, our first task is to define these polynomials for p-variable functions and establish their most important properties. This proves to be surprisingly simple. The notation in the multivariable case is necessarily more complicated, but the notion of the Taylor polynomials, as well as their basic properties, is basically the same as in the single-variable case.

By a **monomial** we mean a product of the form $c \cdot x_1^{s_1} \cdots x_p^{s_p}$, where c is a nonzero real number and the exponents $s_1, \ldots, s_p$ are nonnegative integers.

The **degree** of the monomial $c \cdot x_1^{s_1} \cdots x_p^{s_p}$ is $s_1 + \ldots + s_p$. Every p-variable polynomial can be written as the sum of monomials. Obviously, if a polynomial is not the constant zero function, then it can be written in a way that the p-element sequences of the exponents of its corresponding monomials are distinct. By induction on p one can easily prove that this representation of the polynomials is unique. We call it the **canonical form** of the polynomial.

We say that the **degree** of a nonidentically zero polynomial is the highest degree of the monomials in its canonical form. The constant zero polynomial does not have a degree. Still, when we speak about the set of polynomials of degree at most n, we will include the identically zero polynomial among them.

Lemma 1.92. *Let*

$$g(x) = \sum_{\substack{s_1, \ldots, s_p \geq 0 \\ s_1 + \ldots + s_p \leq n}} c_{s_1 \ldots s_p} \cdot (x_1 - a_1)^{s_1} \cdots (x_p - a_p)^{s_p}. \tag{1.24}$$

Then $g(a) = c_{0 \ldots 0}$, *and furthermore, for every* $k \leq n$ *and* $1 \leq i_1, \ldots, i_k \leq p$ *we have*

$$D_{i_1 \ldots i_k} g(a) = s_1! \cdots s_p! \cdot c_{s_1 \ldots s_p}, \tag{1.25}$$

[17] Brook Taylor (1685–1731), English mathematician.

where $s_1, \ldots, s_p$ *denotes the number of indices of* $1, \ldots, p$ *in the sequence* $(i_1, \ldots, i_k)$.

Proof. The equality $g(a) = c_{0\ldots0}$ is obvious. Let the indices $1 \le i_1, \ldots, i_k \le p$ be fixed, with $k \le n$. For simplicity, we write $\mathcal{D}$ instead of $D_{i_1 \ldots i_k}$. It is easy to see that if g_1 and g_2 are polynomials, then $\mathcal{D}(g_1 + g_2) = \mathcal{D}g_1 + \mathcal{D}g_2$ and $\mathcal{D}(\lambda g_1) = \lambda \cdot \mathcal{D}g_1$ for every $\lambda \in \mathbb{R}$. Thus, the kth-order partial derivative $\mathcal{D}g(a)$ can be computed by applying $\mathcal{D}$ to each of the terms on the right-hand side of (1.24) and summing the values of the resulting partial derivatives at the point a. Consider the kth-order partial derivative

$$\mathcal{D}(x_1 - a_1)^{t_1} \cdots (x_d - a_d)^{t_p} \tag{1.26}$$

and its value at a:

$$\left(\mathcal{D}(x_1 - a_1)^{t_1} \cdots (x_d - a_d)^{t_p}\right)(a). \tag{1.27}$$

It is easy to see that if the index i is present in the sequence $(i_1, \ldots, i_k)$ more than t_i times, then 1.26 is constant and equal to zero. On the other hand, if there is an index i such that i is present in the sequence $(i_1, \ldots, i_k)$ fewer than t_i times, then the polynomial 1.26 is divisible by $x_i - a_i$, and thus the value of (1.27) is zero. Therefore, in applying $\mathcal{D}$ to the right-hand side of (1.24) and taking its value at a, we get a nonzero term only if $(t_1, \ldots, t_p) = (s_1, \ldots, s_p)$.

Furthermore, since $\mathcal{D}(x_1 - a_1)^{s_1} \cdots (x_d - a_d)^{s_p}$ is equal to the constant function $s_1! \cdots s_p!$, it follows that (1.25) holds. □

Let f be n times differentiable at a. By Theorem 1.91, if $n \le k$, then the kth-order partial derivative $D_{i_1 \ldots i_k} f(a)$ does not depend on the order of the indices $i_1, \ldots, i_k$, and only on the number of times these indices are present in the sequence $(i_1, \ldots, i_k)$. Let $s_1, \ldots, s_p$ be nonnegative integers, with $s_1 + \ldots + s_p \le n$. We denote by $D^{s_1 \ldots s_p} f(a)$ the number $D_{i_1 \ldots i_k} f(a)$, where the indices $1, \ldots, p$ are present in the sequence $(i_1, \ldots, i_k)$ exactly $s_1, \ldots, s_p$ times, respectively. Let $D^{0 \ldots 0} f(a) = f(a)$.

Theorem 1.93. *Suppose that the function* f *is* n *times differentiable at* $a = (a_1, \ldots, a_p) \in \mathbb{R}^p$, *and let*

$$t_n(x) = \sum_{\substack{s_1, \ldots, s_p \ge 0 \\ s_1 + \ldots + s_p \le n}} \frac{1}{s_1! \cdots s_p!} \cdot D^{s_1 \ldots s_p} f(a) \cdot (x_1 - a_1)^{s_1} \cdots (x_p - a_p)^{s_p}. \tag{1.28}$$

The polynomial t_n *is the only polynomial of degree at most* n *such that* $t_n(a) = f(a)$, *and*

$$D_{i_1 \ldots i_k} t_n(a) = D_{i_1 \ldots i_k} f(a) \tag{1.29}$$

for every $1 \le k \le n$ *and* $1 \le i_1, \ldots, i_k \le p$.

Proof. It follows from Lemma 1.92 that $t_n(a) = f(a)$ and (1.29) holds for the polynomial t_n.

Let g be a polynomial of degree at most n, and suppose that g satisfies $g(a) = f(a)$ and $D_{i_1 \ldots i_k} g(a) = D_{i_1 \ldots i_k} f(a)$ for every $k \leq n$ and $1 \leq i_j \leq p$ $(1 \leq j \leq k)$. Then the polynomial $q = g(x_1 + a_1, \ldots, x_p + a_p)$ has degree at most n. Write q as the sum of the monomials $c \cdot x_1^{s_1} \cdots x_p^{s_p}$ (with $c \neq 0$). Then $s_1 + \ldots + s_p \leq n$ holds for each term. If we replace x_i by $x_i - a_i$ in g for every $i = 1, \ldots, p$, then we get that (1.24) is true for suitable coefficients $c_{s_1 \ldots s_p}$. Then by Lemma 1.92 we have

$$s_1! \cdots s_p! \cdot c_{s_1 \ldots s_p} = D_{i_1 \ldots i_k} g(a) = D_{i_1 \ldots i_k} f(a)$$

for every $(i_1, \ldots, i_k)$, i.e., $g = t_n$. □

We can see that

$$t_1(x) = f(a) + D_1 f(a) \cdot (x_1 - a_1) + \ldots + D_p f(a) \cdot (x_p - a_p),$$

i.e., the graph of the polynomial t_1 is the tangent plane of graph f at $(a, f(a))$.

The polynomial t_2 in the cases $p = 2$ and $p = 3$ can be written as follows:

$$\begin{aligned} t_2(x, y) = {} & f(a, b) + f'_x(a, b) \cdot (x - a) + f'_y(a, b) \cdot (y - b) + \\ & + \frac{1}{2} \cdot f''_{xx}(a, b) \cdot (x - a)^2 + f''_{xy}(a, b) \cdot (x - a)(y - b) + \frac{1}{2} \cdot f''_{yy}(a, b) \cdot (y - b)^2, \end{aligned}$$

or

$$\begin{aligned} t_2(x, y, z) = {} & f(a, b, c) + f'_x(a, b, c) \cdot (x - a) + f'_y(a, b, c) \cdot (y - b) + \\ & + f'_z(a, b, c) \cdot (z - c) + \\ & + \frac{1}{2} \cdot f''_{xx}(a, b, c) \cdot (x - a)^2 + \\ & + \frac{1}{2} \cdot f''_{yy}(a, b, c) \cdot (y - b)^2 + \frac{1}{2} \cdot f''_{zz}(a, b, c) \cdot (z - c)^2 + \\ & + f''_{xy}(a, b, c) \cdot (x - a)(y - b) + f''_{xz}(a, b, c) \cdot (x - a)(z - c) + \\ & + f''_{yz}(a, b) \cdot (y - b)(z - c), \end{aligned}$$

respectively.

Remark 1.94. If the function f is n times differentiable at a, then the polynomial in (1.28) can be written in the following alternative form:

$$\begin{aligned} t_n(x) = {} & f(a) + \sum_{i=1}^{p} D_i f(a) \cdot (x_i - a_i) + \\ & + \frac{1}{2!} \sum_{i_1, i_2 = 1}^{p} D_{i_1 i_2} f(a) \cdot (x_{i_1} - a_{i_1})(x_{i_2} - a_{i_2}) + \ldots + \\ & + \frac{1}{n!} \sum_{i_1, \ldots, i_n = 1}^{p} D_{i_1 \ldots i_n} f(a) \cdot (x_{i_1} - a_{i_1}) \cdots (x_{i_n} - a_{i_n}). \end{aligned} \tag{1.30}$$

Indeed, suppose that the index i occurs in the sequence $(i_1, \ldots, i_k)$ exactly s_i times $(i = 1, \ldots, p)$. Then $s_1, \ldots, s_p$ are nonnegative integers with $s_1 + \ldots + s_p = k$. It is well known (and easy to show) that the number of possible permutations of the sequence $(i_1, \ldots, i_k)$ is $\frac{k!}{s_1! \cdots s_p!}$. Using the notation of Theorem 1.93, we can see that the term $D^{s_1 \ldots s_p}(x_1 - a_1)^{s_1} \cdots (x_p - a_p)^{s_p}$ occurs $\frac{k!}{s_1! \cdots s_p!}$ times on the right-hand side of (1.30). This proves that (1.28) and (1.30) define the same polynomial.

Definition 1.95. We call the polynomial t_n defined by (1.28) (or by (1.30)) the *nth Taylor polynomial* of the function f at the point a.

The following notion makes it possible to represent the multivariable Taylor polynomials in a simple form similar to that in the single-variable case.

Definition 1.96. If the function f is n times differentiable at $a \in \mathbb{R}^p$, then we call the polynomial

$$\sum_{\substack{s_1, \ldots, s_p \geq 0 \\ s_1 + \ldots + s_p = k}} \frac{k!}{s_1! \cdots s_p!} \cdot D^{s_1 \ldots s_p} f(a) \cdot x_1^{s_1} \cdots x_p^{s_p} =$$

$$= \sum_{i_1, \ldots, i_k = 1}^{p} D_{i_1 \ldots i_k} f(a) \cdot x_{i_1} \cdots x_{i_k} \tag{1.31}$$

the *kth differential* of the function f at a, and we use the notation $d^k f(a)$ $(k \leq n)$. Thus $d^k f(a)$ is not a real number, but a p-variable polynomial. If $b = (b_1, \ldots, b_p) \in \mathbb{R}^p$, then $d^k f(a)(b)$ is the value the polynomial $d^k f(a)$ takes at b; that is,

$$d^k f(a)(b) = \sum_{i_1, \ldots, i_k = 1}^{p} D_{i_1 \ldots i_k} f(a) \cdot b_{i_1} \cdots b_{i_k}.$$

For $p = 2$ and $k = 2$ we have

$$d^2 f(a)(b) = f''_{xx}(a) b_1^2 + 2 f''_{xy}(a) b_1 b_2 + f''_{yy}(a) b_2^2.$$

We can write the nth Taylor polynomial in the form

$$t_n(x) = f(a) + d^1 f(a)(x - a) + \frac{1}{2!} d^2 f(a)(x - a) + \ldots + \frac{1}{n!} d^n f(a)(x - a)$$

using differentials. Again, $d^k f(a)(x - a)$ is the value $d^k f(a)$ takes at $x - a$.

Theorem 1.97. (Taylor's formula) *Let the function f be $(n+1)$ times differentiable at the points of the segment $[a, b]$, where $a, b \in \mathbb{R}^p$. Then there exists a point $c \in [a, b]$ such that*

$$f(b) = t_n(b) + \frac{1}{(n+1)!} d^{n+1} f(c)(b-a). \quad (1.32)$$

Lemma 1.98. *Let the function f be n times differentiable at the points of the segment $[a, b]$, where $a, b \in \mathbb{R}^p$. If $F(t) = f(a + t \cdot (b - a))$ $(t \in [0, 1])$, then the function F is n times differentiable on the interval $[0, 1]$, and*

$$F^{(k)}(t) = d^k f(a + t(b-a))(b-a) \quad (1.33)$$

for every $k \le n$ and $t \in [0, 1]$.

Proof. We prove the lemma by induction on k. If $k = 0$, then the statement is true, since $F^{(0)}(t) = F(t) = f(a + t(b - a))$, and $d^0 f(a + t(b - a))$ is the constant polynomial $f(a + t(b - a))$. If $k = 1$, then (1.33) is exactly part (i) of Theorem 1.79.

Let $1 \le k < n$, and suppose that (1.33) is true for every $t \in [0, 1]$. By the definitions of the differential $d^k f$, we have

$$F^{(k)}(t) = \sum_{i_1,\ldots,i_k=1}^{p} D_{i_1\ldots i_k} f(a + t(b-a)) \cdot (b_{i_1} - a_{i_1}) \cdots (b_{i_k} - a_{i_k}) \quad (1.34)$$

for every $t \in [0, 1]$. Since f is $n > k$ times differentiable at the points of $[a, b]$, every kth-order partial derivative $D_{i_1\ldots i_k} f$ is differentiable there. By part (i) of Theorem 1.79, the function $t \mapsto D_{i_1\ldots i_k} f(a + t(b - a))$ is differentiable at $[0, 1]$, and its derivative is

$$\sum_{i=1}^{p} D_{i,i_1\ldots i_k} f(a + t(b-a)) \cdot (b_i - a_i).$$

This holds for every term on the right-hand side of (1.34). Thus $F^{(k)}$ is differentiable at $[0, 1]$, and its derivative is

$$F^{(k+1)}(t) = \sum_{i_1,\ldots,i_{k+1}=1}^{p} D_{i_1\ldots i_{k+1}} f(a + t(b-a)) \cdot (b_{i_1} - a_{i_1}) \cdots (b_{i_{k+1}} - a_{i_{k+1}}).$$

Therefore, (1.33) holds for $(k + 1)$, and (1.33) has been proved for every $k \le n$. □

Proof of Theorem 1.97. Let $F(t) = f(a + t \cdot (b - a))$, for every $t \in [0, 1]$. By Lemma 1.98, F is $(n + 1)$ times differentiable on the interval $[0, 1]$, and (1.33) holds for every $k \le n + 1$ and $t \in [0, 1]$. If we apply (the single-variable version of) Taylor's formula with Lagrange remainder (see [7, 13.7]), we get (1.32). □

Theorem 1.99. *Let the function f be n times differentiable at $a = (a_1, \ldots, a_p) \in \mathbb{R}^p$, and let t_n be the nth Taylor polynomial of f at a. Then*

$$\lim_{x \to a} \frac{f(x) - t_n(x)}{|x - a|^n} = 0. \tag{1.35}$$

Conversely, if a polynomial q with degree at most n satisfies

$$\lim_{x \to a} \frac{f(x) - q(x)}{|x - a|^n} = 0, \tag{1.36}$$

then $q = t_n$. (In other words, among the polynomials of degree at most n, t_n is the one that approximates the function f best locally at the point a.)

Proof. For $n = 1$, equation (1.35) is exactly the definition of differentiability of f at a. Thus, we may assume that $n \geq 2$.

Let f be n times differentiable at a. The function $g = f - t_n$ is also n times differentiable at a, and by Theorem 1.93, the partial derivatives of g of order at most n are all zero at a. The $(n-1)$st-order partial derivatives of g are differentiable at a, and for the same reason as we mentioned above, both their values at a and the values of their partial derivatives at a are zero. By the definition of differentiability, for every $\varepsilon > 0$ there exists $\delta > 0$ such that if $|x - a| < \delta$, then

$$|D_{i_1 \ldots i_{n-1}} g(x)| \leq \varepsilon \cdot |x - a| \tag{1.37}$$

for every $1 \leq i_j \leq p\,(j = 1, \ldots, n-1)$. Let us apply the $(n-2)$nd Taylor formula for g. We find that for every $x \in B(a, \delta)$ there exists $c \in [a, x]$ such that

$$g(x) = \frac{1}{(n-1)!} d^{n-1} g(c)(x - a) =$$
$$= \frac{1}{(n-1)!} \cdot \sum_{i_1, \ldots, i_{n-1} = 1}^{p} D_{i_1 \ldots i_{n-1}} g(c)(x_{i_1} - a_{i_1}) \cdots (x_{i_{n-1}} - a_{i_{n-1}}).$$

Since $|c - a| < \delta$, it follows from (1.37) that

$$|g(x)| \leq \frac{p^{n-1}}{(n-1)!} \cdot \varepsilon \cdot |c - a| \cdot |x - a|^{n-1} \leq \frac{p^{n-1}}{(n-1)!} \cdot \varepsilon \cdot |x - a|^n.$$

This implies

$$\frac{|f(x) - t_n(x)|}{|x - a|^n} \leq \frac{p^{n-1}}{(n-1)!} \cdot \varepsilon$$

for every $0 < |x - a| < \delta$. Since ε was arbitrary, (1.35) has been proved.

Now let's assume that (1.36) holds for a polynomial q with degree at most n. Then $r = q - t_n$ is a polynomial of degree at most n, and

$$\lim_{x \to a} r(x)/|x - a|^n = 0. \tag{1.38}$$

We need to prove that r is the constant zero function. If $p = 1$, then (1.38) implies that a is a root of r with multiplicity at least $(n+1)$. Since the degree of r is at most n, this is possible only if r is identically zero.

Let $p > 1$ and let $s(t) = r(a + ty)$ $(t \in \mathbb{R})$, where y is a fixed p-dimensional nonzero vector. It is obvious that s is a polynomial in the variable t of degree at most n. Applying Theorem 1.49 on the limit of composite functions, we obtain $\lim_{t\to 0} s(t)/|ty|^n = 0$ and $\lim_{t\to 0} s(t)/|t|^n = 0$. As we saw above, this implies that $s(t) = 0$ for every t. Then $r(a + y) = s(1) = 0$ for every vector $y \in \mathbb{R}^p$, $y \neq 0$. Thus $r \equiv 0$, since r is continuous at the point a. □

Let f be a function of one variable, and suppose that f is twice differentiable at the point $a \in \mathbb{R}$. It is well known that if $f'(a) = 0$ and $f''(a) > 0$, then f has a strict local minimum at the point a, and if $f'(a) = 0$ and $f''(a) < 0$, then f has a strict local maximum at the point a. (See [7, Theorem 12.60].) This implies that if f has a local minimum at the point a, then necessarily $f''(a) \geq 0$. The following application of Taylor's formula gives a generalization of these results to multivariable functions.

To state our theorem, we need to introduce a few concepts from the field of algebra. We say that a p-variable polynomial is a **quadratic form** if every term of its canonical form is of degree two. In other words, a polynomial is a quadratic form if it can be written as $\sum_{i,j=1}^{p} c_{ij} x_i x_j$. Note that if f is twice-differentiable at a, then the second differential $d^2 f(a)$ is a quadratic form, since $d^2 f(a)(x) = \sum_{i,j=1}^{p} D_{ij} f(a) \cdot x_i x_j$.

Definition 1.100. A quadratic form q is *positive (negative) definite* if $q(x) > 0$ $(q(x) < 0)$ for every $x \neq 0$.

A quadratic form q is *positive (negative) semidefinite* if $q(x) \geq 0$ $(q(x) \leq 0)$ for every $x \in \mathbb{R}^p$.

A quadratic form q is *indefinite* if it takes both positive and negative values.

Theorem 1.101. *Let f be twice differentiable at $a \in \mathbb{R}^p$, and let $D_i f(a) = 0$ for every $i = 1, \ldots, p$.*

(i) *If f has a local minimum (maximum) at a, then the quadratic form $d^2 f(a)$ is positive (negative) semidefinite.*

(ii) *If the quadratic form $d^2 f(a)$ is positive (negative) definite, then f has a strict local minimum (maximum) at a.*

Proof. (i) We prove the result by contradiction. Let f have a local minimum at a, and suppose that there exists a point x_0 such that $d^2 f(a)(x_0) < 0$. Since $D_i f(a) = 0$ for every $i = 1, \ldots, p$, we have $d^1 f(a) = 0$, and $t_2(x) = f(a) + \frac{1}{2} \cdot d^2 f(a)(x - a)$ for every x. According to Theorem 1.99,

$$\lim_{x\to a} \frac{f(x) - t_2(x)}{|x - a|^2} = 0. \tag{1.39}$$

For t small enough, (1.39) implies

$$|f(a+tx_0) - t_2(a+tx_0)| < \frac{|d^2 f(a)(x_0)|}{2} \cdot t^2.$$

On the other hand,

$$t_2(a+tx_0) = f(a) + \frac{t^2}{2} \cdot d^2 f(a)(x_0),$$

and thus

$$f(a+tx_0) < t_2(a+tx_0) + \frac{|d^2 f(a)(x_0)|}{2} \cdot t^2 =$$
$$= f(a) + \frac{t^2}{2} \cdot d^2 f(a)(x_0) + \frac{|d^2 f(a)(x_0)|}{2} \cdot t^2 = f(a),$$

for every t small enough. This means that if $d^2 f(a)$ takes a negative value, then f takes a value less than $f(a)$ in every neighborhood of a, which is a contradiction.

We can prove similarly that if f has a local maximum at a, then $d^2 f(a)$ is negative semidefinite. Thus (i) is proved.

Now let $d^2 f(a)$ be positive definite. The function $d^2 f(a)$ is positive and continuous on the set $S(0,1) = \{x \in \mathbb{R}^p \colon |x| = 1\}$. Since $S(0,1)$ is bounded and closed, Theorem 1.51 implies that $d^2 f(a)$ takes a least value on $S(0,1)$. Let this value be m; then $m > 0$ and $d^2 f(a)(x) \geq m$ for every $x \in S(0,1)$. If $x \neq 0$, then $x/|x| \in S(0,1)$, and thus

$$d^2 f(a)(x) = |x|^2 \cdot d^2 f(a)(x/|x|) \geq m \cdot |x|^2. \tag{1.40}$$

By (1.39), there exists $\delta > 0$ such that $|f(x) - t_2(x)| < (m/2) \cdot |x-a|^2$ for every $0 < |x-a| < \delta$. If $0 < |x-a| < \delta$ then (1.40) implies

$$f(x) > t_2(x) - (m/2) \cdot |x-a|^2 \geq f(a) + \tfrac{1}{2} \cdot m \cdot |x-a|^2 - (m/2) \cdot |x-a|^2 = f(a).$$

This proves that f has a strict local minimum at a. Similarly, if $d^2 f(a)$ is negative definite, then f has a strict local maximum at a, which proves (ii). □

Remark 1.102. **1.** For $p = 1$, we have $d^2 f(a)(x) = f''(a) \cdot x^2$, which is positive definite if $f''(a) > 0$, negative definite if $f''(a) < 0$, positive semidefinite if $f''(a) \geq 0$, and negative semidefinite if $f''(a) \leq 0$. (For single-variable functions every quadratic form is semidefinite; there are no indefinite quadratic forms.) Thus, (i) of Theorem 1.101 gives the statement we quoted above: if $f'(a) = 0$ and $f''(a) > 0$, then f has a strict local minimum at the point a.

Note that for $p > 1$, there exist indefinite quadratic forms (e.g., $x_1 x_2$).

2. We show that neither of the converses of the statements of Theorem 1.101 is true. Obviously, every first- and second-order partial derivative of the polynomial

$f(x_1, \ldots, x_p) = x_1^3$ is zero at the origin. Thus the quadratic form $d^2 f(0)$ is constant and equal to zero. Consequently, it is positive semidefinite. Still, the function f does not have a local minimum at the origin, since it takes negative values in every neighborhood of the origin.

Now consider the polynomial $g(x_1, \ldots, x_p) = x_1^4 + \ldots + x_p^4$, which has a strict local minimum at the origin. Since every second-order partial derivative of g is zero at the origin, the quadratic form $d^2 g(0)$ is constant and equal to zero, and is therefore *not* positive definite.

3. The quadratic form $ax^2 + bxy + cy^2$ is positive definite if and only if $a > 0$ and $b^2 - 4ac < 0$. A classic theorem of abstract algebra states that for every quadratic form (of an arbitrary number of variables) an appropriate matrix (or rather the signs of its subdeterminants) formed from the coefficients of the quadratic form can tell us whether the quadratic form is positive (negative) definite, or positive (negative) semidefinite. For a mathematically precise statement see [6, Theorem 7.3.4].

A single-variable differentiable function f is convex on an interval if and only if each of the tangents of graph f is under the graph of the function (see [7, Theorem 12.64]). Also, a twice-differentiable function is convex on an interval if and only if its second derivative is nonnegative everywhere on the interval (see [7, Theorem 12.65]). Both statements can be generalized in the multivariable case.

Definition 1.103. We say that the set $H \subset \mathbb{R}^p$ is *convex* if H contains every segment whose endpoints are in H.

Every ball is convex. Indeed, if $x, y \in B(a, r)$, then

$$\begin{aligned} |x + t(y - x) - a| &= |(1 - t)(x - a) + t(y - a)| \leq \\ &\leq (1 - t)|x - a| + t|y - a| < \\ &< (1 - t)r + tr = r \end{aligned}$$

for every $t \in [0, 1]$, i.e., every point of the segment $[x, y]$ is in $B(a, r)$.

A similar argument shows that every closed ball is convex. It is also easy to see that every open or closed box is also convex.

Definition 1.104. Let $H \subset \mathbb{R}^p$ be convex. We say that the function $f \colon H \to \mathbb{R}$ is *convex on the set* H if for every $x, y \in H$, the single-variable function $t \mapsto f(x + t(y - x))$ is convex on the interval $[0, 1]$. That is, f is convex on H if

$$f((1 - t)x + ty) \leq (1 - t)f(x) + tf(y)$$

for every $x, y \in H$ and $t \in [0, 1]$.

We say that the function $f \colon H \to \mathbb{R}$ is *concave on the set* H if $-f$ is convex on H.

Figure 1.18 shows an example of a convex function.

Theorem 1.105. *Let f be differentiable on the convex and open set $G \subset \mathbb{R}^p$. The function f is convex on G if and only if the graph of f is above the tangent hyperplane at the point $(a, f(a))$ for every $a \in G$. In other words, f is convex on G if and only if*

$$f(x) \geq f(a) + \langle f'(a), x - a \rangle \tag{1.41}$$

for every $a, x \in G$.

Proof. Let f be convex on G, and let a and x be different points of G. By Theorem 1.79, the single-variable function $F(t) = f(a + t(x - a))$ is differentiable at $[0,1]$, and $F'(t) = \langle f'(a + t(x - a)), x - a \rangle$ for every $t \in [0,1]$. Since F is convex on $[0,1]$ (by our assumption), we have

$$f(x) = F(1) \geq F(0) + F'(0) = f(a) + \langle f'(a), x - a \rangle,$$

which is exactly (1.41). (We applied here [7, Theorem 12.64]).

Now suppose (1.41) for every $a, x \in G$. Let F be the same function as above. We have to prove that F is convex on $[0,1]$. By [7, Theorem 12.65], it is enough to show that $F(t) \geq F(t_0) + F'(t_0)(t - t_0)$ for every $t, t_0 \in [0,1]$. Since $F'(t) = \langle f'(a + t(x - a)), x - a \rangle$, we have

$$f(a + t(x - a)) \geq f(a + t_0(x - a)) + \langle f'(a + t_0(x - a)), (t - t_0) \cdot (x - a) \rangle.$$

However, this follows from (1.41) if we apply it with $a + t_0(x - a)$ and $a + t(x - a)$ in place of a and x, respectively. □

Theorem 1.106. *Let f be twice differentiable on the convex and open set $G \subset \mathbb{R}^p$. The function f is convex on G if and only if the quadratic form $d^2 f(a)$ is positive semidefinite for every $a \in G$.*

Proof. Let f be convex on G, and let a and b be different points of G. By Lemma 1.98, the function $F(t) = f(a + t(b - a))$ is twice differentiable on the interval $[0,1]$, and $F''(0) = d^2 f(a)(b - a)$. Since F is convex on $[0,1]$ (by our assumption), we have $F''(0) = d^2 f(a)(b - a) \geq 0$. This is true for every $b \in G$, showing that $d^2 f(a)$ is positive semidefinite. Indeed, since G is open, we must have $B(a,r) \subset G$ for a suitable $r > 0$. For every $x \in \mathbb{R}^p$ we have $a + tx \in B(a,r)$ if t is small enough, i.e., $d^2 f(a)(tx) \geq 0$ for every t small enough. Since $d^2 f(a)(tx) = t^2 \cdot d^2 f(a)(x)$, it follows that $d^2 f(a)(x) \geq 0$, and $d^2 f(a)$ is positive semidefinite.

Now let $d^2 f(a)$ be positive semidefinite for every $a \in G$. Let a and b be distinct points of G, and let $F(t) = f(a + t(b - a))$ $(t \in [0,1])$. By Lemma 1.98, F is twice differentiable on the interval $[0,1]$, and $F''(t) = d^2 f(a + t(b - a))(b - a) \geq 0$, since $d^2 f(a + t(b - a))$ is positive semidefinite. This implies that F is convex on $[0,1]$. Since this is true for every $a, b \in G$, $a \neq b$, this means that f is convex on G. □

Remark 1.107. It is clear how to change the conditions of Theorems 1.105 and 1.106 in order to get necessary and sufficient conditions for concavity of a function.

Example 1.108. Let $p = 2$. The graph of the polynomial $f(x, y) = x^2 + y^2$ is a **rotated paraboloid**, since it can be obtained by rotating the graph of the single-variable function $z = x^2$ around the z-axis. We show that f is convex in the plane.
For every $(a, b) \in \mathbb{R}^2$ we have

$$D_{1,1}f(a, b) = 2,$$
$$D_{2,1}f(a, b) = D_{1,2}f(a, b) = 0,$$
$$\text{and } D_{2,2}f(a, b) = 2.$$

Thus $d^2 f(a, b)(x, y) = 2x^2 + 2y^2$. Since this quadratic form is positive definite, it follows from Theorem 1.106 that f is convex.

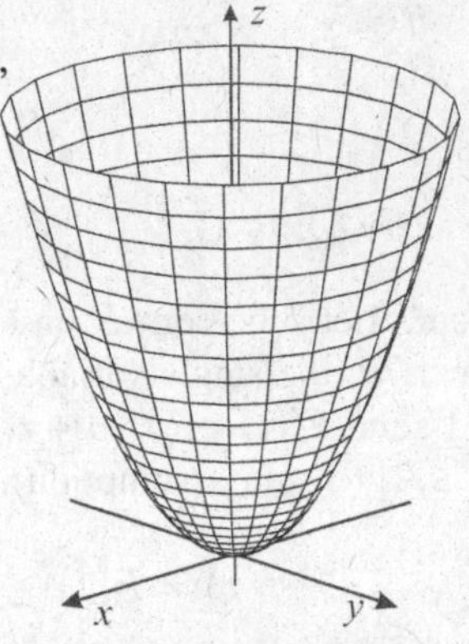

1.18. Figure

Exercises

1.102. What are the third Taylor polynomials of the following functions?

(a) x/y at $(1, 1)$;
(b) $x^3 + y^3 + z^3 - 3xyz$ at $(1, 1, 1)$;
(c) $\sin(x + y)$ at $(0, 0)$;
(d) x^y at $(1, 0)$.

1.103. Find the local extremum points and also the least and greatest values (if they exist) of the following two-variable functions:

(a) $x^2 + xy + y^2 - 3x - 3y$;
(b) $x^3y^2(2 - x - y)$;
(c) $x^3 + y^3 - 9xy$;
(d) $x^4 + y^4 - 2x^2 + 4xy - 2y^2$.

1.104. Let $H \subset \mathbb{R}^p$ be convex. Show that the function $f\colon H \to \mathbb{R}$ is convex if and only if the set

$$\{(x, y) \in \mathbb{R}^{p+1} \colon x \in H,\ y \geq f(x)\} \subset \mathbb{R}^{p+1}$$

is convex.

1.105. Let $G \subset \mathbb{R}^p$ be convex and open. Show that if $f\colon G \to \mathbb{R}$ is convex, then it is continuous.

1.106. Let $G \subset \mathbb{R}^p$ be convex and open. Show that the function $f\colon G \to \mathbb{R}$ is convex if and only if it is continuous and if

$$f\left(\frac{x+y}{2}\right) \le \frac{f(x)+f(y)}{2}$$

holds for every $x, y \in G$.

1.11 Appendix: Tangent Lines and Tangent Planes

In our previous investigations we introduced the notions of tangent lines and tangent planes in connection with approximations by linear functions. However, the intuitive notion of tangent lines also involves the idea that tangents are the "limits of the secant lines." Let, for example, f be a one-variable function differentiable at a. The slope of the line (the "secant") intersecting the graph of f at the points $(a, f(a))$ and $(x, f(x))$ is $(f(x) - f(a))/(x - a)$. This slope converges to $f'(a)$ as $x \to a$, and thus the secant "converges" to the line with slope $f'(a)$ that contains point $(a, f(a))$, i.e., to the tangent line. More precisely, if x converges to a from the right or from the left, then the half-line with endpoint $(a, f(a))$ that intersects $(x, f(x))$ "converges" to one of the half-lines that are subsets of the tangent and lie above $[a, \infty)$ or $(-\infty, a]$, respectively. This property will be used for a more general definition of the tangent.

Let x_0 and x be different points of $\mathbb{R}^p$. The half-line $\overrightarrow{x_0 x}$ with endpoint x_0 and passing through x consists of the points $x_0 + t(x - x_0)$ $(t \in \mathbb{R},\ t \ge 0)$. We say that the unit vector $(x - x_0)/|x - x_0|$ is the **direction vector** of this half-line. Let $x_n \to x_0$ and $x_n \ne x_0$, for every n, and let $(x_n - x_0)/|x_n - x_0| \to v$. In this case we say that **the sequence of half-lines $\overrightarrow{x_0 x_n}$ converges to the half-line** $\{x_0 + tv\colon t \ge 0\}$.

Let $H \subset \mathbb{R}^p$, and let $x_0 \in H'$. If $x_n \in H \setminus \{x_0\}$ and $x_n \to x_0$, then by the Bolzano–Weierstrass theorem (Theorem 1.9), the sequence of unit vectors $(x_n - x_0)/|x_n - x_0|$ has a convergent subsequence. We say that the **contingent** of the set H at x_0 is the set of vectors v for which there exists a sequence $x_n \in H \setminus \{x_0\}$ such that $x_n \to x_0$ and $(x_n - x_0)/|x_n - x_0| \to v$. We denote the contingent of the set H at x_0 by $\operatorname{Cont}(H; x_0)$. It is clear that $\operatorname{Cont}(H; x_0) \ne \emptyset$ for every $x_0 \in H'$.

In the next three examples we investigate the contingents of curves. By a **curve** we mean a map $g\colon [a, b] \to \mathbb{R}^p$ (see [7, p. 380]).

Example 1.109. **1.** If the single-variable function f is differentiable at a, then $\operatorname{Cont}(\operatorname{graph} f; (a, f(a)))$ contains exactly two unit vectors, namely the vector

$$\left(\frac{1}{\sqrt{1+(f'(a))^2}}, \frac{f'(a)}{\sqrt{1+(f'(a))^2}}\right)$$

with slope $f'(a)$ and its negative.

2. Let $g\colon [a,b] \to \mathbb{R}^p$ be a curve, and let g be differentiable at $t_0 \in (a,b)$ with $g'(t_0) \neq 0$. The contingent of the set $\Gamma = g([a,b])$ at $g(t_0)$ contains the unit vectors $\pm g'(t_0)/|g'(t_0)|$. Indeed, if $t_n \to t_0$, then

$$\frac{g(t_n) - g(t_0)}{t_n - t_0} \to g'(t_0).$$

We have

$$\left|\frac{g(t_n) - g(t_0)}{t_n - t_0}\right| \to |g'(t_0)|,$$

which implies

$$\frac{g(t_n) - g(t_0)}{|g(t_n) - g(t_0)|} = \frac{(g(t_n) - g(t_0))/(t_n - t_0)}{|g(t_n) - g(t_0)|/(t_n - t_0)} \to \frac{g'(t_0)}{|g'(t_0)|}$$

if $t_n > t_0$. Therefore, $g'(t_0)/|g'(t_0)| \in \operatorname{Cont}(\Gamma, g(t_0))$. If t_n converges to t_0 from the left-hand side, we get $-g'(t_0)/|g'(t_0)| \in \operatorname{Cont}(\Gamma, g(t_0))$ in the same way.
3. Let g be a curve that passes through the point $g(t_0)$ only once, i.e., $g(t) \neq g(t_0)$ for every $t \neq t_0$. It is easy to see that $g(t_n) \to g(t_0)$ is true only if $t_n \to t_0$. If we also assume that $g'(t_0) \neq 0$, then we obtain that the contingent $\operatorname{Cont}(\Gamma, g(t_0))$ consists of the unit vectors $\pm g'(t_0)/|g'(t_0)|$.

The examples above motivate the following definition of the tangent.

Definition 1.110. Let $x_0 \in H'$, and let $|v| = 1$. We say that the line $\{x_0 + tv\colon t \in \mathbb{R}\}$ is *the tangent line of the set H at the point x_0* if $\operatorname{Cont}(H; x_0) = \{v, -v\}$.

By this definition, the graph of the function f has a tangent line at the point $(a, f(a))$ not only when f is differentiable at a, but also when $f'(a) = \infty$ or $f'(a) = -\infty$. On the other hand, if $f'_-(a) = -\infty$ and $f'_+(a) = \infty$, then graph f does not have a tangent line at $(a, f(a))$.
We can easily generalize Definition 1.110 to tangent planes.

Definition 1.111. Let $x_0 \in H'$, and let S be a plane containing the origin (i.e., let S be a two-dimensional subspace). We say that a plane $\{x_0 + s\colon s \in S\}$ is the *tangent plane of the set H at the point x_0* if $\operatorname{Cont}(H; x_0)$ consists of exactly the unit vectors of S.

Let the function $f\colon \mathbb{R}^2 \to \mathbb{R}$ be differentiable at $(a,b) \in \mathbb{R}^2$. It is not very difficult to show (though some computation is involved) that the contingent of the set graph f at the point $(a, b, f(a,b))$ consists of the unit vectors $(v_1, v_2, v_3) \in \mathbb{R}^3$ for which $v_3 = D_1 f(a,b) v_1 + D_2 f(a,b) v_2$.

Then by Definition 1.74, the plane

$$z = D_1 f(a, b)(x - a) + D_2 f(a, b)(y - b) + f(a, b)$$

is the tangent plane of the graph of f at $(a, b, f(a, b))$. One can see that this plane is the tangent plane of the graph of f according to Definition 1.111 as well.

We can define the notion of tangent hyperplanes in $\mathbb{R}^p$ similarly. One can show that for a graph of a function, the notion of tangent hyperplane according to Definition 1.75 corresponds to Definition 1.111, generalized to $\mathbb{R}^p$.

Chapter 2
Functions from $\mathbb{R}^p$ to $\mathbb{R}^q$

Consider a function $f\colon H \to \mathbb{R}^q$, where H is an arbitrary set, and let the coordinates of the vector $f(x)$ be denoted by $f_1(x), \dots, f_q(x)$ for every $x \in H$. In this way we define the functions $f_1, \dots, f_q$, where $f_i\colon H \to \mathbb{R}$ for every $i = 1, \dots, q$. We call f_i the ith **coordinate function** or **component** of f.

The above defined concept is a generalization of the coordinate functions introduced by Definition 1.47. Indeed, let f be the identity function on $\mathbb{R}^p$, i.e., let $f(x) = x$ for every $x \in \mathbb{R}^p$. Then the coordinate functions of $f\colon \mathbb{R}^p \to \mathbb{R}^p$ are nothing but thc functions $x = (x_1, \dots, x_p) \mapsto x_i$ (with $i = 1, \dots, p$).

Now let $f\colon H \to \mathbb{R}^q$ with $H \subset \mathbb{R}^p$. The coordinate functions of f are real-valued functions defined on the set H; therefore, they are p-variable real-valued functions. The limits, continuity, and differentiability of the function f could be defined using the corresponding properties of f's coordinate functions. However, it is easier, shorter, and more to the point to define these concepts directly for f, just copying the corresponding definitions for real-valued functions. Fortunately, as we shall see, the two approaches are equivalent to each other.

2.1 Limits and Continuity

Definition 2.1. Let $H \subset E \subset \mathbb{R}^p$, and let a be a limit point of H. The *limit of the function* $f\colon E \to \mathbb{R}^q$ *at* a *restricted to* H is $b \in \mathbb{R}^q$ if for every $\varepsilon > 0$ there exists $\delta > 0$ such that $|f(x) - b| < \varepsilon$ whenever $x \in H$ and $0 < |x - a| < \delta$. *Notation:* $\lim_{x \to a,\, x \in H} f(x) = b$.

If the domain of the function f is equal to H (i.e., it is not greater than H), then we omit the part "restricted to the set H" of the definition above, and we simply say that the limit of f at a is b, with notation $\lim_{x \to a} f(x) = b$ or $f(x) \to b$, if $x \to a$.

M. Laczkovich and V.T. Sós, *Real Analysis*, Undergraduate Texts in Mathematics, https://doi.org/10.1007/978-1-4939-7369-9_2

Obviously, $\lim_{x\to a,\, x\in H} f(x) = b$ if and only if for every neighborhood V of b there exists a punctured neighborhood $\dot{U}$ of a such that $f(x) \in V$ if $x \in H \cap \dot{U}$.

Theorem 2.2. *Let $H \subset E \subset \mathbb{R}^p$, let a be a limit point of H, and let $b = (b_1, \ldots, b_q) \in \mathbb{R}^q$. For every function $f\colon E \to \mathbb{R}^q$, we have $\lim_{x\to a,\, x\in H} f(x) = b$ if and only if $\lim_{x\to a,\, x\in H} f_i(x) = b_i$ $(i = 1, \ldots, q)$ holds for every coordinate function f_i of f.*

Proof. The statement follows from the definitions, using the fact that for every point $y = (y_1, \ldots, y_q) \in \mathbb{R}^q$, we have $|y - b| \le |y_1 - b_1| + \ldots + |y_q - b_q|$ and $|y_i - b_i| \le |y - b|$, for each $i = 1, \ldots, q$. □

The **transference principle** follows from the theorem above: $\lim_{x\to a,\, x\in H} f(x) = b$ *if and only if for every sequence $x_n \in H \setminus \{a\}$,* we have that *$x_n \to a$ implies $f(x_n) \to b$.* (This statement is a generalization of the corresponding one dimensional theorem [7, Theorem 10.19].)

It is clear from Theorems 1.40 and 2.2 that *if $\lim_{x\to a,\, x\in H} f(x) = b$ and $\lim_{x\to a,\, x\in H} g(x) = c$, where $b, c \in \mathbb{R}^q$, then $\lim_{x\to a,\, x\in H} \lambda f(x) = \lambda b$ for every $\lambda \in \mathbb{R}$. Furthermore, $\lim_{x\to a,\, x\in H}(f(x) + g(x)) = b + c$ and $\lim_{x\to a,\, x\in H}\langle f(x), g(x)\rangle = \langle b, c\rangle$.*

Definition 2.3. Let $a \in H \subset E \subset \mathbb{R}^p$. We say that the function $f\colon E \to \mathbb{R}^q$ is *continuous at a restricted to the set H* if for every $\varepsilon > 0$ there exists $\delta > 0$ such that if $x \in H$ and $|x - a| < \delta$, then $|f(x) - f(a)| < \varepsilon$.

If the domain of f is equal to H, then we can omit the part "restricted to the set H" from the definition.

If f is continuous at every point $a \in H$, then we say that f is continuous on the set H.

The following theorem follows from Theorem 2.2.

Theorem 2.4. *The function f is continuous at a point a restricted to the set H if and only if this is true for every coordinate function of f.* □

Clearly, f is continuous at a restricted to H if and only if one of the following two conditions holds:

(i) *a is an isolated point of H;*
(ii) *$a \in H \cap H'$ and $\lim_{x\to a,\, x\in H} f(x) = f(a)$.*

The **transference principle for continuity** can be easily verified: *the function $f\colon H \to \mathbb{R}^p$ is continuous at the point $a \in H$ restricted to the set H if and only if $f(x_n) \to f(a)$ holds for every sequence $x_n \in H$ with $x_n \to a$.*

This implies the following statement: *if the functions f and g are continuous at the point a restricted to the set H, then so are the functions $f + g$, $\langle f, g\rangle$ and λf for every $\lambda \in \mathbb{R}$.*

A theorem about the limit of composite functions follows.

Theorem 2.5. *Suppose that*

(i) *$H \subset \mathbb{R}^p$, $g\colon H \to \mathbb{R}^q$ and $\lim_{x\to a} g(x) = c$, where a is a limit point of H;*

(ii) $g(H) \subset E \subset \mathbb{R}^q$, $f\colon E \to \mathbb{R}^s$ *and* $\lim_{x\to c} f(x) = b$;

(iii) $g(x) \neq c$ *in a punctured neighborhood of* a, *or* $c \in E$ *and* f *is continuous at* c *restricted to the set* $g(H)$.

Then

$$\lim_{x\to a} f(g(x)) = b. \square \tag{2.1}$$

Corollary 2.6. *If* g *is continuous at* a *restricted to* H, *and* f *is continuous at the point* $g(a)$ *restricted to the set* $g(H)$, *then* $f \circ g$ *is also continuous at* a *restricted to* H. □

If we wish to generalize Weierstrass's theorem (Theorem 1.51) to functions mapping to $\mathbb{R}^q$, we have to keep in mind that for $q > 1$ there is no natural ordering of the points of $\mathbb{R}^q$. Therefore, we cannot speak about the largest or smallest value of a function. However, the statement on the boundedness still holds; moreover, we can state more.

Theorem 2.7. *Let* $H \subset \mathbb{R}^p$ *be bounded and closed, and let* $f\colon H \to \mathbb{R}^q$ *be continuous. Then the set* $f(H)$ *is bounded and closed in* $\mathbb{R}^q$.

Proof. Applying Weierstrass's theorem (Theorem 1.51) to the coordinate functions of f yields that the set $f(H)$ is bounded.

In order to prove that $f(H)$ is also closed we will use part (iii) of Theorem 1.17. Suppose that $y_n \in f(H)$ and $y_n \to b$. For every n we can choose a point $x_n \in H$ such that $f(x_n) = y_n$. The sequence (x_n) is bounded (since H is bounded). Thus, by the Bolzano–Weierstrass theorem, (x_n) has a convergent subsequence (x_{n_k}). If $x_{n_k} \to a$, then $a \in H$, because the set H is closed. Since the function f is continuous, it follows that

$$b = \lim_{k\to\infty} y_{n_k} = \lim_{k\to\infty} f(x_{n_k}) = f(a),$$

and thus $b \in f(H)$. Then, by Theorem 1.17, the set $f(H)$ is closed. □

Recall the definition of injective functions. A mapping is **injective** (or **one-to-one**) if it takes on different values at different points of its domain. The following theorem states another important property of continuous functions with bounded and closed domains.

Theorem 2.8. *Let* $H \subset \mathbb{R}^p$ *be bounded and closed, and let* $f\colon H \to \mathbb{R}^q$ *be continuous. If* f *is injective on the set* H, *then* f^{-1} *is continuous on the set* $f(H)$.

Proof. Let $y_n \in f(H)$ and $y_n \to b \in f(H)$. Then we have $b = f(a)$ for a suitable $a \in H$. Let $x_n = f^{-1}(y_n)$ for every n; we need to prove that $x_n \to f^{-1}(b) = a$.

We prove by contradiction. Let us assume that the statement is not true. Then there exists $\varepsilon > 0$ such that $x_n \notin B(a, \varepsilon)$, i.e., $|x_n - a| \geq \varepsilon$ for infinitely many n. We may assume that this holds for every n, for otherwise, we could delete the terms of the sequence for which it does not hold. The sequence (x_n) is bounded (since H is bounded), and then, by the Bolzano–Weierstrass theorem, it has a convergent subsequence (x_{n_k}). If $x_{n_k} \to c$. And then $c \in H$, since H is closed. Also, $c \neq a$, since

$$|c-a| = \lim_{k\to\infty} |x_{n_k} - a| \geq \varepsilon.$$

Since the function f is continuous, it follows that

$$f(c) = \lim_{k\to\infty} f(x_{n_k}) = \lim_{k\to\infty} y_{n_k} = b = f(a),$$

which contradicts the assumption that f is injective. □

Remark 2.9. The condition of the boundedness of the set H cannot be omitted from the previous theorem, i.e., the inverse of a continuous and injective function on a closed domain is not necessarily continuous. Consider the following example. Let $p = q = 1$, $H = \mathbb{N}$ and let $f\colon \mathbb{N} \to \mathbb{R}$ be the function with $f(0) = 0$ and $f(n) = 1/n$ for every $n = 1, 2, \ldots$. The set H is closed (since every convergent sequence of H is constant begining from some index), the function f is continuous on H (since every point of H is an isolated point), and f is injective. On the other hand, f^{-1} is not continuous, since

$$0 = f^{-1}(0) \neq \lim_{n\to\infty} f^{-1}(1/n) = \lim_{n\to\infty} n = \infty.$$

(The condition of closedness of H cannot be omitted from the theorem either; see Exercise 2.2.)

Uniform continuity can be defined in the same way as in the case of real-valued functions.

Definition 2.10. We say that the function f is *uniformly continuous* on the set $H \subset \mathbb{R}^p$ if for every $\varepsilon > 0$ there exists $\delta > 0$ such $|f(x) - f(y)| < \varepsilon$ holds whenever $x, y \in H$ and $|x - y| < \delta$ (where δ is independent of x and y).

Heine's theorem remains valid: *if $H \subset \mathbb{R}^p$ is a bounded and closed set and the function $f\colon H \to \mathbb{R}^q$ is continuous, then f is uniformly continuous on H.*

2.2 Differentiability

To define differentiability for an $\mathbb{R}^q$-valued function, we proceed as in the cases of limits and continuity; that is, we simply copy Definition 1.63. However, since we are dealing with functions that map from $\mathbb{R}^p$ to $\mathbb{R}^q$, we need to define linear maps from $\mathbb{R}^p$ to $\mathbb{R}^q$. For this reason we recall some basic notions of linear algebra.

We say that a function $A\colon \mathbb{R}^p \to \mathbb{R}^q$ is a **linear mapping** or a **linear transformation** if $A(x+y) = A(x) + A(y)$ and $A(\lambda x) = \lambda A(x)$ hold for every $x, y \in \mathbb{R}^p$

and $\lambda \in \mathbb{R}$. Clearly, the mapping $A: \mathbb{R}^p \to \mathbb{R}^q$ is linear if and only if each of its coordinate functions is linear.

Let $a_{i1}x_1 + \ldots + a_{ip}x_p$ be the ith coordinate function of the mapping A ($i = 1, \ldots, q$). We call

$$\begin{pmatrix} a_{11} & a_{12} \ldots a_{1p} \\ a_{21} & a_{22} \ldots a_{2p} \\ \vdots & \vdots \quad\;\; \vdots \\ a_{q1} & a_{q2} \ldots a_{qp} \end{pmatrix} \tag{2.2}$$

the matrix of the mapping A. The matrix has q rows and p columns, and the ith row contains the coefficients of the ith coordinate function of A.

It is easy to see that if $x = (x_1, \ldots, x_p) \in \mathbb{R}^p$, then the vector $y = A(x)$ is the product of the matrix of A and the column vector consisting of the coordinates of x. That is,

$$A(x) = \begin{pmatrix} a_{11} & a_{12} \ldots a_{1p} \\ a_{21} & a_{22} \ldots a_{2p} \\ \vdots & \vdots \quad\;\; \vdots \\ a_{q1} & a_{q2} \ldots a_{qp} \end{pmatrix} \begin{pmatrix} x_1 \\ x_2 \\ \vdots \\ x_p \end{pmatrix} = \begin{pmatrix} y_1 \\ y_2 \\ \vdots \\ y_q \end{pmatrix}. \tag{2.3}$$

In other words, the ith coordinate of $A(x)$ is the scalar product of the ith row of A and x.

Definition 2.11. Let $H \subset \mathbb{R}^p$ and $a \in \operatorname{int} H$. We say that *the function $f: H \to \mathbb{R}^q$ is differentiable at the point a* if there exists a linear mapping $A: \mathbb{R}^p \to \mathbb{R}^q$ such that

$$f(x) = f(a) + A(x - a) + \varepsilon(x) \cdot |x - a| \tag{2.4}$$

for every $x \in H$, where $\varepsilon(x) \to 0$ if $x \to a$. (Here $\varepsilon: H \to \mathbb{R}^q$.)

Remark 2.12. Since the function ε can be defined to be 0 at the point a, *the differentiability of the function f is equivalent to* (2.4) *for an appropriate linear mapping A, where $\varepsilon(a) = 0$ and ε is continuous at a.*

We can formulate another equivalent condition: *for an appropriate linear mapping A we have $(f(x) - f(a) - A(x - a))/|x - a| \to 0$* as $x \to a$.

Theorem 2.13. *The function $f: H \to \mathbb{R}^q$ ($H \subset \mathbb{R}^p$) is differentiable at the point $a \in \operatorname{int} H$ if and only if every coordinate function f_i of f is differentiable at a. The jth entry of the ith row of the matrix of A from (2.4) is equal to the partial derivative $D_j f_i(a)$ for every $i = 1, \ldots, q$ and $j = 1, \ldots, p$.*

Proof. Suppose that (2.4) holds for every $x \in H$, where $\varepsilon(x) \to 0$ as $x \to a$. Since the vectors of the two sides of (2.4) are equal, their corresponding coordinates are equal as well. Thus, $f_i(x) = f_i(a) + A_i(x - a) + \varepsilon_i(x) \cdot |x - a|$ for every $x \in H$ and $i = 1, \ldots, q$, where f_i, A_i, ε_i denote the ith coordinate functions of f, A, and ε, respectively. Since A_i is linear and $\varepsilon_i(x) \to 0$ as $x \to a$ (following from the fact that $|\varepsilon_i(x)| \le |\varepsilon(x)|$ for every x), we get that f_i is differentiable at a. By Theorem 1.67, the jth coefficient of the linear function A_i is the $D_j f_i(a)$ partial derivative, which also proves the statement about the matrix A.

Now suppose that every coordinate function f_i of f is differentiable at a. By Theorem 1.67, $f_i(x) = f_i(a) + A_i(x-a) + \varepsilon_i(x)$, where $A_i(x) = D_1 f_i(a)x_1 + \ldots + D_p f_i(a)x_p$ and $\varepsilon_i(x) \to 0$ as $x \to a$. Let $A(x) = (A_1(x), \ldots, A_q(x))$ for every $x \in \mathbb{R}^p$, and let $\varepsilon(x) = (\varepsilon_1(x), \ldots, \varepsilon_q(x))$ for every $x \in H$. The mapping $A\colon \mathbb{R}^p \to \mathbb{R}^q$ is linear, and $\varepsilon(x) \to 0$ as $x \to a$ by Theorem 2.2. In addition, (2.4) holds for every $x \in H$. This proves that f is differentiable at the point a. □

Corollary 2.14. *If f is differentiable at a, then the linear mapping A from (2.4) is unique.* □

Definition 2.15. Let $f\colon H \to \mathbb{R}^q$ with $H \subset \mathbb{R}^p$, and let f be differentiable at $a \in \operatorname{int} H$. We say that the linear mapping $A\colon \mathbb{R}^p \to \mathbb{R}^q$ from (2.4) is *the derivative of the function f at the point a*, and we use the notation $f'(a)$. We call the matrix of the linear mapping $f'(a)$, i.e., the matrix of the partial derivatives $D_j f_i(a)$ ($j = 1, \ldots, p$, $i = 1, \ldots, q$)

$$\begin{pmatrix} D_1 f_1(a) & D_2 f_1(a) \ldots & D_p f_1(a) \\ D_1 f_2(a) & D_2 f_2(a) \ldots & D_p f_2(a) \\ \vdots & \vdots & \vdots \\ D_1 f_q(a) & D_2 f_q(a) \ldots & D_p f_q(a), \end{pmatrix}$$

the *Jacobian matrix*[1] of the function f at the point a.

The following statements are clear from Theorems 1.66, 1.67, 1.71, and 2.13.

Theorem 2.16.

(i) *If the function f is differentiable at the point a, then f is continuous at a. Furthermore, every partial derivative of every coordinate function of f exists and is finite at a.*

(ii) *If every partial derivative of every coordinate function of f exists and is finite in a neighborhood of the point a and is continuous at a, then f is differentiable at a.* □

Example 2.17. Consider the mapping

$$f(x,y) = (e^x \cos y, e^x \sin y) \qquad ((x,y) \in \mathbb{R}^2).$$

[1] Carl Jacobi (1804–1851), German mathematician.

The partial derivatives of f's coordinate functions are

$$D_1 f_1(x,y) = e^x \cos y, \quad D_2 f_1(x,y) = -e^x \sin y,$$
$$D_1 f_2(x,y) = e^x \sin y, \quad D_2 f_2(x,y) = e^x \cos y$$

for every $(x,y) \in \mathbb{R}^2$. Since these partial derivatives are continuous everywhere, it follows from Theorem 2.16 that f is differentiable at every point (a,b) in the plane, and f's Jacobian matrix is

$$\begin{pmatrix} e^a \cos b & -e^a \sin b \\ e^a \sin b & e^a \cos b \end{pmatrix}.$$

Thus, the derivative of f at (a,b) is the linear mapping

$$A(x,y) = ((e^a \cos b)x - (e^a \sin b)y, (e^a \sin b)x + (e^a \cos b)y).$$

Remark 2.18. Let us summarize the different objects we obtain by differentiating different kinds of mappings.

The derivative of a single-variable real function at a fixed point is a real number, namely the limit of the differential quotients.

The derivative of a curve $g\colon [a,b] \to \mathbb{R}^q$ at a given point is a vector of $\mathbb{R}^q$ whose coordinates are the derivatives of g's coordinate functions (see [7, Remark 16.22]).

The derivative of a p-variable real function is a vector of $\mathbb{R}^p$ (the gradient vector) whose components are the partial derivatives of the function at a given point.

Definition 2.15 takes another step toward further abstraction: the derivative of a map $\mathbb{R}^p \to \mathbb{R}^q$ is neither a number nor a vector, but a mapping.

As a consequence of this diversity, the derivative of a function $f\colon \mathbb{R} \to \mathbb{R}$ is a real number (if we consider f a function) but also a vector of dimension one (if we consider f a curve mapping into $\mathbb{R}$).

What's worse, the derivative of a mapping $\mathbb{R}^p \to \mathbb{R}^q$ is a vector for $q = 1$, but it is also a linear mapping, and for $p = q = 1$ it is a real number as well.

We should realize, however, that the essence of the derivative is the linear mapping with which we approximate the function, and the way we represent this linear mapping is less important. For a single-variable function f, the approximating linear function is $f(a) + f'(a)(x-a)$ defining the tangent line. This function is uniquely characterized by the coefficient $f'(a)$ (since it has to take the value $f(a)$ at a). Similarly, a linear function approximating a p-variable real function is the function $f(a) + \sum_{i=1}^p D_i f(a)(x_i - a_i)$ defining the tangent hyperplane. This can be characterized by the vector of its coefficients.

We could have circumvented these inconsistencies by defining the derivative of a function $f\colon \mathbb{R}^p \to \mathbb{R}^q$ not by a linear mapping, but by its matrix (i.e., its Jacobian matrix).[2] In most cases it is much more convenient to think of the derivative as a mapping and not as a matrix, which we will see in the next section. When we talk about mappings between more general spaces (called normed linear spaces), the linear mappings do not always have a matrix. In these cases we have to define the derivative as the linear mapping itself.

We have to accept the fact that the object describing the derivative depends on the dimensions of the corresponding spaces. Fortunately enough, whether we consider the derivative to be a number, a vector, or a mapping will always be clear from the context.

2.3 Differentiation Rules

Theorem 2.19. *If the functions f and g mapping to $\mathbb{R}^q$ are differentiable at the point $a \in \mathbb{R}^p$, then the functions $f+g$ and λf are also differentiable at a. Furthermore, $(f+g)'(a) = f'(a) + g'(a)$ and $(\lambda f)'(a) = \lambda f'(a)$ for every $\lambda \in \mathbb{R}$.*

Proof. The statement is obvious from Theorem 2.13. □

The following theorem concerns the differentiability of a composite function and its derivative.

Theorem 2.20. *Suppose that*

(i) *$H \subset \mathbb{R}^p$, $g \colon H \to \mathbb{R}^q$, and g is differentiable at the point $a \in \operatorname{int} H$;*

(ii) *$g(a) \in \operatorname{int} E \subset \mathbb{R}^q$, $f \colon E \to \mathbb{R}^s$, and f is differentiable at the point $g(a)$.*

Then the composite function $f \circ g$ is differentiable at a, with

$$(f \circ g)'(a) = f'(g(a)) \circ g'(a).$$

To prove this theorem we first need to show that every linear mapping has the Lipschitz property.

Lemma 2.21. *For every linear mapping $A \colon \mathbb{R}^p \to \mathbb{R}^q$ there exists a $K \geq 0$ such that $|A(x) - A(y)| \leq K \cdot |x - y|$ for every $x, y \in \mathbb{R}^p$.*

Proof. Let $e_1, \ldots, e_p$ be a basis of $\mathbb{R}^p$, and let $M = \max_{1 \leq i \leq p} |A(e_i)|$. Then, for every $x = (x_1, \ldots, x_p) \in \mathbb{R}^p$ we have

$$|A(x)| = \left| \sum_{i=1}^{p} x_i \cdot A(e_i) \right| \leq \sum_{i=1}^{p} |x_i| \cdot M \leq Mp \cdot |x|.$$

Thus $|A(x) - A(y)| = |A(x - y)| \leq Mp \cdot |x - y|$ for every $x, y \in \mathbb{R}^p$, and hence $K = Mp$ satisfies the requirements of the lemma. □

---

[2] However, the inconsistencies would not have disappeared entirely. For $p = 1$ (i.e., for curves mapping to $\mathbb{R}^q$) the Jacobian matrix is a $1 \times q$ matrix, in other words, it is a column vector, while the derivative of the curve is a row vector.

Let $\mathcal{K}_A$ denote the set of numbers $K \geq 0$ that satisfy the conditions of Lemma 2.21. Obviously, the set $\mathcal{K}_A$ has a smallest element. Indeed, if $K_0 = \inf \mathcal{K}_A$, then $|A(x) - A(y)| \leq K_0 \cdot |x - y|$ also holds for every $x, y \in \mathbb{R}^p$, and thus $K_0 \in \mathcal{K}_A$.

Definition 2.22. The smallest number, K, satisfying the conditions of Lemma 2.21 is called the *norm* of A, and is denoted by $\|A\|$.

Proof of Theorem 2.20. Let $g'(a) = A$ and $f'(g(a)) = B$. We know that if x is close to a, then $g(a) + A(x - a)$ approximates $g(x)$ well, and if y is close to $g(a)$, then $f(g(a)) + B(y - g(a))$ approximates $f(y)$ well. Therefore, intuitively, if x is close to a, then

$$f(g(a)) + B(g(a) + A(x - a) - g(a)) = f(g(a)) + (BA)(x - a)$$

approximates $f(g(x))$ well; i.e., $(f \circ g)'(a) = BA$. Below we make this argument precise.

Since $g'(a) = A$, it follows that

$$g(x) = g(a) + A(x - a) + \varepsilon(x) \cdot |x - a|, \tag{2.5}$$

where $\lim_{x \to a} \varepsilon(x) = 0$. Let us choose $\delta > 0$ such that $|x - a| < \delta$ implies $x \in H$ and $|\varepsilon(x)| < 1$. Then

$$\begin{aligned} |g(x) - g(a)| &\leq |A(x - a)| + |\varepsilon(x)| \cdot |x - a| \leq \|A\| \cdot |x - a| + |x - a| = \\ &= (\|A\| + 1) \cdot |x - a| \end{aligned} \tag{2.6}$$

for every $|x - a| < \delta$. On the other hand, $f'(g(a)) = B$ implies

$$f(y) = f(g(a)) + B(y - g(a)) + \eta(y) \cdot |y - g(a)|, \tag{2.7}$$

where $\lim_{y \to g(a)} \eta(y) = \eta(g(a)) = 0$. Now g is continuous at the point a by (2.6) (or by Theorem 2.16), whence $g(x) \in E$ if x is close enough to a. Applying (2.7) with $y = g(x)$ and using also (2.5), we get

$$\begin{aligned} f(g(x)) - f(g(a)) &= B(g(x) - g(a)) + \eta(g(x)) \cdot |g(x) - g(a)| = \\ &= B(A(x - a)) + B(\varepsilon(x)) \cdot |x - a| + \eta(g(x)) \cdot |g(x) - g(a)| = \\ &= (B \circ A)(x - a) + r(x), \end{aligned} \tag{2.8}$$

where $r(x) = B(\varepsilon(x)) \cdot |x - a| + \eta(g(x)) \cdot |g(x) - g(a)|$. Then, by (2.6),

$$|r(x)| \leq \|B\| \cdot |\varepsilon(x)| \cdot |x - a| + |\eta(g(x))| \cdot (\|A\| + 1) \cdot |x - a| = \theta(x) \cdot |x - a|,$$

where

$$\theta(x) = \|B\| \cdot |\varepsilon(x)| + (\|A\| + 1) \cdot |\eta(g(x))| \to 0$$

if $x \to a$, since $\eta(g(a)) = 0$ and η is continuous at $g(a)$. Therefore, (2.8) implies that the function $f \circ g$ is differentiable at a, and $(f \circ g)'(a) = B \circ A$. □

Corollary 2.23. (Differentiation of composite functions)
Suppose that the real-valued function f is differentiable at the point $b=(b_1, \ldots, b_q) \in \mathbb{R}^q$, and the real-valued functions $g_1, \ldots, g_q$ are differentiable at the point $a \in \mathbb{R}^p$, where $g_i(a) = b_i$ for every $i = 1, \ldots, q$. Then the function $F(x) = f(g_1(x), \ldots, (g_q(x))$ is differentiable at the point a, and

$$D_j F(a) = \sum_{i=1}^{q} D_i f(b) \cdot D_j g_i(a) \tag{2.9}$$

holds for every $j = 1, \ldots, p$.

Proof. Let $g_1, \ldots, g_q$ be defined in $B(a, \delta)$, and let $G(x) = (g_1(x), \ldots, g_q(x))$ for every $x \in B(a, \delta)$. By Theorem 2.13, the mapping $G\colon B(a, \delta) \to \mathbb{R}^q$ is differentiable at a. Since $F = f \circ G$, Theorem 2.20 implies that F is differentiable at a and its Jacobian matrix (i.e., the vector $F'(a)$) is equal to the product of the Jacobian matrix of f at the point b (i.e., the vector $f'(b)$) and the Jacobian matrix of G at the point a. The jth coordinate of the vector $F'(a)$ is equal to $D_j F(a)$. On the other hand (by the rules of matrix multiplication), the jth coordinate of the vector $F'(a)$ is equal to the scalar product of the vector $f'(b)$ and the jth column of the Jacobian matrix of G. This is exactly equation (2.9). □

Remark 2.24. The formula (2.9) is easy to memorize in the following form. Let $y_1, \ldots, y_q$ denote the variables of f, and let us write also y_i instead of g_i. We get

$$\frac{\partial F}{\partial x_j} = \frac{\partial f}{\partial y_1} \cdot \frac{\partial y_1}{\partial x_j} + \frac{\partial f}{\partial y_2} \cdot \frac{\partial y_2}{\partial x_j} + \ldots + \frac{\partial f}{\partial y_q} \cdot \frac{\partial y_q}{\partial x_j}.$$

The differentiability of products and fractions follows easily from Corollary 2.23.

Theorem 2.25. *Let f and g be real-valued functions differentiable at the point $a \in \mathbb{R}^p$. Then $f \cdot g$, and assuming $g(a) \neq 0$, f/g is also differentiable at a.*

Proof. The function $\varphi(x, y) = x \cdot y$ is differentiable everywhere on $\mathbb{R}^2$. Since $f(x) \cdot g(x) = \varphi(f(x), g(x))$, Corollary 2.23 gives the differentiability of $f \cdot g$ at a. The differentiability of f/g follows similarly, using the fact that the rational function x/y is differentiable on the set $\{(x, y) \in \mathbb{R}^2 : y \neq 0\}$. □

Note that the partial derivatives of $f \cdot g$ and f/g can be obtained using (2.9) (or using the rules of differentiating single-variable functions). (See Exercise 1.92.)

The differentiation rule for the inverse of one-variable functions (see [7, Theorem 12.20]) can be generalized to multivariable functions as follows.

Theorem 2.26. *Suppose that $H \subset \mathbb{R}^p$, the function $f\colon H \to \mathbb{R}^p$ is differentiable at the point $a \in \operatorname{int} H$, and the mapping $f'(a)$ is invertible. Let $f(a) = b$, $\delta > 0$, and* *let* $g\colon B(b,\delta)\to\mathbb{R}^p$ *be a continuous function that satisfies* $g(b)=a$ *and* $f(g(x))=x$ *for every* $x\in B(b,\delta)$.

Then the function g is differentiable at b, and $g'(b)=(f'(a))^{-1}$*, where* $(f'(a))^{-1}$ *is the inverse of the linear mapping* $f'(a)$.

Proof. Without loss of generality, we may assume that $a=b=0$ (otherwise, we replace the functions f and g by $f(x+a)-b$ and $g(x+b)-a$, respectively).

First we also assume that $f'(0)$ is the identity mapping. Then $|f(x)-x|/|x|\to 0$ as $x\to 0$. Since $\lim_{x\to 0} g(x)=0$ and $g\neq 0$ on the set $B(0,\delta)\setminus\{0\}$, it follows from Theorem 2.5 on the limit of composite functions that $|f(g(x))-g(x)|/|g(x)|\to 0$ as $x\to 0$. Since $f(g(x))=x$, we find that $|x-g(x)|/|g(x)|\to 0$ as $x\to 0$.

Now we prove that $g'(0)$ is also the identity mapping, i.e., $\lim_{x\to 0}|g(x)-x|/|x|=0$. First note that $|x-g(x)|\leq |g(x)|/2$ for every $x\in B(0,\delta')$ for a small enough δ'. Thus $x\in B(0,\delta')$ implies

$$|g(x)|\leq |g(x)-x|+|x|\leq (|g(x)|/2)+|x|,$$

whence $|g(x)|\leq 2|x|$, and

$$\frac{|x-g(x)|}{|x|}=\frac{|x-g(x)|}{|g(x)|}\cdot\frac{|g(x)|}{|x|}\leq 2\cdot\frac{|x-g(x)|}{|g(x)|}.$$

Therefore, $\lim_{x\to 0}|g(x)-x|/|x|=0$ holds. We have proved that g is differentiable at the origin, and its derivative is the identity mapping there.

Now we consider the general case (still assuming $a=b=0$). Let $f'(0)=A$. By Theorem 2.20, $f_1=A^{-1}\circ f$ is differentiable at the origin, and its derivative is the linear mapping $A^{-1}\circ A$, which is the identity. The function $g_1=g\circ A$ is continuous in a neighborhood of the origin, with $f_1(g_1(x))=x$ in this neighborhood. Thus, the special case proved above implies that $g_1'(0)$ is also the identity mapping. Since $g=g_1\circ A^{-1}$, Theorem 2.20 on the differentiability of composite functions implies that g is differentiable at the origin, and its derivative is A^{-1} there. □

Exercises

2.1. Let $H\subset\mathbb{R}^p$. Show that the mapping $f\colon H\to\mathbb{R}^q$ is continuous on H if and only if for every open set $V\subset\mathbb{R}^q$ there is an open set $U\subset\mathbb{R}^p$ such that $f^{-1}(V)=H\cap U$.

2.2. Give an example of a bounded set $H\subset\mathbb{R}^p$ and a continuous, injective function $f\colon H\to\mathbb{R}^q$ such that f^{-1} is not continuous on the set $f(H)$.

2.3. Show that if $A = (a_{ij})$ $(i = 1, \ldots, q,\ j = 1, \ldots, p)$, then

$$\|A\| \le \sqrt{\sum_{i=1}^{q} \sum_{j=1}^{p} a_{ij}^2}.$$

Give an example when strict inequality holds.

2.4. Show that if $A = (a_{ij})$ $(i = 1, \ldots, q, j = 1, \ldots, p)$, then

$$\max_{1 \le i \le q,\ 1 \le j \le p} |a_{ij}| \le \|A\|,$$

furthermore,

$$\max_{1 \le i \le q} \sqrt{\sum_{j=1}^{p} a_{ij}^2} \le \|A\|.$$

Give an example when strict inequality holds.

2.5. Let the linear mapping $A\colon \mathbb{R}^p \to \mathbb{R}^p$ be invertible. Show the existence of some $\delta > 0$ and $K \ge 0$ such that $\|B^{-1} - A^{-1}\| \le K \cdot \|B - A\|$ for every B that satisfies $\|B - A\| < \delta$.

2.6. Let $1 \le i \le q$ and $1 \le j \le p$ be fixed. Show that a_{ij} is a continuous (furthermore, Lipschitz) function of A, i.e., there exists K such that $|a_{ij} - b_{ij}| \le K \cdot \|A - B\|$. (Here a_{ij} and b_{ij} are the jth entries of the ith row of the matrices A and B, respectively.)

2.7. Find all differentiable functions $f\colon \mathbb{R}^2 \to \mathbb{R}$ that satisfy $D_1 f \equiv D_2 f$. (S)

2.8. Let the function $f\colon \mathbb{R}^2 \to \mathbb{R}$ be differentiable on the plane, and let $D_1 f(x, x^2) = D_2 f(x, x^2) = 0$ for every x. Show that $f(x, x^2)$ is constant.

2.9. Let $f\colon \mathbb{R}^2 \to \mathbb{R}$ be differentiable on the plane. Let $f(0,0) = 0$, $D_1 f(x, x^3) = x$ and $D_2 f(x, x^3) = x^3$ for every x. Find $f(1,1)$.

2.10. Let $H \subset \mathbb{R}^p$, and let $f\colon H \to \mathbb{R}$ be differentiable at the point $a \in \operatorname{int} H$. We call the set $S = \{x \in \mathbb{R}^p : f(x) = f(a)\}$ the **contour line** corresponding to a. Show that the contour line is perpendicular to the gradient $f'(a)$ in the following sense: if $g\colon (c, d) \to \mathbb{R}^p$ is a differentiable curve whose graph lies in S and $g(t_0) = a$ for some $t_0 \in (c, d)$, then $g'(t_0)$ and $f'(a)$ are perpendicular to each other. (The zero vector is perpendicular to every vector.)

2.11. We say that the function $f\colon \mathbb{R}^p \setminus \{0\} \to \mathbb{R}$ is a **homogeneous function with degree** k (where k is a fixed real), if $f(tx) = t^k \cdot f(x)$ holds for every $x \in \mathbb{R}^p \setminus \{0\}$ and $t \in \mathbb{R}$, $t > 0$. **Euler's theorem**[3] states that if $f\colon \mathbb{R}^p \setminus \{0\} \to \mathbb{R}$ is

[3] Leonhard Euler (1707–1783), Swiss mathematician.

differentiable and homogeneous with degree k, then $x_1 \cdot D_1 f + \ldots + x_p \cdot D_p f = k \cdot f$ for every $x = (x_1, \ldots, x_p) \in \mathbb{R}^p \setminus \{0\}$.
Double-check the theorem for some particular functions (e.g., $xy/\sqrt{x^2+y^2}$, $xy/(x^2+y^2)$, $\sqrt{x^2+y^2}$, etc.).

2.12. Prove Euler's theorem.

2.13. Let the function $f\colon \mathbb{R}^p \to \mathbb{R}^q$ be differentiable at the points of the segment $[a, b]$, where $a, b \in \mathbb{R}^p$. True or false? There exists a point $c \in [a, b]$ such that $f(b) - f(a) = f'(c)(b - a)$. (I.e., can we generalize the mean value theorem (Theorem 1.79) for vector valued functions?) (H S)

2.4 Implicit and Inverse Functions

Solving an equation means that the unknown quantity, given only implicitly by the equation, is made explicit. For example, x is defined implicitly by the quadratic equation $ax^2 + bx + c = 0$, and as we solve this equation, we express x explicitly in terms of the parameters a, b, c. In order to make the nature of this problem more transparent, let's write x_1, x_2, x_3 in place of a, b, c and y in place of x. Then we are given the function $f(x_1, x_2, x_3, y) = x_1 y^2 + x_2 y + x_3$ of four variables, and we have to find a function $\varphi(x_1, x_2, x_3)$ satisfying

$$f(x_1, x_2, x_3, \varphi(x_1, x_2, x_3)) = 0. \tag{2.10}$$

In this case we say that the function $y = \varphi(x_1, x_2, x_3)$ is the solution of equation (2.10). As we know, there is no solution on the set $A = \{(x_1, x_2, x_3)\colon x_2^2 - 4x_1x_3 < 0\} \subset \mathbb{R}^3$, and there are continuous solutions on the set $B = \{(x_1, x_2, x_3) : x_1 \neq 0,\ x_2^2 - 4x_1x_3 \geq 0\} \subset \mathbb{R}^3$, namely each of the functions

$$\varphi_1 = (-x_2 + \sqrt{x_2^2 - 4x_1x_3})/(2x_1), \qquad \varphi_2 = (-x_2 - \sqrt{x_2^2 - 4x_1x_3})/(2x_1)$$

is a continuous solution on B.

Finding the inverse of a function means solving an equation as well. A function φ is the inverse of the function g exactly when the unique solution of the equation $x - g(y) = 0$ is $y = \varphi(x)$.

In general, we cannot expect that the solution y can be given by a (closed) formula of the parameters. Even $f(x, y)$ is not always defined by a closed formula. However, even assuming that $f(x, y)$ is given by a formula, we cannot ensure that y belongs to the same family of functions that we used to express f. For example, $f(x, y) = x - y^3$ is a polynomial, but the solution $y = \sqrt[3]{x}$ of the equation $f(x, y) = 0$ cannot. Based on this observation, it is not very surprising that there exists a function $f(x, y)$ such that f can be expressed by elementary functions, but the solution y of the equation $f(x, y) = 0$ is not. Consider the function $g(x) = x + \sin x$. Then g is strictly monotonically increasing and continuous on the real line, and furthermore it assumes every real value, and thus it has an inverse on $\mathbb{R}$. It can be shown that the inverse of g cannot be expressed by elementary functions only. That is, the equation $x - y - \sin y = 0$ has a unique solution, but the solution is not an elementary function.

The same phenomenon is illustrated by a famous theorem of algebra stating that the roots of a general quintic polynomial cannot be obtained from the coefficients by rational operations and by extractions of roots. That is, there does not exist a function $y = \varphi(x_1, \ldots, x_6)$ defined only by the basic algebraic operations and extraction of roots of the coordinate functions $x_1, \ldots, x_6$ such that y is the solution of the equation $x_1 y^5 + \ldots + x_5 y + x_6 = 0$ on a nonempty, open subset of $\mathbb{R}^6$.

Therefore, solving y explicitly does not necessarily mean expressing y by a (closed) formula; it means only establishing the existence or nonexistence of the solution and describing its properties when it does exist. The simplest related theorem is the following.

Theorem 2.27. *Let f be a two-variable real function such that f is zero at the point $(a, b) \in \mathbb{R}^2$ and continuous on the square $[a - \eta, a + \eta] \times [b - \eta, b + \eta]$ for an appropriate $\eta > 0$. If the section f_x is strictly monotone at every $x \in [a - \eta, a + \eta]$, then there exists a positive real δ such that*

(i) *for every $x \in (a - \delta, a + \delta)$, there exists a unique $\varphi(x) \in (b - \eta, b + \eta)$ such that $f(x, \varphi(x)) = 0$, and furthermore,*
(ii) *the function φ is continuous on the interval $(a - \delta, a + \delta)$.*

Proof. We know that the section f_a is strictly monotonically. Without loss of generality, we may assume that f_a is strictly monotone increasing (the proof of the other case is exactly the same), and thus $f_a(b - \eta) < f_a(b) = f(a, b) = 0 < f_a(b + \eta)$.

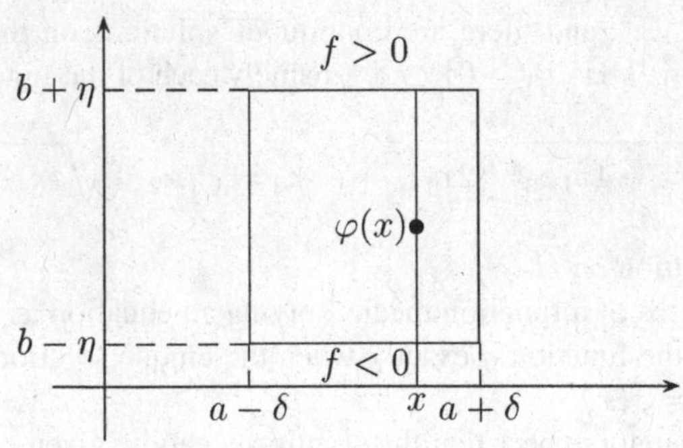

2.1. Figure

Let $\varepsilon > 0$ be small enough to imply $f_a(b - \eta) < -\varepsilon$ and $\varepsilon < f_a(b + \eta)$.

Since f is continuous at the points $(a, b - \eta)$, $(a, b + \eta)$, there exists $0 < \delta < \eta$ such that $|f(x, b - \eta) - f(a, b - \eta)| < \varepsilon$ and $|f(x, b + \eta) - f(a, b + \eta)| < \varepsilon$ for every $x \in (a - \delta, a + \delta)$. That is, if $x \in (a - \delta, a + \delta)$, then

$$f(x, b - \eta) < 0 < f(x, b + \eta).$$

Since f_x is strictly monotone and continuous on the interval $[b-\eta, b+\eta]$, it follows from Bolzano's theorem that there is a unique $\varphi(x) \in (b-\eta, b+\eta)$ such that $f(x, \varphi(x)) = 0$. Thus, we have proved statement (i).

Let $x_0 \in (a-\delta, a+\delta)$ and $\varepsilon > 0$ be fixed. Choose positive numbers δ_1 and $\eta_1 < \varepsilon$ such that

$$(x_0 - \delta_1, x_0 + \delta_1) \subset (a-\delta, a+\delta)$$

and

$$(\varphi(x_0) - \eta_1, \varphi(x_0) + \eta_1) \subset (b-\eta, b+\eta)$$

hold. Following the steps of the first part, we end up with a number $0 < \delta' < \delta_1$ such that for every $x \in (x_0 - \delta', x_0 + \delta')$ there exists a unique

$$y \in (\varphi(x_0) - \eta_1, \varphi(x_0) + \eta_1) \subset (b-\eta, b+\eta)$$

with $f(x, y) = 0$. By (i), $\varphi(x)$ is the only such number, and hence $y = \varphi(x)$. Thus, for $|x - x_0| < \delta'$ we have $|\varphi(x) - \varphi(x_0)| < \eta_1 < \varepsilon$. Therefore, φ is continuous at x_0. □

Corollary 2.28. (Implicit function theorem for single-variable functions) *Suppose that the two-variable function f is zero at the point $(a, b) \in \mathbb{R}^2$ and continuous in a neighborhood of (a, b). Let the partial derivative $D_2 f$ exist and be finite and nonzero in a neighborhood of (a, b). Then there exist positive numbers δ and η such that*

(i) *for every $x \in (a-\delta, a+\delta)$ there exists a unique number $\varphi(x) \in (b-\eta, b+\eta)$ with $f(x, \varphi(x)) = 0$, furthermore,*
(ii) *the function φ is continuous in the interval $(a-\delta, a+\delta)$.*

Proof. It follows from the assumptions that there is a rectangle $(a_1, a_2) \times (b_1, b_2)$ containing (a, b) in its interior such that f is continuous, $D_2 f$ exists and is finite and nonzero in $(a_1, a_2) \times (b_1, b_2)$. The section f_x is strictly monotone in the interval (b_1, b_2) for every $x \in (a_1, a_2)$, since it is differentiable and, by Darboux's theorem[4] [7, Theorem 13.44], its derivative must be everywhere positive or everywhere negative in the interval (b_1, b_2). Then an application of Theorem 2.27 to the rectangle $(a_1, a_2) \times (b_1, b_2)$ finishes the proof. □

Remark 2.29. We will see later that if f is continuously differentiable at (a, b), then the function φ is continuously differentiable at the point a (see Theorem 2.40).

For the single-variable case, it is not difficult to show that the differentiability of f at (a, b) and $D_2 f(a, b) \neq 0$ implies the differentiability of φ at a (see Exercise 2.15). We can calculate $\varphi'(a)$ by applying the differentiation rule of composite functions. Since $f(x, \varphi(x)) = 0$ in a neighborhood of the point a, its derivative is also zero there. Thus,

---

[4] Jean Gaston Darboux (1842–1917), French mathematician. Darboux's theorem states that if $f\colon [a, b] \to \mathbb{R}$ is differentiable, then f' takes on every value between $f'(a)$ and $f'(b)$.

$$D_1 f(a,b) \cdot 1 + D_2 f(a,b) \cdot \varphi'(a) = 0$$

holds, from which we obtain $\varphi'(a) = -D_1 f(a,b)/D_2 f(a,b)$.

Example 2.30. The function $f(x,y) = x^2 + y^2 - 1$ is continuous and (infinitely) differentiable everywhere. If $a^2 + b^2 = 1$ and $-1 < a < 1$, then $D_2 f(a,b) = 2b \neq 0$, and the conditions of Corollary 2.28 are satisfied. Thus, there exists some function ϕ such that ϕ is continuous in a neighborhood of a, $\varphi(a) = b$, and $x^2 + \varphi(x)^2 - 1 = 0$. Namely, if $b > 0$, then the function $\varphi(x) = \sqrt{1-x^2}$ on the interval $(-1,1)$ is such a function. If, however, $b < 0$, then the function $\varphi(x) = -\sqrt{1-x^2}$ satisfies the conditions on the interval $(-1,1)$.

On the other hand, if $a = 1$, then there is no such function in any neighborhood of a, since $x > 1$ implies $x^2 + y^2 - 1 > 0$ for every y. The conditions of Corollary 2.28 are not satisfied here, since $a = 1$ implies $b = 0$ and $D_2 f(1,0) = 0$. The same happens in the $a = -1$ case.

Our next goal is to generalize Corollary 2.28 to multivariable functions.

Corollary 2.28 gives a sufficient condition for the existence of the inverse of a function—at least locally. The inverse of an arbitrary function g is given by the solution of the equation $f(x,y) = 0$, where $f(x,y) = x - g(y)$. Let $g(b) = a$; thus $f(a,b) = 0$. By Corollary 2.28, if g is differentiable in a neighborhood of b such that $g'(x) \neq 0$ in this neighborhood, then there exists a continuous function φ in a neighborhood $(a-\delta, a+\delta)$ of a such that $\varphi(a) = b$ and $g(\varphi(x)) = x$ on $(a-\delta, a+\delta)$.

We expect that a generalization of Corollary 2.28 to multivariable functions would also give a sufficient condition for the existence of the inverse locally. Therefore, we first consider the question of the existence of the inverse function.

Proving the existence of the inverse of a multivariable function is substantially more difficult than for one-variable functions; this is a case in which the analogy with the single-variable case exists but is far from being sufficient. The question is how to decide whether or not a given function is injective on a given set. For a continuous single-variable, real function defined on an interval, the answer is quite simple: the function is injective if it is strictly monotone. (This follows from the Bolzano–Darboux theorem,[5] see [7, Theorem 10.57].) It is not clear, however, how to generalize this condition to continuous multivariable, or vector-valued functions.

Yet another problem is related to the existence of a "global" inverse. Let $f: I \to \mathbb{R}$ be continuous, where $I \subset \mathbb{R}$ is an interval. Given that every point of I has a neighborhood on which the function f is injective, we can easily show that f is injective on the whole interval. Thus, global injectivity follows from local injectivity for single-variable continuous real functions. However, this does not hold for vector-valued or multivariable functions! Let $g: \mathbb{R} \to \mathbb{R}^2$ be a curve with $g(t) = (\cos t, \sin t)$ for every $t \in \mathbb{R}$. The mapping g is injective on every interval

[5] The Bolzano–Darboux's theorem states that if $f: [a,b] \to \mathbb{R}$ is continuous, then f takes on every value between $f(a)$ and $f(b)$.

shorter than 2π (and maps to the unit circle), but g is not injective globally, since it is periodic with period 2π. Similarly, let $f(x, y) = (e^x \cos y, e^x \sin y)$ for every $(x, y) \in \mathbb{R}^2$. The mapping $f: \mathbb{R}^2 \to \mathbb{R}^2$ is injective on every disk of the plane with radius less than π, but f is not injective globally, since $f(x, y + 2\pi) = f(x, y)$ for every $(x, y) \in \mathbb{R}^2$.

Unfortunately, we cannot help this; it seems that there are no natural, simple sufficient conditions for the global injectivity of a vector-valued or multivariable function. Thus, we have to restrict our investigations to the question of local injectivity.

Let the mapping f be differentiable in a neighborhood of the point a. Since the mapping $f(a) + f'(a)(x - a)$ approximates f well locally, we might think that given the injectivity of the linear mapping $f'(a)$, f will also be injective on a neighborhood of a. However, this is not always so, not even in the simple special case of $p = q = 1$. There are everywhere differentiable functions $f: \mathbb{R} \to \mathbb{R}$ such that $f'(0) \neq 0$, but f is nonmonotone on every neighborhood of 0. (See [7, Remark 12.45.4].) Let $f'(0) = b$. By the general definition of the derivative, $f'(0)$ is the linear mapping $x \mapsto b \cdot x$, which is injective. Nonetheless, f is not injective on any neighborhood of 0.

Thus, we need to have stricter assumptions if we wish to prove the local injectivity of f. One can show that if *the linear mapping $f'(x)$ is injective for every x in a neighborhood of a, then f is injective in a neighborhood of a*. The proof involves more advanced topological tools, and hence it is omitted here. We will prove only the special case in which the partial derivatives of f are continuous at a.

Definition 2.31. Let $H \subset \mathbb{R}^p$ and $f: H \to \mathbb{R}^q$. We say that the mapping f is *continuously differentiable* at the point $a \in \operatorname{int} H$ if f is differentiable in a neighborhood of a, and the partial derivatives of the coordinate functions of f are continuous at a.

Theorem 2.32. (Local injectivity theorem) *Let $H \subset \mathbb{R}^p$ and $f: H \to \mathbb{R}^q$, with $p \leq q$. If f is continuously differentiable at the point $a \in \operatorname{int} H$ and the linear mapping $f'(a): \mathbb{R}^p \to \mathbb{R}^q$ is injective, then f is injective in a neighborhood of a.*

Lemma 2.33. *Let $H \subset \mathbb{R}^p$, and let the function $f: H \to \mathbb{R}^q$ be differentiable at the points of the segment $[a, b] \subset H$. If $|D_j f_i(x)| \leq K$ for every $i = 1, \ldots, q$, $j = 1, \ldots, p$ and $x \in [a, b]$, then $|f(b) - f(a)| \leq Kpq \cdot |b - a|$.*

Proof. Applying the mean value theorem (Theorem 1.79) to the coordinate function f_i yields

$$f_i(b) - f_i(a) = \sum_{j=1}^{p} D_j f_i(c_i)(b_i - a_i)$$

for an appropriate point $c_i \in [a, b]$. Thus,

$$|f_i(b) - f_i(a)| \leq \sum_{j=1}^{p} K \cdot |b_i - a_i| \leq Kp \cdot |b - a|$$

for every i, and

$$|f(b) - f(a)| \leq \sum_{i=1}^{q} |f_i(b) - f_i(a)| \leq Kpq \cdot |b - a|. \qquad \square$$

Proof of Theorem 2.32. First, we assume that $p = q$ and $f'(a)$ is the identity map, i.e., $f'(a)(x) = x$ for every $x \in \mathbb{R}^p$. By the definition of the derivative this means that $\lim_{x \to a} |g(x)|/|x - a| = 0$, where $g(x) = f(x) - f(a) - (x - a)$ for every $x \in H$. Obviously, g is continuously differentiable at the point a, and $g'(a)$ is the constant zero mapping. It follows that $D_j g_i(a) = 0$ for every $i, j = 1, \ldots, p$. Since g is continuously differentiable at a, we can choose some $\delta > 0$ such that $|D_j g_i(x)| \leq 1/(2p^2)$ holds for every $x \in B(a, \delta)$ and every $i, j = 1, \ldots, p$. By Lemma 2.33, we have $|g(x) - g(y)| \leq |x - y|/2$ for every $x, y \in B(a, \delta)$. If $x, y \in B(a, \delta)$ and $x \neq y$, then $f(x) \neq f(y)$; otherwise, $f(y) = f(x)$ would imply $g(y) - g(x) = x - y$, which is impossible. We have proved that f is injective on the ball $B(a, \delta)$.

Consider the general case. Let A denote the injective linear mapping $f'(a)$. Let the range of A be V; it is a linear subspace of $\mathbb{R}^q$ (including the case $V = \mathbb{R}^q$). Let $B(y) = A^{-1}(y)$ for every $y \in V$. Obviously, B is a well-defined linear mapping from V to $\mathbb{R}^p$. Extend B linearly to $\mathbb{R}^q$, and let us denote this extension by B as well. (The existence of such an extension is easy to show.) Then the mapping $B \circ A$ is the identity map on $\mathbb{R}^p$.

Clearly, the derivative of the linear mapping B is itself B everywhere. Then, it follows from Theorem 2.20 on the differentiation rules of composite functions that $B \circ f$ is differentiable in a neighborhood of a with $(B \circ f)'(x) = B \circ f'(x)$ there. Then the Jacobian matrix of $B \circ f$ at the point x is equal to the (matrix) product of the matrices of B and $f'(x)$. Thus, every partial derivative of every coordinate function of $B \circ f$ is a linear combination of the partial derivatives $D_j f_i$. This implies that $B \circ f$ is continuously differentiable at the point a. Since $(B \circ f)'(a) = B \circ f'(a) = B \circ A$ is the identity, the already proven special case implies the injectivity of $B \circ f$ in a neighborhood of a. Then f itself has to be injective in this neighborhood. $\square$

Remarks 2.34. **1.** Let $A\colon \mathbb{R}^p \to \mathbb{R}^q$ be a linear mapping. It is well known that A cannot be injective if $p > q$. Indeed, in this case the dimension of the null space of A, i.e, the linear subspace $\{x \in \mathbb{R}^p \colon A(x) = 0\}$, is $p - q > 0$, and thus there exists a point $x \neq 0$ such that $A(x) = 0$. This implies that A can be injective only when $p \leq q$.

2. The local injectivity theorem turns the question of a function's local injectivity into a question of the injectivity of a linear mapping. The latter is easy to answer. A linear mapping $A\colon \mathbb{R}^p \to \mathbb{R}^q$ is injective if and only if $A(x) \neq 0$ for every vector $x \in \mathbb{R}^p$, $x \neq 0$. Furthermore, it is well known that A is injective if and only if the rank of its matrix is p. This means that the matrix of A has p linearly independent rows, or equivalently, the matrix has a nonzero $p \times p$ subdeterminant.

A linear mapping $A: \mathbb{R}^p \to \mathbb{R}^q$ is called **surjective** , if its range is $\mathbb{R}^q$. Since the range of A can be at most p-dimensional, A can be surjective only if $p \geq q$. The following statement is the dual of Theorem 2.32.

Theorem 2.35. *Let $H \subset \mathbb{R}^p$ and let $f: H \to \mathbb{R}^q$, where $p \geq q$. If f is continuously differentiable at the point $a \in \operatorname{int} H$ and the linear mapping $f'(a): \mathbb{R}^p \to \mathbb{R}^q$ is surjective, then the range of f contains a neighborhood of $f(a)$.*

We need to show that if b is close to $f(a)$, then the equation $f(x) = b$ has a solution. We prove this with the help of iterates, which are useful in several cases of solving equations[6].

The most widely used version of this method is given by the following theorem.

We say that the mapping $f: H \to H$ has a **fixed point** at $x \in H$ if $f(x) = x$. Let $f: H \to \mathbb{R}^q$, where $H \subset \mathbb{R}^p$. The mapping f is called a **contraction**, if there exists a number $\lambda < 1$ such that $|f(y) - f(x)| \leq \lambda \cdot |y - x|$ for every $x, y \in H$. (That is, f is contraction if it is Lipschitz with a constant less than 1.)

Theorem 2.36. (Banach's[7] fixed-point theorem) *If $H \subset \mathbb{R}^p$ is a nonempty closed set, then every contraction $f: H \to H$ has a fixed point.*

Proof. Let $|f(y) - f(x)| \leq \lambda \cdot |y - x|$ for every $x, y \in H$, with $0 < \lambda < 1$. Let $x_0 \in H$ be an arbitrarily chosen point, and consider the sequence of points x_n defined by the recurrence $x_n = f(x_{n-1})$ $(n = 1, 2, \ldots)$. (Since f maps H into itself, x_n is defined for every natural number n.) We prove that the sequence x_n is convergent and tends to a fixed point of f.

Let $|x_1 - x_0| = d$. By induction, we get $|x_{n+1} - x_n| \leq \lambda^n d$ for every $n \geq 0$. Indeed, this is clear for $n = 0$, and if it holds for $(n-1)$, then

$$|x_{n+1} - x_n| = |f(x_n) - f(x_{n-1})| \leq \lambda \cdot |x_n - x_{n-1}| \leq \lambda \cdot \lambda^{n-1} d = \lambda^n d.$$

Now we show that (x_n) satisfies the Cauchy criterion (Theorem 1.8). Indeed, for every $\varepsilon > 0$, the convergence of the infinite series $\sum \lambda^n$ implies the existence of some index N such that $|\lambda^n + \ldots + \lambda^m| < \varepsilon$ holds for every $N \leq n < m$. For $N \leq n < m$ we have

$$\begin{aligned}|x_m - x_n| &\leq |x_{n+1} - x_n| + |x_{n+2} - x_{n+1}| + \ldots + |x_m - x_{m-1}| \leq \\ &\leq |\lambda^n + \ldots + \lambda^{m-1}| d < \varepsilon d.\end{aligned}$$

Thus, by Theorem 1.8, (x_n) is convergent. If $x_n \to c$, then $c \in H$ follows from the fact that H is closed. Since $|x_{n+1} - f(c)| = |f(x_n) - f(c)| \leq \lambda \cdot |x_n - c|$, we have $x_{n+1} \to f(c)$, which implies $f(c) = c$, i.e., c is a fixed point of f. □

---

[6] Regarding the solution of equations using iterates, see Exercises 6.4 and 6.5 of [7]. In (a)–(d) of Exercise 6.4 the equations $x = \sqrt{a + x}$, $x = 1/(2 - x)$, $x = 1/(4 - x)$, $x = 1/(1 + x)$ are solved using iterates by defining sequences converging to the respective solutions. The solution of the equation $x^2 = a$ using the same method can be found in Exercise 6.5.

[7] Stefan Banach (1892–1945), Polish mathematician.

Proof of Theorem 2.35. We may assume that $a = 0$ and $f(a) = 0$ (otherwise, we replace f by the function $f(x+a) - f(a)$).

First, assume that $p = q$ and $f'(0)$ is the identity map. Let $g(x) = f(x) - x$ for every $x \in H$. As we saw in the proof of Theorem 2.32, there exists a $\delta > 0$ such that $|g(x) - g(y)| \le |x - y|/2$ for every $x, y \in B(0, \delta)$. We may assume that this inequality also holds for every $x, y \in \overline{B}(0, \delta)$, for otherwise, we could choose a smaller δ. We prove that the range of f contains the ball $B(0, \delta/2)$.

Let $b \in B(0, \delta/2)$ be fixed. The mapping $h(x) = b - g(x)$ maps the closed ball $\overline{B}(0, \delta)$ into itself, since $|x| \le \delta$ implies

$$|h(x)| \le |b| + |g(x)| \le (\delta/2) + |x|/2 \le \delta.$$

Furthermore, since $|h(x) - h(y)| = |g(x) - g(y)| \le |x - y|/2$ for every $x \in \overline{B}(0, \delta)$, it follows that h is a contraction, and then, by Banach's fixed point theorem, it has a fixed-point. If x is such a fixed point, then $x = h(x) = b - g(x) = b + x - f(x)$, i.e., $f(x) = b$.

Now consider the general case $p \le q$. (still assuming $a = 0$ and $f(a) = 0$). Let $e_1, \ldots, e_q$ be a basis of the linear space $\mathbb{R}^q$, and let the points $x_1, \ldots, x_q \in \mathbb{R}^p$ be such that $f'(0)(x_i) = e_i$ $(i = 1, \ldots, q)$. (Such points exist, since the linear mapping $f'(0)$ is surjective.) There exists a linear mapping $A\colon \mathbb{R}^q \to \mathbb{R}^p$ such that $A(e_i) = x_i$ $(i = 1, \ldots, q)$.

Since $0 \in \operatorname{int} H$, we must have $B(0, r) \subset H$ for an appropriate $r > 0$. The mapping A is linear, and thus it is continuous, even Lipschitz by Lemma 2.21. Thus, there exists an $\eta > 0$ such that $|A(x)| < r$ for every $|x| < \eta$. Applying Theorem 2.20 (the differentiation rule for composite functions), we obtain that $f \circ A\colon B(0, \eta) \to \mathbb{R}^q$ is differentiable in the ball $B(0, \eta) \subset \mathbb{R}^q$. We have $(f \circ A)'(0) = f'(0) \circ A$ (since the derivative of the linear mapping A is itself), which is the identity on $\mathbb{R}^q$, by the construction of A. It is easy to see that $f \circ A$ is continuously differentiable at the origin. Thus, by the already proven special case, the range of $f \circ A$ contains a neighborhood of the origin. Then the same is true for f. □

Corollary 2.37. (Open mapping theorem) *Let $H \subset \mathbb{R}^p$ be an open set, and let $f\colon H \to \mathbb{R}^q$ be continuously differentiable at the points of H. If the linear mapping $f'(x)$ is surjective for every $x \in H$, then $f(H)$ is an open set in $\mathbb{R}^q$.*

Proof. If $H \ne \emptyset$, then the assumptions imply $p \ge q$. Let $b \in f(H)$ be arbitrary. Then $b = f(a)$ for a suitable $a \in H$. By Theorem 2.35, $f(H)$ contains a neighborhood of b. Since this is true for every $b \in f(H)$, it follows that $f(H)$ is open. □

The name of Corollary 2.37 comes from the fact that a function $f\colon \mathbb{R}^p \to \mathbb{R}^q$ is called an **open mapping** if $f(G)$ is an open set for every open set $G \subset \mathbb{R}^p$.

Using Theorems 2.32 and 2.35 one obtains a sufficient condition for the existence of a local inverse.

Theorem 2.38. (Inverse function theorem) *Let $H \subset \mathbb{R}^p$ and $a \in \operatorname{int} H$. If $f\colon H \to \mathbb{R}^p$ is continuously differentiable at a and the linear mapping $f'(a)\colon \mathbb{R}^p \to \mathbb{R}^p$ is invertible, then there exist positive numbers δ and η such that*

(i) *for every $x \in B(f(a), \delta)$ there exists a unique $\varphi(x) \in B(a, \eta)$ such that $f(\varphi(x)) = x$,*

(ii) *the function φ defined this way is differentiable on the ball $B(f(a), \delta)$ and is continuously differentiable at the point $f(a)$, and furthermore,*

(iii) *$f'(x)$ is invertible at every $x \in B(a, \eta)$, and $\varphi'(f(x)) = f'(x)^{-1}$ for every $x \in B(f(a), \delta)$.*

If f is continuously differentiable in a neighborhood of a, then we can choose δ and η such that φ is continuously differentiable in $B(f(a), \delta)$.

Proof. By Theorem 2.32, f is injective on some ball $B(a, \eta)$. We may assume that f is differentiable and injective on the closed ball $\overline{B}(a, \eta)$, since otherwise, we could choose a smaller η. Let $K = f(\overline{B}(a, \eta))$. For every $x \in K$ let $\varphi(x)$ denote the unique point in $\overline{B}(a, \eta)$ such that $f(\varphi(x)) = x$. It follows from Theorem 2.8 that the function φ is continuous on the set K.

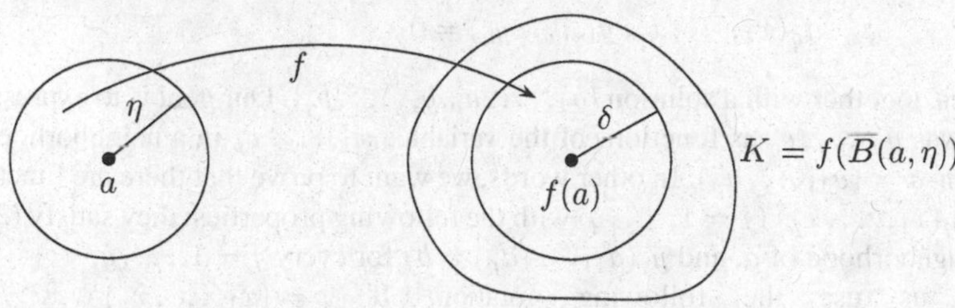

2.2. Figure

Since an invertible linear mapping that maps $\mathbb{R}^p$ into itself is necessarily surjective as well, we find, by Theorem 2.35, that $f(B(a, \eta))$ contains a ball $B(f(a), \delta)$. Obviously, for every point $x \in B(f(a), \delta)$ there exists a unique point in $B(a, \eta)$ whose image by f equals x, namely, the point $\varphi(x)$. This proves (i).

A linear mapping that maps $\mathbb{R}^p$ to itself is injective if and only if the determinant of the mapping's matrix is nonzero. By assumption, the determinant of f's Jacobian matrix at the point a is nonzero. Since the Jacobian matrix is a polynomial in the partial derivatives $D_j f_i$, it follows that the determinant of the Jacobian matrix is continuous at a, implying that it is nonzero in a neighborhood of a. We have proved that the linear mapping $f'(x)$ is injective for every point x close enough to a. By taking a smaller η if necessary, we may assume that $f'(x)$ is injective for every $x \in B(a, \eta)$.

Then it follows from Theorem 2.26 (the differentiation rule for inverse functions) that φ is differentiable in $B(f(a), \delta)$ and (iii) holds on this ball.

We now prove that φ is continuously differentiable at the point $f(a)$. This follows from the equality $\varphi'(f(x)) = f'(x)^{-1}$. Indeed, this implies that the partial derivative $D_j\varphi_i(x)$ equals the jth entry of the ith row in the matrix of the inverse of $f'(\varphi(x))$. Now it is well known that the jth element of the ith row of the inverse of a matrix is equal to A_{ij}/D, where A_{ij} is an appropriate subdeterminant and D is the determinant of the matrix itself (which is nonzero). The point is that the entries of the inverse matrix can be written as rational functions of the entries of the original matrix. Since $D_j f_i(\varphi(x))$, i.e., the entries of the matrix of the mapping $f'(\varphi(x))$ are continuous at $f(a)$, their rational functions are also continuous at $f(a)$.

Thus, if f is continuously differentiable on the ball $B(a, \eta)$, then φ is continuously differentiable on $B(f(a), \delta)$. □

Now we turn to what is called the implicit function theorem, that is, to the generalization of Corollary 2.28 to multivariable functions. Intuitively, the statement of the theorem is the following. Let the equations

$$\begin{aligned} f_1(x_1, \ldots, x_p, y_1, \ldots, y_q) &= 0, \\ f_2(x_1, \ldots, x_p, y_1, \ldots, y_q) &= 0, \\ &\vdots \\ f_q(x_1, \ldots, x_p, y_1, \ldots, y_q) &= 0 \end{aligned} \tag{2.11}$$

be given, together with a solution $(a_1, \ldots, a_p, b_1, \ldots, b_q)$. Our goal is to express the unknowns $y_1, \ldots, y_q$ as functions of the variables $x_1, \ldots, x_p$ in a neighborhood of the point $a = (a_1, \ldots, a_p)$. In other words, we want to prove that there are functions $y_j = y_j(x_1, \ldots, x_p)$ $(j = 1, \ldots, q)$ with the following properties: they satisfy (2.11) in a neighborhood of a, and $y_j(a_1, \ldots, a_p) = b_j$ for every $j = 1, \ldots, q$.

Let us use the following notation. If $x = (x_1, \ldots, x_p) \in \mathbb{R}^p$ and $y = (y_1, \ldots, y_q) \in \mathbb{R}^q$, then (x, y) denotes the vector $(x_1, \ldots, x_p, y_1, \ldots, y_q) \in \mathbb{R}^{p+q}$.

If the function f is defined on a subset of $\mathbb{R}^{p+q}$ and $a = (a_1, \ldots, a_p) \in \mathbb{R}^p$, then f_a denotes the **section function,** obtained by putting $a_1, \ldots, a_p$ in place of $x_1, \ldots, x_p$. That is, f_a is defined at the points $y = (y_1, \ldots, y_q) \in \mathbb{R}^q$ that satisfy $(a, y) \in D(f)$, and $f_a(y) = f(a, y)$ for every such point y. The section f^b can be defined for $b = (b_1, \ldots, b_q) \in \mathbb{R}^q$ in a similar manner. The following lemma is the generalization of the fact that differentiability implies partial differentiability.

Lemma 2.39. *Let $H \subset \mathbb{R}^{p+q}$, and let the function $f\colon H \to \mathbb{R}^s$ be differentiable at the point $(a, b) \in \operatorname{int} H$, where $a \in \mathbb{R}^p$ and $b \in \mathbb{R}^q$. Then the section function f_a is differentiable at the point b, and the section function f^b is differentiable at the point a. If $(f^b)'(a) = A$, $(f_a)'(b) = B$, and $f'(a, b) = C$, then $A(x) = C(x, 0)$ and $B(y) = C(0, y)$ for every $x \in \mathbb{R}^p$ and $y \in \mathbb{R}^q$.*

Proof. Let $r(x, y) = f(x, y) - f(a, b) - C(x - a, y - b)$. Since $f'(a, b) = C$, we have $r(x, y)/|(x, y) - (a, b)| \to 0$ if $(x, y) \to (a, b)$. Since $f(a, y) - f(a, b) - C(0, y - b) = r(a, y)$, it follows that $(f_a)'(b)$ equals the linear mapping $y \mapsto$

$C(0, y)$ $(y \in \mathbb{R}^q)$. A similar argument shows that $(f^b)'(a)(x) = C(x, 0)$ for every $x \in \mathbb{R}^p$. □

Theorem 2.40. (Implicit function theorem) *Let $H \subset \mathbb{R}^{p+q}$ and $(a, b) \in \text{int}\, H$, where $a \in \mathbb{R}^p$ and $b \in \mathbb{R}^q$. Suppose that the function $f \colon H \to \mathbb{R}^q$ vanishes at the point (a, b) (i.e., $f(a, b)$ is the null vector of $\mathbb{R}^q$). If f is continuously differentiable at (a, b) and the linear mapping $(f_a)'(b)$ is injective, then there are positive numbers δ and η such that*

(i) *for every $x \in B(a, \delta)$ there exists a unique point $\varphi(x) \in B(b, \eta)$ such that $f(x, \varphi(x)) = 0$,*

(ii) *the function φ defined this way is differentiable in the ball $B(a, \delta)$ and continuously differentiable at the point a.*

Proof. Let $F(x, y) = (x, f(x, y))$ for every $(x, y) \in H$, where $x \in \mathbb{R}^p$ and $y \in \mathbb{R}^q$. Then F maps the set H into $\mathbb{R}^{p+q}$. We will prove that F is continuously differentiable at the point (a, b), and the linear mapping $F'(a, b)$ is invertible.

Let us proceed with the proof of the theorem, assuming the statements above. Note that $F(a, b) = (a, 0)$. Applying the inverse function theorem to F, we obtain positive numbers δ and η such that $F'(x, y)$ is injective for every $(x, y) \in B((a, b), \eta)$, for every point $(x, z) \in B((a, 0), \delta)$ there exists a unique point $(x, \psi(x, z)) \in B((a, b), \eta)$ such that $F(x, \psi(x, z)) = (x, z)$, and furthermore, the function ψ defined this way is differentiable on the ball $B((a, 0), \delta)$ and is continuously differentiable at the point $(a, 0)$. From the definition of the mapping F it follows that $f(x, \psi(x, z)) = z$ for every point $(x, z) \in B((a, 0), \delta)$.

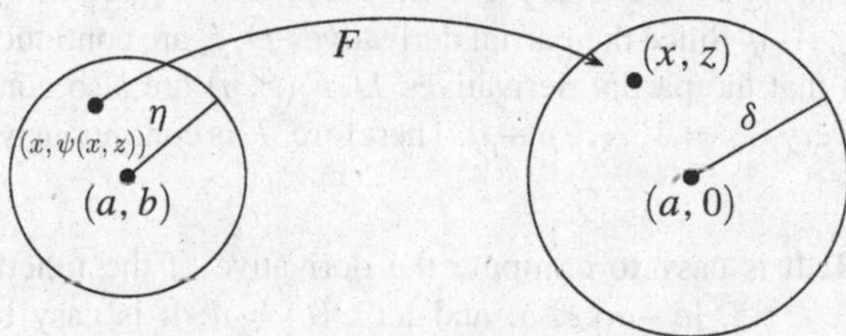

2.3. Figure

Let $\varphi(x) = \psi(x, 0)$ for every point $x \in \mathbb{R}^p$ with $|x - a| < \delta$. The definition makes sense, since $|x - a| < \delta$ implies $(x, 0) \in B((a, 0), \delta)$. It is clear that φ is differentiable on the ball $B(a, \delta)$ of $\mathbb{R}^p$ and continuously differentiable at the point a, and $f(x, \varphi(x)) = 0$ holds for every $x \in B(a, \delta)$.

We now prove the claims on F. First we prove that if f is differentiable at a point (x_0, y_0) and its derivative there is $f'(x_0, y_0) = C$, then F is also differentiable at the given point and $F'(x_0, y_0) = E$, with $E(x, y) = (x, C(x, y))$ for every $x \in \mathbb{R}^p$ and $y \in \mathbb{R}^q$. Indeed, by the definition of the derivative, $\lim_{(x,y)\to(x_0,y_0)} r(x, y)/|(x, y) - (x_0, y_0)| = 0$, where

$$r(x, y) = f(x, y) - f(x_0, y_0) - C(x - x_0, y - y_0)$$

for every $(x,y) \in H$. Thus,

$$\begin{aligned} F(x,y) - F(x_0,y_0) &= (x, f(x,y)) - (x_0, f(x_0,y_0)) = \\ &= (x - x_0, f(x,y) - f(x_0,y_0)) = \\ &= (x - x_0, C(x - x_0, y - y_0) + r(x,y)) = \\ &= E(x - x_0, y - y_0) + t(x,y) \end{aligned}$$

follows, where $t(x,y) = (0, r(x,y))$. Obviously,

$$\lim_{(x,y)\to(x_0,y_0)} t(x,y)/|(x,y) - (x_0,y_0)| = 0,$$

where E is indeed the derivative of the mapping F at the point (x_0,y_0). This proves that F is differentiable in a neighborhood of the point (a,b).

We now prove that if $(f_{x_0})'(y_0)$ is injective, then $F'(x_0,y_0) = E$ is also injective. Since the mapping E is linear, we need to prove that if the vector $(x,y) \in \mathbb{R}^{p+q}$ is nonzero, then $E(x,y) \neq 0$. By Lemma 2.39, $(f_{x_0})'(y_0)$ is equal to the linear mapping $(x,y) \mapsto C(0,y)$ $(x \in \mathbb{R}^p,\ y \in \mathbb{R}^q)$. By assumption, this mapping is injective on $\mathbb{R}^q$, thus $C(0,y) \neq 0$ holds if $y \neq 0$. We know that $E(x,y) = (x, C(x,y))$ for every $x \in \mathbb{R}^p$, $y \in \mathbb{R}^q$. If $x \neq 0$, then $E(x,y) \neq 0$ is clear. On the other hand, if $x = 0$ and $y \neq 0$, then $E(0,y) = (0, C(0,y)) \neq 0$, since $C(0,y) \neq 0$. We have proved that $(x,y) \neq 0$ implies $E(x,y) \neq 0$, i.e., E is injective. Since we assumed the injectivity of $(f_a)'(b)$, t follows that $F'(a,b)$ is also injective.

Let the coordinate functions of F and f be F_i and f_i, respectively. Obviously, $F_i(x,y) = x_i$ for every $i = 1,\ldots,p$, and $F_i(x,y) = f_{i-p}(x,y)$ for every $i = p+1,\ldots,q$. Since the partial derivatives $D_j f_i$ are continuous at the point (a,b), it follows that the partial derivatives $D_j F_i(x,y)$ are also continuous at the point (a,b) for every $i,j = 1,\ldots,p+q$. Therefore, F is continuously differentiable at (a,b). □

Remarks 2.41. **1.** It is easy to compute the derivative of the function φ of Theorem 2.40. Let $c \in \mathbb{R}^p$, $|c-a| < \delta$, and let $\varphi(c) = d$. It is easy to see that the derivative of the mapping $x \mapsto (x, \varphi(x))$ at the point c is the linear mapping $x \mapsto (x, A(x))$ with $\varphi'(c) = A$.

Let $f'(c,d) = C$. It follows from the derivation rule for composite functions that the derivative of the function $f(x,\varphi(x))$ at the point c is the linear function $C(x, A(x))$. Since $f(x,\varphi(x)) = 0$ for every $|x-a| < \delta$, this derivative is zero, i.e.,

$$0 = C(x, A(x)) = C(x,0) + C(0, A(x)).$$

By Lemma 2.39, $C(x,0) = (f^d)'(c)(x)$ and $C(0,y) = (f_c)'(d)(y)$, i.e., the linear mapping $(f^d)'(c) + (f_c)'(d) \circ A$ is identically zero. This implies

$$\varphi'(c) = A = -\left((f_c)'(d)\right)^{-1} \circ (f^d)'(c).$$

We get
$$\varphi'(x) = -\left(f'_x(\varphi(x))\right)^{-1} \circ (f^{\varphi(x)})'(x)$$

for every $x \in B(a,\delta)$.
2. If f satisfies the conditions of Theorem 2.40 and f is continuously differentiable in a neighborhood of the point (a,b), then *we can choose* δ *and* η *such that* φ *is continuously differentiable on the ball* $B(a,\delta)$.

It suffices to choose δ and η such that in addition to parts (i) and (ii) of the theorem, we also require that f be continuously differentiable on the ball $B((a,b),\eta)$. In this case φ will be continuously differentiable at every point (c,d) of the ball $B(a,\delta)$. This follows from Theorem 2.40 applied to the point (c,d) instead of the point (a,b).

As an important application of the implicit function theorem we give a method for finding the conditional extremal points of a function.

Definition 2.42. Let $a \in H \subset \mathbb{R}^p$, $F\colon H \to \mathbb{R}^q$, and let $F(a) = 0$. Let the p-variable real function f be defined in a neighborhood of a, and let $\delta > 0$ be such that $f(x) \le f(a)$ for every point $x \in B(a,\delta)$ that satisfies $F(x) = 0$. Then we say that *the function* f *has a conditional local maximum point at the point* a *with the condition* $F = 0$. Conditional local minima can be defined in a similar manner. If f has a conditional local maximum or minimum at the point a with the condition $F = 0$, then we say that f *has a conditional local extremum at the point* a *with the condition* $F = 0$.

Example 2.43. Suppose we want to find the maximum of the function $f(x,y,z) = x + 2y + 3z$ on the sphere $S = \{(x,y,z) \in \mathbb{R}^3 : x^2 + y^2 + z^2 = 1\}$. By Weierstras's theorem, f has a maximal value on the bounded and closed set S. If f takes on this greatest value at the point a, then f has a conditional local maximum at a with the condition $x^2 + y^2 + z^2 - 1 = 0$.

Theorem 2.44. (Lagrange[8] multiplier method) *Let* $H \subset \mathbb{R}^p$, *and suppose that* $F\colon H \to \mathbb{R}^q$ *vanishes and is continuously differentiable at the point* $a \in \operatorname{int} H$. *Let us denote the coordinate functions of* F *by* $F_1, \ldots, F_q$.

If the p*-variable real function* f *is differentiable at* a *and* f *has a conditional local extremum at the point* a *with the condition* $F = 0$, *then there are real numbers* $\lambda, \lambda_1, \ldots, \lambda_q$ *such that at least one of these numbers is nonzero, and the partial derivatives of the function* $\lambda f + \lambda_1 F_1 + \ldots + \lambda_q F_q$ *are zero at* a.

The $p = 2$, $q = 1$ special case of the theorem above states that the gradients of f and F are parallel to each other at the conditional local extremum points. Intuitively, this can be proved as follows. Condition $F(x,y) = 0$ defines a curve in the plane. If we move along this curve, then we move perpendicularly to the gradient of F at each point of the curve (see Exercise 2.10). As we reach a conditional local

[8] Joseph-Louis Lagrange (1736–1813), Italian-French mathematician.

extremum point of f, we go neither upward nor downward on the graph of f, and thus the gradient of f is also perpendicular to the curve. That is, the two gradients are parallel to each other.

Proof of Theorem 2.44. Consider the matrix

$$\begin{pmatrix} D_1F_1(a) & D_2F_1(a) \dots D_pF_1(a) \\ D_1F_2(a) & D_2F_2(a) \dots D_pF_2(a) \\ \vdots & \vdots \quad \dots \quad \vdots \\ D_1F_q(a) & D_2F_q(a) \dots D_pF_q(a) \\ D_1f(a) & D_2f(a) \dots D_pf(a) \end{pmatrix}. \tag{2.12}$$

We need to prove that the rows of this matrix are linearly dependent. Indeed, in this case there are real numbers $\lambda_1,\dots,\lambda_q,\lambda$ such that at least one of these numbers is nonzero, and the linear combination of the row vectors with coefficients $\lambda_1,\dots,\lambda_q,\lambda$ is zero. Then every partial derivative of $\lambda_1F_1+\dots+\lambda_qF_q+\lambda f$ is zero at a, and this is what we want to prove.

If $p \le q$, then the statement holds trivially. Indeed, the matrix has p columns, and its rank is at most p. Thus, $q+1>p$ row vectors must be linearly dependent.

Therefore, we may assume that $p>q$. We may also assume that the first q row vectors of the matrix (the gradient vectors $F_1'(a),\dots,F_q'(a)$) are linearly independent, since otherwise, there would be nothing to prove.

The vectors $F_1'(a),\dots,F_q'(a)$ are the row vectors of the Jacobian matrix of F at the point a. Since these are linearly independent, the rank of the Jacobian matrix is q, and the matrix has q linearly independent column vectors. Permuting the coordinates of $\mathbb{R}^q$ if necessary, we may assume that the last q columns of the Jacobian matrix are linearly independent.

Let $s=p-q$. Put $b=(a_1,\dots,a_s)\in\mathbb{R}^s$ and $c=(a_{s+1},\dots,a_p)$; then $a=(b,c)$. The Jacobian matrix of the section $F_b\colon \mathbb{R}^q\to\mathbb{R}^q$ at the point c consists of the last q column vectors of the matrix of $F'(a)$. Since these are linearly independent, the linear mapping $(F_b')(c)$ is injective. Therefore, we may apply the implicit function theorem. We obtain $\delta>0$ and a differentiable function $\varphi\colon B(b,\delta)\to\mathbb{R}^q$ such that $\varphi(b)=c$ and $F(x,\varphi(x))=0$ for every $x\in B(b,\delta)$. (Here, $B(b,\delta)$ denotes the ball with center b and radius δ in $\mathbb{R}^s$.)

We know that f has a conditional local extremum point at $a=(b,c)$ with the condition $F=0$. Let us assume that this is a local maximum. This means that if $x\in\mathbb{R}^s$, $y\in\mathbb{R}^q$ and the point (x,y) is close enough to a, and $F(x,y)=0$, then $f(x,y)\le f(a)$. Consequently, if x is close enough to b, then $f(x,\varphi(x))\le f(b,\varphi(b))$. In other words, the function $f(x,\varphi(x))$ has a local maximum at the point b. By Theorem 1.60, the partial derivatives of $f(x,\varphi(x))$ are zero at the point b. If $\varphi_1,\dots,\varphi_q$ are the coordinate functions of φ, then applying Corollary 2.23, we find that for every $i=1,\dots,s$ we have

$$D_i f(a) + \sum_{j=1}^{q} D_{s+j} f(a) \cdot D_i \varphi_j(b) = 0. \tag{2.13}$$

For every $k = 1, \ldots, q$ the function $F_k(x, \varphi(x))$ is constant and equal to zero in a neighborhood of the point b, thus its partial derivatives are zero at b. We get

$$D_i F_k(a) + \sum_{j=1}^{q} D_{s+j} F_k(a) \cdot D_i \varphi_j(b) = 0 \tag{2.14}$$

for every $k = 1, \ldots, q$ and $i = 1, \ldots, s$. Equations (2.13) and (2.14) imply that the first s column vectors of the matrix of (2.12) are linear combinations of the last q column vectors. In other words, the rank of the matrix is at most q. Since the matrix has $q + 1$ rows, they are linearly dependent. □

Remark 2.45. If we want to find the conditional local extremum points a of the function f with condition $F = 0$, then according to Theorem 2.44, we need to find $\lambda, \lambda_1, \ldots, \lambda_q$ such that $\lambda D_i f(a) + \lambda_1 D_i F_1(a) + \ldots + \lambda_q D_i F_q(a) = 0$ for every $i = 1, \ldots, p$. These equations, together with the conditions $F_k(a) = 0$ ($k = 1, \ldots, q$), form a set of $p + q$ equations in $p + q + 1$ unknowns $a_1, \ldots, a_p, \lambda, \lambda_1, \ldots, \lambda_q$. We can also add the equation

$$\lambda^2 + \lambda_1^2 + \ldots + \lambda_q^2 = 1$$

to our system of equations, since instead of $\lambda, \lambda_1, \ldots, \lambda_q$, we could also take $\nu \cdot \lambda, \nu \cdot \lambda_1, \ldots, \nu \cdot \lambda_q$, where $\nu = 1/(\lambda^2 + \lambda_1^2 + \ldots + \lambda_q^2)$. We now have exactly as many equations as unknowns. Should we be lucky enough, these equations are "independent" and they have only a finite number of solutions. Checking these solutions one by one, we can find, in principle, the set of actual conditional local extremum points.

Example 2.46. In Example 2.43 we have seen that the function $f(x, y, z) = x + 2y + 3z$ has a greatest value on the sphere $S = \{(x, y, z) \in \mathbb{R}^3 : x^2 + y^2 + z^2 = 1\}$. If f takes on this greatest value at the point $a = (u, v, w)$, then f has a conditional local maximum at a with the condition $x^2 + y^2 + z^2 - 1 = 0$. Each of the functions mentioned above is continuously differentiable, and thus we can apply Theorem 2.44. We get that there are real numbers λ, μ such that they are not both zero and the partial derivatives of the function $\lambda(x + 2y + 3z) + \mu(x^2 + y^2 + z^2 - 1)$ are zero at the point (u, v, w). Thus, the equations

$$\lambda + 2\mu u = 0, \qquad 2\lambda + 2\mu v = 0, \qquad 3\lambda + 2\mu w = 0, \tag{2.15}$$

and $u^2 + v^2 + w^2 = 1$ hold. The Equations (2.15) imply $\mu \neq 0$, since $\mu = 0$ would imply $\lambda = 0$. Thus, applying (2.15) again gives us $v = 2u$ and $w = 3u$, implying $u^2 + (2u)^2 + (3u)^2 = 1$, $u = \pm 1/\sqrt{14}$, i.e., $(u, v, w) = (1/\sqrt{14}, 2/\sqrt{14}, 3/\sqrt{14})$ or $(u, v, w) = (-1/\sqrt{14}, -2/\sqrt{14}, -3/\sqrt{14})$.

The function $f(x, y, z) = x + 2y + 3z$ also has a least value on the sphere S. Since f is not constant on S, the points where f takes its maximum and its minimum must be different. This means that there are at least two conditional local extremal points. Our calculations above imply that there are exactly two such extremal points, and it is also clear that f assumes its greatest value at the point $(1/\sqrt{14}, 2/\sqrt{14}, 3/\sqrt{14})$ while it takes its least value at the point $(-1/\sqrt{14}, -2/\sqrt{14}, -3/\sqrt{14})$ on S.

Exercises

2.14. Show that in Corollary 2.28 the condition on the finiteness of the partial derivative $D_2 f$ can be omitted. (H)

2.15. Let the function f of Corollary 2.28 be differentiable at the point (a, b). Show directly (i.e., without applying Theorem 2.40) that the function φ is differentiable at the point a and $\varphi'(a) = -D_1 f(a, b)/D_2 f(a, b)$.

2.16. Let $f\colon I \to \mathbb{R}$ be continuous, where $I \subset \mathbb{R}$ is an interval. Show that if every point of I has a neighborhood where f is injective, then f is injective on the whole interval.

2.17. Let $f(x, y) = (e^x \cos y, e^x \sin y)$ for every $(x, y) \in \mathbb{R}^2$.

(a) Show that $f'(a, b)$ is injective at every $(a, b) \in \mathbb{R}^2$.
(b) Show that f is injective in every open disk with radius π.
(c) Let $G = \{(x, y) \in \mathbb{R}^2 \colon x > 0\}$. Define a continuous map $\varphi\colon G \to \mathbb{R}^2$ such that $\varphi(1, 0) = (0, 0)$ and $f \circ \varphi$ is the identity on G. (S)

2.18. Show that a contraction can have at most one fixed point.

2.19. Let $B \subset \mathbb{R}^p$ be an open ball. Show that there exists a contraction $f\colon B \to B$ with no fixed points.

2.20. We call the mapping $f\colon \mathbb{R}^p \to \mathbb{R}^p$ a **similarity with ratio** λ if $|f(x) - f(y)| = \lambda \cdot |x - y|$ holds for every $x, y \in \mathbb{R}^p$. Show that if $0 < \lambda < 1$, then every similarity with ratio λ has exactly one fixed point.

2.21. Find the largest value of $x - y + 3z$ on the ellipsoid $x^2 + \frac{y^2}{2} + \frac{z^2}{3} = 1$.

2.22. Find the largest value of xy with the condition $x^2 + y^2 = 1$.

2.23. Find the largest value of xyz with the condition $x^2 + y^2 + z^2 = 3$.

2.24. Find the largest value of xyz with the condition $x + y + z = 5$, $xy + yz + xz = 8$.

Chapter 3
The Jordan Measure

3.1 Definition and Basic Properties of the Jordan Measure

One of the main goals of mathematical analysis, besides applications in physics, is to compute the measure of sets (arc length, area, surface area, and volume).

We deal with the concepts of area and volume at once; we will use the word *measure* instead. We will actually define measure in every space $\mathbb{R}^p$, and area and volume will be the special cases when $p = 2$ and $p = 3$.

We call the sets A and B **nonoverlapping** if they do not share any interior points.

If we want to convert the intuitive meaning of measure into a precise notion, then we should first list our expectations for the concept. Measure has numerous properties which we consider natural. We choose three out of these:

(a) The measure of the box $R = [a_1, b_1] \times \cdots \times [a_p, b_p]$ equals the product of its sides lengths, that is, $(b_1 - a_1) \cdots (b_p - a_p)$.

(b) If we decompose a set into the union of finitely many nonoverlapping sets, then the measure of the set is the sum of the measures of the parts.

(c) If $A \subset B$ then the measure of A is not greater than the measure of B.

We will see that these requirements naturally determine to which sets we can assign a measure, and what that measure should be.

Definition 3.1. If $R = [a_1, b_1] \times \cdots \times [a_p, b_p]$, then we let $\mu(R)$ denote the product $(b_1 - a_1) \cdots (b_p - a_p)$.

Let A be an arbitrary bounded set in $\mathbb{R}^p$. Cover A in every possible way by finitely many boxes $R_1, \ldots, R_K$, and form the sum $\sum_{i=1}^{K} \mu(R_i)$ for each cover. The *outer measure* of the set A is defined as the infimum of the set of all the sums we obtain this way. We denote the outer measure of the set A by $\overline{\mu}(A)$.

If A does not have an interior point, then we define the *inner measure* to be equal to zero. If A does have an interior point, then choose every combination of finitely

M. Laczkovich and V.T. Sós, *Real Analysis*, Undergraduate Texts in Mathematics, https://doi.org/10.1007/978-1-4939-7369-9_3

many boxes $R_1, \ldots, R_K$ in A such that they are pairwise nonoverlapping, and form the sum $\sum_{i=1}^{K} \mu(R_i)$ each time. The inner measure of A is defined as the supremum of the set of all such sums. The inner measure of the set A will be denoted by $\underline{\mu}(A)$.

It is intuitively clear that for any bounded set A, the values $\underline{\mu}(A)$ and $\overline{\mu}(A)$ are finite, moreover $0 \le \underline{\mu}(A) \le \overline{\mu}(A)$. (We shall prove these statements shortly.) Now by restrictions (a) and (c) above, it is clear that the measure of the set A should fall between $\underline{\mu}(A)$ and $\overline{\mu}(A)$. If $\underline{\mu}(A) < \overline{\mu}(A)$, then without further inspection, it is not clear which number (between $\underline{\mu}(A)$ and $\overline{\mu}(A)$) we should consider the measure of A to be. Therefore, when speaking about sets having measure, we will restrict ourselves to sets for which $\underline{\mu}(A) = \overline{\mu}(A)$, and this shared value will be called the measure of A.

Definition 3.2. We call the bounded set $A \subset \mathbb{R}^d$ *Jordan*[1] *measurable* if $\underline{\mu}(A) = \overline{\mu}(A)$. The *Jordan measure* of the set A (the measure of A, for short) is the common value $\underline{\mu}(A) = \overline{\mu}(A)$, which we denote by $\mu(A)$.

If $p \ge 3$ then instead of Jordan measure we can say **volume**, if $p = 2$ then **area**, and if $p = 1$ then we can say **length** as well. If we want to emphasize that we are talking about the inner, outer, or Jordan measure of a p dimensional set, then instead of $\underline{\mu}(A)$, $\overline{\mu}(A)$, or $\mu(A)$ we may write $\underline{\mu}_p(A)$, $\overline{\mu}_p(A)$, or $\mu_p(A)$.

Before proceeding with the investigation of measurable sets and the calculation of their measure, we will now consider a different approach to define measure.

Finding the (approximate) area of a plane figure can be done by covering the plane by a very fine square-grid, and counting the number of small squares intersecting the figure. Our next goal is translating this idea into a precise notion. From now on, $|V|$ denotes the cardinality of a finite set V.

We call the box $R = [a_1, b_1] \times \ldots \times [a_p, b_p] \subset \mathbb{R}^p$ a **cube with side length** s, if $b_1 - a_1 = \ldots = b_p - a_p = s$. (For $p = 1$ and $p = 2$ the cubes with side length s are nothing else than the closed intervals of length s, and the squares with side length s, respectively.)

We denote by $\mathcal{K}_n$ the set of cubes $\left[\frac{i_1-1}{n}, \frac{i_1}{n}\right] \times \ldots \times \left[\frac{i_p-1}{n}, \frac{i_p}{n}\right]$ with side length $1/n$, where $i_1, \ldots, i_p$ are arbitrary integers. These cubes are mutually nonoverlapping, and their union covers the whole $\mathbb{R}^p$ space.

For every set $A \subset \mathbb{R}^p$, the cubes of $\mathcal{K}_n$ belong to one of three separate classes. The cube K is an **interior cube,** or an **exterior cube**, if $K \subset \operatorname{int} A$ or $K \subset \operatorname{ext} A$, respectively. If a cube is neither an interior, nor an exterior cube, it is called a **boundary cube**. Since the sets $\operatorname{int} A$, $\operatorname{ext} A$ and ∂A are pairwise disjoint, every cube that intersects the boundary of A is necessarily a boundary cube.

[1] Camille Jordan (1838–1922), French mathematician.

In fact, the boundary cubes are exactly the ones with $K \cap \partial A \neq \emptyset$. In order to prove this statement, let us assume that a boundary cube K does not intersect ∂A. Then $K \subset \operatorname{int} A \cup \operatorname{ext} A$. Since K is neither an interior nor an exterior cube, hence it contains some points x and y such that $x \in \operatorname{int} A$ and $y \in \operatorname{ext} A$. Since K is a convex set, it contains the whole segment $[x, y]$. We know that every segment connecting a point of A and another point of $\mathbb{R}^p \setminus A$ always contains a boundary point of A (see Theorem 1.19). Thus $K \cap \partial A \neq \emptyset$, and we reached contradiction.

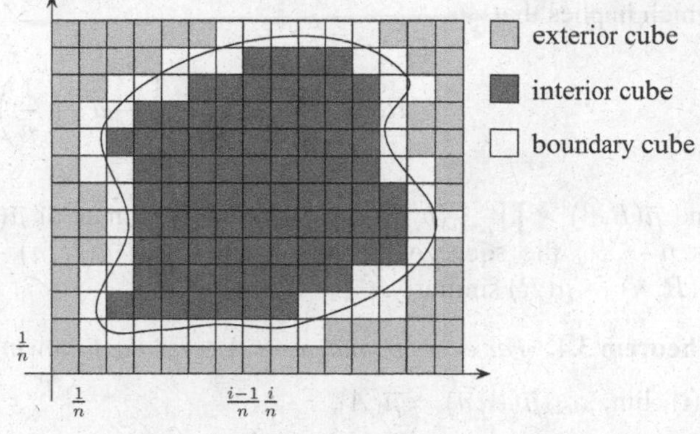

3.1. Figure

Obviously, a cube is an interior or a boundary cube exactly when it intersects the set $(\operatorname{int} A) \cup (\partial A)$, i.e., the closure of A.

We denote the total measure of interior boxes by $\underline{\mu}(A, n)$. In other words,

$$\underline{\mu}(A, n) = \frac{|\{K \in \mathcal{K}_n \colon K \subset \operatorname{int} A\}|}{n^p}.$$

We denote the total measure of interior and boundary boxes by $\overline{\mu}(A, n)$. That is,

$$\overline{\mu}(A, n) = \frac{|\{K \in \mathcal{K}_n \colon K \cap \operatorname{cl} A \neq \emptyset\}|}{n^p}.$$

We will now show that for every bounded set A, the sequences $\underline{\mu}(A, n)$ and $\overline{\mu}(A, n)$ converge to the inner measure and the outer measure of A, respectively.

Lemma 3.3. *For every box $R = [a_1, b_1] \times \ldots \times [a_p, b_p]$ we have $\lim_{n\to\infty} \overline{\mu}(R, n) = \lim_{n\to\infty} \underline{\mu}(R, n) = \prod_{j=1}^{p} (b_j - a_j)$.*

Proof. Let n be fixed. There are integers p_j, q_j such that

$$(p_j - 1)/n < a_j \le p_j/n \quad \text{and} \quad (q_j - 1)/n \le b_j < q_j/n \; (j = 1, \ldots, p).$$

One can easily see that a cube $\left[\frac{i_1-1}{n}, \frac{i_1}{n}\right] \times \ldots \times \left[\frac{i_p-1}{n}, \frac{i_p}{n}\right]$ intersects the closure of R (i.e., R itself) if $p_j \le i_j \le q_j$ $(j = 1, \ldots, p)$. Therefore,

$$\overline{\mu}(R, n) = n^{-p} \cdot \prod_{j=1}^{p} (q_j - p_j + 1),$$

which implies that

$$\overline{\mu}(R,n) \leq \prod_{j=1}^{p} \left(b_j + \frac{1}{n} - a_j + \frac{1}{n} \right),$$

and $\overline{\mu}(R,n) \geq \prod_{j=1}^{p} (b_j - a_j)$. Since both estimate of $\overline{\mu}(R,n)$ converge to $\mu(R)$ as $n \to \infty$, the squeeze theorem implies that $\overline{\mu}(R,n) \to \mu(R)$. We can prove $\underline{\mu}(R,n) \to \mu(R)$ similarly. □

Theorem 3.4. *For every bounded set $A \in \mathbb{R}^p$ the following hold.*

(i) $\lim_{n\to\infty} \overline{\mu}(A,n) = \overline{\mu}(A)$,

(ii) $\lim_{n\to\infty} \underline{\mu}(A,n)(A) = \underline{\mu}(A)$,

(iii) $\underline{\mu}(A) \leq \overline{\mu}(A)$.

Proof. (i) Let $\varepsilon > 0$ be fixed. Then there are boxes $R_1, \ldots, R_N$ covering A such that $\sum_{i=1}^{N} \mu(R_i) < \overline{\mu}(A) + \varepsilon$. The union of the cubes of $\mathcal{K}_n$ intersecting the closure of A cover A itself, thus the definition of $\overline{\mu}(A)$ implies $\overline{\mu}(A,n) \geq \overline{\mu}(A)$. On the other hand, a cube intersecting the closure of A also intersects one of the boxes R_i, thus

$$\overline{\mu}(A,n) \leq \sum_{i=1}^{N} \overline{\mu}(R_i,n).$$

By Lemma 3.3, $\lim_{n\to\infty} \sum_{i=1}^{N} \overline{\mu}(R_i,n) = \sum_{i=1}^{N} \mu(R_i) < \overline{\mu}(A) + \varepsilon$, and thus there exists an integer n_0 such that $\overline{\mu}(A,n) \leq \sum_{i=1}^{N} \overline{\mu}(R_i,n) < \overline{\mu}(A) + \varepsilon$ for every $n > n_0$. We obtain that $\overline{\mu}(A) \leq \overline{\mu}(A,n) < \overline{\mu}(A) + \varepsilon$ for $n > n_0$. Since ε was arbitrary, we get $\lim_{n\to\infty} \overline{\mu}(A,n) = \overline{\mu}(A)$.

(ii) If the interior of the set A is empty, then $\underline{\mu}(A) = 0$ and $\underline{\mu}(A,n) = 0$ for every n, thus $\underline{\mu}(A,n) \to \underline{\mu}(A)$. Suppose $\operatorname{int} A \neq \emptyset$, and let $\varepsilon > 0$ be fixed. There exist nonoverlapping boxes $R_1, \ldots, R_N$ in A such that $\sum_{i=1}^{N} \mu(R_i) > \underline{\mu}(A) - \varepsilon$. Then

$$\underline{\mu}(A,n) \geq \sum_{i=1}^{N} \underline{\mu}(R_i,n), \tag{3.1}$$

since if a cube is in the interior of R_i then it has to be an interior cube of A as well. By Lemma 3.3,

$$\lim_{n\to\infty} \sum_{i=1}^{N} \underline{\mu}(R_i,n) = \sum_{i=1}^{N} \mu(R_i) > \underline{\mu}(A) - \varepsilon,$$

and thus there exists an integer n_0 such that $\sum_{i=1}^{N} \underline{\mu}(R_i,n) > \underline{\mu}(A) - \varepsilon$ for every $n > n_0$. Now (3.1) implies $\underline{\mu}(A,n) > \underline{\mu}(A) - \varepsilon$ for $n > n_0$. On the other hand

$\underline{\mu}(A,n) \le \underline{\mu}(A)$, since the total volume of the cubes of $\mathcal{K}_n$ in the interior of A is at most $\underline{\mu}(A)$ by the definition of $\underline{\mu}(A)$. We proved that $\underline{\mu}(A,n) \to \underline{\mu}(A)$.

Statement (iii) is obvious from (i), (ii), and from the inequality $\underline{\mu}(A,n) \le \overline{\mu}(A,n)$. □

Remark 3.5. We should realize that if R is a box, then $\overline{\mu}(R) \le \mu(R)$ and $\underline{\mu}(R) \ge \mu(R)$ follow immediately from the definition, and thus, by $\underline{\mu}(R) \le \overline{\mu}(R)$ we have $\overline{\mu}(R) = \underline{\mu}(R) = \mu(R)$. As a result, R is measurable, and Definition 3.2 gives the same value for $\mu(R)$ as the original definition of $\mu(R)$.

Theorem 3.6. *Let A and B be bounded sets. Then*

(i) $\overline{\mu}(A \cup B) \le \overline{\mu}(A) + \overline{\mu}(B)$,

(ii) *if A and B are nonoverlapping, then $\underline{\mu}(A \cup B) \ge \underline{\mu}(A) + \underline{\mu}(B)$, and*

(iii) *if $A \subset B$, then $\underline{\mu}(A) \le \underline{\mu}(B)$ and $\overline{\mu}(A) \le \overline{\mu}(B)$.*

Proof. It is clear that $\overline{\mu}(A \cup B, n) \le \overline{\mu}(A,n) + \overline{\mu}(B,n)$ holds for every n. Then we get (i) by letting $n \to \infty$. If A and B are nonoverlapping, then $\underline{\mu}(A \cup B, n) \ge \underline{\mu}(A,n) + \underline{\mu}(B,n)$, which yields (ii). Suppose $A \subset B$. Every box in A is also in B, thus $\underline{\mu}(A) \le \underline{\mu}(B)$. Finally, if the union of a set of boxes cover B it also covers A, which implies $\overline{\mu}(B) \ge \overline{\mu}(A)$. □

Theorem 3.7. *For every bounded set A,*

$$\overline{\mu}(A) = \overline{\mu}(\operatorname{cl} A) = \underline{\mu}(A) + \overline{\mu}(\partial A).$$

Proof. Recall that $\overline{\mu}(A,n)$ denotes the total volume of the interior and boundary cubes of $\mathcal{K}_n$. Since ∂A is closed, the cubes intersecting the closure of ∂A are the same as the cubes intersecting ∂A, and thus the total volume of the boundary cubes is $\overline{\mu}(\partial A, n)$. We obtain

$$\overline{\mu}(A) = \overline{\mu}(\operatorname{cl} A, n) = \underline{\mu}(A,n) + \overline{\mu}(\partial A, n).$$

Then, letting $n \to \infty$ yields the desired equality. □

Definition 3.8. We say that a set $A \subset \mathbb{R}^p$ is a *null set,* if $\overline{\mu}(A) = 0$.

Theorem 3.9. *A set $A \subset \mathbb{R}^p$ is measurable if and only if it is bounded and its boundary is a null set.*

Proof. The statement is clear from $\overline{\mu}(A) - \underline{\mu}(A) = \overline{\mu}(\partial A)$. □

For our further discussion of the theory of measurable sets, it is necessary to give conditions for being a null set.

Definition 3.10. The *diameter* of a bounded set $A \subset \mathbb{R}^p$ is the number

$$\operatorname{diam} A = \sup\{|x - y| : x, y \in A\}.$$

The diameter of the empty set is zero by definition.

Example 3.11. The diameter of the cubes of $\mathbb{R}^p$ with side length s is $s \cdot \sqrt{p}$. Indeed, let $R = [a_1, b_1] \times \ldots \times [a_p, b_p] \subset \mathbb{R}^p$, where $b_1 - a_1 = \ldots = b_p - a_p = s$. If $x = (x_1, \ldots, x_p) \in R$ and $y = (y_1, \ldots, y_p) \in R$, then $x_i, y_i \in [a_i, b_i]$ and $y_i - x_i \leq b_i - a_i = s$, for every $i = 1, \ldots, p$, and

$$|x - y| = \sqrt{\sum_{i=1}^{p} (x_i - y_i)^2} \leq \sqrt{p \cdot s^2} = s \cdot \sqrt{p}$$

follows. Thus $\operatorname{diam} R \leq s \cdot \sqrt{p}$. On the other hand, the distance of the points $a = (a_1, \ldots, a_p)$ and $b = (b_1, \ldots, b_p)$ is exactly $s \cdot \sqrt{p}$, which implies $\operatorname{diam} R = s \cdot \sqrt{p}$.

Lemma 3.12. *Let a positive number δ be given. Then every box can be decomposed into finitely many nonoverlapping boxes with diameter smaller than δ.*

Proof. Let the box R be fixed. If $n > \sqrt{p}/\delta$, then every cube with side length $1/n$ has diameter smaller than δ. Consider the cubes $K \cap R$, where $K \in \mathcal{K}_n$ and $\operatorname{int} K \cap \operatorname{int} R \neq \emptyset$ (there is only finitely many such cubes). These cubes are nonoverlapping, and their (common) diameter is smaller than δ. Since their union covers the interior of R and it is closed, it also covers R. □

Theorem 3.13. *Let $H \subset \mathbb{R}^p$ be bounded and closed, and let $f : H \to \mathbb{R}$ be continuous. Then* $\operatorname{graph} f$ *has measure zero in* $\mathbb{R}^{p+1}$.

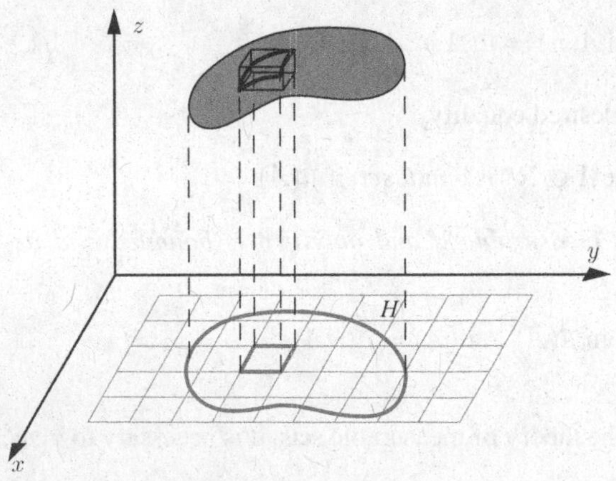

3.2. Figure

Proof. Let $\varepsilon > 0$ be fixed. By Heine's theorem (Theorem 1.53), there exists $\delta > 0$ such that $|f(x) - f(y)| < \varepsilon$ for every $x, y \in H$ with $|x - y| < \delta$.

Since H is bounded, it can be covered by a box R. Using Lemma 3.12, we can decompose R into finitely many nonoverlapping boxes $R_1, \ldots, R_n$ with diameter less than δ. Let I denote the set of indices i

with $R_i \cap H \neq \emptyset$. If $i \in I$, then the set $A_i = R_i \cap H$ is bounded, closed, and nonempty. Then, by Weierstrass' theorem, f has a smallest and a largest value on A_i. If $m_i = \min_{x \in A_i} f(x)$ and $M_i = \max_{x \in A_i} f(x)$, then $\operatorname{diam} A_i < \delta$ implies $M_i - m_i < \varepsilon$.

The union of the boxes $R_i \times [m_i, M_i]$ $(i \in I)$ covers graph f. Thus

$$\overline{\mu}_{p+1}(\operatorname{graph} f) \leq \sum_{i \in I} \mu_p(R_i) \cdot (M_i - m_i) \leq \sum_{i \in I} \mu_p(R_i) \cdot \varepsilon \leq \mu_p(R) \cdot \varepsilon, \quad (3.2)$$

since R_i are nonoverlapping, and then $\sum_{i=1}^n \mu_p(R_i) \leq \mu_p(R)$ by statement (ii) of Theorem 3.6. (Here $\overline{\mu}_{p+1}(\operatorname{graph} f)$ and $\mu_p(R_i)$ are the outer measures of the set graph f in $\mathbb{R}^{p+1}$ and the measure of R_i in $\mathbb{R}^p$, respectively.) Since ε was arbitrary, graph f has measure zero. □

Corollary 3.14. *Every ball is measurable.*

Proof. The ball $B(a, r)$ is bounded, with boundary $S(a, r)$. It is easy to see that $S(a, r)$ is the union of the graphs of the functions $f(x_1, \ldots, x_{p-1}) = \sqrt{r^2 - \sum_{i=1}^{p-1} x_i^2}$ and $g(x_1, \ldots, x_{p-1}) = -\sqrt{r^2 - \sum_{i=1}^{p-1} x_i^2}$ defined on the set

$$A = \left\{ (x_1, \ldots, x_{p-1}) \in \mathbb{R}^{p-1} \colon \sum_{i=1}^{p-1} (x_i - a_i)^2 \leq r^2 \right\}.$$

Since A is bounded and closed in $\mathbb{R}^{p-1}$, furthermore f and g are continuous on A, hence Theorem 3.13 and part (i) of Theorem 3.6 imply that $S(a, r)$ has measure zero. Thus $B(a, r)$ is measurable by Theorem 3.9. □

Intuitively, it is clear (and also not too hard to prove), that a hyperplane does not have interior points. Thus, the inner measure of a bounded subset of a hyperplane is always zero. We show that the outer measure of such a set is also zero.

Lemma 3.15. *Every bounded subset of a hyperplane of $\mathbb{R}^p$ has measure zero in $\mathbb{R}^p$.*

Proof. Let our hyperplane be the set

$$H = \{(x_1, \ldots, x_p) \in \mathbb{R}^p \colon a_1 x_1 + \ldots + a_p x_p = b\},$$

where not every a_i is zero. We may assume that $a_p \neq 0$. Indeed, in the case of $a_p = 0$ and $a_i \neq 0$ we could simply swap the corresponding coordinates. Obviously, the mapping

$$(x_1, \ldots, x_i, \ldots, x_p) \mapsto (x_1, \ldots, x_p, \ldots, x_i)$$

does not change the volume of a box, thus it also leaves the inner and outer measures of every set unchanged as well. The image of a hyperplane is a hyperplane and the image of a bounded set is a bounded set, i.e., we get that the assumption of $a_p \neq 0$ is indeed justified.

Let $A \subset H$ be a bounded set. The set A can be covered by some box R. Let $R = R_1 \times [c, d]$, where R_1 is a box in $\mathbb{R}^{p-1}$. With these assumptions, A is a subset of the graph of the function

$$f(x_1, \ldots, x_{p-1}) = \left(b - \sum_{i=1}^{p-1} a_i x_i \right) / a_p \qquad ((x_1, \ldots, x_{p-1}) \in R_1),$$

which has measure zero, by Theorem 3.13. □

Since every polyhedron is bounded and its boundary can be covered by finitely many number of hyperplanes, it follows that *every polyhedron is measurable.*

The $p = 2$ case yields that *every polygon is measurable.*

One can show that *every bounded and convex set is also measurable.* The proof can be found in the appendix.

We continue with a closer inspection of the measurable sets.

Theorem 3.16. *If A and B are measurable sets, then $A \cup B$, $A \cap B$, and $A \setminus B$ are also measurable.*

Proof. Since A and B are bounded, so are the sets listed in the statement of the theorem. Thus, it is enough to show that the boundaries of these sets have measure zero. Since A and B are measurable, we have $\overline{\mu}(\partial A) = \overline{\mu}(\partial B) = 0$, and thus $\overline{\mu}((\partial A) \cup (\partial B)) \leq \overline{\mu}(\partial A) + \overline{\mu}(\partial B) = 0$, that is $(\partial A) \cup (\partial B)$ has measure zero. Then every subset of $(\partial A) \cup (\partial B)$ also has measure zero. Therefore, it is enough to show that the boundaries of the sets $A \cup B$, $A \cap B$, and $A \setminus B$ are subsets of $(\partial A) \cup (\partial B)$. This is easy to check using the definition of the boundary (see Exercise 1.10). □

We denote the set of all Jordan-measurable sets in $\mathbb{R}^p$ by $\mathcal{J}$. The previous theorem states that if $A, B \in \mathcal{J}$, then $A \cup B$, $A \cap B$, $A \setminus B$ are also in $\mathcal{J}$. The statements of the following theorem can be summarized as follows: the function $\mu \colon \mathcal{J} \to \mathbb{R}$ is *non-negative, additive, translation-invariant, and normalized.*

Theorem 3.17.

(i) *$\mu(A) \geq 0$ for every $A \in \mathcal{J}$.*

(ii) *If $A, B \in \mathcal{J}$ are nonoverlapping then $\mu(A \cup B) = \mu(A) + \mu(B)$.*

(iii) *If $A \in \mathcal{J}$ and B is a translation of the set A (i.e., there exists a vector v such that $B = A + v = \{x + v \colon x \in A\}$), then $B \in \mathcal{J}$ and $\mu(B) = \mu(A)$.*

(iv) *$\mu([0, 1]^p) = 1$.*

Proof. Part (i) is obvious, since $\underline{\mu}(A) \geq 0$ for every bounded set. If $A, B \in \mathcal{J}$, then (without additional conditions)

$$\mu(A \cup B) = \overline{\mu}(A \cup B) \leq \overline{\mu}(A) + \overline{\mu}(B) = \mu(A) + \mu(B).$$

(The first equality follows from Theorem 3.16.) If A and B are nonoverlapping, then

$$\mu(A \cup B) = \underline{\mu}(A \cup B) \geq \underline{\mu}(A) + \underline{\mu}(B) = \mu(A) + \mu(B)$$

also holds. The two inequalities together imply (ii).

Part (iii) follows from the fact that if the set R' is a translation of the box R, then R' is also a box, and $\mu(R') = \mu(R)$, obviously. Thus, if the set B is a translation of the set A, then $\overline{\mu}(B) = \overline{\mu}(A)$, since $\overline{\mu}(B)$ and $\overline{\mu}(A)$ are the infima of the same set of numbers. Similarly, $\underline{\mu}(B) = \underline{\mu}(A)$. If A is measurable, then $\overline{\mu}(B) = \overline{\mu}(A) = \underline{\mu}(A) = \underline{\mu}(B)$, and B is also measurable with $\mu(B) = \mu(A)$. Finally, part (iv) follows from the fact that $[0,1]^p$ is a box. □

According to the following theorem, the function μ is the only set function satisfying these four conditions.

Theorem 3.18. *Let the function $t\colon \mathcal{J} \to \mathbb{R}$ be non-negative, additive, translation-invariant, and normalized. (I.e., let t satisfy the conditions of the previous theorem.) Then $t(A) = \mu(A)$ for every set $A \in \mathcal{J}$.*

Proof. First we show that t is monotone; that is, $A, B \in \mathcal{J}$, $A \subset B$ implies $t(A) \leq t(B)$. Indeed, the additivity and non-negativity of t imply

$$t(B) = t(A \cup (B \setminus A)) = t(A) + t(B \setminus A) \geq t(A).$$

The cubes of $\mathcal{K}_n$ are translations of each other, thus the translation-invariance of t implies that $t(K) = t(K')$ for every $K, K' \in \mathcal{K}_n$. Since $[0,1]^p$ is the union of n^p nonoverlapping cubes of $\mathcal{K}_n$, hence the additivity of t and the fact that t is normalized imply

$$1 = t([0,1]^p) = n^p \cdot t(K),$$

i.e., $t(K) = 1/n^p$ for every $K \in \mathcal{K}_n$. Let $A \in \mathcal{J}$ be arbitrary, and let B_n be the union of the cubes of $\mathcal{K}_n$ which lie in the interior of A. We have $\mu(B_n) = \underline{\mu}(A, n)$. Since t is monotone, additive, and t is the same as μ on the cubes of $\mathcal{K}_n$, it follows that $t(A) \geq t(B_n) = \mu(B_n) = \underline{\mu}(A, n)$. Let C_n be the union of the cubes of $\mathcal{K}_n$ intersecting the closure of A. We have $\mu(C_n) = \overline{\mu}(A, n)$, and again, since t is monotone, additive, and t is the same as μ on the cubes of $\mathcal{K}_n$, we have $t(A) \leq t(C_n) = \mu(C_n) = \overline{\mu}(A, n)$. Thus, $\underline{\mu}(A, n) \leq t(A) \leq \overline{\mu}(A, n)$ for every n. Since $\overline{\mu}(A, n) \to \mu(A)$ and $\underline{\mu}(A) \to \mu(A)$ as $n \to \infty$, it follows that $t(A) = \mu(A)$. □

Later we will see that the Jordan measure is not only translation-invariant, but it also is isometry-invariant (Theorem 3.36)[2]. This is easy to show for some special isometries.

[2] By an isometry we mean a distance preserving bijection from $\mathbb{R}^p$ onto itself (see page 115).

Let $\phi_a(x) = 2a - x$ for every $x, a \in \mathbb{R}^p$. The mapping ϕ_a is called the **reflection through the point** a.

Lemma 3.19. *For every bounded set $A \subset \mathbb{R}^p$ and for every point $a \in \mathbb{R}^p$ we have $\overline{\mu}(\phi_a(A)) = \overline{\mu}(A)$ and $\underline{\mu}(\phi_a(A)) = \underline{\mu}(A)$. If $A \subset \mathbb{R}^p$ is measurable, then $\phi_a(A)$ is also measurable with $\mu(\phi_a(A)) = \mu(A)$.*

Proof. If $R' = \phi_a(R)$, where R is a box, then R' is also a box and $\mu(R') = \mu(R)$. It follows that if $B = \phi_a(A)$, then $\overline{\mu}(B) = \overline{\mu}(A)$, since $\overline{\mu}(B)$ and $\overline{\mu}(A)$ are the infima of the same set of numbers. We get $\underline{\mu}(B) = \underline{\mu}(A)$ similarly. If A is measurable, then $\overline{\mu}(B) = \overline{\mu}(A) = \underline{\mu}(A) = \underline{\mu}(B)$, thus B is also measurable, with $\mu(B) = \mu(A)$. □

For every positive number λ and for every point $a \in \mathbb{R}^p$ we say that the mapping $\psi_{\lambda,a}(x) = \lambda x + a$ ($x \in \mathbb{R}^p$) is a **homothetic transformation** with ratio λ.

Lemma 3.20. *For every bounded set $A \subset \mathbb{R}^p$ and for every point $a \in \mathbb{R}^p$ we have $\underline{\mu}(\psi_{\lambda,a}(A)) = \lambda^p \cdot \underline{\mu}(A)$ and $\overline{\mu}(\psi_{\lambda,a}(A)) = \lambda^p \cdot \overline{\mu}(A)$. If $A \subset \mathbb{R}^p$ is measurable, then $\psi_{\lambda,a}(A)$ is also measurable with $\mu(\psi_{\lambda,a}(A)) = \lambda^p \cdot \mu(A)$.*

Proof. For every box R, $\psi_{\lambda,a}(R)$ is also a box, with side-lengths λ times the corresponding side lengths of R. Thus $\mu(\psi_{\lambda,a}(R)) = \lambda^p \cdot \mu(R)$.

Now, $A \subset R_1 \cup \ldots \cup R_n$ if and only if $\psi_{\lambda,a}(A) \subset \psi_{\lambda,a}(R_1) \cup \ldots \cup \psi_{\lambda,a}(R_n)$, which implies $\overline{\mu}(\psi_{\lambda,a}(A)) = \lambda^p \cdot \overline{\mu}(A)$. The inequality $\underline{\mu}(\psi_{\lambda,a}(A)) = \lambda^p \cdot \underline{\mu}(A)$ can be proven similarly. The case of measurable sets should be clear. □

Exercises

3.1. For every $0 \le a \le b$ find a set $H \subset \mathbb{R}^2$ such that $\underline{\mu}(H) = a$ and $\overline{\mu}(H) = b$.

3.2. (a) Is there a non-measurable set whose boundary is measurable?
(b) Does the measurability of the closure, the interior, and the boundary of a set imply the measurability of the set itself?
(c) Does the measurability of a set imply the measurability of its closure, interior, and boundary?

3.3. Let (r_n) be an enumeration of all rational number in $[0, 1]$. Is $\bigcup_{n=1}^{\infty} ([r_n, r_n + (1/n)] \times [0, 1/n])$, as a subset of $\mathbb{R}^2$ measurable?

3.4. Show that if A is bounded, then $\underline{\mu}(A) = \underline{\mu}(\text{int}\, A)$.

3.5. Prove that for every bounded set $A \subset \mathbb{R}^p$ and for every $\varepsilon > 0$ there is an open set G such that $A \subset G$ and $\overline{\mu}(G) < \overline{\mu}(A) + \varepsilon$. (S)

3.6. (a) Show that if A and B are bounded and nonoverlapping sets, then $\overline{\mu}(A \cup B) \geq \overline{\mu}(A) + \underline{\mu}(B)$.
(b) Give an example of two sets A and B with the property $\overline{\mu}(A \cup B) > \overline{\mu}(A) + \underline{\mu}(B)$.

3.7. Let the function $f\colon [a,b] \to \mathbb{R}$ be non-negative, bounded, and let A be the domain under the graph of f; i.e., $A = \{(x,y)\colon x \in [a,b],\ 0 \leq y \leq f(x)\}$. Show that $\underline{\mu}(A)$ and $\overline{\mu}(A)$ are equal to the lower and upper integral of f, respectively.

3.8. Let $f\colon [a,b] \to \mathbb{R}$ be bounded. Is it true that if the graph of f is measurable, then f is integrable?

3.9. Construct a function $f\colon [0,1] \to [0,1]$ whose graph is not measurable.

3.10. Let $H \subset R$, where R is a box. Show that $\underline{\mu}(H) = \mu(R) - \overline{\mu}(R \setminus H)$.

3.11. Show that if A is measurable and $H \subset A$, then $\underline{\mu}(H) = \mu(A) - \overline{\mu}(A \setminus H)$.

3.12. Let A be a bounded set. Show that A is measurable if and only if, for every bounded set H we have $\overline{\mu}(H) = \overline{\mu}(H \cap A) + \overline{\mu}(H \setminus A)$.

3.13. Let (a_n, b_n) $(n = 1, 2, \ldots)$ be open intervals, with a bounded union. Show that $\underline{\mu}\left(\cup_{n=1}^{\infty}(a_n, b_n)\right) \leq \sum_{n=1}^{\infty}(b_n - a_n)$. (S)

3.14. For every $\varepsilon > 0$, construct a bounded, open subset G of the real line such that $\underline{\mu}(G) < \varepsilon$ and $\overline{\mu}(G) \geq 1$. (S)

3.15. Construct a bounded and closed subset F of the real line such that $\underline{\mu}(F) = 0$ and $\overline{\mu}(F) \geq 1$. (S)

3.16. Let $m(H) = (\overline{\mu}(H) + \underline{\mu}(H))/2$ for every bounded set $H \subset \mathbb{R}^2$. Show that m is not additive.

3.17. Show that omitting any one of the conditions being additive, translation-invariant, normalized, or non-negative, the remaining properties do not imply that a function defined on $\mathcal{J}$ with these properties is necessarily the Jordan-measure. ($*$ S)

3.18. Let A and B be bounded sets. Show that the following statements are equivalent.

(a) $\overline{\mu}(A \cup B) = \overline{\mu}(A) + \overline{\mu}(B)$.
(b) For every $\varepsilon > 0$ there exist M, N measurable sets such that $A \subset M$, $B \subset N$ and $\overline{\mu}(M \cap N) < \varepsilon$.
(c) $(\partial A) \cap (\partial B)$ has measure zero.
(d) There exists a measurable set M such that $A \subset M$ and $\overline{\mu}(M \cap B) = 0$. ($*$)

3.19. Let A and B be bounded sets such that $\underline{\mu}(A) = 1$, $\overline{\mu}(A) = 6$, $\underline{\mu}(B) = 2$, $\overline{\mu}(B) = 4$, $\overline{\mu}(A \cup B) = 10$. What values can $\underline{\mu}(A \cup B)$ take?

3.20. Let $A_1, \ldots, A_n$ be measurable sets in the unit cube, and let the sum of their measures be larger then 100. Show that there exists a point which belongs to more than 100 of these sets.

3.21. Let $F_1 \supset F_2 \supset \ldots$ be bounded and closed sets, and let $\bigcap_{n=1}^{\infty} F_n$ consist of a single point. Show that $\overline{\mu}(F_n) \to 0$.

3.22. Let $A_1 \subset A_2 \subset \ldots$ be measurable sets in $\mathbb{R}^p$ with a bounded union. Which of the following statements is true?
(a) $\overline{\mu}\left(\bigcup_{n=1}^{\infty} A_n\right) = \lim_{n\to\infty} \mu(A_n)$; (b) $\underline{\mu}\left(\bigcup_{n=1}^{\infty} A_n\right) = \lim_{n\to\infty} \mu(A_n)$.

3.23. Let $A_1 \supset A_2 \supset \ldots$ be measurable sets in $\mathbb{R}^p$. Which of the following statements is true?
(a) $\overline{\mu}\left(\bigcap_{n=1}^{\infty} A_n\right) = \lim_{n\to\infty} \mu(A_n)$; (b) $\underline{\mu}\left(\bigcap_{n=1}^{\infty} A_n\right) = \lim_{n\to\infty} \mu(A_n)$.

3.24. Let A_n $(n = 1, 2, \ldots)$ be nonoverlapping measurable sets in $\mathbb{R}^p$ with a bounded union. Which of the following statements is true?

(a) $\overline{\mu}\left(\bigcup_{n=1}^{\infty} A_n\right) \le \sum_{n=1}^{\infty} \overline{\mu}(A_n)$; (b) $\overline{\mu}\left(\bigcup_{n=1}^{\infty} A_n\right) \ge \sum_{n=1}^{\infty} \overline{\mu}(A_n)$;
(c) $\underline{\mu}\left(\bigcup_{n=1}^{\infty} A_n\right) \le \sum_{n=1}^{\infty} \underline{\mu}(A_n)$; (d) $\underline{\mu}\left(\bigcup_{n=1}^{\infty} A_n\right) \ge \sum_{n=1}^{\infty} \underline{\mu}(A_n)$.

3.2 The measure of a Few Particular Sets

Example 3.21. Let f be a non-negative bounded function on the interval $[a, b]$, and let $A_f = \{(x, y)\colon x \in [a, b],\ 0 \le y \le f(x)\}$ be the set under the graph of f. We show that if f is integrable on $[a, b]$, then the set A_f is Jordan measurable and $\mu(A_f) = \int_a^b f\,dx$.

Indeed, let $F : a = x_0 < x_1 < \ldots < x_n = b$ be an arbitrary partition of the interval $[a, b]$, and let $m_i = \inf\{f(x)\colon x_{i-1} \le x \le x_i\}$ and $M_i = \sup\{f(x)\colon x_{i-1} \le x \le x_i\}$ for every $i = 1, \ldots, n$. Since the rectangles $R_i = [x_{i-1}, x_i] \times [0, m_i]$ $(i = 1, \ldots, n)$ are nonoverlapping and are subsets of A_f, it follows that

$$s_F = \sum_{i=1}^{n} m_i(x_i - x_{i-1}) = \sum_{i=1}^{n} \mu(R_i) \le \underline{\mu}(A_f).$$

On the other hand, the rectangles $T_i = [x_{i-1}, x_i] \times [0, M_i]$ $(i = 1, \ldots, n)$ cover A_f, and thus

$$S_F = \sum_{i=1}^{n} M_i(x_i - x_{i-1}) = \sum_{i=1}^{n} \mu(T_i) \ge \overline{\mu}(A_f).$$

Therefore, we have $s_F \le \underline{\mu}(A_f) \le \overline{\mu}(A_f) \le S_F$ for every partition F. The supremum of the set of numbers s_F, when F runs through all partitions of $[a, b]$ equals $\int_a^b f\,dx$, and thus $\int_a^b f\,dx \le \underline{\mu}(A_f)$. On the other hand, the infimum of the set of numbers S_F also equals $\int_a^b f\,dx$, and thus $\int_a^b f\,dx \ge \overline{\mu}(A_f)$. This implies $\underline{\mu}(A_f) = \overline{\mu}(A_f) = \int_a^b f\,dx$; that is, A_f is measurable, and $\mu(A_f) = \int_a^b f\,dx$.

The converse of the statement above is also true: if A_f is measurable, then f is integrable. This follows from Theorem 3.25 to be proved presently. Indeed, applying the theorem for $p = 2$ and with the coordinates x and y interchanged, we obtain that if A_f is measurable, then the function $x \mapsto \mu((A_f)_x) = f(x)$ is integrable on $[a, b]$.

Example 3.22. **(The Cantor set)** We start with a well-known subset of the real line. Remove from the interval $[0, 1]$ its open middle third, i.e., cut the interval $(1/3, 2/3)$ out. Remove the open middle thirds of the remaining closed intervals, i.e., cut the intervals $(1/9, 2/9)$ and $(7/9, 8/9)$ out. Continue the process infinitely many times, each time cutting out the open middle thirds of the remaining closed intervals. The set of the remaining points is the **Cantor set,** denoted by C.

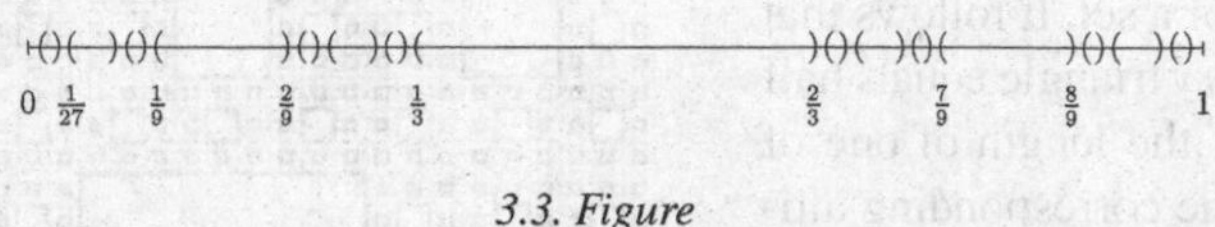

3.3. Figure

The set C is closed, since the $[0, 1] \setminus C$ is a union of open interval, thus $[0, 1] \setminus C$ is open. The elements of C can be described using the ternary representation of numbers. In these representations every digit is one of $0, 1$ and 2. It is easy to see that during the first step of the construction of C we removed those points whose first digit is 1 (in both ternary representations, if there are more than one). During the second step we remove the numbers whose second digit is 1 (in both representations, if there are more than one), and so on. Thus, $x \in [0, 1]$ is in C if and only if all digits of its ternary representation are 0 or 2. (If there are two representations of x then one of them must be of this form.) It follows that the cardinality of C is continuum. Indeed, we can define a one-to-one mapping between the subsets of $\mathbb{N}^+$ and C, if we map every set $H \subset \mathbb{N}^+$ into the ternary representation $0, a_1 a_2 \ldots$ such that $a_i = 2$ if $i \in H$ and $a_i = 0$ if $i \notin H$. Since the cardinality of the system of all subsets of $\mathbb{N}^+$ is continuum, we get that the cardinality of C is continuum as well.

Now we show that C has measure zero. After the nth step of the construction we are left with 2^n closed intervals of length 3^{-n}, whose union covers C. Thus, the outer measure of C is at most $(2/3)^n$. Since $(2/3)^n \to 0$ as $n \to \infty$, it follows that $\overline{\mu}(C) = 0$. Therefore, the cardinality of the Cantor set is continuum, yet it has measure zero. (In other words, C, as a subset of the real line, has the largest possible cardinality and the smallest possible measure.)

Example 3.23. **(Triangles and polygons)** Let H be a triangle whose AB side is parallel to the x axis. As Figure 3.4 shows, if neither of the angles at the vertices A and B is obtuse, then H can be decomposed into three pieces that can be rearranged to form a rectangle, whose base is AB and whose altitude is half the length of the altitude of the triangle.

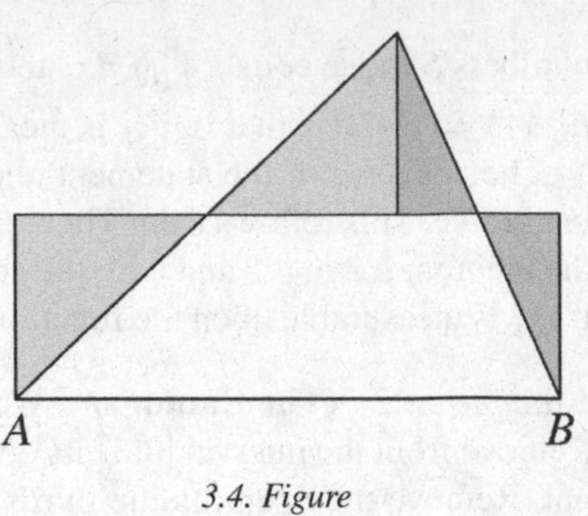

3.4. Figure

The pieces of the triangle are measurable sets (since they are polygons), and the applied transformations (the identity, and two reflections through points) do not change the measurability of these sets, and leave their area unchanged. Therefore, *the area of H is equal to the product of the length of AB and half the length of the altitude with foot AB*. (If one of the angles at A or B is obtuse, then we can represent H as the difference of two triangles with no such obtuse angles, and deduce the same formula for its area.)

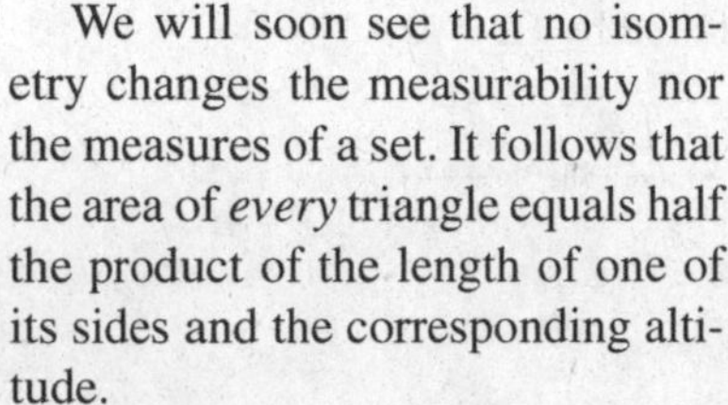

We will soon see that no isometry changes the measurability nor the measures of a set. It follows that the area of *every* triangle equals half the product of the length of one of its sides and the corresponding altitude.

It is also easy to show that every polygon can be cut up into finitely many nonoverlapping triangles (see Exercises 3.30 and 3.31). This implies that the Jordan measure of a polygon is the same as its area as defined in geometry.

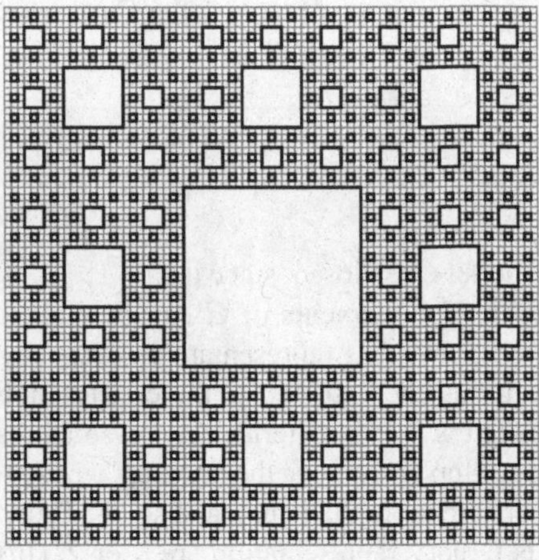

3.5. Figure

Example 3.24. **(Sierpiński's[3] carpet)** The Sierpiński carpet, one of the analogues of the Cantor set in the plane, is defined as follows. Let us start with the closed unit square, i.e., the set $[0, 1] \times [0, 1]$. Divide this set into 9 equal squares, and remove the center open square, i.e., cut out the set $(1/3, 2/3) \times (1/3, 2/3)$. Repeat this step on each of the remaining 8 closed squares, then keep repeating infinitely many times. The set of the remaining points is the **Sierpiński carpet** (Figure 3.5). The outer measure of this carpet is zero, since it can be covered by 8^n squares with area 9^{-n} for every n, and $(8/9)^n \to 0$ if $n \to \infty$.

The most important tool for computing the measure of sets is provided by the next theorem.

If $A \subset \mathbb{R}^p$ and $x \in \mathbb{R}$, then we denote by A^y the set

$$\{(x_1, \ldots, x_{p-1} \in \mathbb{R}^{p-1} \colon (x_1, \ldots, x_{p-1}, y) \in A\},$$

and call it **the section of A with height y**.

---

[3] Wacław Sierpiński (1882–1969), Polish mathematician.

Theorem 3.25. *Let $A \subset R \subset \mathbb{R}^p$, where A is measurable and $R = [a_1, b_1] \times \ldots \times [a_p, b_p]$. The functions $y \mapsto \overline{\mu}_{p-1}(A^y)$ and $y \mapsto \underline{\mu}_{p-1}(A^y)$ are integrable on $[a_p, b_p]$, and*

$$\mu_p(A) = \int_{a_p}^{b_p} \overline{\mu}_{p-1}(A^y)\, dy = \int_{a_p}^{b_p} \underline{\mu}_{p-1}(A^y)\, dy. \tag{3.3}$$

Proof. We will prove the theorem in the case of $p = 2$. The proof in the general case is similar.

Let $A \subset \mathbb{R}^2$ be a measurable set such that $A \subset [a, b] \times [c, d]$. We have to prove that the functions $y \mapsto \overline{\mu}_1(A^y)$ and $x \mapsto \underline{\mu}_1(A^y)$ are integrable in $[a, b]$, and

$$\mu_2(A) = \int_a^b \overline{\mu}_1(A^y)\, dx = \int_a^b \underline{\mu}_1(A^y)\, dx.$$

As we saw in the proof of Lemma 3.15, the role of the coordinates is symmetric; interchanging the x- and y-coordinates does not affect the measure of sets. Therefore, it is enough to prove the following: if $A \subset [a, b] \times [c, d]$ is measurable, then the functions $x \mapsto \overline{\mu}_1(A_x)$ and $x \mapsto \underline{\mu}_1(A_x)$ are integrable in $[a, b]$, and

$$\mu_2(A) = \int_a^b \overline{\mu}_1(A_x)\, dx = \int_a^b \underline{\mu}_1(A_x)\, dx. \tag{3.4}$$

Since $A \subset [a, b] \times [c, d]$, we have $A_x \subset [c, d]$ for all $x \in [a, b]$. It follows that if $x \in [a, b]$ then $\underline{\mu}_1(A_x) \le \overline{\mu}_1(A_x) \le d - c$, so the functions $\underline{\mu}_1(A_x)$ and $\overline{\mu}_1(A_x)$ are bounded in $[a, b]$.

Let $\varepsilon > 0$ be given, and pick rectangles $T_i = [a_i, b_i] \times [c_i, d_i]$ $(i = 1, \ldots, n)$ such that $A \subset \bigcup_{i=1}^n T_i$ and $\sum_{i=1}^n \mu_2(T_i) < \mu_2(A) + \varepsilon$. We may assume that $[a_i, b_i] \subset [a, b]$ for all $i = 1, \ldots, n$. Let

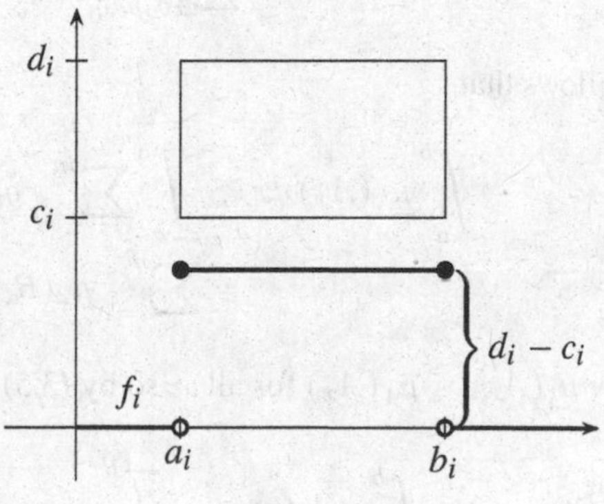

3.6. Figure

$$f_i(x) = \begin{cases} 0, & \text{if } x \notin [a_i, b_i], \\ d_i - c_i, & \text{if } x \in [a_i, b_i] \end{cases} \qquad (i = 1, \ldots, n).$$

Then f_i is integrable in $[a, b]$, and $\int_a^b f_i\, dx = \mu_2(T_i)$. For arbitrary $x \in [a, b]$ the sections A_x are covered by those intervals $[c_i, d_i]$ which correspond to indices i for which $x \in [a_i, b_i]$. Thus, by the definition of the outer measure,

$$\overline{\mu}_1(A_x) \le \sum\nolimits_{x\in[a_i,b_i]} (d_i - c_i) = \sum_{i=1}^{n} f_i(x).$$

It follows that

$$\overline{\int_a^b} \overline{\mu}_1(A_x)\,dx \le \overline{\int_a^b} \sum\nolimits_{i=1}^{n} f_i\,dx = \int_a^b \sum\nolimits_{i=1}^{n} f_i\,dx = \\ = \sum\nolimits_{i=1}^{n} \mu_2(T_i) < \mu_2(A) + \varepsilon. \tag{3.5}$$

Now let $R_i = [p_i,q_i] \times [r_i,s_i]$ $(i = 1,\ldots,m)$ be non-overlapping rectangles such that $A \supset \bigcup_{i=1}^{m} R_i$ and $\sum_{i=1}^{m} \mu_2(R_i) > \mu_2(A) - \varepsilon$. Then $[p_i,q_i] \subset [a,b]$ for all $i = 1,\ldots,m$. Let

$$g_i(x) = \begin{cases} 0, & \text{if} \quad x \notin [p_i,q_i], \\ s_i - r_i, & \text{if} \quad x \in [p_i,q_i] \end{cases} \qquad (i = 1,\ldots,m).$$

Then g_i is integrable in $[a,b]$, and $\int_a^b g_i\,dx = \mu_2(R_i)$. If $x \in [a,b]$ then the section A_x contains all the intervals $[r_i,s_i]$ whose indices i satisfy $x \in [a_i,b_i]$. We can also easily see that if x is distinct from all points p_i, q_i, then these intervals are non-overlapping. Then by the definition of the inner measure

$$\underline{\mu}_1(A_x) \ge \sum\nolimits_{x\in[p_i,q_i]} (s_i - r_i) = \sum\nolimits_{i=1}^{m} g_i(x).$$

It follows that

$$\underline{\int_a^b} \underline{\mu}_1(A_x)\,dx \ge \underline{\int_a^b} \sum\nolimits_{i=1}^{m} g_i\,dx = \int_a^b \sum\nolimits_{i=1}^{m} g_i\,dx = \\ = \sum\nolimits_{i=1}^{m} \mu_2(R_i) > \mu_2(A) - \varepsilon. \tag{3.6}$$

Now $\underline{\mu}_1(A_x) \le \overline{\mu}_1(A_x)$ for all x, so by (3.5) and (3.6) we get that

$$\mu_2(A) - \varepsilon < \underline{\int_a^b} \underline{\mu}_1(A_x)\,dx \le \overline{\int_a^b} \underline{\mu}_1(A_x)\,dx \le \overline{\int_a^b} \overline{\mu}_1(A_x)\,dx < \mu_2(A) + \varepsilon.$$

Since this holds for all ε, $\underline{\int_a^b} \underline{\mu}_1(A_x)\,dx = \overline{\int_a^b} \underline{\mu}_1(A_x)\,dx = \mu_2(A)$, which means that the function $x \mapsto \underline{\mu}_1(A_x)$ is integrable on $[a,b]$ with integral $\mu_2(A)$. We obtain $\int_a^b \overline{\mu}_1(A_x)\,dx = \mu_2(A)$ the same way. □

We call the set $A \subset \mathbb{R}^p$ a **cone with base** H **and vertex** c if $H \subset \mathbb{R}^{p-1}$, $c \in \mathbb{R}^p$, and A is the union of the segments $[x,c]$ $(x \in H \times \{0\})$.

Theorem 3.26. *Let the set $H \subset \mathbb{R}^{p-1}$ be bounded and convex, and let $c = (c_1, \ldots, c_p) \in \mathbb{R}^p$, where $c_p > 0$. Then the cone with base H and vertex c is measurable, and $\mu_p(A) = \frac{1}{p} \cdot \mu_{p-1}(H) \cdot c_p$.*

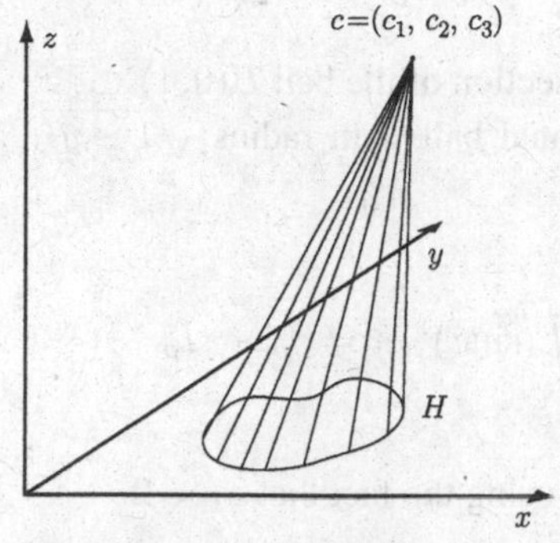

3.7. Figure

Proof. It is easy to see that A is bounded and convex. Thus, by Theorem 3.37 of the appendix, A is measurable.

Let $0 \le y < 1$ and $x \in H$. The vector $(1-y) \cdot (x, 0) + y \cdot c$ is in A, and its last coordinate is $y \cdot c_p$. Conversely, if $(v, y \cdot c_p) \in A$, then $v = (1-y) \cdot x + y \cdot d$, where $x \in H$ and $d = (c_1, \ldots, c_{p-1})$. This implies

$$A^{y \cdot c_p} = \{(1-y) \cdot x + y \cdot d \colon x \in H\}.$$

In other words, the section $A^{y \cdot c_p}$ can be obtained by applying a homothetic transformation with ratio $(1-y)$ to H, then translating the resulting set by the vector $y \cdot d$. By Theorem 3.17 and Lemma 3.20, the measure of this set is $(1-y)^{p-1} \cdot \mu_{p-1}(H)$. Applying Theorem 3.25 yields the measure of A:

$$\mu_p(A) = \int_0^{c_p} \mu_{p-1}(A^u)\, du = \int_0^1 \mu_{p-1}\left(A^{y \cdot c_p}\right) \cdot c_p\, dy =$$

$$= \int_0^1 (1-y)^{p-1} \cdot \mu_{p-1}(H) \cdot c_p\, dt = \mu_{p-1}(H) \cdot c_p \cdot \frac{1}{p}. \quad \square$$

Remark 3.27. Since c_p is the height of the cone, hence the volume of the cone is the product of the area of its base and its height, divided by the dimension. This yields the formula for the area of a triangle in the case of $p = 2$, and the well-known formula of the volume of cones of the three dimensional space in the case of $p = 3$.

Our next aim is to compute the measure of balls. The unit balls of $\mathbb{R}^p$ are translations of each other, thus, by Theorem 3.17, their measure is the same. Let us denote this measure by γ_p. A ball with radius r can be obtained by applying a homothetic transformation with ratio r to the unit ball, thus, by Lemma 3.20, $\mu_p(B(x, r)) = \gamma_p \cdot r^p$ for every $x \in \mathbb{R}^p$ and $r > 0$. It is enough to find the constants γ_p.

Theorem 3.28.

(i) $\gamma_{2k} = \frac{\pi^k}{k!}$ *for every positive integer k, and*

(ii) $\gamma_{2k+1} = \frac{\pi^k \cdot 2^{2k+1} \cdot k!}{(2k+1)!}$ *for every non-negative integer k.*

Proof. Let $I_n = \int_0^\pi \sin^n x\,dx$. It is well-known that

$$I_{2k} = \pi \cdot \frac{1}{2} \cdot \frac{3}{4} \cdot \ldots \cdot \frac{2k-1}{2k} \quad \text{and} \quad I_{2k+1} = 2 \cdot \frac{2}{3} \cdot \frac{4}{5} \cdot \ldots \cdot \frac{2k}{2k+1}.$$

(See [7, Theorem 15.12].) If $-1 \le y \le 1$, then the section of the ball $B(0,1) \subset \mathbb{R}^p$ with height y, i.e., $B(0,1)^y$ is a $(p-1)$-dimensional ball with radius $\sqrt{1-y^2}$. Then, by Theorem 3.25,

$$\gamma_p = \int_{-1}^{1} \gamma_{p-1} \cdot \left(\sqrt{1-y^2}\right)^{p-1} dy = \gamma_{p-1} \int_0^\pi (\sin t)^p\,dt = \gamma_{p-1} \cdot I_p.$$

The statement of the theorem follows by induction, using the fact that $\gamma_1 = 2$. □

Remarks 3.29. **1.** It is easy to see that the sequence I_n is strictly decreasing, and $I_5 = 16/15 > 1 > I_6 = 10\pi/32$. It follows that

$$\gamma_1 < \gamma_2 < \gamma_3 < \gamma_4 < \gamma_5 > \gamma_6 > \gamma_7 > \ldots,$$

thus the volume of the 5-dimensional unit ball is the largest.

2. The sequence γ_p converges to zero at a rate faster than exponential. By applying **Stirling's[4] formula**[5] one can check that

$$\gamma_p \sim \left(\frac{2\pi e}{p}\right)^{p/2} \cdot (\pi p)^{-1/2}$$

as $p \to \infty$. This is surprising, since the smallest box containing the unit ball has volume 2^p, which converges to infinity at an exponential rate. This phenomenon can be formulated as follows: *in high dimensions the ball only covers a very small part of the box that contains it.*

Next we compute the measure of parallelepipeds.

Definition 3.30. Let $x_1, \ldots, x_k \in \mathbb{R}^p$ be vectors ($k \le p$). We call the set

$$\{\lambda_1 x_1 + \ldots + \lambda_k x_k \colon 0 \le \lambda_1, \ldots, \lambda_k \le 1\}$$

the *parallelepiped* spanned by the vectors x_i, and use the notation $P(x_1, \ldots, x_k)$. If $k < p$ or $k = p$ and the vectors $x_1, \ldots, x_p$ are linearly dependent, then we say that the parallelepiped $P(x_1, \ldots, x_k)$ is *degenerated.* If the vectors $x_1, \ldots, x_p$ are linearly independent, then the parallelepiped $P(x_1, \ldots, x_p)$ is *non-degenerated.*

---

[4] James Stirling (1692–1770), Scottish mathematician.

[5] Stirling's formula is the statement $n! \sim (n/e)^n \cdot \sqrt{2\pi n}$ $(n \to \infty)$. See [7, Theorem 15.15].

Since the parallelepipeds are bounded and convex sets, they are measurable by Theorem 3.37 of the appendix. The next theorem gives the geometric interpretation of the determinant.

Theorem 3.31. *If $x_i = (a_{i,1}, \dots, a_{i,p})$ $(i = 1, \dots, p)$, then the volume of the parallelepiped $P(x_1, \dots, x_p)$ is the absolute value of the determinant* $\begin{vmatrix} a_{1,1} & a_{1,2} & \dots & a_{1,p} \\ \vdots & \vdots & \dots & \vdots \\ a_{p,1} & a_{p,2} & \dots & a_{p,p} \end{vmatrix}$.

Proof. Let $D(x_1, \dots, x_p)$ denote the value of the determinant. If $D(x_1, \dots, x_p) = 0$, then the parallelepiped is degenerated. It is easy to see that every degenerated parallelepiped can be covered by a hyperplane. Thus, by Lemma 3.15, every degenerated parallelepiped has measure zero. Therefore, the statement of the theorem holds in this case.

Thus we may assume that $D(x_1, \dots, x_p) \neq 0$, i.e., the vectors $x_1, \dots, x_p$ are linearly independent. The statement of the theorem is obvious for $p = 1$, thus we may also assume that $p \geq 2$.

We know that the value of $D(x_1, \dots, x_p)$ does not change if we add a constant multiple of one of its rows to another row. Now we show that these operations also leave the volume of parallelepiped $P(x_1, \dots, x_p)$ unchanged. What we prove is that the parallelepiped $P(x_1 + \lambda x_2, x_2, \dots, x_p)$ can decomposed into pieces which can be reassembled to give the original parallelepiped $P(x_1, \dots, x_p)$, implying the equality of the two volumes. We denote by $A + a$ the translation of the set $A \subset \mathbb{R}^p$ by a vector $a \in \mathbb{R}^p$. That is, $A + a = \{x + a \colon x \in A\}$.

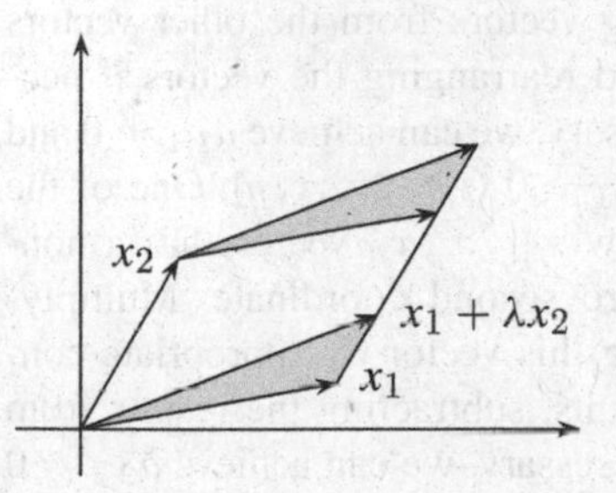

3.8. Figure

First consider the case of $p = 2$. For every $u, v \in \mathbb{R}^2$ let $T(u, v)$ denote the triangle with vertices 0, u and v. Let the vectors $x_1, x_2 \in \mathbb{R}^2$ be linearly independent, and let $\lambda > 0$. It is easy to see that

$$P(x_1 + \lambda x_2, x_2) \cup T(x_1, x_1 + \lambda x_2) = \\ = P(x_1, x_2) \cup (T(x_1, x_1 + \lambda x_2) + x_2), \tag{3.7}$$

with non-overlapping sets on both sides. Since

$$\mu\left(T(x_1, x_1 + \lambda x_2) + x_2\right) = \mu\left(T(x_1, x_1 + \lambda x_2)\right),$$

hence $\mu\left(P(x_1 + \lambda x_2, x_2)\right) = \mu\left(P(x_1, x_2)\right)$, and we are done.

For the $p > 2$ case, let us use the notation $A + B = \{x + y \colon x \in A,\ y \in B\}$. Let $x_1, \dots, x_p \in \mathbb{R}^p$ be linearly independent vectors, and let $\lambda > 0$. It is easy to see that

$$P(x_1,\ldots,x_p)=P(x_1,x_2)+P(x_3,\ldots,x_p),$$

furthermore, each element of $P(x_1,\ldots,x_p)$ can by uniquely written as $x+y$, where $x\in P(x_1,x_2)$ and $y\in P(x_3,\ldots,x_p)$. Thus, by (3.7) we obtain

$$P(x_1+\lambda x_2,x_2,x_3,\ldots,x_p)\cup[T(x_1,x_1+\lambda x_2)+P(x_3,\ldots,x_p)]=$$
$$=P(x_1,\ldots,x_p)\cup([T(x_1,x_1+\lambda x_2)+P(x_3,\ldots,x_p)]+x_2),$$

with nonoverlapping convex polyhedra on both sides. Since the volume is translation-invariant, hence the measures of the second sets of both sides are equal, implying

$$\mu\left(P(x_1+\lambda x_2,x_2,x_3,\ldots,x_p)\right)=\mu\left(P(x_1,x_2,x_3,\ldots,x_p)\right). \tag{3.8}$$

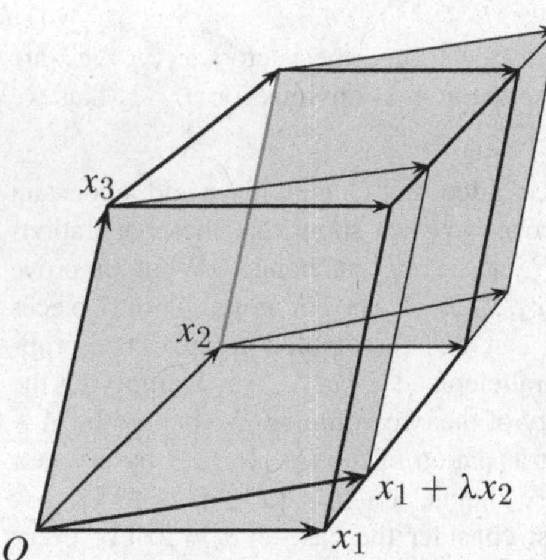

3.9. Figure

Replacing x_i by $x_i+\lambda x_j$ for $\lambda>0$ yields a similar results. If $\lambda<0$, then plugging $\lambda\mapsto-\lambda$, $x_1\mapsto x_1+\lambda x_2$ into (3.8) gives us the desired result.

One of the vectors $x_1,\ldots,x_p$ has a non-zero first coordinate. Multiplying this vector by appropriate constants, subtracting the resulting vectors from the other vectors and rearranging the vectors if necessary, we can achieve $a_{1,1}\neq 0$ and $a_{i,1}=0$ $(i=2,\ldots,p)$. One of the new $x_2,\ldots,x_p$ vectors has a non-zero second coordinate. Multiplying this vector by appropriate constants, subtracting the results from the other vectors and rearranging the vectors if necessary, we can achieve $a_{2,2}\neq 0$ and $a_{i,2}=0$ $(i\neq 2)$. Repeating this process for each of the other coordinates results in a system that satisfies $a_{i,j}\neq 0 \iff i=j$. This system satisfies the statement of the theorem, since the box $P(x_1,\ldots,x_p)$ has volume $\prod_{j=1}^{p}|a_{j,j}|$, and the value of the determinant is $\prod_{j=1}^{p}a_{j,j}$. Since neither the value of the determinant, nor the measure of the parallelepiped were changed by our operations, the theorem is proved. □

Exercises

3.25. Let C be the Cantor set. Show, that $\{x+y\colon x,y\in C\}=[0,2]$ and $\{x-y\colon x,y\in C\}=[-1,1]$.

3.26. Are there points of the Cantor set (apart from 0 and 1) with finite decimal expansion? (S)

3.27. We define a function $f\colon C \to [0,1]$ as follows. If $x \in C$ and the ternary representation of x is $0, a_1 a_2 \ldots$, where $a_i = 0$ or $a_i = 2$ for every i, then we define

$$f(x) = \sum_{i=1}^{\infty} \frac{a_i}{2^{i+1}}.$$

In other words, the value of $f(x)$ is obtained by dividing the digits of the ternary representation of x by 2, and reading the result as a binary representation. Show that

(a) the function f maps C onto $[0,1]$;
(b) the function f is monotone increasing on C;
(c) if $a, b \in C$, $a < b$, and $(a,b) \cap C = \emptyset$, then $f(a) = f(b)$.

3.28. Extend f to the interval $[0,1]$ such that if $a, b \in C$, $a < b$, and $(a,b) \cap C = \emptyset$, then let f be equal to the constant $f(a) = f(b)$ on the interval (a,b). Denote this new function by f, also. The function f defined above is the **Cantor function.** Show that

(a) the function f is monotone increasing on $[0,1]$;
(b) the function f is continuous on $[0,1]$. (H)

3.29. Let D_7 denote the set of numbers $x \in [0,1]$ whose decimal (in the scale of 10) do not have a digit 7. Show that D_7 is closed and has measure zero.

3.30. Show that every polygon can be decomposed into finitely many nonoverlapping triangles. (H)

3.31. Show that every polygon can be decomposed into finitely many nonoverlapping triangles with the added condition that the set of vertices of each triangle of the decomposition is a subsets of the set of vertices of the polygon. (∗)

3.32. Let $H \subset \mathbb{R}^p$ be convex, and let $c \in \mathbb{R}^p$. Show that the union of the segments $[x,c]$ $(x \in H)$ is convex.

3.33. Let the set $H \subset \mathbb{R}^{p-1}$ be measurable (not necessarily convex), and let $c = (c_1, \ldots, c_p) \in \mathbb{R}^p$, with $c_p > 0$. Show that the cone A with base H and vertex c is measurable, and $\mu_p(A) = \frac{1}{p} \cdot \mu_{p-1}(H) \cdot c_p$. (H)

3.3 Linear Transformations and the Jordan Measure

Our next aim is to prove that the measure is not only translation-invariant, but is invariant under all isometries. We prove this in two steps. First we show that every isometry can be written as the composition of a special linear transformation and a translation, then we figure out how linear transformations change the measure of a set. The isometry invariance of the Jordan measure will follow from these two steps.

First we summarize the basics on isometries. We say that a mapping $f\colon \mathbb{R}^p \to \mathbb{R}^p$ is **distance preserving**, if $|f(x) - f(y)| = |x - y|$ for every $x, y \in \mathbb{R}^p$. The mapping $f\colon \mathbb{R}^p \to \mathbb{R}^p$ is an **isometry,** if it is a distance preserving bijection of $\mathbb{R}^p$ onto itself. We denote the set of all isometries of $\mathbb{R}^p$ by G_p. (Therefore, a mapping f is an isometry if it is distance preserving, and its range is the whole space $\mathbb{R}^p$. We will show presently that every distance preserving mapping is an isometry.) The sets A and B are called **congruent** if there is an isometry f such that $B = f(A)$.

It is easy to see that the inverse of an isometry is also an isometry, and the composition of two isometries is also an isometry. (In other words, G_p forms a group with respect to the composition operation.) We denote the set of all translations of $\mathbb{R}^p$ (i.e., the mappings $x \mapsto x + c$ $(x \in \mathbb{R}^p)$) by T_p. It is clear that every translation is an isometry; i.e., $T_p \subset G_p$.

We say that a linear transformation A mapping $\mathbb{R}^p$ into itself is **orthogonal,** if it preserves the scalar product, i.e., if $\langle Ax, Ay\rangle = \langle x, y\rangle$ for every $x, y \in \mathbb{R}^p$. Since $\langle Ax, Ay\rangle = \langle x, A^T Ay\rangle$ (where A^T denotes the transpose of A), hence A is orthogonal if and only if $A^T A = I$, were I is the identity transformation. Therefore, A is orthogonal if and only if the column vectors of the matrix of A are **orthonormal**; that is, they are pairwise orthogonal unit vectors. The conditions $A^T A = I$ and $AA^T = I$ are equivalent, hence A is orthogonal if and only if the row vectors of the matrix of A are orthonormal. We denote the set of all the orthogonal linear transformations of $\mathbb{R}^p$ by O_p.

If $A \in O_p$, then $|Ax|^2 = \langle Ax, Ax\rangle = \langle x, x\rangle = |x|^2$, thus $|Ax| = |x|$ for every $x \in \mathbb{R}^p$. Consequently, $|Ax - Ay| = |A(x - y)| = |x - y|$ for every $x, y \in \mathbb{R}^p$. Thus every orthogonal linear transformation is an isometry: $O_p \subset G_p$.

Lemma 3.32. *Let $a, b \in \mathbb{R}^p$ be distinct points. Then the set $\{x \in \mathbb{R}^p\colon |x - a| = |x - b|\}$ is a hyperplane (called the* **orthogonal bisector hyperplane of the points** *a* **and** *b.*

Proof. For every $x \in \mathbb{R}^p$ we have

$$\begin{aligned}
|x-a| = |x-b| &\iff |x-a|^2 = |x-b|^2 \iff \\
&\iff \langle x-a, x-a\rangle = \langle x-b, x-b\rangle \iff \\
&\iff |x|^2 - 2\langle a, x\rangle + |a|^2 = |x|^2 - 2\langle b, x\rangle + |b|^2 \iff \\
&\iff \langle 2(b-a), x\rangle = |b|^2 - |a|^2 \iff \\
&\iff \sum_{i=1}^{p} c_i x_i = |b|^2 - |a|^2,
\end{aligned}$$

where $2(b - a) = (c_1, \ldots, c_p)$. □

Lemma 3.33. *Let $f\colon \mathbb{R}^p \to \mathbb{R}^p$ be distance preserving, and let $g \in G_p$. If the values of f and g are equal at $p+1$ points of general position (i.e., if the points cannot be covered by a single hyperplane), then $f \equiv g$.*

Proof. Suppose that $f(x) \neq g(x)$. If $f(y) = g(y)$, then $|y-x| = |f(y) - f(x)| = |g(y) - f(x)| = |y - (g^{-1} \circ f)(x)|$. This implies that the set $\{y\colon f(y) = g(y)\}$ is a subset of the orthogonal bisector hyperplane of the points x and $(g^{-1} \circ f)(x)$, which contradicts our assumption. □

Theorem 3.34.

(i) *The mapping $f\colon \mathbb{R}^p \to \mathbb{R}^p$ is an orthogonal linear transformation if and only if $f(0) = 0$ and f is distance preserving.*

(ii) *Every distance preserving map $f\colon \mathbb{R}^p \to \mathbb{R}^p$ is an isometry.*

(iii) $G_p = \{f \circ g\colon g \in O_p,\ f \in T_p\}$.

Proof. (i) We have already proved that an orthogonal linear transformation f is an isometry (and thus, distance preserving). Clearly, $f(0) = 0$ also holds for every linear map. Now we show that if $f\colon \mathbb{R}^p \to \mathbb{R}^p$ is distance preserving and $f(0) = 0$, then $f \in O_p$. Let $e_i = (0,\ldots,0,\underset{i}{1},0,\ldots,0)$ and $v_i = f(e_i)$ for every $i = 1,\ldots,p$. Let g denote the linear transformation whose matrix has the column vectors $v_1,\ldots,v_p$.

These column vectors are orthonormal. Indeed, on the one hand

$$|v_i - 0| = |f(e_i) - f(0)| = |e_i - 0| = 1$$

for every i. On the other hand, for $i \neq j$ we have

$$|v_i - v_j| = |f(e_i) - f(e_j)| = |e_i - e_j| = \sqrt{2},$$

and $-2\langle v_i, v_j\rangle = |v_i - v_j|^2 - |v_i|^2 - |v_j|^2 = 0$. Thus $g \in O_p$.

Now, the distance preserving map f and the isometry g are equal at the points $0, e_1,\ldots,e_p$. It is easy to see that these points are of general position and thus, by Lemma 3.33, $f \equiv g$.

(ii) Let $h\colon \mathbb{R}^p \to \mathbb{R}^p$ be distance preserving. Let f be the translation by $h(0)$, and let $g(x) = h(x) - h(0)$ $(x \in \mathbb{R}^p)$. Then g is distance preserving and $g(0) = 0$, thus $g \in O_p$ by (i). Therefore, g is an isometry and, since $h = f \circ g$, we find that h is also an isometry. This also proves (iii). □

Let $\Lambda\colon \mathbb{R}^p \to \mathbb{R}^p$ be a linear transformation, and let the determinant of Λ be $\det \Lambda$. The following theorem gives the measure theoretic meaning of this determinant.

Theorem 3.35. *For every bounded $A \subset \mathbb{R}^p$ and for every linear transformation $\Lambda\colon \mathbb{R}^p \to \mathbb{R}^p$ we have $\underline{\mu}(\Lambda(A)) = |\det \Lambda| \cdot \underline{\mu}(A)$ and $\overline{\mu}(\Lambda(A)) = |\det \Lambda| \cdot \overline{\mu}(A)$. If A is measurable, then $\Lambda(A)$ is also measurable, and $\mu(\Lambda(A)) = |\det \Lambda| \cdot \mu(A)$.*

Proof. First we compute the measure of $\Lambda(R)$, where $R = [0, a_1] \times \ldots \times [0, a_p]$. Obviously, $\Lambda(R)$ is the parallelepiped spanned by the vectors $\Lambda(a_1 e_1), \ldots, \Lambda(a_p e_p)$, where $e_i = (0, \ldots, 0, \underset{i}{1}, 0, \ldots, 0)$ $(i = 1, \ldots, p)$. Since the determinant of a matrix with row vectors $\Lambda(a_i e_i) = a_i \Lambda(e_i)$ is $a_1 \cdots a_p \cdot \det \Lambda$, Theorem 3.31 gives

$$\mu(\Lambda(R)) = |a_1 \cdots a_p \cdot \det \Lambda| = |\det \Lambda| \cdot \mu(R).$$

Then, using the translation invariance of the Jordan measure and the linearity of Λ we get that $\mu(\Lambda(R)) = |\det \Lambda| \cdot \mu(R)$ for every box R.

Let $\det \Lambda$ be denoted by D, and let A be a bounded set. If $D = 0$, then the range of Λ is a proper linear subspace of $\mathbb{R}^p$. Every such subspace can be covered by a $(p-1)$ dimensional subspace, that is, by a hyperplane. Thus $\Lambda(A) \subset \Lambda(\mathbb{R}^p)$ is part of a hyperplane. By Lemma 3.15 we obtain that $\underline{\mu}(\Lambda(A)) = 0 = |D| \cdot \underline{\mu}(A)$ and $\overline{\mu}(\Lambda(A)) = 0 = |D| \cdot \overline{\mu}(A)$.

Suppose now $D \neq 0$. Then Λ is invertible. Let $\{R_1, \ldots, R_n\}$ be a system of nonoverlapping boxes in A. The parallelepipeds $\Lambda(R_i)$ $(i = 1, \ldots, n)$ are also nonoverlapping[6], and are subsets of $\Lambda(A)$. Thus

$$\underline{\mu}(\Lambda(A)) \geq \sum_{i=1}^{n} \mu(\Lambda(R_i)) = \sum_{i=1}^{n} |D| \cdot \mu(R_i) = |D| \cdot \sum_{i=1}^{n} \mu(R_i). \tag{3.9}$$

Since $\underline{\mu}(A)$ is the supremum of the set of numbers $\sum_{i=1}^{n} \mu(R_i)$, hence (3.9) implies $\underline{\mu}(\Lambda(A)) \geq |D| \cdot \underline{\mu}(A)$.

The linear transformation Λ^{-1} maps the set $\Lambda(A)$ into the set A, thus switching $\Lambda(A)$ and A in the previous argument gives $\underline{\mu}(A) \geq |\det \Lambda^{-1}| \cdot \underline{\mu}(\Lambda(A)) = |D^{-1}| \cdot \underline{\mu}(\Lambda(A))$. Thus, $\underline{\mu}(A) \cdot |D| \geq \underline{\mu}(\Lambda(A))$, and $\underline{\mu}(\Lambda(A)) = |D| \cdot \underline{\mu}(A)$.

We get $\overline{\mu}(\Lambda(A)) = |D| \cdot \overline{\mu}(A)$ by a similar argument.

If A is measurable, then A is bounded and $\mu(A) = \underline{\mu}(A) = \overline{\mu}(A)$. In this case, $\Lambda(A)$ is also bounded, and $\underline{\mu}(\Lambda(A)) = |D| \cdot \underline{\mu}(A) = |D| \cdot \overline{\mu}(A) = \overline{\mu}(\Lambda(A))$. It follows that $\Lambda(A)$ is measurable and $\mu(\Lambda(A)) = |D| \cdot \mu(A)$. □

Theorem 3.36. *Let the bounded sets $A, B \subset \mathbb{R}^p$ be congruent. Then $\overline{\mu}(A) = \overline{\mu}(B)$ and $\underline{\mu}(A) = \underline{\mu}(B)$.*

If A is measurable, then B is also measurable and $\mu(A) = \mu(B)$.

Proof. We know that translations do not change the outer and inner measures of sets. By part (iii) of Theorem 3.34, it is enough to prove the statement for orthogonal linear transformations.

If $\Lambda \in O_p$, then $(\det \Lambda)^2 = \det(\Lambda^T \Lambda) = \det I = 1$, thus $\det \Lambda = \pm 1$. Therefore, the statement follows from the previous theorem. □

---

[6] This needs some consideration, see Exercise 3.37.

Exercises

3.34. Show that if $A \subset \mathbb{R}^p$ is bounded, then among the open balls contained by A there is one with maximum radius.

3.35. Let $A \subset \mathbb{R}^p$ be a bounded set with a non-empty interior. Let $B_1 = B(a_1, r_1)$ be (one of) the open ball(s) in A with maximum radius. Suppose we have already chosen the balls $B_1, \ldots, B_{n-1}$, and that the interior of $A \setminus (B_1 \cup \ldots \cup B_{n-1})$ is non-empty. Then let $B_n = B(a_n, r_n)$ be (one of) the open ball(s) in $A \setminus (B_1 \cup \ldots \cup B_{n-1})$ with maximum radius. Suppose we chose infinitely many balls during this process.
Prove the following statements.

(a) $\operatorname{int} A \subset \bigcup_{n=1}^{N} B(a_n, r_n) \cup \bigcup_{n=N+1}^{\infty} B(a_n, 2 \cdot r_n)$ for every $N \geq 1$. (*)
(b) $\underline{\mu}\left(\bigcup_{n=1}^{\infty} B_n\right) = \underline{\mu}(A)$.
(c) If A is measurable, then $\overline{\mu}\left(\bigcup_{n=1}^{\infty} B_n\right) = \mu(A)$.

3.36. Using the results of the previous exercise, give a new proof of the isometry invariance of the inner and outer measure of sets.

3.37. Let $A, B \subset \mathbb{R}^p$ be nonoverlapping sets. Show that $\Lambda(A)$ and $\Lambda(B)$ are also nonoverlapping for every linear transformation $\Lambda\colon \mathbb{R}^p \to \mathbb{R}^p$. (S)

3.4 Appendix: The Measurability of Bounded Convex Sets

Our aim is to prove the following theorem.

Theorem 3.37. *Every bounded convex set is measurable.*

Lemma 3.38. *If $F \subset G \subset \mathbb{R}^p$, where F is closed and G is bounded and open, then $\overline{\mu}(F) \leq \underline{\mu}(G)$.*

Proof. For every $x \in F$ we have $x \in G$, thus there exists $r(x) > 0$ such that $B(x, r(x)) \subset G$. The open balls $B(x, r(x))$ cover F. Then, by Borel's theorem (Theorem 1.31), finitely many of these balls also cover F. Let H be the union of these (finitely many) balls. The set H is measurable and $F \subset H \subset G$, hence $\overline{\mu}(F) \leq \overline{\mu}(H) = \underline{\mu}(H) \leq \underline{\mu}(G)$. □

Remark 3.39. In general $F \subset G$ does not imply $\overline{\mu}(F) \leq \underline{\mu}(G)$. E.g., if $F = G$ is bounded and non-measurable, then $\overline{\mu}(F) > \underline{\mu}(G)$.

Lemma 3.40. *Let $A \subset \mathbb{R}^p$ be convex. If $a \in \operatorname{cl} A$ and $b \in \operatorname{int} A$, then the points of segment $[a, b]$ are in $\operatorname{int} A$ (with the possible exception of a).*

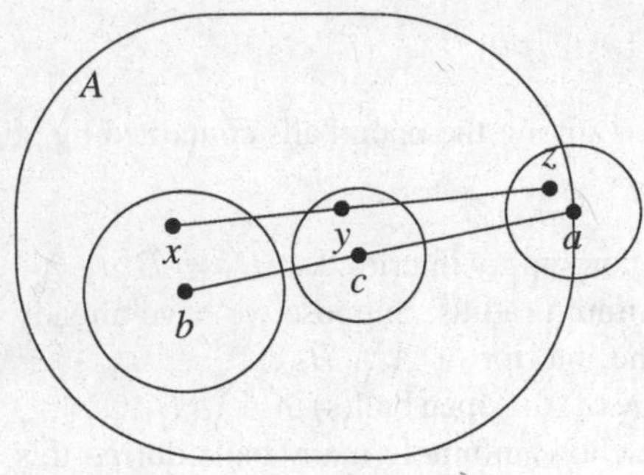

3.10. Figure

Proof. Let $c \in [a,b] \setminus \{a\}$ be arbitrarily. Then $c = (1-t)a + tb$, where $0 < t \le 1$. Since $b \in \operatorname{int} A$, there exists $r > 0$ such that $B(b,r) \subset A$. Let $\delta = tr/2$; we show that $B(c,\delta) \subset A$, which will prove that $c \in \operatorname{int} A$.

Let $y \in B(c,\delta)$ be arbitrary; we show that $y \in A$. Since $a \in \operatorname{cl} A$, the set $B(a,\delta) \cap A$ is non-empty. Pick a point $z \in B(a,\delta) \cap A$. Clearly, there exists a unique point x such that $y = (1-t)z + tx$. We show that $x \in B(b,r)$, i.e., $|x-b| < r$.

Indeed, $tx = y - (1-t)z$ and $tb = c - (1-t)a$. Subtracting the two equations from each other and taking the absolute value of both sides we find

$$t|x-b| = |tx - tb| = |y - c - ((1-t)z - (1-t)a)| \le$$
$$* \le |y-c| + (1-t)|z-a| < \delta + (1-t)\delta \le 2\delta = tr,$$

thus $|x-b| < r$, and $x \in B(b,r) \subset A$. Since $z \in A$ and A is convex, it follows that $y = tx + (1-t)z \in A$. □

Lemma 3.41. *Let $A \subset \mathbb{R}^p$ be convex. If* $\operatorname{int} A = \emptyset$*, then A can be covered by a hyperplane.*

Proof. We may assume $0 \in A$, since otherwise we can take an appropriate translated copy of A. Let V be the linear subspace of $\mathbb{R}^p$ generated by A. It is enough to show that $V \ne \mathbb{R}^p$, since in this case V is the subset of a $(p-1)$-dimensional linear subspace, which is a hyperplane containing A.

Suppose $V = \mathbb{R}^p$. Then A is a generating system in $\mathbb{R}^p$, and then it contains the linearly independent vectors $u_1, \ldots, u_p$. Since A is convex and $0 \in A$, hence $t_1u_1 + \ldots + t_pu_p \in A$ for $t_1, \ldots, t_n \ge 0$ and $t_1 + \ldots + t_p \le 1$. It follows that A contains the parallelepiped $P = P(u_1/p, \ldots, u_p/p)$. The parallelepiped P is measurable with a positive measure, thus its interior is non-empty. Since $\operatorname{int} P \subset \operatorname{int} A$, hence the interior of A is also non-empty, which contradicts the assumption. □

Proof of Theorem 3.37. Let $A \subset \mathbb{R}^p$ be bounded and convex. We distinguish between two cases.

I: $\operatorname{int} A \ne \emptyset$. We may assume that $0 \in \operatorname{int} A$, otherwise we could take an appropriate translated copy of A. For every $t \in \mathbb{R}$ and $x \in \mathbb{R}^p$, let $\phi_t(x) = t \cdot x$. Now $\phi_t(\operatorname{cl} A) \subset \operatorname{int} A$ holds for every $0 < t < 1$. Indeed, if $x \in \operatorname{cl} A$, then $tx = tx + (1-t)0 \in \operatorname{int} A$, since $0 \in \operatorname{int} A$, and we can apply Lemma 3.40.

The mapping ϕ_t is continuous, and the set $\operatorname{cl} A$ is bounded and closed. Then, by Theorem 2.7, the set $\phi_t(\operatorname{cl} A)$ is also bounded and closed. Thus, by Lemma 3.38, we have $\overline{\mu}(\phi_t(\operatorname{cl} A)) \leq \underline{\mu}(\operatorname{int} A)$. Since $A \subset \operatorname{cl} A$ and $\operatorname{int} A \subset A$, it follows that $\overline{\mu}(\phi_t(A)) \leq \underline{\mu}(A)$. Applying Lemma 3.20 we find that $t^p \cdot \overline{\mu}(A) \leq \underline{\mu}(A)$. This holds for every $0 < t < 1$, thus $\overline{\mu}(A) \leq \underline{\mu}(A)$, i.e., A is measurable.

II: $\operatorname{int} A = \emptyset$. Then, by Lemma 3.41, A is the subset of a hyperplane. By Lemma 3.15 it follows that the set A has measure zero and, consequently, it is measurable. □

Chapter 4
Integrals of Multivariable Functions I

4.1 The Definition of the Multivariable Integral

The concept of the integral of a multivariable function arose as an attempt to solve some problems in mathematics, physics, and in science in general, similarly to the case of the integral of a single-variable function. We give an example from physics.

Finding the weight of an object via its density. Given is a rectangular plate made of an inhomogeneous material, whose density is known everywhere. That is, we know the ratio of its weight and its area in a small neighborhood of each of its points. Our job is to find the weight of the plate.

We assume that the weight is a monotone function of the density, which means that if we change the material of the plate in such a way that the density does not decrease at any of its points, then the weight of the whole plate will not decrease either.

Let the plate be expressed in coordinates as $R = [a, b] \times [c, d]$, and let $f(x, y)$ be its density at the point $(x, y) \in R$. Let $a = x_0 < x_1 < \ldots < x_n = b$ and $c = y_0 < y_1 < \ldots < y_k = d$ be arbitrary partitions, and let ρ_{ij} denote the weight of the region $R_{ij} = [x_{i-1}, x_i] \times [y_{j-1}, y_j]$ for every $1 \le i \le n$ and $1 \le j \le k$. Let

$$m_{ij} = \inf\{f(x, y) \colon (x, y) \in R_{ij}\} \quad \text{and} \quad M_{ij} = \sup\{f(x, y) \colon (x, y) \in R_{ij}\}.$$

If the density of the region R_{ij} was m_{ij} at every point of R_{ij}, the weight of R_{ij} would be $m_{ij} \cdot \mu(R_{ij})$ (by the definition of density). The monotonicity condition implies that $\rho_{ij} \ge m_{ij} \cdot \mu(R_{ij})$. Similarly, we get that $\rho_{ij} \le M_{ij} \cdot \mu(R_{ij})$ for every $1 \le i \le n$ and $1 \le j \le m$. Thus, if the weight of R is ρ, then we have

$$\sum_{i=1}^{n} \sum_{j=1}^{k} m_{ij} \cdot \mu(R_{ij}) \le \rho \le \sum_{i=1}^{n} \sum_{j=1}^{k} M_{ij} \cdot \mu(R_{ij}).$$

M. Laczkovich and V.T. Sós, *Real Analysis*, Undergraduate Texts in Mathematics, https://doi.org/10.1007/978-1-4939-7369-9_4

These inequalities hold for every partition $a = x_0 < x_1 < \ldots < x_n = b$ and $c = y_0 < y_1 < \ldots < y_k = d$. If we are lucky, only one number satisfies these inequalities, and that will be the value of the weight.

The argument above is similar to the reasoning that led to the concept of the Riemann[1] integral of a single-variable function. Accordingly, the definition of the integral of a multivariable function is obtained as an immediate generalization of the Riemann integral of a single-variable function.

Definition 4.1. The *partition* of a rectangle $R = [a,b] \times [c,d]$ is a system of rectangles $R_{ij} = [x_{i-1}, x_i] \times [y_{j-1}, y_j]$, where $a = x_0 < x_1 < \ldots < x_n = b$ and $c = y_0 < y_1 < \ldots < y_k = d$. We call the points x_i and y_j the *base points*, and the rectangles R_{ij} the *division rectangles* of the partition.

Let $f \colon R \to \mathbb{R}$ be a bounded function and let

$$m_{ij} = \inf\{f(x,y) \colon (x,y) \in R_{ij}\}$$

and $\quad M_{ij} = \sup\{f(x,y) \colon (x,y) \in R_{ij}\}$

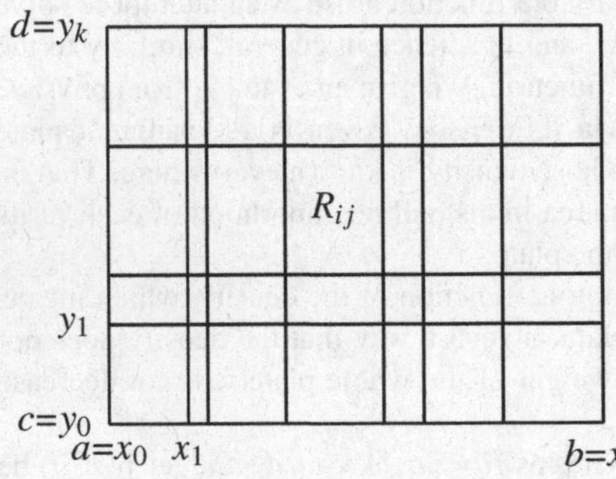

4.1. Figure

for every $1 \le i \le n$ and $1 \le j \le k$. We call the sums

$$s_F(f) = \sum_{i=1}^{n} \sum_{j=1}^{k} m_{ij} \cdot \mu(R_{ij})$$

and

$$S_F(f) = \sum_{i=1}^{n} \sum_{j=1}^{k} M_{ij} \cdot \mu(R_{ij})$$

the *lower* and *upper sums* of the function f with partition $F = \{R_{ij}\}$. If f is given, and it is obvious which function we are talking about, we will use the notation s_F and S_F instead of $s_F(f)$ and $S_F(f)$, respectively.

Similarly to the case of single-variable Riemann integration, we say that a function is integrable if there exists only one number between its lower and upper sums. First we show that *for every bounded function f, there exists a number between every lower and every upper sum of f.*

The proof goes similarly to the single-variable case. We say that a partition F' is a **refinement** of the partition F if every base point of F is also a base point of F'.

---

[1] Georg Friedrich Bernhard Riemann (1826–1866), German mathematician.

Lemma 4.2. *Let $f: R \to \mathbb{R}$ be bounded, and let the partition F' be a refinement of the partition F. Then we have $s_F \le s_{F'}$ and $S_F \ge S_{F'}$.*

Proof. First we show that if F' is obtained by adding one new base point to F, then $s_{F'} \ge s_F$. This follows from the fact that if a division rectangle R_{ij} of F is cut into two rectangles by the new partition, then the infimum of f is at least $m_{ij} = \inf\{f(x): x \in R_{ij}\}$ on both of the new rectangles, and thus the corresponding contribution of the lower sum to these two pieces is at least $m_{ij} \cdot t(R_{ij})$.

Then the statement of the lemma is proved by induction on the number of new base points, since every added base point either increases the lower sum or leaves it unchanged.

The proof of the inequality on the upper sums is similar. □

Lemma 4.3. *Let $f: R \to \mathbb{R}$ be bounded. If F_1 and F_2 are arbitrary partitions of $[a, b]$, then $s_{F_1} \le S_{F_2}$.*

Proof. Let F be the union of the partitions F_1 and F_2, i.e., let the set of the base points of F consist of the base points of F_1 and F_2. The partition F is a refinement of both F_1 and F_2. Clearly, we have $s_F \le S_F$ (since $m_{ij} \le M_{ij}$ for every i, j). Then, by Lemma 4.2, we obtain $s_{F_1} \le s_F \le S_F \le S_{F_2}$. □

Let $\mathcal{F}$ denote the set of partitions of the rectangle R. The lemma above states that for every partition $F_2 \in \mathcal{F}$ the upper sum S_{F_2} is an upper bound of the set $\{s_F: F \in \mathcal{F}\}$. Therefore, the least upper bound of this set, i.e., $\sup_{F \in \mathcal{F}} s_F$, is not larger than S_{F_2} for all $F_2 \in \mathcal{F}$. In other words, $\sup_{F \in \mathcal{F}} s_F$ is a lower bound of the set $\{S_F: F \in \mathcal{F}\}$, and we get

$$\sup_{F \in \mathcal{F}} s_F \le \inf_{F \in \mathcal{F}} S_F. \tag{4.1}$$

It is clear that for every real number I we have $s_F \le I \le S_F$ for every partition F if and only if

$$\sup_{F \in \mathcal{F}} s_F \le I \le \inf_{F \in \mathcal{F}} S_F. \tag{4.2}$$

This proves that for every bounded function f there exists a number between the set of its lower sums and the set of its upper sums.

Definition 4.4. Let $f: R \to \mathbb{R}$ be a bounded function. The function f is called *integrable* on the rectangle R if $\sup_{F \in \mathcal{F}} s_F = \inf_{F \in \mathcal{F}} S_F$. We call the number $\sup_{F \in \mathcal{F}} s_F = \inf_{F \in \mathcal{F}} S_F$ the *integral* of the function f on the rectangle R, and denote it by $\int_R f(x, y)\, dxdy$.

We introduce new notation for the numbers $\sup_{F \in \mathcal{F}} s_F$ and $\inf_{F \in \mathcal{F}} S_F$.

Definition 4.5. Let $f\colon R \to \mathbb{R}$ be a bounded function. We call $\sup_{F \in \mathcal{F}} s_F$ the *lower integral* of f and denote it by $\underline{\int_R} f(x,y)\,dxdy$. We call $\inf_{F \in \mathcal{F}} S_F$ the *upper integral* of f and denote it by $\overline{\int_R} f(x,y)\,dxdy$.

We can summarize (4.1) and (4.2) with the help of the new notation.

Theorem 4.6.

(i) *For every bounded function* $f\colon R \to \mathbb{R}$ *we have* $\underline{\int_R} f(x,y)\,dx\,dy \le \overline{\int_R} f(x,y)\,dx\,dy$.

(ii) *For every real number* I, *the inequalities* $s_F \le I \le S_F$ *hold for every partition* F *if and only if* $\underline{\int_R} f(x,y)\,dx\,dy \le I \le \overline{\int_R} f(x,y)\,dx\,dy$.

(iii) *The bounded function* f *is integrable on* R *if and only if* $\underline{\int_R} f(x,y)\,dx\,dy = \overline{\int_R} f(x,y)\,dx\,dy$, *and then*

$$\int_R f(x,y)\,dx\,dy = \overline{\int_R} f(x,y)\,dx\,dy = \underline{\int_R} f(x,y)\,dx\,dy.$$

□

The definitions, theorems, and arguments used in the case of the integral of a single-variable function can be copied almost word by word for the integrals of two-variable functions. Moreover, these notions and theorems can be easily generalized to p-variable functions as well.

Definition 4.7. Let $R = [a_1, b_1] \times \ldots \times [a_p, b_p] \subset \mathbb{R}^p$ be a box. If $a_i = x_{i,0} < x_{i,1} < \ldots < x_{i,n_i} = b_i$ for every $i = 1, \ldots, p$, then we call the system of boxes

$$R_{j_1 \ldots j_p} = [x_{1,j_1-1}, x_{1,j_1}] \times \ldots \times [x_{p,j_p-1}, x_{p,j_p}]$$

(where $1 \le j_i \le n_i$ for every $i = 1, \ldots, p$) a *partition* of the box R. If f is bounded on R, we can define its *lower and upper sums* in the same way that we did in the two-variable case.

The proof of $s_{F_1} \le S_{F_2}$ for a pair of arbitrary partitions F_1 and F_2 is exactly the same as in the two-variable case. With these in hand, we can define the *lower and upper integrals, integrability, and the value of the integral* of a function f in the same way that we did in Definitions 4.5 and 4.4. We denote the integral of the function f on the box R by $\int_R f(x_1, \ldots, x_p)\,dx_1 \cdots dx_p$, or $\int_R f(x)\,dx$ and $\int_R f\,dx$ for short.

Below we give a list of theorems on integrals of multivariable functions whose proofs closely follow the arguments of their corresponding counterparts in the single-variable case. (As for the latter, see, e.g., the theorems of Section 14.3 of [7].) We suggest the reader check these proofs again in this more general context.

This could be useful for more than one reason: it helps in understanding the new notions, and it also makes clear that the ideas used in the multivariable case are essentially the same as those applied to functions of one variable.

A bounded function $f: R \to \mathbb{R}$ is integrable and its integral equals I if and only if for every $\varepsilon > 0$ there exists a partition F such that

$$I - \varepsilon < s_F \le S_F < I + \varepsilon.$$

A bounded function $f: R \to \mathbb{R}$ is integrable if and only if for every $\varepsilon > 0$ there exists a partition F with $S_F - s_F < \varepsilon$.

We introduce a new notation for $S_F - s_F$, just as in the single-variable case. Let H be a nonempty set, and let $f: H \to \mathbb{R}$ be a bounded function. We call the quantity

$$\omega(f; H) = \sup f(H) - \inf f(H) = \sup\{|f(x) - f(y)| : x, y \in H\}$$

the **oscillation** of the function f on H.

The **oscillatory sum** of a bounded function $f: R \to \mathbb{R}$ corresponding to the partition F is

$$\Omega_F(f) = \sum \omega(f; R_{j_1 \ldots j_p}) \cdot \mu\left(R_{j_1 \ldots j_p}\right),$$

where $R_{j_1 \ldots j_p}$ runs through the division boxes of the partition F. Obviously, $\Omega_F(f) = S_F - s_F$.

A bounded function $f: R \to \mathbb{R}$ is integrable if and only if for every $\varepsilon > 0$ there exists a partition F such that $\Omega_F < \varepsilon$.

The **approximating sums** of a bounded function f corresponding to the partition F are the sums

$$\sigma_F(f; (c_{j_1 \ldots j_p})) = \sum f\left(c_{j_1 \ldots j_p}\right) \cdot \mu\left(R_{j_1 \ldots j_p}\right),$$

for every choice of the points $c_{j_1 \ldots j_p} \in R_{j_1 \ldots j_p}$.

For every bounded function $f: R \to \mathbb{R}$ and partition F,

$$\inf_{(c_1, \ldots, c_n)} \sigma_F = s_F \qquad \text{and} \qquad \sup_{(c_1, \ldots, c_n)} \sigma_F = S_F.$$

That is, the infimum and the supremum of the approximating sums (over all possible choices of the points c_i) are s_F and S_F, respectively.

A bounded function $f\colon R \to \mathbb{R}$ is integrable and its integral equals I if and only if for every $\varepsilon > 0$ there exists a partition F such that every approximating sum σ_F has $|\sigma_F - I| < \varepsilon$.

If f is continuous on the box R, then f is integrable on R.

If f is integrable on the box R, then the function cf is also integrable on R, and $\int_R cf\,dx = c\int_R f\,dx$.

If f and g are integrable on the box R, then $f+g$ is also integrable on R, and $\int_R (f+g)\,dx = \int_R f\,dx + \int_R g\,dx$.

If f and g are integrable on the box R, then the functions $|f|$ and $f \cdot g$ are also integrable on R, and furthermore, if $|g(x)| \ge \delta > 0$ for every $x \in R$, then f/g is also integrable on R.

Let g be integrable on the box R, and let f be a continuous real-valued function on an interval $[\alpha, \beta]$ containing the range of g (i.e., containing the set $g(R)$). Then the function $f \circ g$ is also integrable on R.

Exercise

4.1. Let $A \subset R \subset \mathbb{R}^p$, where R is a box, and let

$$f(x) = \begin{cases} 1, & \text{if } x \in A, \\ 0, & \text{if } x \in R \setminus A \end{cases}.$$

Show that

(a) $\underline{\int_R} f\,dx = \underline{\mu}(A)$ and $\overline{\int_R} f\,dx = \overline{\mu}(A)$, and furthermore,

(b) f is integrable on R if and only if A is measurable, and then $\int_R f\,dx = \mu(A)$.

4.2 The Multivariable Integral on Jordan Measurable Sets

So far we have defined the multivariable integral of functions only on boxes. However, the definition of the integral and most of our previous theorems hardly used the fact that the underlying sets and the parts of their partitions are boxes. Since we often encounter problems in which those conditions are not met, it will be useful to generalize the definition of the integral to a more general situation.

Definition 4.8. Let $A \subset \mathbb{R}^p$ be a Jordan measurable set. A system of sets $F = \{A_1, \ldots, A_n\}$ is called a *partition* of the set A if $A_1, \ldots, A_n$ are nonoverlapping and nonempty measurable sets whose union is A.

We say that the partition $\{B_1, \ldots, B_m\}$ is a *refinement* of the partition $\{A_1, \ldots, A_n\}$ if for every $j \in \{1, \ldots, m\}$ there is an $i \in \{1, \ldots, n\}$ such that $B_j \subset A_i$.

Let $f\colon A \to \mathbb{R}$ be a bounded function. We say that the *lower sum* of f corresponding to the partition F is the sum $s_F = \sum_{i=1}^n m_i \cdot \mu(A_i)$, where $m_i = \inf\{f(x)\colon x \in A_i\}$ $(i = 1, \ldots, n)$. The *upper sum* of the function f corresponding to the partition F is the sum $S_F = \sum_{i=1}^n M_i \cdot \mu(A_i)$, where $M_i = \sup\{f(x)\colon x \in A_i\}$ $(i = 1, \ldots, n)$.

Lemma 4.9. *Let $A \subset \mathbb{R}^p$ be nonempty and Jordan measurable, and let $f\colon A \to \mathbb{R}$ be bounded.*

(i) *If F_1 and F_2 are partitions of A and F_2 is a refinement of F_1, then $s_{F_2} \ge s_{F_1}$ and $S_{F_2} \le S_{F_1}$.*
(ii) *If F_1 and F_2 are arbitrary partitions of A, then $s_{F_1} \le S_{F_2}$.*

Proof. (i) Let $F_1 = \{A_1, \ldots, A_n\}$ and $F_2 = \{B_1, \ldots, B_m\}$. Clearly, $\inf\{f(x)\colon x \in B_j\} \ge \inf\{f(x)\colon x \in A_i\}$ $(i = 1, \ldots, n)$ whenever $B_j \subset A_i$. If F_2 is a refinement of F_1, then each A_i is the union of the sets B_j that are subsets of A_i. From these observations the inequality $s_{F_2} \ge s_{F_1}$ easily follows. The inequality $S_{F_2} \le S_{F_1}$ is obtained similarly.

(ii) The sets $A_i \cap B_j$ $(i = 1, \ldots, n,\ j = 1, \ldots, m)$ are nonoverlapping, and their union is also A. Let $C_1, \ldots, C_k$ be an enumeration of the sets $A_i \cap B_j$ that are nonempty. Then the partition $F = \{C_1, \ldots, C_k\}$ is a common refinement of F_1 and F_2. Then by (i), we have $s_{F_1} \le s_F \le S_F \le S_{F_2}$. □

Definition 4.10. Let $A \subset \mathbb{R}^p$ be nonempty and Jordan measurable, and let $\mathcal{F}$ denote the set of all partitions of A. If $f\colon A \to \mathbb{R}$ is bounded, then the number $\sup_{F \in \mathcal{F}} s_F$ is said to be the *lower integral* of f and is denoted by $\underline{\int_A} f\,dx$. Similarly, we say that $\inf_{F \in \mathcal{F}} S_F$ is the *upper integral* of f and denote it by $\overline{\int_A} f\,dx$.

As a corollary of Lemma 4.9 we have $\underline{\int_A} f\,dx \le \overline{\int_A} f\,dx$ for every bounded function $f\colon A \to \mathbb{R}$. We say that a function f is *integrable* on the set A if $\underline{\int_A} f\,dx = \overline{\int_A} f\,dx$. We call $\underline{\int_A} f\,dx = \overline{\int_A} f\,dx$ the *integral* of the function f on the set A, and denote it by $\int_A f\,dx$ or $\int_A f\,dx_1 \ldots dx_p$.

If $A = \emptyset$, then we define $\int_A f\,dx$ to be zero.

Note that if the set A is a box and $f\colon A \to \mathbb{R}$ is bounded, then we have defined the integrability and the integral of f *twice*, first in Definition 4.7 using partitions of boxes, and then in Definition 4.10 using partitions of measurable sets. We will see presently that the two definitions are equivalent (see Remark 4.13).

We call the sum

$$\Omega_F = S_F - s_F = \sum_{i=1}^{n} \omega(f; A_i) \cdot t(A_i)$$

the **oscillatory sum** of the bounded function $f\colon A \to \mathbb{R}$ corresponding to a partition $F = \{A_1, \ldots, A_n\}$.

The proofs of the following theorems are the same as their respective counterparts for the single-variable case.

A bounded function f is integrable on A if and only if for every $\varepsilon > 0$ there exists a partition F such that $\Omega_F < \varepsilon$.

If a function is integrable on the set A, then it is also integrable on every nonempty Jordan measurable subset of A.

Let f be defined on $A \cup B$, where A, B are nonoverlapping Jordan measurable sets. If f is integrable on both A and B, then f is integrable on $A \cup B$, and

$$\int_{A\cup B} f(x)\,dx = \int_A f(x)\,dx + \int_B f(x)\,dx. \tag{4.3}$$

Using the last two theorems, we can reduce integration on an arbitrary measurable set A to integration on boxes. Indeed, pick a box R containing A. We extend the given function $f\colon A \to \mathbb{R}$ to R by setting it zero everywhere on $R \setminus A$:

$$\overline{f}(x) = \begin{cases} f(x), & \text{if } x \in A, \\ 0, & \text{if } x \in R \setminus A. \end{cases}$$

Clearly, f is integrable on A if and only if $\overline{f}$ is integrable on R, and then $\int_A f\,dx = \int_R \overline{f}\,dx$.

Since the integral of the constant function equal to 1 on A is $\mu(A)$, we have the following theorem.

Theorem 4.11. *Let $A \subset R \subset \mathbb{R}^p$, where A is measurable and R is a box. Then the function*

$$f(x) = \begin{cases} 1, & \text{if } x \in A, \\ 0, & \text{if } x \in R \setminus A, \end{cases}$$

is integrable on R, and its integral equals $\mu(A)$. □

See Exercise 4.1. for the converse of this theorem.

Let $f\colon [a, b] \to \mathbb{R}$ be integrable with integral I. Then for every $\varepsilon > 0$ there exists $\delta > 0$ such that for every partition F finer than δ we have $I - \varepsilon < s_F \le I \le S_F <$

$I + \varepsilon$ (see [7, Theorem 14.23]). This fact can be generalized to multivariable functions.

Recall that the diameter of a nonempty set A is $\operatorname{diam} A = \sup\{|x - y| : x, y \in A\}$. We say that the **mesh** of a partition $F = \{A_1, \ldots, A_n\}$ is $\delta(F) = \max_{1 \le i \le n} \operatorname{diam} A_i$. The **partition F is finer than** η if $\delta(F) < \eta$.

Theorem 4.12.

(i) *Let $A \subset \mathbb{R}^p$ be Jordan measurable, and let $f : A \to \mathbb{R}$ be bounded. For every partition F_0 of the set A and for every $\varepsilon > 0$, there exists $\delta > 0$ such that*

$$s_{F_0} - \varepsilon < s_F \le S_F < S_{F_0} + \varepsilon$$

for every partition F finer than δ.

(ii) *Let f be integrable on the set A and let $\int_A f\,dx = I$. For every $\varepsilon > 0$ there exists $\delta > 0$ such that $I - \varepsilon < s_F \le S_F < I + \varepsilon$ for every partition F finer than δ.*

We will prove Theorem 4.12 in the first appendix.

Remarks 4.13. **1.** If the set A is a box and the function $f : A \to \mathbb{R}$ is bounded, then the integrability of f and its integral are defined in both of Definitions 4.7 and 4.10, first with the help of subdividing boxes, then with the help of subdividing measurable sets. We will now prove that these two definitions are equivalent.

Let us call Definition 4.7 the box partition definition, and call Definition 4.10 the general definition. Clearly, it is enough to prove that the upper and the lower integrals are the same in the two cases.

Since every partition of the box partition definition is also a partition of the general definition, the upper integral of f based on the box partition definition cannot be smaller then the upper integral of f based on the general definition.

Let $F_0 = \{A_1, \ldots, A_N\}$ be a partition of the general definition. It is enough to prove that for every $\varepsilon > 0$ there exists a box partition F such that $S_F < S_{F_0} + \varepsilon$. By (i) of Theorem 4.12, $S_F < S_{F_0} + \varepsilon$ for F fine enough. Since by Lemma 3.12 the box A has an arbitrarily fine partition, Definition 4.7 proves our claim.

The equality of the lower integrals can be proved similarly.

2. Do we really need both definitions? The question is only natural, since we have just proved their equivalence. As a matter of fact, we do not need both, and we could get by using either. However, what justifies the first definition is the simplicity of the box partition definition and the fact that it is a natural generalization of the single-variable integral in that boxes are the natural generalizations of intervals. On the other hand, the existence of the general definition is justified by the fact that it is independent of the definition of boxes and, consequently, of the choice of coordinate system. Also, some arguments and ideas are simpler and more natural in the general context.

It is well known that if f is bounded on $[a, b]$ and continuous there except at finitely many points, then f is integrable on $[a, b]$. (See [7, Theorem 14.43].) We generalize this theorem below. Note that this theorem gives a much stronger statement even in the one-dimensional case than the theorem quoted above.

Theorem 4.14. *Let $A \subset \mathbb{R}^p$ be nonempty and Jordan measurable. If the function $f\colon A \to \mathbb{R}$ is bounded and continuous on A except at the points of a set of measure zero, then f is integrable on A.*

Proof. Let $|f(x)| \le K$ for every $x \in A$. Since A is Jordan measurable, ∂A has measure zero. By assumption, the set $D = \{x \in A\colon f \text{ is not continuous at } x\}$ also has measure zero, and thus $\mu((\partial A) \cup D) = 0$.

Let $\varepsilon > 0$ be fixed. By Exercise 3.5, there exists an open set G such that $(\partial A) \cup D \subset G$ and $\overline{\mu}(G) < \varepsilon$. Since $\partial A \subset G$, we have $A \setminus G = (\mathrm{cl}\, A) \setminus G = \mathrm{cl}\, A \cap (\mathbb{R}^p \setminus G)$, and thus $A \setminus G$ is closed. The function f is continuous at every point of $A \setminus G$, since $D \subset G$. Since $A \setminus G$ is bounded and closed, it follows from Heine's theorem (see page 70) that f is uniformly continuous on $A \setminus G$, i.e., that there exists $\delta > 0$ such that $|f(x) - f(y)| < \varepsilon$ for every $x, y \in A \setminus G$ with $|x - y| < \delta$. Let the partition $\{F_1, \ldots, F_k\}$ of $A \setminus G$ be finer than δ. (We can construct such a partition by choosing $n > \delta/\sqrt{p}$ and taking the intersections $K \cap (A \setminus G)$, where $K \in \mathcal{K}_n$ and $K \cap (A \setminus G) \ne \emptyset$.)

Consider the partition $F = \{F_1, \ldots, F_k, A \cap G\}$ of the set A. By the choice of the sets F_i, we have $\omega(f; F_i) < \varepsilon$ for every $i = 1, \ldots, k$. Thus,

$$\begin{aligned}\Omega_F &= \sum_{i=1}^{k} \omega(f; F_i) \cdot \mu(F_i) + \omega(f; A \cap G) \cdot \mu(A \cap G) \le \\ &\le \varepsilon \cdot \mu\left(\bigcup_{i=1}^{k} F_i\right) + 2K \cdot \mu(A \cap G) \le \\ &\le \varepsilon \cdot \mu(A) + 2K \cdot \varepsilon = (\mu(A) + 2K) \cdot \varepsilon.\end{aligned}$$

Since ε was arbitrary, this proves the integrability of f on A. □

Remarks 4.15. **1.** As a corollary of the theorem, we can see that if $f\colon [0, 1] \to \mathbb{R}$ is bounded and is continuous everywhere outside of the points of the Cantor set, then f is integrable on $[0, 1]$, since the Cantor set is of measure zero (see Example 3.22). Such a function is, e.g., the function with $f(x) = 1$ for every $x \in C$ and $f(x) = 0$ for every $x \in [0, 1] \setminus C$. Since the cardinality of the Cantor set is that of the continuum, we can see that there exist integrable functions that are not continuous at uncountably many points.

2. The converse of the theorem does not hold: an integrable function is not necessarily continuous everywhere outside of a set of measure zero. Consider, for example, the **Riemann function**, which is defined as follows. If $x \in \mathbb{R}$ is irrational, then we define $f(x) = 0$. If $x \in \mathbb{R}$ is rational and $x = p/q$, where p, q are coprime integers and $q > 0$, then we define $f(x) = 1/q$. It is well known that the Riemann function is integrable on every interval $[a, b]$. See [7, Example 14.45]. The Riemann function is integrable on $[0, 1]$, yet it is discontinuous at every rational point, and the set $\mathbb{Q} \cap [0, 1]$ is not of measure zero.

The integral of a nonnegative function of one variable gives the area under the graph of the function (see Example 3.21.1). More generally, the area of a normal domain (see below) is the difference between the integrals of the functions defining the domain (see [7, Theorem 16.5]). This result can be generalized to multivariable integrals as well.

Let f and g be integrable functions defined on a nonempty measurable set $B \subset \mathbb{R}^p$ such that $f(x) \le g(x)$ for every $x \in B$. The set

$$A = \{(x, y) \in \mathbb{R}^{p+1} : x \in B,\ f(x) \le z \le g(x)\} \tag{4.4}$$

is the **normal domain** defined by f and g.

It is easy to see that every ball is a normal domain. One can prove that every bounded, closed, and convex set is also a normal domain (see Exercise 4.3).

Theorem 4.16.

(i) *If $B \subset \mathbb{R}^p$ is nonempty and measurable, f and g are integrable on B, and $f(x) \le g(x)$ for every $x \in B$, then the normal domain defined by* (4.4) *is measurable, and its measure is $\int_B (g - f)\,dx$.*

(ii) *Let the function $f : B \to \mathbb{R}$ be nonnegative and bounded. The set*

$$A_f = \{(x, y) : x \in B,\ 0 \le y \le f(x)\}$$

is measurable if and only if f is integrable on B, and the measure of A_f is $\int_B f(x)\,dx$.

Proof. (i) First we assume that B is a box. For a given $\varepsilon > 0$ choose partitions (into boxes) F_1 and F_2 of B such that $\Omega_{F_1}(f) < \varepsilon$ and $\Omega_{F_2}(g) < \varepsilon$. Let the partition F be a refinement of F_1 and F_2, and let the division boxes of the partition F be $R_1, \ldots, R_n$. Then we have $\Omega_F(f) < \varepsilon$ and $\Omega_F(g) < \varepsilon$. Let $m_i(f)$, $m_i(g)$, $M_i(f)$, and $M_i(g)$ be the infimum and supremum of the functions f and g respectively on the box R_i. Then the boxes $R_i \times [m_i(f), M_i(g)]$ $(i = 1, \ldots, n)$ cover the set A, so

$$\begin{aligned}
\overline{\mu}_{p+1}(A) &\le \sum_{i=1}^{n} (M_i(g) - m_i(f)) \cdot \mu_p(R_i) = \\
&= S_F(g) - s_F(f) < \\
&< \int_B g\,dx + \varepsilon - \left(\int_B f\,dx - \varepsilon \right) = \\
&= \int_B (g - f)\,dx + 2\varepsilon.
\end{aligned} \tag{4.5}$$

Let I denote the set of indices i that satisfy $M_i(f) \le m_i(g)$. Then the boxes $R_i \times [M_i(f), m_i(g)]$ $(i \in I)$ are contained in A and are nonoverlapping, so

$$\begin{aligned}\underline{\mu}_{p+1}(A) &\geq \sum_{i\in I}(m_i(g) - M_i(f)) \cdot \mu_p(R_i) \geq \\ &\geq \sum_{i=1}^{n}(m_i(g) - M_i(f)) \cdot \mu_2(R_i) = \\ &= s_F(g) - S_F(f) > \\ &> \int_B g\,dx - \varepsilon - \int_B f\,dx - \varepsilon = \\ &= \int_B (g-f)\,dx - 2\varepsilon. \end{aligned} \tag{4.6}$$

Since ε was arbitrary, by (4.5) and (4.6) it follows that A is measurable and has volume $\int_B (g-f)\,dx$. This proves (i) in the case that B is a rectangle.

If B is measurable, then let R be a box containing B, and extend f and g to R by putting $f(x) = g(x) = 0$ $(x \in R \setminus B)$. Then f, g are integrable on R, and thus by what we proved above, the normal domain

$$A' = \{(x, y) \in \mathbb{R}^{p+1} : x \in R,\ f(x) \le y \le g(x)\}$$

is measurable, and its volume is $\int_R (g-f)\,dx = \int_B (g-f)\,dx$. The set $A' \setminus A$ is a bounded subset of the hyperplane $\{(x_1, \ldots, x_{p+1}) : x_{p+1} = 0\}$. Consequently, $\mu_{p+1}(A' \setminus A) = 0$ by Lemma 3.15. Therefore, $A = A' \setminus (A' \setminus A)$ is measurable and $\mu_{p+1}(A) = \mu_{p+1}(A')$, which completes the proof of (i).

(ii) Let f be nonnegative and integrable on B. Then A_f is the normal domain determined by the functions 0 and f. Therefore, by (i), the set A_f is measurable, and its volume equals $\int_B f(x, y)\,dx\,dy$.

Finally, if A_f is measurable, then the integrability of f follows from Theorem 3.25. □

Exercise

4.2. Let $A \subset \mathbb{R}^p$ be measurable and let $f : A \to \mathbb{R}$ be nonnegative and bounded. Show that if $\int_A f\,dx = 0$, then $\mu(\{x \in A : f(x) \ge a\}) = 0$ for every $a > 0$. Does the converse of this statement also hold?

4.3. Prove that every bounded, closed, and convex set is a normal domain. (∗)

Consider a set $A \subset \mathbb{R}^p$ made of a homogeneous material. Then the weight of every measurable subset of A is d times the volume of the subset, where d is a positive constant (the density). Let $\{A_1, \ldots, A_n\}$ be a partition of A, and let the points $c_i \in A_i$ be arbitrary. If the partition is fine enough, then concentrating the weight $d \cdot \mu(A_i)$ at the point c_i for every i creates a system of points with weight distribution close to that of A. We expect the center of mass of this system to be close to the center of mass of the original set.

The center of mass of the system of points c_i is $\frac{1}{\mu(A)} \cdot \sum_{i=1}^n \mu(A_i) \cdot c_i$. If the partition is fine enough, then the jth coordinate of this point is close to $\frac{1}{\mu(A)} \cdot \int_A x_j \, dx$.

This motivates the following definition: the **center of mass** of a measurable set $A \subset \mathbb{R}^p$ with positive measure is the point

$$\left(\frac{1}{\mu(A)} \int_A x_1 \, dx, \ldots, \frac{1}{\mu(A)} \int_A x_p \, dx \right).$$

4.4. Let $s(A)$ denote the center of mass of the measurable set A of positive measure. Show that if A and B are nonoverlapping measurable sets with positive measure, then $s(A \cup B) = \frac{\mu(A)}{\mu(A)+\mu(B)} \cdot s(A) + \frac{\mu(B)}{\mu(A)+\mu(B)} \cdot s(B)$.

4.5. Suppose a point $r(A) \in \mathbb{R}^p$ is assigned to every Jordan measurable set with positive measure $A \subset \mathbb{R}^p$ and that the following conditions are satisfied:

(i) if $A \subset R$, where R is a box, then $r(A) \in R$; furthermore,
(ii) if A and B are nonoverlapping measurable sets with positive measure, then $r(A \cup B) = \frac{\mu(A)}{\mu(A)+\mu(B)} \cdot r(A) + \frac{\mu(B)}{\mu(A)+\mu(B)} \cdot r(B)$.

Show that $r(A)$ equals the center of mass of A for every measurable set A with positive measure.

4.3 Calculating Multivariable Integrals

The most important method of calculating multivariable integrals is provided by the next theorem. It states that every integral can be reduced to lower-dimensional integrals.

Let us use the following notation. If $x = (x_1, \ldots, x_p) \in \mathbb{R}^p$ and $y = (y_1, \ldots, y_q) \in \mathbb{R}^q$, then let (x, y) be the vector[2] $(x_1, \ldots, x_p, y_1, \ldots, y_q) \in \mathbb{R}^{p+q}$.

Let $A \subset \mathbb{R}^p$, $B \subset \mathbb{R}^q$ and $f\colon (A \times B) \to \mathbb{R}$. Recall that the sections of f are defined by $f_x(y) = f^y(x) = f(x, y)$. More precisely, this means that for every $x \in A$ the function f_x is defined on B and $f_x(y) = f(x, y)$ for every $y \in B$. Similarly, for every $y \in B$ the function f^y is defined on A and $f^y(x) = f(x, y)$ for every $x \in A$.

Theorem 4.17. *Let $A \subset \mathbb{R}^p$ and $B \subset \mathbb{R}^q$ be a pair of boxes and let $f\colon (A \times B) \to \mathbb{R}$ be integrable on the box $A \times B$. Then*

[2] We already used this notation in the implicit function theorem.

(i) *the functions* $y \mapsto \overline{\int_A} f^y\,dx$ *and* $y \mapsto \underline{\int_A} f^y\,dx$ *are integrable on* B, *and*

$$\int_{A\times B} f\,dx\,dy = \int_B \left(\overline{\int_A} f^y\,dx\right)\,dy = \int_B \left(\underline{\int_A} f^y\,dx\right)\,dy;$$

(ii) *the functions* $x \mapsto \overline{\int_B} f_x\,dy$ *and* $x \mapsto \underline{\int_B} f_x\,dy$ *are integrable on* A, *and*

$$\int_{A\times B} f\,dx\,dy = \int_A \left(\overline{\int_B} f_x\,dy\right)\,dx = \int_A \left(\underline{\int_B} f_x\,dy\right)\,dx.$$

Proof. Since (i) and (ii) are transformed into each other if we switch the roles of x and y, it is enough to prove (ii).

Let $\int_{A\times B} f\,dx\,dy = I$. For $\varepsilon > 0$ fixed, there exists a partition of the box $A \times B$ (in the sense of Definition 4.7) such that $I - \varepsilon < s_F \le S_F < I + \varepsilon$. From the definition of partition it is clear that there exist partitions $F_1 = \{A_1, \ldots, A_n\}$ and $F_2 = \{B_1, \ldots, B_m\}$ of the boxes A and B, respectively, such that F consists of the boxes $A_i \times B_j$ ($i = 1, \ldots, n,\ j = 1, \ldots, m$). Let

$$m_{ij} = \inf\{f(x,y) \colon (x,y) \in A_i \times B_j\} \quad \text{and} \quad M_{ij} = \sup\{f(x,y) \colon (x,y) \in A_i \times B_j\}$$

for every $1 \le i \le n$ and $1 \le j \le m$. If $x \in A_i$, then the upper sum of the function $f_x \colon B \to \mathbb{R}$ corresponding to the partition F_2 is at most $\sum_{j=1}^m M_{ij} \cdot \mu(B_j)$, since $f_x(y) = f(x,y) \le M_{ij}$ for every $y \in B_j$. This implies that

$$\overline{\int_B} f_x\,dy \le S_{F_2}(f_x) \le \sum_{j=1}^m M_{ij} \cdot \mu(B_j)$$

for every $x \in A_i$. In other words, the right-hand side is an upper bound on the values of the function $x \mapsto \overline{\int_B} f_x\,dy$ on the set A_i for every $i = 1, \ldots, n$. Therefore, the upper sum of this function corresponding to the partition F_1 is at most $\sum_{i=1}^n \left(\sum_{j=1}^m M_{ij} \cdot \mu(B_j)\right) \cdot \mu(A_i)$. Thus

$$\overline{\int_A}\left(\overline{\int_B} f_x\,dy\right)\,dx \le \sum_{i=1}^n \sum_{j=1}^m M_{ij} \cdot \mu(A_i) \cdot \mu(B_j) = S_F < I + \varepsilon. \tag{4.7}$$

On the other hand, if $x \in A_i$, then the lower sum of the function $f_x \colon B \to \mathbb{R}$ corresponding to the partition F_2 is at least $\sum_{j=1}^m m_{ij} \cdot \mu(B_j)$, since $f_x(y) = f(x,y) \ge m_{ij}$ for every $y \in B_j$. This implies that

$$\underline{\int}_B f_x\,dy \geq s_{F_2}(f_x) \geq \sum_{j=1}^{m} m_{ij} \cdot \mu(B_j)$$

for every $x \in A_i$. That is, the right-hand side is a lower bound for the values of the function $x \mapsto \underline{\int}_B f_x\,dy$ on the set A_i for every $i = 1, \ldots, n$; hence the lower sum of this function corresponding to the partition F_1 is at least $\sum_{i=1}^{n} \left(\sum_{j=1}^{m} m_{ij} \cdot \mu(B_j) \right) \cdot \mu(A_i)$. This proves

$$\underline{\int}_A \left(\underline{\int}_B f_x\,dy \right) dx \geq \sum_{i=1}^{n} \sum_{j=1}^{m} m_{ij} \cdot \mu(A_i) \cdot \mu(B_j) = s_F > I - \varepsilon. \tag{4.8}$$

Combining the inequalities (4.7) and (4.8), we get

$$I - \varepsilon < \underline{\int}_A \left(\underline{\int}_B f_x\,dy \right) dx \leq \overline{\int}_A \left(\overline{\int}_B f_x\,dy \right) dx < I + \varepsilon$$

for every $\varepsilon > 0$. Consequently, we have

$$\underline{\int} \left(\underline{\int}_B f_x\,dy \right) dx = \overline{\int}_A \left(\overline{\int}_B f_x\,dy \right) dx = I, \tag{4.9}$$

which also implies

$$\underline{\int}_A \left(\underline{\int}_B f_x\,dy \right) dx = \overline{\int}_A \left(\underline{\int}_B f_x\,dy \right) dx = I.$$

Therefore, the function $x \mapsto \underline{\int}_B f_x\,dy$ is integrable on A and its integral is I.

We obtain from (4.9) by a similar argument that $x \mapsto \overline{\int}_B f_x\,dy$ is also integrable on A and its integral is I. □

As an application, we can reduce the integrals of functions defined on normal domains to lower-dimensional integrals.

Theorem 4.18. *Suppose that*

$$A = \{(x, y, z) \colon (x, y) \in B,\ f(x, y) \leq z \leq g(x, y)\},$$

where $B \subset \mathbb{R}^2$ is measurable, f and g are integrable on B, and $f(x, y) \leq g(x, y)$ for every $(x, y) \in B$. If h is continuous on A, then

$$\int_A h(x, y, z)\,dx\,dy\,dz = \int_B \left(\int_{f(x,y)}^{g(x,y)} h(x, y, z)\,dz \right) dx\,dy. \tag{4.10}$$

Proof. Since B is bounded and the functions f, g are bounded on B, there exists a box $R = [a_1, b_1] \times [a_2, b_2] \times [a_3, b_3]$ containing A. Let h be zero everywhere on the set $\big([a_1, b_1] \times [a_2, b_2]\big) \setminus B$. If $(x, y) \in \big([a_1, b_1] \times [a_2, b_2]\big) \setminus B$, then $h(x, y, z) = 0$ for every $z \in [a_3, b_3]$. If, however, $(x, y) \in B$, then

$$\int_{a_3}^{b_3} h(x, y, z)\, dz = \int_{f(x,y)}^{g(x,y)} h(x, y, z)\, dz.$$

Thus, part (ii) of Theorem 4.17 applied to the case $p = 2,\ q = 1$ gives (4.10). □

Note that Theorem 4.16 is a special case of our previous theorem applied to the function $h \equiv 1$. It is clear that both the notion of normal domains and the previous theorem can be generalized to higher dimensions.

Applying Theorem 4.17 successively, we obtain the following corollary.

Corollary 4.19. (Theorem of successive integration) *Let f be integrable on the box $R = [a_1, b_1] \times \ldots \times [a_p, b_p] \subset \mathbb{R}^p$. Then*

$$\int_R f\, dx = \int_{a_p}^{b_p} \ldots \left(\int_{a_2}^{b_2} \left(\int_{a_1}^{b_1} f(x_1, \ldots, x_p)\, dx_1 \right) dx_2 \right) \ldots\, dx_p,$$

assuming that the corresponding sections are integrable. □

Example 4.20. The function e^{x+y} is integrable on the square $[0, 1] \times [0, 1]$, since it is continuous. By the theorem of successive integration, its integral is

$$\int_{[0,1]\times[0,1]} e^{x+y}\, dx\, dy = \int_0^1 \left(\int_0^1 e^{x+y}\, dy \right) dx = \int_0^1 e^x \cdot \left(\int_0^1 e^y\, dy \right) dx =$$

$$= \int_0^1 e^x \cdot (e - 1)\, dx = (e - 1)^2.$$

Remarks 4.21. **1.** By generalizing Example 4.20, we can show that if the single-variable functions $f\colon [a, b] \to \mathbb{R}$ and $g\colon [c, d] \to \mathbb{R}$ are integrable, then the function $f(x) \cdot g(y)$ is integrable on $[a, b] \times [c, d]$ and its integral is $\left(\int_a^b f(x)\, dx\right) \cdot \left(\int_c^d g(y)\, dy\right)$ (see Exercise 4.8).

2. The existence of the integrals $\int_B f_x\, dy$ and $\int_A f^y\, dx$ (for every y and x, respectively) does not necessarily follow from the integrability of the function f. In other words, the lower and upper integrals in Theorem 4.17 cannot be replaced by integrals. Take the following example.

Let f be the Riemann function. Since $\int_0^1 f(x)\,dx = 0$, it follows from statement (ii) of Theorem 4.16 that the set $A = \{(x,y)\colon 0 \le y \le f(x)\}$ has measure zero. Let B be the set of points $(x,y) \in A$ whose coordinates are rational. Since B has measure zero, it is measurable. By Theorem 4.11, the function

$$g(x,y) = \begin{cases} 1, & \text{if } (x,y) \in B, \\ 0, & \text{if } (x,y) \notin B, \end{cases} \tag{4.11}$$

is integrable on $[0,1] \times [0,1]$, and its integral is zero. However, the section g_x is not integrable on $[0,1]$ if $x \in [0,1]$ and x is rational. Indeed, let $x = p/q$. Then $g_x(y) = 0$ for every irrational number y and $g_x(y) = 1$ for every rational number $y \in [0,1/q]$. Consequently, $\underline{\int_0^1} g_x\,dy = 0$ and $\overline{\int_0^1} g_x\,dy > 0$, and thus g_x is not integrable.

However, one can prove that if f is integrable on the rectangle $[a,b] \times [c,d]$, then the set of points $x \in [a,b]$ where f_x is integrable on $[c,d]$ is everywhere dense in $[a,b]$, and the set of points $y \in [c,d]$ where f^y is integrable on $[a,b]$ is everywhere dense in $[c,d]$ (see Exercise 4.9).

3. Let f be integrable on the rectangle $[a,b] \times [c,d]$. If f_x is integrable on $[c,d]$ for every $x \in [a,b]$, and f^y is integrable on $[a,b]$ for every $y \in [c,d]$, then Theorem 4.17 implies

$$\int_a^b \left(\int_c^d f_x\,dy \right) dx = \int_c^d \left(\int_a^b f^y\,dx \right) dy. \tag{4.12}$$

We emphasize that without the assumption of the integrability of f, (4.12) is not necessarily true, not even if every integral in (4.12) exists. For example, let $f(x,y) = (x^2 - y^2)/(x^2+y^2)^2$ and $f(0,0) = 0$, and let $[a,b] = [c,d] = [0,1]$. Then the left-hand side of (4.12) is $\pi/4$, while the right-hand side is $-\pi/4$. Another example is the following. Let $f(x,y) = (x-y)/(x+y)^3$ if $x + y \neq 0$ and $f(x,y) = 0$ if $x + y = 0$, and let $[a,b] = [c,d] = [0,1]$. Then the two sides of (4.12) are $-1/2$ and $1/2$, respectively (see Exercise 4.11).

4. It can also happen that f is not integrable, despite the fact that (4.12) holds. It is not difficult to construct a set $A \subset [0,1] \times [0,1]$ that contains a point from every box but does not contain three collinear points. If $f(x,y) = 1$, where $(x,y) \in A$ and $f(x,y) = 0$ otherwise, then both sides of (4.12) are zero, but f is not integrable on $[0,1] \times [0,1]$ (see Exercise 4.12).

5. Theorem 4.17 holds for arbitrary measurable sets $A \subset \mathbb{R}^p$, $B \subset \mathbb{R}^q$ in place of boxes. This can be proved either by repeating the original proof of Theorem 4.17 almost verbatim or by reducing the statement to Theorem 4.17. Indeed, let $A \subset \mathbb{R}^p$, $B \subset \mathbb{R}^q$ be measurable, and let f be integrable on $A \times B$. Let $R \subset \mathbb{R}^p$ and $S \subset \mathbb{R}^q$ be boxes with $A \subset R$ and $B \subset S$, and let f be defined as zero everywhere on $(R \times S) \setminus (A \times B)$. Applying Theorem 4.17 to the box $R \times S$, we obtain the desired statement by (4.3).

Similarly to the single-variable case, using appropriate substitutions to find a more easily computable integral is an important method of finding the value of a multivariable integral. The theorem of integration by substitution in this context is as follows.

Theorem 4.22. (Integration by substitution) *Let $G \subset \mathbb{R}^p$ be open, and let $g\colon G \to \mathbb{R}^p$ be continuously differentiable. If H is measurable, $\operatorname{cl} H \subset G$, and g is injective on $\operatorname{int} H$, then $g(H)$ is also measurable, and*

$$\mu(g(H)) = \int_H |\det g'(x)|\, dx. \tag{4.13}$$

Furthermore, if $f\colon g(H) \to \mathbb{R}$ is bounded, then

$$\int_{g(H)} f\, dt = \int_H f(g(x)) \cdot |\det g'(x)|\, dx \tag{4.14}$$

in the sense that if either the left-hand side or the right-hand side exists, then the other side exists as well and they are equal.

The proof is given in the second appendix.

Remarks 4.23. **1.** The right-hand side of the formulas (4.13) and (4.14) contain the absolute value of g's Jacobian determinant (i.e., the determinant of its Jacobian matrix). This might look surprising at first, since the integration by substitution formula for single-variable functions is

$$\int_{g(a)}^{g(b)} f\, dt = \int_a^b f(g(x)) \cdot g'(x)\, dx, \tag{4.15}$$

and it has g' instead of $|g'|$. To resolve this "paradox," let us consider (4.14) in the case that $p = 1$ and $H = [a, b]$.

Let $g\colon [a, b] \to \mathbb{R}$ be continuously differentiable on an open interval containing $[a, b]$, and let g be injective on (a, b). It is easy to see that g has to be strictly monotone on $[a, b]$, and thus the sign of g' does not change on this interval. If g' is nonnegative on $[a, b]$, then g is monotonically increasing with $g(H) = [g(a), g(b)]$. Then (4.14) gives (4.15).

If, however, g' is nonpositive on $[a, b]$, then g is monotonically decreasing, and thus $g(H) = [g(b), g(a)]$. Then (4.14) implies

$$\int_{g(b)}^{g(a)} f\, dt = \int_a^b f(g(x)) \cdot (-g'(x))\, dx.$$

Multiplying both sides by -1, we get (4.15).

We can see from this latter case that if we omitted the absolute value, (4.14) would give the wrong result.

2. If the mapping $g\colon \mathbb{R}^p \to \mathbb{R}^p$ is linear, then the function $|\det g'(x)|$ is constant, and (4.13) turns into the statement of Theorem 3.35.

Remarks 4.24. **1.** An important step in the proof of Theorem 4.22 is to show that if $H \subset \mathbb{R}^p$ is measurable, $G \subset \mathbb{R}^p$ is open, $\operatorname{cl} H \subset G$, and $g\colon G \to \mathbb{R}^p$ is continuously differentiable, then $g(H)$ is measurable (see Theorem 4.30). Let's examine to what extent we can relax the conditions on g in this theorem.

The following simple example shows that the continuous image of a measurable set is not necessarily measurable. Let $A = \{1/n\colon n \in \mathbb{N}^+\}$. Then A is a measurable subset of the real line (and it has measure zero). Since every point of A is an isolated point, every function $g\colon A \to \mathbb{R}$ is continuous on A. Let $g(1/n) = r_n$, where (r_n) is an enumeration of the set of rational numbers in $[0,1]$. Then $g(A) = [0,1] \cap \mathbb{Q}$, and thus g is not measurable.

Slightly modifying the example above, we can make g differentiable on an open set containing A. Choose mutually disjoint open intervals around the points $1/n$ (e.g., $I_n = \left(\frac{1}{n} - \frac{1}{3n^2}, \frac{1}{n} + \frac{1}{3n^2}\right)$ $(n \in \mathbb{N}^+)$ will work). Then $G = \bigcup_{n=1}^{\infty} I_n$ is an open set containing A. Let $g(x) = r_n$ for every $x \in I_n$ and $n = 1, 2, \ldots$. Obviously, g is differentiable at every point of G (and its derivative is zero everywhere), but $g(A) = [0,1] \cap \mathbb{Q}$ is not measurable.

The next example gives a continuous mapping defined on a closed interval and a measurable subset of the interval whose image is not measurable. Let C be the Cantor set and f the Cantor function (see Exercise 3.28). We know that f is continuous on $[0,1]$, and $f(C) = [0,1]$. Let B be an arbitrary nonmeasurable subset of $[0,1]$. The set $A = C \cap f^{-1}(B)$ is measurable, since it has measure zero. On the other hand, $f(A) = B$ is not measurable. In this example we can choose B to be a closed set (see Exercise 3.15). This makes A also closed, since f is continuous.

2. Let us show some positive results now. One can prove that *if $H \subset \mathbb{R}^p$ is measurable, $G \subset \mathbb{R}^p$ is open, $\operatorname{cl} H \subset G$, and $g\colon G \to \mathbb{R}^p$ is differentiable, then $g(H)$ is also measurable.* That is, we can omit the condition on the continuity of the derivative g' from Theorem 4.30; the differentiability of g is sufficient. The proof requires some advanced topological and measure-theoretic tools. The same holds for the following theorem: *if $H \subset \mathbb{R}^p$ is measurable and $g\colon H \to \mathbb{R}^p$ has the Lipschitz property, then $g(H)$ is also measurable.* In the case $p = 1$ both statements are provable using tools that are already at our disposal (see Exercises 4.15 and 4.16).

Below, we present an important application of the theorem on integration by substitution.

Theorem 4.25. (Substitution by polar coordinates)*Let $P(r,\varphi) = (r\cos\varphi, r\sin\varphi)$ for every $r, \varphi \in \mathbb{R}$. If the set $A \subset [0,\infty) \times [0,2\pi]$ is measurable, then $P(A)$ is also measurable, and $\mu(P(A)) = \int_A r\,dr\,d\varphi$. Furthermore, if $f\colon P(A) \to \mathbb{R}$ is bounded, then*

$$\int_{P(A)} f(x,y)\,dx\,dy = \int_A f(r\cos\varphi, r\sin\varphi)\cdot r\,dr\,d\varphi \tag{4.16}$$

holds in the sense that if either the left-hand side or the right-hand side exist, then the other side exists as well, and they are equal.

Proof. Consider the mapping

$$P(x,y) = (x\cos y, x\sin y) \qquad ((x,y)\in\mathbb{R}^2).$$

Obviously, P is continuously differentiable on $\mathbb{R}^2$. We show that P is injective on the open set $G = \{(x,y)\in\mathbb{R}^2 \colon x>0,\ 0<y<2\pi\}$.

Let $(x_1,y_1), (x_2,y_2)\in G$ and $P(x_1,y_1) = P(x_2,y_2)$. Then

$$x_1 = |P(x_1,y_1)| = |P(x_2,y_2)| = x_2,$$

and thus $\cos y_1 = \cos y_2$ and $\sin y_1 = \sin y_2$. Using $0 < y_1, y_2 < 2\pi$, we get $y_1 = y_2$.

The Jacobian determinant of the mapping P is

$$\det P'(x,y) = \begin{vmatrix} \cos y & -x\sin y \\ \sin y & x\cos y \end{vmatrix} = x.$$

Applying Theorem 4.22 with $g = P$ and using the notation $x = r$, $y = \varphi$, we obtain the statements of the theorem. □

Examples 4.26. **1.** Let B_R be the closed disk of radius R, centered at the origin. We have $B_R = P([0,R]\times[0,2\pi])$. By Theorem 4.25 the area of the disk B_R is

$$\mu(B_R) = \int_{[0,R]\times[0,2\pi]} r\,dr\,d\varphi.$$

The integral can be calculated easily using successive integration:

$$\int_{[0,R]\times[0,2\pi]} r\,dr\,d\varphi = \int_0^{2\pi}\left(\int_0^R r\,dr\right)d\varphi = \int_0^{2\pi}(R^2/2)\,d\varphi = R^2\pi.$$

2. According to Theorem 4.25, for every bounded function $f\colon B_R\to\mathbb{R}$ we have

$$\int_{B_R} f(x,y)\,dx\,dy = \int_{[0,R]\times[0,2\pi]} f(r\cos\varphi, r\sin\varphi)\cdot r\cdot dr\,d\varphi \tag{4.17}$$

in the sense that if either the right-hand side or the left-hand side exists, then so does the other, and they are equal. For example, let $f(x,y) = \sqrt{R^2 - x^2 - y^2}$. The integral of f is the volume of a hemisphere (cf. Corollary 3.14). Applying (4.17), we obtain

$$\int\limits_{B_R} \sqrt{R^2 - x^2 - y^2}\, dx\, dy = \int\limits_{[0,R]\times[0,2\pi]} \sqrt{R^2 - r^2} \cdot r \cdot dr\, d\varphi =$$

$$= 2\pi \cdot \int\limits_0^R \sqrt{R^2 - r^2} \cdot r \cdot dr = 2\pi \cdot \left[-\frac{1}{3}(R^2 - r^2)^{3/2} \right]_0^R = \frac{2R^3\pi}{3},$$

and thus the volume of a ball with radius R is $4R^3\pi/3$.

3. It is a well-known fact that the primitive function of e^{-x^2} is not an elementary function (see, e.g., [7, Section 15.5]). Still, one can calculate the value of the improper integral $\int_0^\infty e^{-x^2}\, dx$ using the theory of the Γ function. (See Example 19.21 and Exercise 19.45 of [7]. As for the Γ function, see Definition 8.39 and the subsequent discussion in Chapter 8 of this volume). Now we present a direct method of finding the value of this integral. We show that $\int_{-\infty}^\infty e^{-x^2}\, dx = \sqrt{\pi}$.

By applying (4.17) to the function $f(x,y) = e^{-x^2-y^2}$ we get

$$\int\limits_{B_R} e^{-x^2-y^2}\, dx\, dy = \int\limits_{[0,R]\times[0,2\pi]} e^{-r^2} \cdot r\, dr\, d\varphi =$$

$$= 2\pi \cdot \int\limits_0^R e^{-r^2} r\, dr = \tag{4.18}$$

$$= 2\pi \cdot \left[-\frac{1}{2} e^{-r^2} \right]_0^R = \pi \cdot \left(1 - e^{-R^2}\right).$$

Since $[-R/2, R/2]^2 \subset B_R \subset [-R, R]^2$ and the function $e^{-x^2-y^2}$ is everywhere positive, it follows that

$$\int\limits_{[-R/2,R/2]^2} e^{-x^2-y^2}\, dx\, dy \le \int\limits_{B_R} e^{-x^2-y^2}\, dx\, dy \le \int\limits_{[-R,R]^2} e^{-x^2-y^2}\, dx\, dy. \tag{4.19}$$

Now, by Remark 4.21.1 we have

$$\int\limits_{[-R,R]^2} e^{-x^2-y^2}\, dx\, dy = \left(\int\limits_{-R}^R e^{-x^2}\, dx \right)^2$$

and

$$\int\limits_{[-R/2,R/2]^2} e^{-x^2-y^2}\,dx\,dy = \left(\int\limits_{-R/2}^{R/2} e^{-x^2}\,dx\right)^2.$$

Comparing (4.19) and (4.18) gives

$$\left(\int\limits_{-R/2}^{R/2} e^{-x^2}\,dx\right)^2 \le \pi\cdot\left(1-e^{-R^2}\right) \le \left(\int\limits_{-R}^{R} e^{-x^2}\,dx\right)^2 \tag{4.20}$$

for every $R > 0$. We know that the improper integral $\int_{-\infty}^{\infty} e^{-x^2}\,dx$ is convergent. Thus, if R converges to infinity in (4.20), then

$$\left(\int\limits_{-\infty}^{\infty} e^{-x^2}\,dx\right)^2 \le \pi \le \left(\int\limits_{-\infty}^{\infty} e^{-x^2}\,dx\right)^2,$$

and thus $\int_{-\infty}^{\infty} e^{-x^2}\,dx = \sqrt{\pi}$.

Exercises

4.6. Compute the following integrals:

(a) $\int_A (x^2+y^2)\,dx\,dy,\ A = \{(x,y)\colon x,y \ge 0,\ x+y \le 1\}$;
(b) $\int_A \sqrt{x^2+y^2}\,dx\,dy,\ A = \{(x,y)\colon x,y > 0,\ x^2+y^2 \le x\}$;
(c) $\int_A \sqrt{y-x^2}\,dx\,dy,\ A = \{(x,y)\colon x^2 \le y \le 4\}$;
(d) $\int_A \sin(x^2+y^2)\,dx\,dy,\ A = \{(x,y)\colon \pi^2 \le x^2+y^2 \le 4\pi^2\}$;
(e) $\int_A \frac{1}{\sqrt{2a-x}}\,dx\,dy$, where A is the part of the disk of center (a,a) and radius a that lies in the half-plane $x \le a$ $(a > 0)$;
(f) $\int_A \frac{\sin x\cdot\sqrt{1+e^{x^2y^2}}}{\operatorname{ch} x\cdot\operatorname{ch} y}\,dx\,dy$, where A is the disk with center at the origin and radius R;
(g) $\int_A |xyz|\,dx\,dy\,dz$, with $A = \{(x,y,z)\colon x^2+\frac{y^2}{4}+\frac{z^2}{9} \le 1\}$. (H)

4.7. Let f be the Riemann function. Which of the following integrals exists?

(a) $\int_{[0,1]\times[0,1]} f(x)\,dx\,dy$;
(b) $\int_0^1 \left(\int_0^1 f(x)\,dy\right)\,dx$;
(c) $\int_0^1 \left(\int_0^1 f(x)\,dx\right)\,dy$.

4.8. Show that if the single-variable functions $f\colon [a,b] \to \mathbb{R}$ and $g\colon [c,d] \to \mathbb{R}$ are integrable, then the function $f(x) \cdot g(y)$ is integrable on the rectangle $[a,b] \times [c,d]$, and its integral is $\left(\int_a^b f(x)\,dx\right) \cdot \left(\int_c^d g(y)\,dy\right)$.

4.9. Show that if f is integrable on the rectangle $[a,b] \times [c,d]$, then the set of points $x \in [a,b]$ where f_x is integrable on $[c,d]$ is everywhere dense in $[a,b]$, and the set of points $y \in [c,d]$ where f^y is integrable on $[a,b]$ is everywhere dense in $[c,d]$. (H)

4.10. Let f be twice continuously differentiable on the rectangle R. Find $\int_R D_{12} f \, dx\,dy$.

4.11. Double-check that equation (4.12) does not hold on the square $[0,1] \times [0,1]$ for the functions below:
(a) $f(x,y) = (x^2 - y^2)/(x^2 + y^2)^2$, if $|x| + |y| \neq 0$ and $f(0,0) = 0$;
(b) $f(x,y) = (x-y)/(x+y)^3$ if $x + y \neq 0$ and $f(x,y) = 0$ if $x + y = 0$.

4.12. Suppose that the set $A \subset [0,1] \times [0,1]$ has a point in every box but does not contain three collinear points. Show that if $f(x,y) = 1$ for every $(x,y) \in A$ and $f(x,y) = 0$ otherwise, then both sides of (4.12) are zero, but f is not integrable on $[0,1] \times [0,1]$.

4.13. Find the center of mass of the following sets:
(a) $\{(x,y)\colon a \le x \le b,\ 0 \le y \le f(x)\}$, where f is nonnegative and integrable on $[a,b]$;
(b) $\{(x,y)\colon x,y \ge 0,\ y \le x^2,\ x + y \le 1\}$;
(c) $\{(x,y)\colon x,y \ge 0,\ \sqrt{x} + \sqrt{y} \le 1\}$;
(d) $\{(x,y)\colon (x^2+y^2)^3 \le 4x^2y^2\}$;
(e) $\{(r,\varphi)\colon r \le R,\ \varphi \in [\alpha,\beta]\}$;
(f) $\{(r,\varphi)\colon r \le 1 + \cos\varphi,\ \varphi \in [0,\pi/2]\}$;
(g) $\{(x,y,z)\colon x,y \ge 0,\ x^2 + y^2 \le z \le x + y\}$.

4.14. Let $0 \le \alpha < \beta \le 2\pi$, and let $f\colon [\alpha,\beta] \to \mathbb{R}$ be nonnegative and integrable. Prove, using Theorem 4.25, that the **sector-like region**

$$\{(r\cos\varphi,\ r\sin\varphi) : 0 \le r \le r(\varphi),\ \alpha \le \varphi \le \beta\}$$

is measurable, and its area is $\frac{1}{2}\int_\alpha^\beta r^2(\varphi)\,d\varphi$.

4.15. Let $f\colon [a,b] \to \mathbb{R}$ have the Lipschitz property. Show that
(a) if $A \subset [a,b]$ has measure zero, then $f(A)$ also has measure zero, and
(b) if $A \subset [a,b]$ is measurable, then $f(A)$ is also measurable.

4.16. Let $f\colon [a,b] \to \mathbb{R}$ be differentiable. Show that

(a) if $A \subset [a,b]$ has measure zero, then $f(A)$ also has measure zero, and
(b) if $A \subset [a,b]$ is measurable, then $f(A)$ is also measurable. (∗∗)

4.17. Let T be the trapezoid bounded by the lines $y = a - x$, $y = b - x$, $y = \alpha x$, $y = \beta x$, with $a < b$ and $0 < \alpha < \beta$. Find the area of T by representing T in the form $f([a, b] \times [\alpha, \beta])$, where f is the inverse of the mapping $(x, y) \mapsto (x + y, y/x)$, and applying the first statement of Theorem 4.22.

4.18. Let D be the region bounded by the hyperbolas $xy = a^2$, $xy = 2a^2$ and the lines $y = x$, $y = 2x$. Find the area of D by representing it in the form $f([a^2, 2a^2] \times [1, 2])$, where f is the inverse of the mapping $(x, y) \mapsto (xy, y/x)$.

4.19. Let $N = [a, b] \times [c, d]$ with $0 < a < b$ and $0 < c < d$, and let $f(x, y) = (y^2/x, \sqrt{xy})$. Find the area of $f(N)$.

4.20. Prove that

(a) $1 + \frac{1}{3^2} + \frac{1}{5^2} + \ldots = \frac{\pi^2}{8}$ and
(b) $1 + \frac{1}{2^2} + \frac{1}{3^2} + \ldots = \frac{\pi^2}{6}$,

using the following exercises. Let T denote the open triangle with vertices $(0, 0)$, $(0, \pi/2)$, $(\pi/2, 0)$, and let $f(x, y) = (\sin x/\cos y, \sin y/\cos x)$ for every $(x, y) \in T$. Show that

(c) f is one-to-one, mapping T onto the open square $N = (0, 1) \times (0, 1)$;
(d) f^{-1} is continuously differentiable on N, and $\det(f^{-1})'(x, y) = \frac{1}{1 - x^2y^2}$ for every $(x, y) \in N$;
(e) if $N_n = (0, 1 - (1/n)) \times (0, 1 - (1/n))$, then $f^{-1}(N_n)$ is measurable with area $\int_{N_n} 1/(1 - x^2y^2)\, dx\, dy$;
(f) $\mu(f^{-1}(N_n)) \to \mu(T) = \pi^2/8$ as $n \to \infty$;
(g) $\displaystyle\lim_{n\to\infty} \int_{N_n} \frac{1}{1 - x^2y^2}\, dx\, dy = 1 + \frac{1}{3^2} + \frac{1}{5^2} + \ldots$.

4.4 First Appendix: Proof of Theorem 4.12

If $A \subset \mathbb{R}^p$ and $\delta > 0$, we call the set

$$U(A, \delta) = \bigcup_{x \in A} B(x, \delta)$$

the **neighborhood of the set** A **with radius** δ. In other words, $U(A, \delta)$ is the set of points y for which there exists a point $x \in A$ such that $|x - y| < \delta$. Since $U(A, \delta)$ is the union of open sets, it is itself open.

Lemma 4.27. *If $A \subset \mathbb{R}^p$ is bounded, then for every $\varepsilon > 0$ there exists $\delta > 0$ such that $\overline{\mu}(U(A, \delta)) < \overline{\mu}(A) + \varepsilon$, where $\overline{\mu}$ is the (Jordan) outer measure.*

Proof. Let $\varepsilon > 0$ be fixed, and choose the boxes $R_i = [a_{i,1}, b_{i,1}] \times \ldots \times [a_{i,p}, b_{i,p}]$ $(i=1,\ldots,n)$ such that $A \subset \bigcup_{i=1}^n R_i$ and $\sum_{i=1}^n \mu(R_i) < \overline{\mu}(A) + \varepsilon$. Let $R_i(\delta) = [a_{i,1} - \delta, b_{i,1} + \delta] \times \ldots \times [a_{i,p} - \delta, b_{i,p} + \delta]$ for every $\delta > 0$. Obviously, for δ small enough, $\sum_{i=1}^n \mu(R_i(\delta)) < \overline{\mu}(A) + \varepsilon$. Fix such a δ. Since $U(A, \delta) \subset \bigcup_{i=1}^n R_i(\delta)$, it follows that $\overline{\mu}(U(A, \delta)) < \overline{\mu}(A) + \varepsilon$. □

Proof of (i) of Theorem 4.12. We show that $S_F < S_{F_0} + \varepsilon$ for every F fine enough. We may assume that f is nonnegative; otherwise, we add a large enough constant c to f. It is easy to see that by adding a constant c to f we increase the value of every upper sum by the same number (namely, by $c \cdot \mu(A)$). Thus, if $S_F < S_{F_0} + \varepsilon$ holds for the function $f + c$, then the same inequality holds for f as well.

Let $0 \le f(x) \le K$ for every $x \in A$, and let $F_0 = \{A_1, \ldots, A_N\}$. Let $F = \{B_1, \ldots, B_n\}$ be a partition finer than δ, and let's calculate how much larger S_F can be, compared to S_{F_0}.

Let $M_i = \sup\{f(x) \colon x \in A_i\}$ $(i = 1, \ldots, N)$ and $M'_j = \sup\{f(x) \colon x \in B_j\}$ $(j = 1, \ldots, n)$. Now, $S_{F_0} = \sum_{i=1}^N M_i \cdot \mu(A_i)$ and $S_F = \sum_{j=1}^n M'_j \cdot \mu(B_j)$.

We partition the set of indices j into two classes, based on whether the set B_j is or is not a subset of one of the sets A_i. Let J_1 and J_2 denote the two classes, respectively. If $j \in J_1$ and $B_j \subset A_i$, then clearly $M'_j \le M_i$. The sum of the products $M'_j \cdot \mu(B_j)$ for which $B_j \subset A_i$ is at most $M_i \cdot \mu\left(\bigcup\{B_j \colon B_j \subset A_i\}\right)$. Now, $M_i \ge 0$ implies that the sum is at most $M_i \cdot \mu(A_i)$. By summing these upper estimates for every $i = 1, \ldots, N$ we get $\sum_{j \in J_1} M'_j \cdot \mu(B_j) \le S_{F_0}$.

We now show that

$$\bigcup_{j \in J_2} B_j \subset U\left(\bigcup_{i=1}^N \partial A_i, \delta\right). \tag{4.21}$$

Indeed, if $j \in J_2$, then B_j is not a subset of any of the sets A_i. For every $x \in B_j$, x is in one of the sets A_i, because $x \in A = \bigcup_{i=1}^N A_i$. Since B_j is not a subset of A_i, there exists a point $y \in B_j$ such that y is not in A_i. We know that the segment $[x, y]$ intersects the boundary of A_i; let $z \in [x, y] \cap \partial A_i$. By assumption, F is finer than δ, i.e., $\operatorname{diam} B_j < \delta$. Thus, we have $|x - y| < \delta$, which implies $|x - z| < \delta$. We have proved that if $j \in J_2$, then every point of B_j is closer than δ to a boundary point of at least one of the A_i, which is exactly (4.21).

The sets A_i are measurable, and thus, by Theorems 3.9 and 3.6, the set $\bigcup_{i=1}^N \partial A_i$ has measure zero. Then, by Lemma 4.27, we can choose $\delta > 0$ such that the right-hand side of (4.21) is less than ε.

Therefore, if the partition F is finer than δ, then

$$\sum_{j \in J_2} M'_j \cdot \mu(B_j) \le K \cdot \sum_{j \in J_2} \mu(B_j) = K \cdot \mu\left(\bigcup_{j \in J_2} B_j\right) \le K \cdot \varepsilon,$$

and

$$S_F = \sum_{j \in J_1} M'_j \cdot \mu(B_j) + \sum_{j \in J_2} M'_j \cdot \mu(B_j) \le S_{F_0} + K \cdot \varepsilon.$$

A similar argument proves that $S_F \geq s_{F_0} - K \cdot \varepsilon$ for every fine enough partition F. □

Proof of (ii) of Theorem 4.12. Since f is integrable, there is a partition F_0 such that $\Omega_{F_0} < \varepsilon/2$. Then we have

$$I - (\varepsilon/2) < s_{F_0} \leq S_{F_0} < I + (\varepsilon/2).$$

Therefore, an application of statement (i) completes the proof. □

4.5 Second Appendix: Integration by Substitution (Proof of Theorem 4.22)

We know that if Q is a cube and A is a linear map, then the volume of the parallelepiped $A(Q)$ is $|\det A| \cdot \mu(Q)$. (See Theorem 3.31.) It seems plausible that if the mapping g is close to the linear map A on the cube Q, then the measure of $g(Q)$ is close to $|\det A| \cdot \mu(Q)$. In the next lemma we show that the outer measure of $g(Q)$ is not much larger than $|\det A| \cdot \mu(Q)$.

Lemma 4.28. *Let $A\colon \mathbb{R}^p \to \mathbb{R}^p$ be a linear map, let $Q \subset \mathbb{R}^p$ be a cube, $c \in Q$, $0 < \delta < 1$, and let $g\colon Q \to \mathbb{R}^p$ be a mapping such that $|g(x) - g(c) - A(x-c)| < \delta|x - c|$ for every $x \in Q$, $x \neq c$. Then*

$$\overline{\mu}(g(Q)) \leq (|\det A| + C\delta) \cdot \mu(Q), \tag{4.22}$$

where the constant C depends only on p and A.

Proof. Let $P = A(Q - c) + g(c)$. Then P is a (possibly degenerate) parallelepiped. If $x \in Q$, then $y = A(x - c) + g(c) \in P$ and

$$|g(x) - y| = |g(x) - g(c) - A(x - c)| < \delta|x - c| \leq \sqrt{p} \cdot h\delta,$$

where h is the side length of Q. (See Example 3.11.) Thus $g(x) \in U(P, \sqrt{p} \cdot h\delta)$ for every $x \in Q$, i.e.,

$$g(Q) \subset U(P, r), \quad \text{where} \quad r = \sqrt{p} \cdot h\delta. \tag{4.23}$$

We show that

$$U(P, r) \subset P \cup U(\partial P, r). \tag{4.24}$$

Indeed, if the points $x \in U(P, r) \setminus P$ and $y \in P$ satisfy $|x - y| < r$, then $[x, y] \cap \partial P \neq \emptyset$, since $x \neq P$ and $y \in P$. If $z \in [x, y] \cap \partial P$, then $|x - z| < r$, and thus $x \in U(\partial P, r)$.

The boundary of the parallelepiped P can be covered by $2p$ hyperplanes. Namely, if $P = P(u_1, \ldots, u_p)$, then the hyperplanes $S_i^j = \{t_1u_1 + \ldots + t_pu_p : t_i = j\}$ ($i = 1, \ldots, p$, $j = 0, 1$) (these are the images of the sides of the cube Q) cover ∂P-t. Therefore,

$$U(P, r) \subset P \cup U(\partial P, r) \subset P \cup \bigcup_{i=1}^{p} \left[U(S_i^0 \cap P, r) \cup U(S_i^1 \cap P, r)\right]. \tag{4.25}$$

Next we prove

$$\overline{\mu}(U(S_i^j \cap P, r)) \leq (M + 2r)^{p-1} \cdot 2r \tag{4.26}$$

for every $i = 1, \ldots, p$ and $j = 0, 1$, where $M = \operatorname{diam} P$. Indeed, for i and j fixed, the set $H = S_i^j \cap P$ lies in a hyperplane, and $\operatorname{diam} H \leq \operatorname{diam} P = M$. Since the outer measure is invariant under isometries, we may assume that $H \subset \mathbb{R}^{p-1} \times \{0\}$. There exists a $(p-1)$-dimensional cube N with side length M such that $P \subset N \times \{0\}$. Thus, $U(P, r) \subset N' \times [-r, r]$, where N' is a $(d-1)$-dimensional cube with side length $M + 2r$, which implies (4.26).

If $x, y \in Q$, then $|Ax - Ay| \leq \|A\| \cdot |x - y| \leq \|A\|\sqrt{p} \cdot h$, which implies $M = \operatorname{diam} P \leq \|A\|\sqrt{p} \cdot h$. Comparing (4.23)–(4.26), we get

$$\begin{aligned}
\overline{\mu}(g(Q)) \leq \overline{\mu}(U(P, r)) &\leq \mu(P) + 4p(M + 2r)^{p-1} \cdot r = \\
&= |\det A| \cdot h^p + 4p(M + 2\sqrt{p} \cdot h\delta)^{p-1} \cdot \sqrt{p} \cdot h\delta \leq \\
&\leq |\det A| \cdot h^p + 4p(\sqrt{p} \cdot h)^p \cdot (\|A\| + 2\delta)^{p-1} \cdot \delta \leq \\
&\leq \left(|\det A| + 4(\|A\| + 2)^{p-1} p^{(p/2)+1} \cdot \delta\right) \cdot \mu(Q),
\end{aligned}$$

which proves (4.22). □

Lemma 4.29. *Let $H \subset G \subset \mathbb{R}^p$, where H is a bounded and closed set and G is open. Suppose that $g : G \to \mathbb{R}^p$ is differentiable at the points of H, and $|\det g'(x)| \leq K$ for every $x \in H$. Then $\overline{\mu}(g(H)) \leq K \cdot \overline{\mu}(H)$.*

Proof. It is enough to prove that $\overline{\mu}(g(Q \cap H)) \leq K \cdot \mu(Q)$ for every cube $Q \in \mathcal{K}_{2^n}$. Indeed, from this we obtain

$$\overline{\mu}(g(H)) \leq \sum_{\substack{Q \in \mathcal{K}_{2^n} \\ Q \cap H \neq \emptyset}} \overline{\mu}(g(Q \cap H)) \leq \sum_{\substack{Q \in \mathcal{K}_{2^n} \\ Q \cap H \neq \emptyset}} K \cdot \mu(Q) = K \cdot \overline{\mu}(H, 2^n),$$

and the right-hand side converges to $K \cdot \overline{\mu}(H)$ as $n \to \infty$. (Here, $\overline{\mu}(H, 2^n)$ denotes the sum of the volumes of the cubes of $\mathcal{K}_{2^n}$ that intersect H.)

We prove by contradiction. Assume that the statement does not hold. Then there exist $n \in \mathbb{N}$, $Q_1 \in \mathcal{K}_{2^n}$, and $0 < \eta < 1$ such that $\overline{\mu}(g(Q_1 \cap H)) > (K + \eta) \cdot \mu(Q_1)$. Since

$$\sum_{\substack{Q\in\mathcal{K}_{2n+1}\\ Q\subset Q_1}} (K+\eta)\cdot\mu(Q) = (K+\eta)\cdot\mu(Q_1) < \overline{\mu}(g(Q_1\cap H)) \le$$
$$\le \sum_{\substack{Q\in\mathcal{K}_{2n+1}\\ Q\subset Q_1}} \overline{\mu}(g(Q\cap H)),$$

there is a cube $Q_2 \in \mathcal{K}_{2^{n+1}}$ such that $\overline{\mu}(g(Q_2\cap H)) > (K+\eta)\cdot\mu(Q_2)$. Repeating this argument, we obtain a sequence of nested cubes $Q_i \in \mathcal{K}_{2^{n+i-1}}$ such that

$$\overline{\mu}(g(Q_i\cap H)) > (K+\eta)\cdot\mu(Q_i) \tag{4.27}$$

for every i. Let $\bigcap_{i=1}^{\infty} Q_i = \{c\}$. Since (4.27) implies $Q_i\cap H \neq \emptyset$ and H is closed, it follows that $c\in H$. By assumption, f is differentiable at c, and $|\det g'(c)| \le K$. Fix $0<\delta<\eta/C$, where C is the constant appearing in (4.22), depending only on $g'(c)$ and p. Since $\operatorname{diam} Q_i \to 0$, we have $Q_i\subset G$ and

$$|g(x)-g(c)-g'(c)(x-c)| \le \delta\cdot|x-c| \qquad (x\in Q_i) \tag{4.28}$$

for every i large enough. By Lemma 4.28, it follows that

$$\overline{\mu}(g(Q_i\cap H)) \le \overline{\mu}(g(Q_i)) \le (K+C\cdot\delta)\cdot\mu(Q_i) < (K+\eta)\cdot\mu(Q_i),$$

which contradicts (4.27). □

Theorem 4.30. *Let $G\subset\mathbb{R}^p$ be open, and let $g\colon G\to\mathbb{R}^p$ be continuously differentiable. If H is measurable and $\operatorname{cl} H\subset G$, then $g(H)$ is also measurable, and*

$$\mu(g(H)) \le \int_H |\det g'(x)|\,dx. \tag{4.29}$$

Proof. First we prove the measurability of $g(H)$. Since g is continuous and $\operatorname{cl} H$ is bounded and closed, it follows from Theorem 2.7, that $g(\operatorname{cl} H)$ is also bounded and closed. Thus $g(H)$ is bounded as well. It is enough to prove that $\partial(g(H))$ has measure zero.

Let $X=\{x\in G\colon \det g'(x)=0\}$. We show that

$$\partial(g(H)) \subset g(\partial H)\cup g(X\cap\operatorname{cl} H). \tag{4.30}$$

Since $g(\operatorname{cl} H)$ is closed, $\operatorname{cl} g(H)\subset g(\operatorname{cl} H)$ and $\partial(g(H))\subset \operatorname{cl} g(H)\subset g(\operatorname{cl} H)$. Let $y\in\partial(g(H))$, i.e., $y=g(x)$ for a suitable $x\in\operatorname{cl} H$. If $x\in\partial H$, then $g(x)\in g(\partial H)$. If, however, $x\notin\partial H$, then $x\in\operatorname{int} H$. In this case, $\det g'(x)=0$, since if $\det g'(x)\neq 0$, then by the open mapping theorem (Corollary 2.37), $y=g(x)\in\operatorname{int} g(H)$, which is impossible. Thus $x\in X$, which proves (4.30).

Now, ∂H is closed and $\det g'$ is bounded on ∂H, since it is continuous. By Lemma 4.29 it follows that $\mu(g(\partial H)) = 0$, since $\mu(\partial H) = 0$ by the measurability of H. The set $X \cap \operatorname{cl} H$ is also closed, since $\det g'$ is continuous on $\operatorname{cl} H$. Since $|\det g'(x)| = 0$ for every $x \in X$, it follows from Lemma 4.29 that $\mu(g(X \cap \operatorname{cl} H)) = 0$. Then we have $\mu(\partial(g(H))) = 0$ by (4.30), which proves the measurability of $g(H)$.

We turn now to the proof of (4.30). Since the function $|\det g'|$ is continuous and bounded, it is integrable on H. Let $\operatorname{cl} H = \bigcup_{i=1}^n H_i$ be a partition, where the sets H_i are closed. If $M_i = \sup\{|\det g'(x)| : x \in H_i\}$, then

$$\mu(g(H)) \le \mu(g(\operatorname{cl} H)) \le \sum_{i=1}^n \mu(g(H_i)) \le \sum_{i=1}^n M_i \cdot \mu(H_i)$$

by Lemma 4.29. The right-hand side will be arbitrarily close to $\int_H |\det g'|$ if the partition is fine enough, which proves (4.30). □

Proof of Theorem 4.22. First we assume that $g\colon G \to \mathbb{R}^p$ is a continuously differentiable injective map with $\det g'(x) \ne 0$ for every $x \in G$. By the open mapping theorem it follows that $g(G)$ is open, and by the inverse function theorem, the map $g^{-1}\colon g(G) \to G$ is continuously differentiable. Therefore, if $A \subset G$ and $x \in G$, then $x \in \operatorname{int} A \iff g(x) \in \operatorname{int} g(A)$ and $x \in \partial A \iff g(x) \in \partial(g(A))$.

Let H be measurable with $\operatorname{cl} H \subset G$. Then $g(H)$ is also measurable by Theorem 4.30. Let f be a nonnegative and bounded function on $g(H)$. Let $\varepsilon > 0$ be fixed and let $F\colon g(H) = \bigcup_{i=1}^n A_i$ be a partition such that the sets $A_1, \dots, A_n$ are mutually disjoint, and $s_F(f) > \underline{\int}_{g(H)} f - \varepsilon$. The sets $H_i = g^{-1}(A_i)$ $(i = 1, \dots, n)$ form a partition of H into disjoint sets. (The measurability of the sets H_i follows from Theorem 4.30 applied to the mapping g^{-1}.) Let $m_i = \inf\{f(x) : x \in A_i\} = \inf\{f(g(x)) : x \in H_i\}$, and let γ denote the function defined by $\gamma(x) = m_i$ for $x \in H_i$. By Theorem 4.30 we have

$$\begin{aligned}
\left(\underline{\int}_{g(H)} f\right) - \varepsilon < s_F(f) &= \sum_{i=1}^n m_i \cdot \mu(A_i) = \sum_{i=1}^n m_i \cdot \mu(g(H_i)) \le \\
&\le \sum_{i=1}^n m_i \cdot \int_{H_i} |\det g'(x)|\, dx = \\
&= \int_H \gamma(x) \cdot |\det g'(x)|\, dx \le \\
&\le \underline{\int}_H f(g(x)) \cdot |\det g'(x)|\, dx,
\end{aligned} \tag{4.31}$$

since $\gamma \le f \circ g$. This is true for every ε, whence

$$\underline{\int}_{g(H)} f \le \underline{\int}_{H} f(g(x)) \cdot |\det g'(x)|\, dx. \tag{4.32}$$

The function $\varphi = g^{-1}$ is again continuously differentiable, and we also know that

$$\det g'(\varphi(x)) \cdot \det \varphi'(x) = 1 \tag{4.33}$$

for every $x \in g(G)$. (This is a corollary of the differentiation rule for composite functions (Theorem 2.20), using the fact that $g(\varphi(x)) = x$ for every $x \in g(G)$.) If we apply (4.32) with φ in place of g, with $g(H)$ in place H, and with $(f \circ g) \cdot |\det g'|$ in place of f, then we get

$$\underline{\int}_{H} f(g(x)) \cdot |\det g'(x)|\, dx \le \underline{\int}_{g(H)} f.$$

(Note that $(f \circ g) \cdot |\det g'|$ is bounded and nonnegative on $g(H)$, and $(f \circ g)(\varphi(x)) = f(x)$. We also used (4.33).) Comparing this with (4.32), we get

$$\underline{\int}_{g(H)} f = \underline{\int}_{H} f(g(x)) \cdot |\det g'(x)|\, dx. \tag{4.34}$$

Applying this to the function $f \equiv 1$, we obtain (4.13). It follows that if (4.34) holds for a function f, then it also holds for $f + c$ with an arbitrary choice of the constant c. Thus, (4.34) holds for every bounded function. Applying (4.34) for $-f$ and multiplying both sides by (-1) gives

$$\overline{\int}_{g(H)} f = \overline{\int}_{H} f(g(x)) \cdot |\det g'(x)|\, dx \tag{4.35}$$

for every bounded function f. Obviously, if either the left-hand side or the right-hand side of (4.14) exists, then the other side exists as well and they are equal. Thus, we have proved the theorem assuming that g is injective on G and $\det g' \neq 0$.

If we assume that g is injective only on $\operatorname{int} H$ and we also allow the case $\det g' = 0$, then we argue as follows. Let $X = \{x \in G\colon \det g'(x) = 0\}$. Since $\det g'$ is uniformly continuous on $\operatorname{cl} H$, we have that for a fixed $\varepsilon > 0$ there exists $\delta > 0$ such that $x, y \in \operatorname{cl} H$ and $|x - y| < \delta$ implies $|\det g'(x) - \det g'(y)| < \varepsilon$. For n fixed, let

$$A = \bigcup\{Q \in \mathcal{K}_n\colon Q \subset (\operatorname{int} H) \setminus X\},$$
$$B = \bigcup\{Q \in \mathcal{K}_n\colon Q \subset \operatorname{int} H,\ Q \cap X \neq \emptyset\},$$
$$D = \bigcup\{Q \in \mathcal{K}_n\colon Q \cap \partial H \neq \emptyset\}.$$

If $n > \sqrt{p}/\delta$, then the diameter of the cubes is less than δ, and $|\det g'(x)| < \varepsilon$ for every $x \in B$. By Lemma 4.29, $\mu(g(B)) \le \varepsilon \cdot \mu(B)) \le \varepsilon \cdot \mu(H)$. Since $\mu(D) = \overline{\mu}(\partial H, n)$, it follows that $\mu(D) < \varepsilon$ also holds for n large enough. This implies $\mu(g(\mathrm{cl}\, H \cap D)) \le M \cdot \varepsilon$, where $M = \max_{x \in \mathrm{cl}\, H} |\det g'(x)|$.

For an arbitrary bounded function $f\colon g(H) \to \mathbb{R}$ we have

$$\underline{\int}_{g(A)} f = \underline{\int}_{A} f(g(x)) \cdot |\det g'(x)|\, dx,$$

since g is injective on the open set $(\mathrm{int}\, H) \setminus X$ and $\det g' \ne 0$ there. Furthermore, $A \subset (\mathrm{int}\, H) \setminus X$, and we already proved the theorem in this case. Since

$$g(H) \setminus g(A) \subset g(H \cap B) \cup g(H \cap D)$$

implies $\mu(g(H) \setminus g(A)) \le (\mu(H) + M) \cdot \varepsilon$, we have that $\underline{\int}_{g(A)} f$ and $\underline{\int}_{g(H)} f$ differ from each other by at most $K \cdot (\mu(H) + M) \cdot \varepsilon$, where $K = \sup_{x \in g(H)} |f(x)|$.

We show that the lower integrals of the function $F = (f \circ g) \cdot |\det g'|$ on the sets H and A differ from each other by at most $K \cdot (\mu(H) + M) \cdot \varepsilon$. Indeed, $H = A \cup B \cup (H \cap D)$ is a partition of the set H, and thus

$$\underline{\int}_{H} F = \underline{\int}_{A} F + \underline{\int}_{B} F + \underline{\int}_{H \cap D} F.$$

Now, $\left|\underline{\int}_B F\right| \le K \cdot \varepsilon \cdot \mu(B)$, since $|\det g'| < \varepsilon$, and $|F| \le K \cdot \varepsilon$ on the set B. Furthermore, $\left|\underline{\int}_{H \cap D} F\right| \le K \cdot M \cdot \mu(H \cap D) < KM\varepsilon$, which implies our statement.

Summing up, we get that the two sides of (4.34) can differ from each other by at most $2K \cdot (\mu(H) + M) \cdot \varepsilon$. Since ε was arbitrary, (4.34) holds for every bounded function. As we have shown above, (4.14) follows, which proves Theorem 4.22. $\square$

Chapter 5
Integrals of Multivariable Functions II

5.1 The Line Integral

The notion of the line integral was motivated by some problems in physics. One of these problems is the computation of the work done by a force that changes while moving a point. The mathematical model describing the situation is the following.

Let $G \subset \mathbb{R}^3$ be an open set, and let a force act at every point x of G. This force is described by $f(x) \in \mathbb{R}^3$ in the sense that the magnitude of the force is $|f(x)|$, and its direction is the same as the direction of the vector $f(x)$. We say that the pair (G, f) describes a **field of force** or briefly a **field**. For example, let us place a point at the origin having mass m. By Newton's[1] law of gravity, this point attracts a unit-weight point $x \neq 0$ with a force of magnitude $\kappa \cdot m/|x|^2$ and of direction opposite to that of x (where κ is the gravitational constant). This field—called the **gravitational field**—is defined by the open set $G = \mathbb{R}^3 \setminus \{0\}$ and by the function $f(x) = -\kappa m \cdot x/|x|^3$.

Suppose that the force moves a point. We want to find the work that the force exerts along the direction of motion. We know that if the motion is linear and the force is constant and acts in the direction of the point's motion, then the work is the product of the force's magnitude and the displacement of the point. If the force acts in the opposite direction to the motion, then the work is the negative of this product.

Now let the motion of the point be linear and the force constant, but let the force act in a direction different from that of the motion. If the motion is perpendicular to the direction of the force, then by a law of physics, there is no work done[2].

---

[1] Isaac Newton (1643–1727), English mathematician, astronomer, and physicist.

[2] The direction of the motion of a planet traveling in a circular orbit around the sun is always perpendicular to the direction of the force acting toward the center of the circle. Therefore, no work is done during the motion of the planet, and this is the reason why the planets can keep on orbiting indefinitely, at least in theory.

M. Laczkovich and V.T. Sós, *Real Analysis*, Undergraduate Texts in Mathematics, https://doi.org/10.1007/978-1-4939-7369-9_5

By a law of physics, work is additive in the following sense: the work done by the sum of two forces is the sum of the works done by each force. Let a point move along the segment $[u, v]$ from u to v while a constant f force acts on it. Let $f = h_1 + h_2$, where h_1 is parallel to $v - u$ and h_2 is perpendicular to $v - u$. The work done is $|h_1| \cdot |v - u| + 0$ if the direction of h_1 is the same as that of the vector $(v - u)$, and $-|h_1| \cdot |v - u| + 0$ if the direction of g is opposite to that of the vector $(v - u)$. We can see that the amount of work done is equal to the scalar product $\langle f, v - u \rangle$ in both cases.

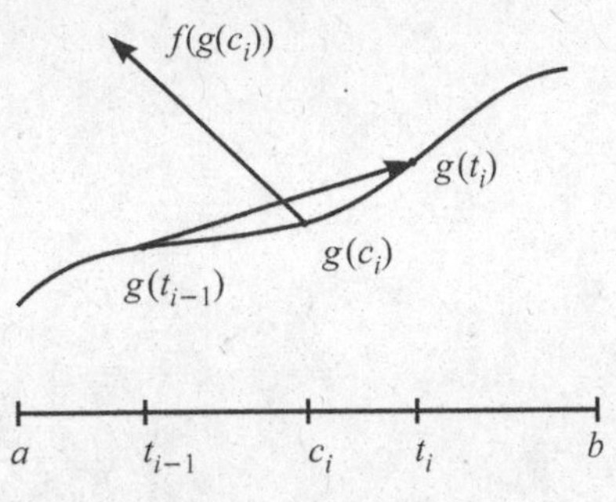

5.1. Figure

In the general case, the motion of the point is described by a curve $g\colon [a, b] \to \mathbb{R}^3$. Suppose that a force of magnitude and direction $f(g(t))$ acts at the point at $g(t)$ for every $t \in [a, b]$. Consider a fine partition $a = t_0 < t_1 < \ldots < t_n = b$, and suppose that the subarc γ_i of g corresponding to the interval $[t_{i-1}, t_i]$ is well approximated by the segment $[g(t_{i-1}), g(t_i)]$. Furthermore, let the force be close to a constant on the arc γ_i. (If the curve is continuously differentiable and the function f is continuous, then these conditions hold for a fine enough partition.) Then the work done by the force on the arc γ_i will be close to the scalar product $\langle f(g(c_i)), g(t_i) - g(t_{i-1}) \rangle$, where $c_i \in [t_{i-1}, t_i]$ is arbitrary. Since the total amount of work done by the force is the sum of the works done along the arcs γ_i, the work can be approximated by the sums $\sum_{i=1}^{n} \langle f(g(c_i)), g(t_i) - g(t_{i-1}) \rangle$. Clearly, if there exists a number I such that these sums get arbitrarily close to I a the partition becomes increasingly finer, then I is the amount of the total work. This motivates the following definition of the line integral.

Definition 5.1. Let $g\colon [a, b] \to \mathbb{R}^p$ be a curve mapping to $\mathbb{R}^p$, and let $f\colon g([a, b]) \to \mathbb{R}^p$. We say that *the line integral* $\int_g \langle f, dx \rangle$ exists *and its value is the number* I if for every $\varepsilon > 0$ there exists $\delta > 0$ such that for every partition $a = t_0 < t_1 < \ldots < t_n = b$ finer than δ and for arbitrary points $c_i \in [t_{i-1}, t_i]$ $(i = 1, \ldots, n)$ we have

$$\left| I - \sum_{i=1}^{n} \langle f(g(c_i)), g(t_i) - g(t_{i-1}) \rangle \right| < \varepsilon. \tag{5.1}$$

Let the coordinate functions of f and g be $f_1, \ldots, f_p$ and $g_1, \ldots, g_p$, respectively. The sum in (5.1) becomes

$$\sum_{i=1}^{n}\sum_{j=1}^{p} f_j(g(c_i))\cdot(g_j(t_i)-g_j(t_{i-1})) =$$

$$= \sum_{j=1}^{p}\left(\sum_{i=1}^{n} f_j(g(c_i))\cdot(g_j(t_i)-g_j(t_{i-1}))\right) \stackrel{\text{def}}{=} \sum_{j=1}^{p} S_j. \qquad (5.2)$$

The sums S_j are nothing but than the approximation sums of the Stieltjes[3] integral[4] $\int_a^b (f_j \circ g)\, dg_j$. Obviously, if every S_j is close to a number I_j if the partition is fine enough, then their sum is close to $I = I_1 + \ldots + I_p$. This observation motivates the following concept.

Definition 5.2. Let $g = (g_1, \ldots, g_p)\colon [a, b] \to \mathbb{R}^p$ be a curve mapping to $\mathbb{R}^p$, let a real function h be defined on the set $g([a, b])$, and let $1 \le j \le p$ be fixed. We say that *the line integral $\int_g h\, dx_j$ exists (with respect to x_j) and it is equal to I*, if the Stieltjes integral $\int_a^b (h \circ g)\, dg_j$ exists and equals I.

For $p = 2$, we may write $\int_g h\, dx$ and $\int_g h\, dy$ instead of $\int_g h\, dx_1$ and $\int_g h\, dx_2$, respectively. For $p = 3$, we may also use the notation $\int_g h\, dz$ instead of $\int_g h\, dx_3$.

Remarks 5.3. **1.** Let $1 \le j \le p$ be fixed and let the coordinate function g_j be continuous and strictly monotone on the parameter interval $[a, b]$. One can show that the existence of the line integral $\int_g h\, dx_j$ is equivalent to the existence of the Riemann integral $\int_{g_j(a)}^{g_j(b)} H(u)\, du$, where the function H is the composition of the mappings

$$[g_j(a), g_j(b)] \to [a, b] \to g([a, b]) \to \mathbb{R}.$$

The first mapping is the inverse of g_j, the second is g, and the third is h. (See [7, Theorem 18.5].)

2. If a segment is parallel to the x_1-axis then among the p line integrals taken on this segment, only the one with respect to dx_1 can be nonzero. Indeed, let $a = (a_1, a_2, \ldots, a_p)$, $b = (b_1, a_2, \ldots, a_p)$, and let $g\colon [a_1, b_1] \to \mathbb{R}^p$ be an arbitrary parametrization of the segment $[a, b]$; i.e., let $g([a_1, b_1]) = [a, b]$. For every function

---

[3] Thomas Joannes Stieltjes (1856–1894), Dutch mathematician.

[4] Let $f, g\colon [a, b] \to \mathbb{R}$. We say that the **Stieltjes integral** $\int_a^b f\, dg$ exists and its value equals I if for every $\varepsilon > 0$ there exists $\delta > 0$ such that if $F\colon a = x_0 < x_1 < \ldots < x_n = b$ is a partition of $[a, b]$ with mesh smaller than δ and $c_i \in [x_{i-1}, x_i]$ $(i = 1, \ldots, n)$ are arbitrary inner points, then $|\sigma_F(f, g; (c_i)) - I| < \varepsilon$, where $\sigma_F(f, g; (c_i)) = \sum_{i=1}^{n} f(c_i)\cdot(g(x_i) - g(x_{i-1}))$. For the basic properties of the Stieltjes integral, see [7, Chapter 18].

$h\colon [a,b] \to \mathbb{R}$ and $j \neq 1$, the line integral $\int_g h\,dx_j$ exists and its value is zero, since $j \neq 1$ implies that the coordinate function g_j is constant, and thus every approximation sum is zero.

Similar statements are true for segments parallel to the other axes.

3. Let $a = (a_1, a_2, \ldots, a_p)$, $b = (b_1, a_2, \ldots, a_p)$, and let $g\colon [a_1, b_1] \to \mathbb{R}^p$ be a one-to-one continuous parametrization of the segment $[a,b]$; i.e., let g be a continuous bijection between $[a_1, b_1]$ and $[a,b]$ with $g(a_1) = a$ and $g(b_1) = b$. (For example, the function $g(t) = (t, a_2, \ldots, a_p)$ $(t \in [a_1, b_1])$ has this property.)

For every function $h\colon [a,b] \to \mathbb{R}$, the line integral $\int_g h\,dx_1$ exists if and only if the section $h^{(a_2,\ldots,a_p)}(t) = h(t, a_2, \ldots, a_p)$ is Riemann integrable on the segment $[a_1, b_1]$, and the line integral is equal to the Riemann integral $\int_{a_1}^{b_1} h^{(a_2,\ldots,a_p)}(t)\,dt$.

This follows from the fact that the line integral and the Riemann integral have the same approximation sums.

4. Based on Definitions 5.1 and 5.2 it is clear that if

$$f = (f_1, \ldots, f_p)\colon g([a,b]) \to \mathbb{R}^p$$

and the line integrals $\int_g f_j\,dx_j$ exist for every $j = 1, \ldots, p$, then the line integral $\int_g \langle f, dx\rangle$ also exists and

$$\int_g \langle f, dx\rangle = \sum_{j=1}^{p} \int_g f_j\,dx_j. \tag{5.3}$$

5. The converse of the statement above is not true, since if f is perpendicular to $g(u) - g(t)$ for every $t, u \in [a,b]$, then the left-hand side of (5.3) exists and it is zero (since every approximating sum is zero), while the line integrals $\int_g f_j\,dx_j$ do not necessarily exist.

For example, let $p = 2$, $g(t) = (t,t)$ for every $t \in [0,1]$, and $f(t,t) = (h(t), -h(t))$ for every $t \in [0,1]$, where $h\colon [0,1] \to \mathbb{R}$ is an arbitrary function. The left-hand side of (5.3) exists and is equal to zero. On the other hand, the line integral $\int_g f_1\,dx_1$ exists if and only if the Riemann integral $\int_0^1 h(t)\,dt$ also exists.

It is well known that if $f\colon [a,b] \to \mathbb{R}$ is continuous and $g\colon [a,b] \to \mathbb{R}$ is of bounded variation, then the Stieltjes integral $\int_a^b f\,dg$ exists. Furthermore, if g is differentiable and g' is integrable on $[a,b]$, then the value of the Stieltjes integral is $\int_a^b f \cdot g'\,dx$. (See [7, Theorems 18.10 and 18.12].) These results together with the connection between line integrals and Stieltjes integrals give the following.

Theorem 5.4. *Let $g = (g_1, \ldots, g_p)\colon [a, b] \to \mathbb{R}^p$ be a continuous and rectifiable*[5] *curve, and let the function $h\colon g([a, b]) \to \mathbb{R}$ be continuous on the set $g([a, b])$. The line integral $\int_g h\, dx_j$ exists for every $j = 1, \ldots, p$.* □

Theorem 5.5. *Let $g = (g_1, \ldots, g_p)\colon [a, b] \to \mathbb{R}^p$ be a differentiable curve, and let g_j' be integrable on $[a, b]$ for every $j = 1, \ldots, p$. If the function $h\colon g([a, b]) \to \mathbb{R}$ is continuous on the set $g([a, b])$, then the line integral $\int_g h\, dx_j$ exists, and it is $\int_a^b h(g(t)) \cdot g_j'(t)\, dt$, for every $j = 1, \ldots, p$.* □

Remark 5.6. If we also assume that g_j is strictly monotone on the parameter interval $[a, b]$ then by Remark 5.3.1, it follows that $\int_g h\, dx_j$ is equal to the Riemann integral $\int_{g_j(a)}^{g_j(b)} H(u)\, du$, where $H = h \circ g \circ (g_j)^{-1}$. It is easy to see that this latter integral turns into $\int_a^b h(g(t)) \cdot g_j'(t)\, dt$, with the substitution $u = g_j(t)$.

Example 5.7. Let $g(t) = (R\cos t, R\sin t)$ $(t \in [0, 2\pi])$ be the usual parametrization of the circle centered at the origin with radius R. Find the line integrals $\int_g x^2 y\, dx$ and $\int_g xy^2\, dy$. Since the conditions of Theorem 5.5 hold, the value of the first integral is

$$\int_0^{2\pi} R^3 \cos^2 t \cdot \sin t \cdot (-R\sin t)\, dt = -\frac{R^4}{4} \int_0^{2\pi} (\sin 2t)^2\, dt =$$

$$= -\frac{R^4}{8} \int_0^{2\pi} (1 - \cos 4t)\, dt = -R^4\pi/4,$$

and similarly $\int_g xy^2\, dy = R^4\pi/4$.

We could have guessed, without any calculation, that the two values need to be the negatives of each other. Indeed, the sum $M = \int_g x^2 y\, dx + \int_g xy^2\, dy$ is the work done by the force $(x^2 y, y^2 x)$ along the circle centered at the origin with radius R. Since the vector $(x^2 y, y^2 x)$ is parallel to (x, y), which is perpendicular to the tangent to the circle at the point (x, y), there is no work done, i.e., $M = 0$.

Combining Theorem 5.5 and (5.3), we obtain the following theorem.

Theorem 5.8. *Let $g = (g_1, \ldots, g_p)\colon [a, b] \to \mathbb{R}^p$ be a differentiable curve, and let g_j' be integrable on $[a, b]$ for every $j = 1, \ldots, p$. If the function $f\colon g([a, b]) \to \mathbb{R}^p$ is continuous on the set $g([a, b])$, then the line integral $\int_g \langle f, dx \rangle$ exists and equals $\int_a^b \langle f(g(t)), g'(t) \rangle\, dt$.* □

Our next aim is to prove the analogue of the Newton–Leibniz[6] formula for line integrals. First, we need to generalize the notion of primitive function to multivariable functions.

---

[5] We say that a curve is rectifiable if its length is finite. See [7, Definition 16.15].

[6] Gottfried Wilhelm (von) Leibniz (1646–1716), German mathematician and philosopher.

Definition 5.9. Let $G \subset \mathbb{R}^p$ be open and let $f = (f_1, \ldots, f_p) \colon G \to \mathbb{R}^p$. We say that *the function $F \colon G \to \mathbb{R}$ is a primitive function of the function f* if F is differentiable on G and $F' = f$, i.e., if $D_j F = f_j$ $(j = 1, \ldots, p)$ on the set G.

Example 5.10. Let $p = 3$, $G = \mathbb{R}^3 \setminus \{(0,0,0)\}$, and let

$$f(x,y,z) = \left(\frac{x}{(x^2+y^2+z^2)^{3/2}}, \frac{y}{(x^2+y^2+z^2)^{3/2}}, \frac{z}{(x^2+y^2+z^2)^{3/2}} \right),$$

for every $(x,y,z) \in G$. The function $F(x,y,z) = -1/\sqrt{x^2+y^2+z^2}$ is a primitive function of f on G.

Theorem 5.11. (The Newton–Leibniz formula (for line integrals)) *Let $G \subset \mathbb{R}^p$ be open and let $F \colon G \to \mathbb{R}$ be a primitive function of the continuous function $f \colon G \to \mathbb{R}^p$. Then for every continuous and rectifiable curve $g \colon [a,b] \to G$ we have $\int_g \langle f, dx \rangle = F(g(b)) - F(g(a))$.*

Proof. The set $K = g([a,b])$ is a bounded and closed subset of the open set G. First, we prove that there exist a bounded and closed set D and a positive number r with the following properties: $K \subset D \subset G$ and $B(z,r) \subset D$ for every $z \in K$, where $B(z,r)$ denotes the ball with center z and radius r.

Choose a positive number r_z for every point $z \in K$ such that $B(z, r_z) \subset G$. The balls $B(z, r_z/3)$ $(z \in K)$ cover the set K. Since K is compact, it follows from Borel's covering theorem (see Theorem 1.31) that finitely many of these balls cover K. Suppose that the balls $B(z_i, r_{z_i}/3)$ $(i = 1, \ldots, N)$ cover K. Let $\overline{B}_i = \overline{B}(z_i, 2r_{z_i}/3)$ for every $(i = 1, \ldots, N)$. The set

$$D = \overline{B}_1 \cup \ldots \cup \overline{B}_N$$

is bounded and closed, it contains K, and it is a subset of G, since

$$\overline{B}_i \subset B(z_i, r_{z_i}) \subset G$$

for every i. Let r denote the minimum of the numbers $r_{z_i}/3$ $(i = 1, \ldots, N)$. Let $z \in K$ be arbitrary. Since the balls $B(z_i, r_{z_i}/3)$ cover K, there exists i such that $z \in B(z_i, r_{z_i}/3)$. Thus by $r \le r_{z_i}/3$ we have

$$B(z,r) \subset B(z_i, r + r_{z_i}/3) \subset B(z_i, 2r_{z_i}/3) \subset D.$$

We have proved that the set D and the number r have the desired properties.

Since D is bounded and closed, it follows from Heine's theorem (Theorem 1.53) that f is uniformly continuous on D. Let $\varepsilon > 0$ be fixed, and choose $0 < \eta < r$ such that $|f(x) - f(y)| < \varepsilon$ holds for every $x, y \in D$ with $|x - y| < \eta$.

The function g is uniformly continuous on $[a,b]$; thus there exists $\delta > 0$ such that $|g(u) - g(v)| < \eta$, whenever $u, v \in [a,b]$ and $|u - v| < \delta$.

By Theorem 5.4, the line integral $\int_g \langle f, dx\rangle$ exists; let its value be I. Choose a partition $a = t_0 < t_1 < \ldots < t_n = b$ finer than δ and such that

$$\left| I - \sum_{i=1}^{n} \langle f(g(c_i)), g(t_i) - g(t_{i-1})\rangle \right| < \varepsilon$$

for arbitrary points $c_i \in [t_{i-1}, t_i]$. Denote the point $g(t_i)$ by y_i for every $i = 1, \ldots, n$. By the choice of δ, we have $|y_i - y_{i-1}| < \eta < r$, and thus the segment $[y_{i-1}, y_i]$ is contained by D, since $y_i \in K$ and $|y - y_i| < r$ for every $y \in [y_{i-1}, y_i]$. Then by the mean value theorem (Theorem 1.79), there is a point $d_i \in [y_{i-1}, y_i]$ such that

$$F(y_i) - F(y_{i-1}) = \langle F'(d_i), y_i - y_{i-1}\rangle = \langle f(d_i), y_i - y_{i-1}\rangle. \qquad (5.4)$$

Since $c_i \in [t_{i-1}, t_i]$, we have that $|c_i - t_{i-1}| < \delta$ and $|c_i - t_i| < \delta$, and then $|g(c_i) - y_{i-1}| < \eta$ and $|g(c_i) - y_i| < \eta$ follow. Therefore, $[y_{i-1}, y_i] \subset B(g(c_i), \eta)$ (since every ball is convex), which implies $d_i \in B(g(c_i), \eta)$, i.e., $|g(c_i) - d_i| < \eta$. Thus $|f(g(c_i)) - f(d_i)| < \varepsilon$. Then by (5.4) we obtain

$$\begin{aligned} |F(y_i) - F(y_{i-1}) - \langle f(g(c_i)), y_i - y_{i-1}\rangle| &= |\langle f(d_i) - f(g(c_i)), y_i - y_{i-1}\rangle| \le \\ &\le |f(d_i) - f(g(c_i))| \cdot |y_i - y_{i-1}| \le \\ &\le \varepsilon \cdot |y_i - y_{i-1}|. \end{aligned}$$

Summing these inequalities yields

$$\left| F(g(b)) - F(g(a)) - \sum_{i=1}^{n} \langle f(g(c_i)), y_i - y_{i-1}\rangle \right| \le \varepsilon \cdot \sum_{i=1}^{n} |y_i - y_{i-1}| \le \varepsilon \cdot L,$$

where L denotes the length of the curve g. Thus $|F(g(b)) - F(g(a)) - I| \le \varepsilon \cdot (L+1)$, and since ε was arbitrary, we get $F(g(b)) - F(g(a)) = I$. □

Remarks 5.12. **1.** If every component g_j of the curve g is differentiable with an integrable derivative on $[a,b]$, then $\int_g \langle f, dx\rangle = F(g(b)) - F(g(a))$ easily follows from Theorem 5.8. Indeed, the differentiation rule of composite functions (Corollary 2.23) gives

$$(F \circ g)' = D_1 F(g) \cdot g_1' + \ldots + D_p F(g) \cdot g_p' = f_1(g) \cdot g_1' + \ldots + f_p(g) \cdot g_p'.$$

Thus the derivative of the single-variable function $F \circ g$ is $\langle f \circ g, g'\rangle$; in other words, $F \circ g$ is a primitive function of $\langle f \circ g, g'\rangle$ on $[a,b]$. Then the (one-variable) Newton–Leibniz formula gives

$$\int_g \langle f, dx\rangle = \int_a^b \langle f(g(t)), g'(t)\rangle\, dt = F(g(b)) - F(g(a)).$$

2. The interpretation of Theorem 5.11 from the point of view of physics is as follows: if the function describing a field has a primitive function, then the work done along any (continuous and rectifiable) curve depends only on the initial point and the endpoint of the curve, and the amount of work done is the increment of the primitive function between these two points. In this context the negative of the primitive function is called the **potential function.**[7]

If a field has a potential function, we call it a **conservative field**. By Example 5.10, the gravitational field of a point mass is conservative. (We will soon see that not every continuous mapping has a primitive function; see Example 5.162.)

Exercises

5.1. Find the following line integrals.[8]

(a) $\int_g (x^2 + y^2)\, dx$, with $g(t) = (t, t)$ $(t \in [0, 1])$;

(b) $\int_g e^x\, dx$, where $g\colon [a, b] \to \mathbb{R}^2$ is an arbitrary continuous and rectifiable curve;

(c) $\int_g e^x\, dy$, where $g(t) = (t, t^2)$ $(t \in [0, 1])$;

(d) $\int_g \sin y\, dy$, where $g\colon [a, b] \to \mathbb{R}^2$ is an arbitrary continuous and rectifiable curve;

(e) $\int_g (x^2 - 2xy)\, dx - \int_g (x^2 - 3y^2)\, dy$, where $g\colon [a, b] \to \mathbb{R}^2$ is an arbitrary continuous and rectifiable curve;

(f) $\int_g (x^2 - 2xy)\, dx$, where $g(t) = (t, t^2)$ $(t \in [0, 1])$;

(g) $\int_g f(x)\, dx + \int_g h(y)\, dy$, where $f, h\colon \mathbb{R} \to \mathbb{R}$ are continuous functions, and $g\colon [a, b] \to \mathbb{R}^2$ is a continuous and rectifiable curve;

(h) $\int_g \operatorname{arc\,tg}(e^x - \sin x)\, dy$, where g is a parametrization of the boundary of the rectangle $[a, b] \times [c, d]$;

(i) $\int_g f(x^2 + y^2) \cdot x\, dx + \int_g f(x^2 + y^2) \cdot y\, dy$, where $f\colon \mathbb{R} \to \mathbb{R}$ is continuous, and $g\colon [a, b] \to \mathbb{R}^2$ is a continuous and rectifiable curve.

---

[7] The work increases the energy, and thus the difference of the values of the potential function between two points is nothing but the increment of the potential energy between the points. This motivates the nomenclature.

[8] A note on notation: we use the notation $\operatorname{tg} x$ and $\operatorname{ctg} x$ for the functions $\sin x/\cos x$ and $\cos x/\sin x$. The inverse of the restriction of $\operatorname{tg} x$ to $(-\pi/2, \pi/2)$ is denoted by $\operatorname{arc\,tg} x$.

5.2. Let $G \subset \mathbb{R}^p$ be a connected open set, and suppose that the mapping $f\colon G \to \mathbb{R}^p$ has a primitive function on G. Show that the difference of any two primitive functions of f is constant. (H)

5.3. Let $a < b < c$, let $u, v \in \mathbb{R}^p$ be nonzero perpendicular vectors, and let $g_1(t) = u$ if $a \le t < b$, $g_1(b) = g_2(b) = 0$, $g_2(t) = v$ if $b < t \le c$. Furthermore, let $g(t) = g_1(t)$ if $t \in [a, b]$, and $g(t) = g_2(t)$ if $t \in [b, c]$. Show that if $f(0) = f(u) = u$ and $f(v) = 0$, then the line integrals $\int_{g_1} \langle f, dx\rangle$ and $\int_{g_2} \langle f, dx\rangle$ exist, but $\int_g \langle f, dx\rangle$ does not exist.

5.4. Let $g_1\colon [a, b] \to \mathbb{R}^p$ and $g_2\colon [b, c] \to \mathbb{R}^p$ be continuous curves with $g_1(b) = g_2(a)$. Let $g(t) = g_1(t)$ if $t \in [a, b]$, and $g(t) = g_2(t)$ if $t \in [b, c]$. Show that if f is bounded and the line integrals $\int_{g_1} \langle f, dx\rangle$ and $\int_{g_2} \langle f, dx\rangle$ exist, then $\int_g \langle f, dx\rangle$ also exists and $\int_g \langle f, dx\rangle = \int_{g_1} \langle f, dx\rangle + \int_{g_2} \langle f, dx\rangle$.

5.5. Let the function $\varphi\colon [a, b] \to \mathbb{R}$ be continuous, and let $g(t) = (t, \varphi(t))$ for every $t \in [a, b]$. Show that for every continuous function $f\colon (\operatorname{graph} \varphi) \to \mathbb{R}$, the line integral $\int_g f\, dx$ exists and equals $\int_a^b f(t, \varphi(t))\, dt$.

5.2 Conditions for the Existence of the Primitive Function

We call a curve $g\colon [a, b] \to \mathbb{R}^p$ a **closed curve** if $g(a) = g(b)$. By Theorem 5.11, if a continuous function f has a primitive function on an open set G, then the line integral of f on every continuous and rectifiable closed curve lying in G is zero. Our next aim is to prove the converse of this statement.

Let the curve g_1 start at the point x and end at the point y, and let the curve g_2 start at the point y and end at the point z. Intuitively it is clear that if we join the curves g_1 and g_2, then the integral along the resulting curve g is equal to the sum of the integrals along the curves g_1 and g_2 (think of the additivity of the work done). We prove that this statement indeed holds under certain extra conditions (but not without them; see Exercise 5.3).

Lemma 5.13. *Let $g_1\colon [a, b] \to \mathbb{R}^p$ and $g_2\colon [b, c] \to \mathbb{R}^p$ be continuous rectifiable curves, with $a < b < c$ and $g_1(b) = g_2(b)$. Let $g(t) = g_1(t)$ if $t \in [a, b]$, and $g(t) = g_2(t)$ if $t \in [b, c]$. Then the curve $g\colon [a, c] \to \mathbb{R}^p$ is also continuous and rectifiable; furthermore, for every continuous function $f\colon g([a, c]) \to \mathbb{R}^p$ we have*

$$\int_g \langle f, dx\rangle = \int_{g_1} \langle f, dx\rangle + \int_{g_2} \langle f, dx\rangle. \tag{5.5}$$

Proof. We leave the proof of the continuity and rectifiability of g to the reader. The integral of (5.5) exists by Theorem 5.4. Let F be a fine partition of the interval $[a, c]$ in which b is one of the division points. Then the approximating sum for F will be close to the left-hand side of (5.5). On the other hand, dividing the sum into two parts gives a pair of approximating sums corresponding to the intervals $[a,b]$ and $[b,c]$, respectively, which are close to the corresponding terms on the right-hand side of (5.5). Hence, equation (5.5) follows. The precise proof is again left to the reader. □

Note that equation (5.5) holds even if we relax some of the conditions (see Exercise 5.4). Sufficient conditions for the existence of the primitive function of a continuous function can now be easily proven.

Theorem 5.14. *Let $G \subset \mathbb{R}^p$ be a nonempty open set, and let $f\colon G \to \mathbb{R}^p$ be continuous. The function f has a primitive function on G if and only if $\int_g \langle f, dx\rangle = 0$ holds for every continuous and rectifiable closed curve lying in G.*

Proof. The necessity of the condition follows from Theorem 5.11. To prove sufficiency, first we assume that G is connected (see Definition 1.21). Let $x_0 \in G$ be fixed. Every point $x \in G$ can be reached from x_0 by a continuous and rectifiable curve of G. (Indeed, it follows from Theorem 1.22 that we can connect x_0 and x by a polygonal line T. It is easy to see that there exists a continuous and rectifiable curve $g\colon [a,b] \to \mathbb{R}^p$ with $g([a,b]) = T$.)

Let $F(x) = \int_{g_1} \langle f, dx\rangle$, where $g_1\colon [a,b] \to G$ is a continuous and rectifiable curve lying in G and such that $g_1(a) = x_0$ and $g_1(b) = x$. We show that F does not depend on the choice of g_1. Let $h\colon [c,d] \to \mathbb{R}$ be another continuous and rectifiable curve in G with $h(c) = x_0$ and $h(d) = x$. Moving along the curve g_1 from x_0 to x, then moving along the curve h from x to x_0 results in a closed curve. Since the line integral along a closed curve is zero, it follows that f has the same line integral along the two curves h and g_1.

More precisely, let $g_2(t) = h(d+b-t)$ for $t \in [b, b+(d-c)]$. (The curve g_2 "goes along" the curve h backward and its parameter interval connects to $[a,b]$.) It is easy to check that the curve $g_2\colon [b, b+(d-c)] \to G$ is continuous and rectifiable, $g_2(b) = x$, $g_2(b+(d-c)) = x_0$, and $\int_{g_2} \langle f, dx\rangle = -\int_h \langle f, dx\rangle$. The last equality follows from the fact that every approximating sum of $\int_{g_2} \langle f, dx\rangle$ is equal to the negative of an approximating sum of $\int_h \langle f, dx\rangle$.

Let $g(t) = g_1(t)$ for $t \in [a,b]$, and $g(t) = g_2(t)$ for $t \in [b, b+(d-c)]$. Then by Lemma 5.13, we have (5.5). The left-hand side is zero, since we integrate along a continuous, rectifiable, and closed curve of G. The right-hand side is $\int_{g_1} \langle f, dx\rangle - \int_h \langle f, dx\rangle$; thus $\int_h \langle f, dx\rangle = \int_{g_1} \langle f, dx\rangle$.

We have shown that $F(x)$ is well defined, i.e., that it depends only on x (if x_0 is fixed). By Lemma 5.13, we have $F(v) - F(u) = \int_g \langle f, dx\rangle$ whenever $u, v \in G$ and g is a continuous rectifiable curve moving from u to v.

Now we show that $D_1 F(y) = f_1(y)$ for every $y \in G$. Since G is open, there exists $r > 0$ such that $\overline{B}(y,r) \subset G$. Let $y = (y_1, \ldots, y_p)$, $u = (y_1 - r, y_2, \ldots, y_p)$, and $v = (y_1 + r, y_2, \ldots, y_p)$. The segment $[u,v]$ is contained by the closed ball $\overline{B}(y,r)$; thus it is also contained by G. For $z = (z_1, y_2, \ldots, y_p) \in [u,v]$ we have $F(z) - F(y) = \int_g \langle f, dx\rangle$, where g is any continuous and rectifiable parametrization of the segment $[y,z]$. By Remark 5.3.1, if this parametrization is one-to-one, then the line integral is equal to the Riemann integral of the section function

$\int_{y_1}^{z_1} f_1^{(y_2,\ldots,y_p)}\,dt$. Since f_1 is continuous on $[u,v]$, it follows that the function $z_1 \mapsto \int_{y_1}^{z_1} f_1^{(y_2,\ldots,y_p)}\,dt$ is differentiable at y_1, and its derivative is $f_1^{(y_2,\ldots,y_p)}(y_1) = f_1(y)$ there. (See [7, Theorem 15.5].) This means that

$$\lim_{z_1\to y_1} \frac{F(z_1, y_2, \ldots, y_p) - F(y)}{z_1 - y_1} = f_1(y),$$

i.e., $D_1 f(y) = f_1(y)$.

The equality $D_j F(y) = f_j(y)$ can be proved similarly for every $j = 2, \ldots, p$. Thus the partial derivatives of F exist everywhere on G. Since they are also continuous, it follows from Theorem 1.71 that F is differentiable everywhere on G, and F is a primitive function of f.

We have proved our theorem in the case that G is connected. In the general case, let us present G as the union of mutually disjoint, nonempty, and connected open sets G_i $(i \in I)$. For every i there exists a function $F_i \colon G_i \to \mathbb{R}$ such that F_i is differentiable on G_i and its derivative is f (restricted to G_i). Let $F(x) = F_i(x)$ for every $x \in G_i$ and $i \in I$. Obviously, F is a primitive function of f on the set G. □

Remark 5.15. In the proof above we could have restricted the definition of $F(x) = \int_g \langle f, dx\rangle$ to integrals along polygonal lines between x_0 and x. That is, *having* $\int_g \langle f, dx\rangle = 0$ *for every closed polygonal line lying in G is a sufficient condition for the existence of the primitive function.* Moreover, having $\int_g \langle f, dx\rangle = 0$ for every *simple* closed polygonal line is also sufficient. (A closed polygonal line is called simple if it does not intersect itself.) Indeed, every closed polygonal line T is the union of finitely many simple closed polygonal lines T_i, and the line integral of every function along T is the sum of its line integrals along T_i (see Exercise 5.6). Therefore, *if f is continuous on an open set G and if* $\int_g \langle f, dx\rangle = 0$ *holds for every simple closed polygonal line lying in G, then f has a primitive function on G.*

Examples 5.16. **1.** Let us consider $g(t) = (R\cos t, R\sin t)$ $(t \in [0, 2\pi])$, the usual parametrization of the circle centered at the origin and of radius R. Let's compute the line integrals $\int_g \frac{x}{x^2+y^2}\,dx$ and $\int_g \frac{y}{x^2+y^2}\,dy$. Applying Theorem 5.5 gives us

$$\int_g \frac{x}{x^2+y^2}\,dx = \int_0^{2\pi} \frac{\cos t}{R} \cdot (-R\sin t)\,dt = 0$$

and

$$\int_g \frac{y}{x^2+y^2}\,dy = \int_0^{2\pi} \frac{\sin t}{R} \cdot (R\cos t)\,dt = 0.$$

Thus the line integral of the function $f(x,y) = \left(\frac{x}{x^2+y^2}, \frac{y}{x^2+y^2}\right)$ is zero along the circle with center 0 and radius R. This has to be so, since the function f has a primitive function on the set $\mathbb{R}^2 \setminus \{(0,0)\}$; namely, the function $\frac{1}{2}\log(x^2+y^2)$ is a primitive function.

2. Find the line integrals $\int_g \frac{-y}{x^2+y^2}\,dx$ and $\int_g \frac{x}{x^2+y^2}\,dy$ along the same circle. Applying Theorem 5.5 gives us

$$\int_g \frac{-y}{x^2+y^2}\,dx = \int_0^{2\pi} -\frac{\sin t}{R}\cdot(-R\sin t)\,dt = \pi$$

and

$$\int_g \frac{x}{x^2+y^2}\,dy = \int_0^{2\pi} \frac{\cos t}{R}\cdot(R\cos t)\,dt = \pi.$$

Thus the line integral of the function

$$f(x,y) = \left(\frac{-y}{x^2+y^2}, \frac{x}{x^2+y^2}\right) \tag{5.6}$$

along a circle with radius R is not zero. Therefore, *the function f does not have a primitive function on the set $\mathbb{R}^2 \setminus \{(0,0)\}$.*

These examples show an important difference between single- and multivariable analysis. We know that every continuous single-variable function always has a primitive function (see [7, Theorem 15.5]). In contrast, the function defined in (5.6) does not have a primitive function on G, even though it is both continuous and differentiable on $G = \mathbb{R}^2 \setminus \{(0,0)\}$.

Another necessary condition for the existence of a primitive function is given by the following theorem.

Theorem 5.17. *Let $f\colon G \to \mathbb{R}^p$ be differentiable on an open set $G \subset \mathbb{R}^p$. If f has a primitive function on G, then*

$$D_i f_j(x) = D_j f_i(x) \tag{5.7}$$

for every $x \in G$ and $i, j = 1, \ldots, p$. In other words, if the differentiable function f has a primitive function on an open set G, then its Jacobian matrix is symmetric at every point $x \in G$.

Proof. Let F be a primitive function of f on G. Then the function F is twice differentiable on G and thus, by Theorem 1.86, we have

$$D_i f_j(x) = D_i D_j F(x) = D_{ij} F(x) = D_{ji} f(x) = D_j D_i F(x) = D_j f_i(x)$$

for every $x \in G$ and $i, j = 1, \ldots, p$. □

Remark 5.18. In general, the conditions of the previous theorem are *not sufficient* for the existence of a primitive function. The function defined by (5.6) is differentiable on the set $G = \mathbb{R}^2 \setminus \{(0,0)\}$, with $D_2 f_1 = D_1 f_2$ (check this!). But as we saw, the function does not have a primitive function on G.

We shall see presently that the cause of this phenomenon is that G "has a hole in it." We will prove that the conditions of Theorem 5.17 are sufficient for the existence of a primitive function for open sets of a simpler structure (what are called 1-connected sets).

First, let us find the linear transformations having a primitive function.

Theorem 5.19. *A linear transformation* $A\colon \mathbb{R}^p \to \mathbb{R}^p$ *has a primitive function if and only if the matrix of* A *is symmetric.*

Proof. Every linear transformation is differentiable, and its derivative is itself everywhere, and thus its Jacobian matrix is the same as its matrix. Then by Theorem 5.17, if a linear transformation A has a primitive function, its matrix is symmetric.

Now let us assume that the matrix of A is symmetric, and let this matrix be (a_{ij}) with $a_{ij} = a_{ji}$ $(i, j = 1, \ldots, p)$. The ith coordinate of the vector $A(x)$ is $\sum_{j=1}^{p} a_{ij}x_j$ for every vector $x = (x_1, \ldots, x_p) \in \mathbb{R}^p$. It is easy to see that the function $F(x) = \frac{1}{2} \cdot \sum_{i=1}^{p} \sum_{j=1}^{p} a_{ij}x_i x_j$ is a primitive function of the mapping A. Indeed, F is differentiable, since it is a polynomial (see Theorem 1.90). For every $1 \le i \le p$, we have $D_i F(x) = \sum_{j=1}^{p} a_{ij}x_j$, i.e., $D_i F(x)$ is the same as the ith coordinate function of $A(x)$. That is, F is a primitive function of the mapping A. □

Lemma 5.20. (Goursat's[9] lemma) *Let* $G \subset \mathbb{R}^p$ *be open, and let* a, b, c *be points of* G *such that the convex hull* H *of the set* $\{a, b, c\}$ *is in* G.[10]

Let g *denote the closed polygonal line* $[a, b] \cup [b, c] \cup [c, a]$. *Suppose that for every* $x \in H$, $f\colon G \to \mathbb{R}^p$ *is differentiable at* x *with a symmetric Jacobian matrix. Then* $\int_g \langle f, dx\rangle = 0$.

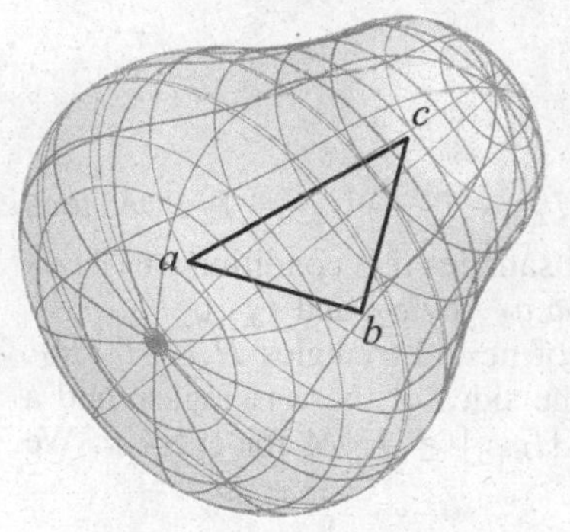

5.2. Figure

For the proof we need the following lemma, called a **trivial estimate**.

Lemma 5.21. *Let* $g\colon [a, b] \to \mathbb{R}^p$ *be a continuous and rectifiable curve, and let the function* $h\colon g([a, b]) \to \mathbb{R}$ *be continuous on* $g([a, b])$. *Then*

$$\left| \int_g \langle h, dx\rangle \right| \le K \cdot s(g),$$

where K *is an upper bound of the function* $|h|$ *on the set* $g([a, b])$, *and* $s(g)$ *is the arc length of* g.

[9] Édouard Jean-Baptiste Goursat (1858–1936), French mathematician.

[10] The convex hull H is nothing other than the triangle with vertices a, b, c.

Proof. For every partition $a = t_0 < t_1 < \ldots < t_n = b$ and for every choice of the inner points $c_i \in [t_{i-1}, t_i]$ $(i = 1, \ldots, n)$ we have

$$\left| \sum_{i=1}^{n} \langle h(g(c_i)), g(t_i) - g(t_{i-1}) \rangle \right| \le \sum_{i=1}^{n} K \cdot |g(t_i) - g(t_{i-1})| \le K \cdot s(g).$$

Thus the statement follows from Definition 5.1. □

Proof of Goursat's lemma. Let $\int_g \langle f, dx \rangle = I_0$; we need to prove that $I_0 = 0$. The midpoints of the sides of the triangle H are $(a+b)/2$, $(b+c)/2$, $(c+a)/2$. The three segments connecting these midpoints cut H into congruent triangles $H_{1,1}, \ldots, H_{1,4}$. Let $g_{1,i}$ denote the polygon whose edges are the sides of the triangle $H_{1,i}$. With an appropriate direction of $g_{1,i}$ we have

$$I_0 = \int_g \langle f, dx \rangle = \sum_{i=1}^{4} \int_{g_{1,i}} \langle f, dx \rangle. \tag{5.8}$$

Indeed, we integrate twice (in opposite directions) along every side of the polygons $g_{1,i}$ contained in the interior of H (i.e., along the segments connecting the midpoints), so their contributions cancel on the right-hand side of (5.8). The union of the remaining segments is exactly g. Thus (given an appropriate direction of the polygons $g_{1,i}$) the sum of their corresponding terms is I_0. Let $\int_{g_{1,i}} \langle f, dx \rangle = I_{1,i}$ $(i = 1, 2, 3, 4)$. Then $I_0 = I_{1,1} + I_{1,2} + I_{1,3} + I_{1,4}$ by (5.8). Therefore, we have $|I_{1,i}| \ge |I_0|/4$ for at least one i. Choose an i that satisfies this condition and let us denote the triangle $H_{1,i}$ by H_1, the closed polygon $g_{1,i}$ by g_1, and $I_{1,i}$ by I_1.

The segments connecting the midpoints of the sides of the triangle H_1 cut H_1 into the isomorphic triangles $H_{2,1}, \ldots, H_{2,4}$. Let $g_{2,i}$ denote the polygon whose edges are the sides of the triangle $H_{2,i}$. With an appropriate direction of $g_{2,i}$ we have

$$I_1 = \int_{g_1} \langle f, dx \rangle = \sum_{i=1}^{4} \int_{g_{2,i}} \langle f, dx \rangle.$$

Let $\int_{g_{2,i}} \langle f, dx \rangle = I_{2,i}$ $(i = 1, 2, 3, 4)$. Then $I_1 = I_{2,1} + I_{2,2} + I_{2,3} + I_{2,4}$, and thus $|I_{2,j}| \ge |I_1|/4$ for at least one j. Pick some j that satisfies this condition and let us denote the triangle $H_{2,j}$ by H_2, the closed polygon $g_{2,j}$ by g_2, and $I_{2,j}$ by I_2.

By repeating this process we get a sequence of nested triangles $H = H_0, H_1, H_2, \ldots$, a sequence of polygons g_k defined by the sides of these triangles, and a sequence $I_k = \int_{g_k} \langle f, dx \rangle$ of numbers such that $|I_{k+1}| \ge |I_k|/4$ for every k. We have

$$|I_0| \le 4 \cdot |I_1| \le 4^2 \cdot |I_2| \le \ldots,$$

i.e., $|I_0| \le 4^k \cdot |I_k|$ for every k.

Let s_k be the length of the perimeter of the triangle H_k, i.e., the arc length of the polygon g_k. Since the triangle H_{k+1} is similar to the triangle H_k with its sized halved (i.e., with ratio $1/2$), we have $s_{k+1} = s_k/2$ for every k, and thus $s_k = s_0/2^k$ $(k = 1, 2, \ldots)$.

Now, $H, H_1, H_2, \ldots$ are nested nonempty closed sets, and thus it follows from Cantor's theorem (Theorem 1.25) that their intersection is nonempty. Let $d \in \bigcap_{k=1}^{\infty} H_k$.

Let $\varepsilon > 0$ be fixed. Since f is differentiable at the point d, there exists $\delta > 0$ such that

$$f(x) = f(d) + f'(d)(x-d) + \eta(x),$$

where $|\eta(x)| < \varepsilon \cdot |x-d|$ for each $|x-d| < \delta$. Now, $d \in H_k$ and the length of H_k's perimeter is $s_0/2^k$; thus every point of g_k is closer to d than $s_0/2^k$. For a large enough k, we have $g_k \subset B(d,\delta)$ and

$$I_k = \int_{g_k} \langle f, dx\rangle = \int_{g_k} \langle f(d) + f'(d)(x-d), dx\rangle + \int_{g_k} \langle \eta, dx\rangle. \qquad (5.9)$$

By assumption, the matrix of the linear transform $f'(d)$ is symmetric. Then, by Theorem 5.19, the mapping $f'(d)(x)$ has a primitive function. Let $c = f(d) - f'(d)(d)$. The constant mapping c also has a primitive function: if $c = (c_1, \ldots, c_p)$, then the function $c_1x_1 + \ldots + c_px_p$ works. Thus the mapping $f(d) + f'(d)(x-d)$ also has a primitive function on $\mathbb{R}^p$. Therefore, by Theorem 5.14, its integral is zero on every continuous, closed, and rectifiable curve. Thus the value of the first integral on the right-hand side of (5.9) is zero. On the other hand, the trivial estimate (Lemma 5.21) gives

$$\left| \int_{g_k} \langle \eta, dx\rangle \right| \le \varepsilon \cdot s_k^2,$$

since every point x of the triangle H_k satisfies $|x-d| \le s_k$, and thus $|\eta(x)| \le \varepsilon \cdot |x-d| < \varepsilon \cdot s_k$. We have proved $|I_k| < \varepsilon \cdot s_0^2 \cdot 4^{-k}$, which implies

$$|I_0| \le 4^k \cdot |I_k| < \varepsilon \cdot s_0^2.$$

This is true for every $\varepsilon > 0$, and thus $I_0 = 0$, which completes the proof. □

Theorem 5.22. *Let $G \subset \mathbb{R}^p$ be a convex open set. A differentiable mapping $f\colon G \to \mathbb{R}^p$ has a primitive function on G if and only if its Jacobian matrix is symmetric at every point $x \in G$.*

Proof. The necessity of the condition follows from Theorem 5.17.

To prove sufficiency, let the Jacobian matrix of f be symmetric at every point of G. In order to prove the existence of a primitive function it is enough to show that the integral of f is zero on every closed polygon in G (see Theorem 5.14 and Remark 5.15). Let $a_0, a_1, \ldots, a_n = a_0$ be points of G; we show that the integral of f along the polygon $p = [a_0, a_1] \cup \ldots \cup [a_{n-1}, a_0]$ is zero. (The polygon p lies in G, since by the convexity of G we have $[a_{i-1}, a_i] \subset G$ for every $i = 1, \ldots, n$.) We now employ induction on n. For $n = 0$, the polygon is reduced to a single point, and the integral of any function on a singleton (i.e., along a constant curve) is zero. For $n = 1$, the integral of f on p is zero, since we integrate along the segment $[a_0, a_1]$ twice, first from a_0 to a_1 and then from a_1 to a_0. The sum of these two integrals is zero.

Let $n > 1$ and suppose that the statement is true for $n-1$. Then the integral of f along the closed polygon $p' = [a_0, a_1] \cup \ldots \cup [a_{n-3}, a_{n-2}] \cup [a_{n-2}, a_0]$ is zero. The integral of f is also zero along the closed polygon $g = [a_{n-2}, a_{n-1}] \cup [a_{n-1}, a_0] \cup [a_0, a_{n-2}]$. Indeed, if H is the convex hull of the points a_{n-2}, a_{n-1}, a_0, then g is a closed polygon consisting of the sides of the triangle H. Since the Jacobian matrix of f is symmetric at every point of H (since it is symmetric at every point of G), it follows from Goursat's lemma that the integral of f along g is zero. Now we have

$$\int_p \langle f, dx \rangle = \int_{p'} \langle f, dx \rangle + \int_g \langle f, dx \rangle. \tag{5.10}$$

Indeed, the union of the segments of p' and g is p together with the segment $[a_{n-2}, a_0]$ counted twice, with two different directions. Since the integrals over a segment with two different directions are the negatives of each other, we obtain (5.10). Since both terms of the right-hand side of (5.10) are zero, we have $\int_p \langle f, dx \rangle = 0$, which proves the theorem. □

Example 5.23. Consider the function f defined by (5.6). Then f is differentiable on the set $G = \mathbb{R}^2 \setminus \{(0,0)\}$ with $D_2 f_1 = D_1 f_2$ there (see Remark 5.18). Thus the Jacobian matrix of f is symmetric at every point $x \in G$, and thus by the previous theorem, f has a primitive function on every convex and open subset of G. We can prove this directly.

Indeed, the function $\operatorname{arc\,tg}(y/x)$ is a primitive function of f on each of the convex open sets $\{(x,y)\colon x > 0\}$ and $\{(x,y)\colon x < 0\}$ (check this fact). On the other hand, the function $-\operatorname{arc\,tg}(x/y)$ is a primitive function of the function f on each of the convex open sets $\{(x,y)\colon y > 0\}$ and $\{(x,y)\colon y < 0\}$ (check this fact as well).

Using these two functions, we can find the primitive function of f on an arbitrary convex and open subset of the domain G as follows. First, note that $\operatorname{arc\,tg} x + \operatorname{arc\,tg}(1/x) = \pi/2$ for $x > 0$, and $\operatorname{arc\,tg} x + \operatorname{arc\,tg}(1/x) = -\pi/2$ for $x < 0$. (These follow from the fact that for $0 < \alpha < \pi/2$ we have $1/\operatorname{tg}\alpha = \operatorname{ctg}\alpha = \operatorname{tg}((\pi/2) - \alpha)$. With the help of these formulas one can easily see that the function

$$F(x,y) = \begin{cases} \operatorname{arc\,tg}(y/x), & \text{if } x > 0, \\ -\operatorname{arc\,tg}(x/y) + \pi/2, & \text{if } y > 0, \\ \operatorname{arc\,tg}(y/x) + \pi, & \text{if } x < 0 \end{cases}$$

is well defined and is differentiable on the open set $G' = \mathbb{R}^2 \setminus \{(0,y)\colon y \le 0\}$, with $F' = f$ there. (That is, F is the primitive function of f on G'. Note that by Exercise 5.2, all the primitive functions of f on G' are of the form $F + c$, where c is a constant.) A similar construction can be applied for the complement of every half-line starting from the origin. Thus f has a primitive function on every such open set.

It is also easy to see that if $H \subset G$ is convex, then there is a half-line L with endpoint at the origin such that $H \subset \mathbb{R}^2 \setminus L$. This shows that f has a primitive function on every convex and open subset of the open set G, in accordance with Theorem 5.22.

This example illustrates another—just as remarkable—fact. By Example 5.16.2, the integral of the function f along *an arbitrary* origin-centered circle is the same number. Next we show that this follows from the fact that f has a primitive function on every disk lying in G.

Definition 5.24. Let $G \subset \mathbb{R}^p$ be open, and let $g_1, g_2 \colon [a, b] \to G$ be continuous closed curves. We say that the curves g_1 and g_2 can be *continuously deformed into each other* or in other words, g_1 and g_2 are *homotopic* curves in G, if there exists a continuous mapping $\varphi\colon ([a, b] \times [0, 1]) \to G$ such that $\varphi(t, 0) = g_1(t)$ and $\varphi(t, 1) = g_2(t)$ for every $t \in [a, b]$. Furthermore, $\varphi(a, u) = \varphi(b, u)$ for every $u \in [0, 1]$.

Example 5.25. Let $p = 2$ and $G = \mathbb{R}^2 \setminus \{(0, 0)\}$. Then

$$g_1(t) = (R_1 \cos t, R_1 \sin t) \qquad (t \in [0, 2\pi])$$

and

$$g_2(t) = (R_2 \cos t, R_2 \sin t) \qquad (t \in [0, 2\pi])$$

are homotopic curves in G for every $R_1, R_2 > 0$. Indeed, the mapping

$$\varphi(t, u) = ((R_1 + (R_2 - R_1)u) \cos t, (R_1 + (R_2 - R_1)u) \sin t)\,,\ t \in [0, 2\pi],\ u \in [0, 1]$$

satisfies the conditions of Definition 5.24.

Theorem 5.26. *Let $G \subset \mathbb{R}^p$ be open, let $f\colon G \to \mathbb{R}^p$ be continuous, and suppose that every $x \in G$ has a neighborhood on which f has a primitive function. Then we have*

$$\int_{\gamma_1} \langle f, dx\rangle = \int_{\gamma_2} \langle f, dx\rangle \tag{5.11}$$

whenever γ_1 and γ_2 are continuous rectifiable homotopic closed curves lying in G.

Proof. Let $T = [a, b] \times [0, 1]$, and let $\varphi\colon T \to G$ be a mapping satisfying the conditions of Definition 5.24.

If a point moves around the perimeter of the rectangle T in the positive direction starting from the vertex $(a, 0)$, then the image of this point by the mapping φ is a continuous closed curve γ that consists of four parts: the curve γ_1, a continuous curve ρ going from the endpoint of γ_1 to the endpoint of γ_2, the curve γ_2 traversed in the opposite direction, and the curve ρ, also traversed in the opposite direction. It follows that

$$\int_{\gamma} \langle f, dx\rangle = \int_{\gamma_1} \langle f, dx\rangle - \int_{\gamma_2} \langle f, dx\rangle. \tag{5.12}$$

Thus it is enough to show that the left-hand side of (5.12) is zero.

The main idea of the proof is the following. Cut T into congruent rectangles T_i ($i = 1, \ldots, n^2$). Let g_i denote the image of the perimeter of T_i by the map φ. Then we have

$$\int_{\gamma} \langle f, dx\rangle = \sum_{i=1}^{n^2} \int_{g_i} \langle f, dx\rangle. \tag{5.13}$$

Indeed, the right-hand side is the sum of the integrals of f along the images of the sides of the rectangles T_i by φ. The integrals corresponding to the sides of T_i that lie inside of T appear twice with opposing signs. Thus these integrals cancel, and the right-hand side of (5.13) is the integral $\int_\gamma \langle f, dx\rangle$.

Now, for n large enough, the diameters of T_i and g_i are small enough for g_i to be covered by a ball in which f has a primitive function. It follows from Theorem 5.14 that every term on the right-had side of (5.13) is zero; thus the left-hand side is also zero.

Now we turn to the precise proof. In the argument we also have to handle the problem of the existence of the integrals on the right-hand side of (5.13), since the maps g_i are continuous, but not necessarily rectifiable. A simple solution of this problem is replacing the nonrectifiable images of the sides of T_i by segments.

Since the set $[a, b] \times [0, 1]$ is bounded and closed, and the mapping φ is continuous, the set $H = \varphi([a, b] \times [0, 1])$ is also bounded and closed (see Theorem 2.7).

Now we prove that for a suitable $\delta > 0$, the function f has a primitive function in the ball $B(x, \delta)$ for every $x \in H$. Suppose there is no such δ. Then for every positive integer n there exists $x_n \in H$ such that f does not have a primitive function in the ball $B(x_n, 1/n)$. Since the set H is bounded, it follows from the Bolzano–Weierstrass theorem (Theorem 1.9) that the sequence (x_n) has a convergent subsequence (x_{n_k}). If $x_{n_k} \to x$, then $x \in H \subset G$ (since H is closed), and there exists $r > 0$ such that f has a primitive function in the ball $B(x, r)$. For k large enough we have $|x_{n_k} - x| < r/2$ and $1/n_k < r/2$. For such a k, we have $B(x_{n_k}, 1/n_k) \subset B(x, r)$, and consequently, f has a primitive function in the ball $B(x_{n_k}, 1/n_k)$. This, however, contradicts the choice of x_{n_k}. This contradiction proves the existence of $\delta > 0$ such that f has a primitive function in the ball $B(x, \delta)$ for every $x \in H$. Fix such a δ.

By Heine's theorem (Theorem 1.53), there exists $\eta > 0$ such that $|\varphi(x) - \varphi(y)| < \delta$ whenever $x, y \in [a, b] \times [0, 1]$ and $|x - y| < \eta$. Let n be large enough to satisfy both $(b - a)/n < \eta/2$ and $1/n < \eta/2$.

Cut T into congruent rectangles T_i $(i = 1, \ldots, n^2)$, and let g_i denote the image of the boundary of T_i by the map φ. By the choice of n, the diameter of T_i is smaller than η, and then by the choice of η, the diameter of g_i is smaller than δ.

Let $x \in g_i$ be arbitrary. Then $g_i \subset B(x, \delta)$, and f has a primitive function in the ball $B(x, \delta)$ by the choice of δ. Thus $\int_{g_i} \langle f, dx\rangle = 0$, assuming that g_i is rectifiable. However, g_i is not necessarily rectifiable, and thus we replace g_i by a rectifiable curve g_i' as follows. Whenever the image of a side $[u, v]$ of some rectangle T_i by φ is not rectifiable, we replace it by the segment $[\varphi(u), \varphi(v)]$. In this way we obtain a rectifiable curve g_i' that is also contained in the ball $B(x, \delta)$, since every ball is convex. Then by Theorem 5.14, $\int_{g_i'} \langle f, dx\rangle = 0$ for every i. On the other hand, a proof similar to that of (5.13) gives

$$\int_\gamma \langle f, dx\rangle = \sum_{i=1}^{n^2} \int_{g_i'} \langle f, dx\rangle.$$

Therefore, we have $\int_\gamma \langle f, dx\rangle = 0$. □

Definition 5.27. Let $G \subset \mathbb{R}^p$ be open, and let $g \colon [a, b] \to G$ be a continuous closed curve. We say that the curve g can be *continuously deformed into a point,* or g *is* *null-homotopic* in G, if there is a point $c \in G$ such that g and the curve with constant value c are homotopic to each other in G.

The open set $G \subset \mathbb{R}^p$ is called *simply connected* or 1-*connected* if it is connected and every continuous closed curve in G is null-homotopic in G.

Remarks 5.28: **1.** Every convex open set is simply connected. Indeed, let $g\colon [a, b] \to G$ be an arbitrary continuous closed curve in G. Choose a point $c \in G$, and let $\varphi(t, u) = u \cdot c + (1 - u) \cdot g(t)$ for every $(t, u) \in [a, b] \times [0, 1]$. Obviously, φ satisfies the conditions of Definition 5.24 on the curves $g_1 = g$ and $g_2 \equiv c$. That is, g is homotopic to the curve with constant value c in G. This is true for every continuous closed curve of G, and thus G is simply connected.
2. One can prove that a connected open set of the plane is simply connected if and only if it is the bijective and continuous image of a convex open set (of the plane).

This statement is not true for higher-dimensional spaces. Consider the open set $G = \{x \in \mathbb{R}^3 \colon r < |x| < R\}$ of three-dimensional space, where $0 < r < R$. It is easy to see that every continuous closed curve of G can be continuously deformed into a point, i.e., G is simply connected. It is clear intuitively that G is not a continuous bijective map of (a three-dimensional) convex open set. However, the proof of this is not easy.

Corollary 5.29. *Let $G \subset \mathbb{R}^p$ be open, let $f\colon G \to \mathbb{R}^p$ be continuous, and let every point $x \in G$ have a neighborhood in which f has a primitive function. Then we have $\int_g \langle f, dx\rangle = 0$ whenever g is a null-homotopic continuous, rectifiable, and closed curve in G.*

Proof. The claim follows trivially from Theorem 5.26, since the integral of every function along a constant curve is zero. □

Corollary 5.30. *Let $G \subset \mathbb{R}^p$ be a simply connected open set, let $f\colon G \to \mathbb{R}^p$ be continuous, and suppose that every $x \in G$ has a neighborhood in which f has a primitive function. Then the function f has a primitive function on G.*

Proof. The statement follows immediately from Corollary 5.29 and Theorem 5.14. □

Remark 5.31. The condition on the continuity of f can be omitted; see Exercise 5.12.

Theorem 5.32. *Let $G \subset \mathbb{R}^p$ be a simply connected open set. A differentiable mapping $f : G \to \mathbb{R}^p$ has a primitive function on G if and only if the Jacobian matrix of f is symmetric at every point $x \in G$.*

Proof. The statement follows immediately from Theorem 5.22 and from Corollary 5.30. □

Exercises

5.6. Show that every closed polygonal line T can be decomposed into finitely many non-self-intersecting polygonal lines T_i. (Therefore, the line integral of an arbitrary function along T equals the sum of the line integrals along T_i.) (H)

5.7. Find the continuously differentiable functions $F\colon \mathbb{R}^2 \to \mathbb{R}$ for which

(a) $\int_g F\,dx = 0$ for every continuous rectifiable closed curve;

(b) $\int_g F\,dx + \int_g F\,dy = 0$ for every continuous rectifiable closed curve.

5.8. Decide whether the following mappings have a primitive function on their respective domains. If a mapping has a primitive function, find one.

(a) $(x+y, x-y)$; (b) $(x^2+y, x+\operatorname{ctg} y)$;

(c) (x^2-2xy, y^2-x^2); (d) $\left(\frac{x^2-y^2}{(x^2+y^2)^2}, \frac{2xy}{(x^2+y^2)^2}\right)$;

(e) $\left(\frac{x}{\sqrt{x^2+y^2}}, \frac{y}{\sqrt{x^2+y^2}}\right)$; (f) $\left(\frac{y}{\sqrt{x^2+y^2}}, \frac{x}{\sqrt{x^2+y^2}}\right)$;

(g) $\left(\frac{-y}{\sqrt{x^2+y^2}}, \frac{x}{\sqrt{x^2+y^2}}\right)$; (h) $\left(\log\sqrt{x^2+y^2}, \operatorname{arc\,tg}(x/y)\right)$;

(i) $\left(\operatorname{arc\,tg}(x/y), -\log\sqrt{x^2+y^2}\right)$;

(j) $\left(\frac{y}{1+x^2}+x, \operatorname{arc\,tg} x + \frac{z}{y}, \log y + z\right)$;

(k) $\left(\frac{z}{x} - \frac{\sin z}{x^2 y}, \frac{z}{y} - \frac{\sin z}{xy^2}, \log(xy) + \frac{\cos z}{xy}\right)$.

5.9. (a) Find the continuously differentiable functions $f\colon \mathbb{R}^2 \to \mathbb{R}$ for which the mapping $(f, f)\colon \mathbb{R}^2 \to \mathbb{R}^2$ has a primitive function.

(b) Find the continuously differentiable functions $f, g\colon \mathbb{R}^2 \to \mathbb{R}$ for which the mappings $(f, g)\colon \mathbb{R}^2 \to \mathbb{R}^2$ and $(g, f)\colon \mathbb{R}^2 \to \mathbb{R}^2$ both have a primitive function.

5.10. Compute the line integral of the mapping $f(x, y) = (\log\sqrt{x^2+y^2}, \operatorname{arc\,tg}(x/y))$ along the curve $g(t) = (\operatorname{sh} t, 1 + \operatorname{ch} t)$ $(t \in [0, 1])$.

5.11. Let $G = \mathbb{R}^2 \setminus \{(0,0)\}$, and let $f = (f_1, f_2)\colon G \to \mathbb{R}^2$ be a continuously differentiable function such that $D_2 f_1 = D_1 f_2$. Let the line integral of f on the unit circle (oriented in the positive direction) be I. Show that the line integral of f along every continuous rectifiable closed curve $g\colon [a, b] \to G$ is $n \cdot I$, for some integer n. What is the intuitive meaning of n?

5.12. Let $G \subset \mathbb{R}^p$ be a simply connected open set, and let $f\colon G \to \mathbb{R}^p$ be a mapping such that every $x \in G$ has a neighborhood on which f has a primitive function. Show that f also has a primitive function in G. (*)

5.13. Let $G = \mathbb{R}^2 \setminus \{(0,0)\}$, and let $f = (f_1, f_2)\colon G \to \mathbb{R}^2$ be differentiable. Show that if $D_2 f_1 = D_1 f_2$ on G, then f has a primitive function on every set that is the complement of a half-line whose endpoint is the origin.

5.3 Green's Theorem

The multivariable variants of the Newton–Leibniz formula are called integral theorems. As we will see presently, these are especially useful and important for applications.

When discussing the topic of integral theorems, we necessarily have to wander into the fields of differential geometry and topology. Consequently, we have to take for granted several facts in these areas and cannot explain some ideas and results that look intuitively clear. The reason is that in order to present a precise proof of several theorems to come, we would need to develop some parts of geometric measure theory, topology, and differential geometry in such depth that would fill an entire book in itself while drifting away from the topic of this book significantly.

Therefore, at times we will need to give up the principle—to which we have kept ourselves so far—of not using anything in our proofs that we have not proved before (except the axioms, of course). The reader is warned either to consider this part of the book a popular introduction or to read the books [12] and [1] for further details.

The topic of simple closed curves is the first on which we have to accept some facts without proofs.

Definition 5.33. We call a curve $g\colon [a,b] \to \mathbb{R}^p$ a *simple closed curve* if g is a continuous closed curve that is injective on the set $[a,b)$. In other words, the curve $g\colon [a,b] \to \mathbb{R}^p$ is a simple closed curve if and only if it is continuous and for every $a \le t < u \le b$ we have $g(t) = g(u)$ if and only if $t = a$ and $u = b$.

It is intuitively clear that if $g\colon [a,b] \to \mathbb{R}^2$ *is a simple closed plane curve, then the open set* $\mathbb{R}^2 \setminus g([a,b])$ *has exactly two components, whose common boundary is the set* $g([a,b])$. *Furthermore, exactly one of these two components is bounded.* This statement is known as the **Jordan curve theorem**. However, the proof of this theorem is far form being simple. The reader can find a proof in [9]. Other proofs can be found in [11] and [13]. From now on, we will take the Jordan curve theorem for granted.

For a simple closed plane curve $g\colon [a,b] \to \mathbb{R}^2$, the bounded component of $\mathbb{R}^2 \setminus g([a,b])$ is called the **bounded domain with boundary** g.

We need to define the direction of a simple closed curve. Intuitively, if we move along the simple closed curve g, then the bounded domain with boundary g is either to our left-hand side or our right-hand side. In the first case we call the direction of the curve positive, and in the second case the direction is negative. The precise definition is the following.

First, we define the directed angle of a pair of nonzero vectors $a = (a_1, a_2)$ and $b = (b_1, b_2)$. Intuitively, the angle is positive or negative according to whether the half-line ℓ_b starting from the origin and going through b can be reached by a positive or negative rotation from the half-line ℓ_a starting from the origin and going through a. We can check that the sign of the determinant $\begin{vmatrix} a_1 & a_2 \\ b_1 & b_2 \end{vmatrix}$ is different in the two cases. This motivates the following definition.

We say that the **undirected angle** of the vectors a and b is the angle $\gamma \in [0, \pi]$ of the half-lines ℓ_a and ℓ_b. (We can also define this angle by the formula $\langle a, b\rangle = |a| \cdot |b| \cdot \cos\gamma$.) We say that the **directed angle** of the vectors a and b is γ if the determinant $\begin{vmatrix} a_1 & a_2 \\ b_1 & b_2 \end{vmatrix}$ is nonnegative, and $-\gamma$ if the determinant $\begin{vmatrix} a_1 & a_2 \\ b_1 & b_2 \end{vmatrix}$ is negative. (Obviously, the directed angle depends on the order of a and b.)

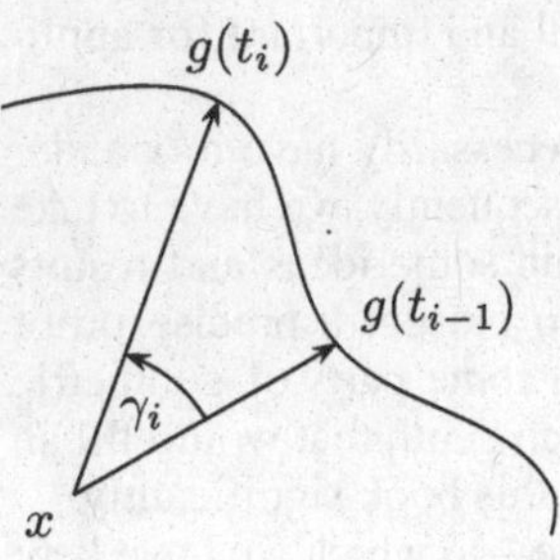

5.3. Figure

Let $g\colon [a,b] \to \mathbb{R}^2$ be a continuous plane curve, let $\Gamma = g([a,b])$, and let $x \in \mathbb{R}^2 \setminus \Gamma$. We define the **winding number** $w(g;x)$ of the curve g around point x as follows. Let $a = t_0 < t_1 < \ldots < t_n = b$ be a partition such that $\operatorname{diam} g([t_{i-1}, t_i]) \le \operatorname{dist}(x, \Gamma)$ holds for every i. Let us denote by γ_i the directed angle of the vectors $g(t_{i-1}) - x$ and $g(t_i) - x$, and let $w(g;x) = \sum_{i=1}^n \gamma_i$. One can prove that the value of $w(g;x)$ is the same for every such partition.

One can also prove that if g is a simple closed curve and A is the bounded domain with boundary g, then either $w(g;x) = 2\pi$ for every $x \in A$ or $w(g;x) = -2\pi$ for every $x \in A$. In the first case we say that the direction of the curve is **positive**, and in the second case we say that the direction of the curve is **negative**.

We can now discuss the integral theorems of the plane. Let f be a two-variable function. From now on we will use the notation $\frac{\partial f}{\partial x}$ instead of $D_1 f$ and $\frac{\partial f}{\partial y}$ instead of $D_2 f$. (This notation makes the theorems easier to memorize.)

Theorem 5.34. (Green's[11] theorem) *Let g be a continuous rectifiable positively oriented simple closed plane curve, and let A be the bounded domain with boundary g. Let $\operatorname{cl} A \subset G$, where G is open, and let $f\colon G \to \mathbb{R}$ be continuous.*

(i) *If $\frac{\partial f}{\partial y}$ exists and is continuous on $\operatorname{cl} A$, then*

$$\int_g f\,dx = -\int_A \frac{\partial f}{\partial y}\,dx\,dy. \tag{5.14}$$

(ii) *If $\frac{\partial f}{\partial x}$ exists and is continuous on $\operatorname{cl} A$, then*

$$\int_g f\,dy = \int_A \frac{\partial f}{\partial x}\,dx\,dy. \tag{5.15}$$

Remark 5.35. Formula (5.15) can be memorized by "deleting" ∂x and dx on the right-hand side (since these "cancel each other out").

Switching x and y turns formula (5.15) into (5.14). Switching the coordinate axes is nothing other than the reflection about the line $y = x$, and reflections change the direction of simple closed curves. Since (5.14) is also about positively oriented curves, we need to take the negative of one of the sides.

We can memorize the negative sign in (5.14) by first switching dx and dy on the right-hand side before the cancellation of ∂y and dy, which causes a negative sign.

These operations can be endowed with precise mathematical meaning using the theory of differential forms. See, e.g., Chapter 10 of the book [12].

[11] George Green (1793–1841), British mathematician and physicist.

Sketch of the proof of Theorem 5.34. First, we prove statement (i) in the special case that A is the interior of a normal domain.

Let φ and ψ be continuous functions on the interval $[c,d]$ and let $\varphi(x) < \psi(x)$, for every $x \in (c,d)$. Let

$$A = \{(x,y)\colon c < x < d\ \varphi(x) < y < \psi(x)\}.$$

Let g be be a continuous positively oriented simple closed curve that parametrizes the boundary of A. This means that g can be divided into the continuous curves g_1, g_2, g_3, g_4, where g_1 parametrizes the graph of the function φ such that the first component of g_1 is strictly monotonically increasing, g_2 is the (possibly degenerate) vertical segment connecting the points $(d,\varphi(d))$ and $(d,\psi(d))$, g_3 parametrizes the graph of the function ψ such that the first component of g_3 is strictly monotonically decreasing, and finally, g_4 is the vertical segment connecting the points $(c,\psi(c))$ and $(c,\varphi(c))$. It is easy to see that

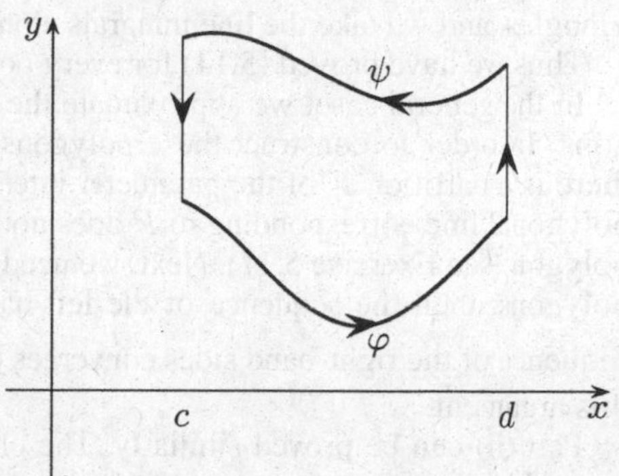

5.4. Figure

$$\int_{g_1} f\,dx = \int_c^d f(x,\varphi(x))\,dx \text{ and } \int_{g_3} f\,dx = -\int_c^d f(x,\psi(x))\,dx$$

(see Exercise 5.5). By Remark 5.3.2, $\int_{g_2} f\,dx = \int_{g_4} f\,dx = 0$, and thus the value of the left-hand side of (5.14) is

$$\int_c^d (f(x,\varphi(x)) - f(x,\psi(x)))\,dx.$$

On the other hand, according to the theorem of successive integration, the value of the right-hand side of (5.14) is

$$-\int_c^d \left(\int_{\varphi(x)}^{\psi(x)} \frac{\partial f}{\partial y}\,dy \right) dx = -\int_c^d [f(x,y)]_{y=\varphi(x)}^{\psi(x)}\,dx =$$

$$= -\int_c^d (f(x,\psi(x)) - f(x,\varphi(x)))\,dx,$$

i.e., (5.14) holds.

Note that every triangle is a normal region, and thus (i) holds for every triangle.

Now we prove (i) for every polygon, that is, for every simple closed polygonal line. It is easy to see that every polygon can be partitioned into nonoverlapping triangles (see Exercise 3.30). We also know that (5.14) holds for every triangle of this partition. Summing these equations yields our statement for polygons. Indeed, it follows from (4.3) that the sum of the right-hand sides equals the right-hand side of (5.14). The line integrals along those segments that lie in the interior of the polygon cancel each other out on the left-hand side, since every such segment belongs to two triangles, and we take the line integrals along them twice, with opposing directions.

Thus we have proved (5.14) for every polygon.

In the general case, we approximate the curve g by a sequence of suitable polygons. In order to construct these polygons we need to show that for every $\delta > 0$ there is a partition F of the parameter interval that is finer than δ and such that the polygonal line corresponding to F does not intersect itself, i.e., it is a simple closed polygon (see Exercise 5.17). Next, we need to prove that if we apply (5.14) to these polygons, then the sequence of the left-hand sides converges to $\int_g f\,dx$, and the sequence of the right-hand sides converges to $-\int_A \frac{\partial f}{\partial y}\,dx\,dy$. We skip the details of this argument.

Part (ii) can be proved similarly. The only difference lies in the very first step; instead of proving the statement for normal domains, we prove it for sets of the form

$$\{(x,y) \colon c < y < d,\ \varphi(y) < x < \psi(y)\},$$

where φ and ψ are continuous functions on $[c,d]$ and $\varphi(y) < \psi(y)$, for every $y \in (c,d)$. □

Remark 5.36. According to Theorem 5.32, if $G \subset \mathbb{R}^p$ is a simply connected open set, the mapping $f \colon G \to \mathbb{R}^p$ is differentiable, and the Jacobian matrix of f is symmetric for every $x \in G$, then f has a primitive function on G. Applying Green's theorem, we can give a new proof for the $p = 2$ special case of this theorem (adding the extra assumption that f is continuously differentiable).

We will need the intuitively obvious fact that a connected open set $G \subset \mathbb{R}^2$ is simply connected if and only if G has "no holes" in it; that is, for every simple closed curve g in G, the bounded domain with boundary g is a subset of G. (The statement is false in higher dimensions; see Remark 5.28.2.)

Let $G \subset \mathbb{R}^2$ be a simply connected open set and let $f = (f_1, f_2) \colon G \to \mathbb{R}^2$ be continuously differentiable. We show that if $D_2 f_1(x,y) = D_1 f_2(x,y)$ holds for every $(x,y) \in G$, then f has a primitive function on G.

It is enough to show that the line integral of f is zero for every polygon S in G. Let $g \colon [a,b] \to G$ be a continuous and rectifiable parametrization of the polygon $S \subset G$. Now, g is a simple closed curve. We may assume that g is positively directed, for otherwise, we could switch to the curve $g_1(t) = g(-t)$ $(t \in [-b,-a])$. The curve g_1 also parametrizes S (in the opposite direction), and if the integral of f along g_1 is zero, then it is also zero along g, since the two integrals are the negatives of each other.

The value of the line integral of f along g is the sum $\int_g f_1\,dx + \int_g f_2\,dy$. Let A denote the bounded domain with boundary g. Since G is simply connected, we have that $A \subset G$, and thus $\mathrm{cl}A = A \cup \partial A = A \cup g([a,b]) \subset G$. Thus the partial derivatives of f_1 and f_2 exist and they are continuous on an open set containing $\mathrm{cl}A$ (namely, on G), and we can apply Green's theorem. We get

$$\int_g f_1\,dx + \int_g f_2\,dy = -\int_A \frac{\partial f_1}{\partial y}\,dx\,dy + \int_A \frac{\partial f_2}{\partial x}\,dx\,dy = 0,$$

since our conditions imply $\frac{\partial f_1}{\partial y} = \frac{\partial f_2}{\partial x}$ everywhere in G. □

An application from physics. An important interpretation of Green's theorem in physics is related to the flow of fluids. Let some fluid flow in a region G of the plane and let the direction and speed of the flow be constant at every point $(x, y) \in G$. By that we mean the following: there exists a unit vector $v = v(x, y)$ such that the fluid passing through the point (x, y) always flows in the direction of v, and there exists a number $c = c(x, y) \geq 0$ such that the amount of fluid flowing during a unit of time through every segment containing (x, y) perpendicular to v with length h short enough is $h \cdot c$. We put $f(x, y) = c(x, y) \cdot v(c, y)$. Then at every point (x, y), the direction of the flow is the same as the direction of $f(x, y)$, while its speed is the absolute value of $f(x, y)$.

Let $g\colon [a, b] \to G$ be a positively oriented simple closed curve, and let the bounded domain with boundary g be A. Let us find the amount of fluid flowing through the boundary of the domain A (i.e., through the set $\Gamma = g([a, b])$) in a unit of time.

Consider a fine partition $a = t_0 < t_1 < \ldots < t_n = b$, and let $c_i \in [t_{i-1}, t_i]$ $(i = 1, \ldots, n)$ be inner points of this partition. Let us assume that the subarc Γ_i of Γ corresponding to the interval $[t_{i-1}, t_i]$ of the partition is close to the segment $J_i = [g(t_{i-1}), g(t_i)]$, and that f is close to the vector $f(d_i)$ on the subarc Γ_i, where $d_i = g(c_i)$. If J_i is perpendicular to $f(d_i)$, then the amount of fluid flowing though J_i in unit time is approximately $m_i = |J_i| \cdot |f(d_i)|$, where $|J_i| = |g(t_i) - g(t_{i-1})|$ is the length of the segment J_i. If J_i is not perpendicular to $f(d_i)$, then it is easy to see that the amount m_i of fluid flowing through J_i equals the amount of fluid flowing through J_i', where J_i' is the projection of J_i to the line perpendicular to $f(d_i)$. Let the coordinate functions of v be v_1 and v_2. We obtain the vector $\tilde{v} = (-v_2, v_1)$ by rotating v by 90 degrees in the positive direction. Thus the length of the segment J_i' is the absolute value of the scalar product $\langle g(t_i) - g(t_{i-1}), \tilde{v}(d_i)\rangle$, and m_i is the absolute value of

$$\langle g(t_i) - g(l_{i-1}), \tilde{v}(d_i)\rangle \cdot |f(c_i)| = \langle g(t_i) - g(t_{i-1}), |f(c_i)| \cdot \tilde{v}(d_i)\rangle =$$
$$= \langle g(t_i) - g(t_{i-1}), \tilde{f}(c_i)\rangle,$$

where $\tilde{f} = (-f_2, f_1)$.

It is easy to see that the scalar product $\langle g(t_i) - g(t_{i-1}), \tilde{f}(c_i)\rangle$ is positive when the flow through the subarcs Γ_i is of outward direction from A, and negative when the flow is of inward direction into A. It follows that the signed sum $\sum_{i=1}^{n}\langle g(t_i) - g(t_{i-1}), \tilde{f}(g(c_i))\rangle$ is approximately equal to the amount of fluid going either into A or out from A. Clearly, if the line integral of the function $\tilde{f}$ along g exists, then the value of this integral,

$$\int_g (-f_2)\,dx + \int_g f_1\,dy, \tag{5.16}$$

is equal to this amount.

Now let us assume that f is continuously differentiable and compute the amount of fluid flowing through a small rectangle $R = [x_1, x_2] \times [y_1, y_2]$ in unit time, using another method.

The amount of fluid flowing through the horizontal segment $[(x_1, y), (x_2, y)]$ in unit time is approximately $(x_2 - x_1)f_2$, while the amount of fluid flowing through the vertical segment $[(x, y_1), (x, y_2)]$ is approximately $(y_2 - y_1)f_1$. It follows that for a rectangle R small enough, the amount of fluid flowing through R's opposite sides is approximately the same.

We are interested in this small difference between the amounts flowing through the vertical sides. The difference is

$$(y_2 - y_1) \cdot f_1\left(x_2, \frac{y_1 + y_2}{2}\right) - (y_2 - y_1) \cdot f_1\left(x_1, \frac{y_1 + y_2}{2}\right) \approx$$
$$\approx (y_2 - y_1)(x_2 - x_1) \cdot \frac{\partial f_1}{\partial x} = t(R) \cdot \frac{\partial f_1}{\partial x},$$

where the partial derivative is taken at some inside point using the single-variable version of Lagrange's mean value theorem. Similarly, the difference between the amounts of fluid flowing through the two horizontal sides is approximately $t(R) \cdot \frac{\partial f_2}{\partial y}$.

Thus the amount of fluid flowing through the sides of the rectangle is approximately $t(R) \cdot \left(\frac{\partial f_1}{\partial x} + \frac{\partial f_2}{\partial y}\right)$. We call

$$\frac{\partial f_1}{\partial x} + \frac{\partial f_2}{\partial y}$$

the **divergence** of f and denote it by $\operatorname{div} f$. It is clear from the argument above that the physical meaning of the divergence is the amount of fluid flowing from an area (as a source) if $\operatorname{div} f > 0$ or flowing into the area (as a sink) when $\operatorname{div} f < 0$) of unit size in unit time. That is, the amount of fluid "created" in the set A is $\int_A \operatorname{div} f \, dx \, dy$.

Comparing this with the amount (5.16), we get that

$$\int_g (-f_2)\, dx + \int_g f_1\, dy = \int_A \left(\frac{\partial f_1}{\partial x} + \frac{\partial f_2}{\partial y}\right) dx\, dy. \tag{5.17}$$

Note that this is nothing other than the two statements of Green's theorem combined, and it formulates the natural physical phenomenon that the amount of fluid flowing from a domain is the same as the amount of fluid flowing through its boundary.

Returning to the physical meaning of the divergence, we should note that if $\operatorname{div} f$ is constant and equal to zero on A, then the amount of fluid flowing through the boundary of the domain A is zero, i.e., the amount of fluid flowing into A is the same as the amount of fluid leaving A.

If $\operatorname{div} f(x_0, y_0) > 0$, then $\operatorname{div} f$ is positive in a small neighborhood of the point (x_0, y_0), and fluid flows from a small neighborhood of the point, i.e., (x_0, y_0) is a **source**. On the other hand, if $\operatorname{div} f(x_0, y_0) < 0$, then fluid flows into a small neighborhood of the point, i.e., (x_0, y_0) is a **sink**.

One can guess that the formulas of Green's theorem are variants of the Newton–Leibniz formula. The analogy, however, is not entirely immediate. It is useful to write the integrals $\int_g f\,dx$ and $\int_g f\,dy$ in another form that makes the analogy clear. To do this, we define a new integral and also illustrate it with an example from physics.

Let a frictional force of magnitude $f(x)$ hinder the motion at every point of the domain G. How much work does a solid do while it moves along the curve $g\colon [a,b] \to G$? Let $a = t_0 < t_1 < \ldots < t_n = b$ be a partition, and let the arc of the curve g corresponding to the division interval $[t_{i-1}, t_i]$ be well approximated by the segment $[g(t_{i-1}), g(t_i)]$. Furthermore, let the force f be nearly constant on the arc γ_i. Since the frictional force is independent of the direction of motion, the amount of work done along the arc γ_i is approximately $f(g(c_i)) \cdot |g(t_i) - g(t_{i-1})|$, where $c_i \in [t_{i-1}, t_i]$ is an arbitrary inner point. Thus the total work can be approximated by the sum $\sum_{i=1}^n f(g(c_i)) \cdot |g(t_i) - g(t_{i-1})|$. If there exists a number I such that this sum approximates I arbitrarily well for a fine enough partition, then I is the total amount of work done.

Definition 5.37. Let $g\colon [a,b] \to \mathbb{R}^p$ be a curve and let the real function f be defined on the set $g([a,b])$. We say that the *line integral with respect to arc length* $\int_g f\,ds$ *exists and its value is* I if for every $\varepsilon > 0$ there exists $\delta > 0$ such that

$$\left| I - \sum_{i=1}^{n} f(g(c_i)) \cdot |g(t_i) - g(t_{i-1})| \right| < \varepsilon$$

holds for every partition $a = t_0 < t_1 < \ldots < t_n = b$ finer than δ and for arbitrary inner points $c_i \in [t_{i-1}, t_i]$ $(i = 1, \ldots, n)$.

The proof of the following theorem on the existence and valuc of the line integral with respect to arc length can be proved similarly to Theorem 16.20 of [7].

Theorem 5.38. *Let g be a continuous and rectifiable curve, and let the function f be continuous on the set $g([a,b])$. Then the line integral with respect to arc length $\int_g f\,ds$ exists. If the components of g are differentiable and their derivatives are integrable on $[a,b]$, then we have*

$$\int_g f\,ds = \int_a^b f(g(t)) \cdot |g'(t)|\,dt.$$

We would like to compress formulas (5.14) and (5.15) into a single formula. This requires the introduction of some new notation. First, we extend the integrals that have been defined for real-valued functions to functions mapping into $\mathbb{R}^q$. The extended integral is evaluated component by component. For example, for a measurable set $H \subset \mathbb{R}^p$ and a mapping $f = (f_1, \cdots, f_q)\colon H \to \mathbb{R}^q$ whose components are integrable on H, let

$$\int_H f\,dt = \left(\int_H f_1\,dt, \ldots, \int_H f_q\,dt \right).$$

Let $g\colon [a,b] \to \mathbb{R}^p$ be an arbitrary curve. The integral with respect to arc length of the function $f\colon g([a,b]) \to \mathbb{R}^q$ is also defined component by component, i.e., we put

$$\int_g f\,ds = \left(\int_g f_1\,ds, \ldots, \int_g f_q\,ds\right).$$

Now let us revisit Green's formulas. Let $g = (g_1, g_2)\colon [a, b] \to \mathbb{R}^2$ be a differentiable positively oriented simple closed curve, and let A be the bounded domain with boundary g. If $g'(t) \neq 0$ at some point t, then the vector $g'(t) = (g_1'(t), g_2'(t))$ has the same direction as the tangent to the curve g at the point $g(t)$. By rotating the tangent-directed unit vector by 90 degrees in the negative direction, we get the **outer normal vector** of the curve, i.e., the unit vector perpendicular to the tangent and pointing outward from the domain A. We denote the outer normal vector at the point $g(t)$ by $n(g(t))$. Again, this is the unit vector with direction $(g_2'(t), -g_1'(t))$, and

$$n(g(t)) = \frac{1}{|g'(t)|} \cdot (g_2'(t), -g_1'(t)).$$

For $f\colon g([a, b]) \to \mathbb{R}$, we have

$$\int_g f n\,ds = \int_a^b \frac{f(g(t))}{|g'(t)|} \cdot (g_2'(t), -g_1'(t)) \cdot |g'(t)|\,dt =$$

$$= \left(\int_a^b f(g(t))g_2'(t)\,dt, -\int_a^b f(g(t))g_1'(t)\,dt\right) =$$

$$= \left(\int_g f\,dy, -\int_g f\,dx\right). \tag{5.18}$$

Comparing this to Green's theorem, we have the following theorem.

Theorem 5.39. *Let $g\colon [a, b] \to \mathbb{R}^2$ be a positively oriented simple closed plane curve that is the union of finitely many continuously differentiable arcs. Let A be the bounded domain with boundary g, and let $G \supset \mathrm{cl}A$ be open. If $f\colon G \to \mathbb{R}$ is continuously differentiable, then*

$$\int_g f n\,ds = \int_A f'\,dx\,dy. \tag{5.19}$$

The formula states that the integral of the derivative of a function f on the set A is the same as the integral with respect to arc length of the mapping fn along the boundary of A.

Note that equality (5.19), using formula (5.18), is equivalent to Green's theorem in the sense that the equality of the two components gives the two statements of Green's theorem.

Remark 5.40. Equality (5.19) can be viewed as the two-dimensional variant of the Newton–Leibniz formula. According to the original Newton–Leibniz formula, the integral of the function f' on the interval $[a, b]$ is equal to the signed "integral" of f along the boundary of $[a, b]$, i.e., the difference $f(b) - f(a)$. Since we can say that the vector (number) 1 is the outer normal at the point b of the interval, and the vector -1 is the outer normal at the point a of the interval, it follows that $f(b) - f(a) = f(a) \cdot n(a) + f(b) \cdot n(b)$, which is the exact analogue of the left-hand side of (5.19).

Exercises

5.14. Let g be a continuous rectifiable positively oriented simple closed plane curve, and let A be the bounded domain with boundary g. Show that both of the line integrals $\int_g x\,dy$ and $-\int_g y\,dx$ are equal to the area of A.

5.15. Test the statement of the previous exercise for the following curves:

(a) $g(t) = (a \cdot \cos t, b \cdot \sin t)$ $(t \in [0, 2\pi])$ (ellipse);
(b) an arbitrary parametrization of the rectangle $[a, b] \times [c, d]$ satisfying the conditions.

5.16. Find the area of the bounded domains with the following boundaries:

(a) $g(t) = (2t - t^2, 2t^2 - t^3)$ $(t \in [0, 2])$;
(b) $g(t) = (a \cdot \cos^3 t, a \cdot \sin^3 t)$ $(t \in [0, 2\pi])$.

5.17. Show that every simple closed plane curve has an arbitrarily fine non-self-intersecting inscribed polygon (that is, there are arbitrarily fine partitions of the parameter interval such that the corresponding polygonal line does not intersect itself). (H S)

5.18. Let g and A satisfy the conditions of Green's theorem. Show that if $f = (f_1, f_2)$ is continuously differentiable on an open set containing cl$A$, then the line integral of f along g is $\int_A (D_1 f_2 - D_2 f_1)\,dx\,dy$.

5.19. Let $g\colon [a, b] \to \mathbb{R}^p$ be a differentiable curve whose coordinate functions are integrable on $[a, b]$. Show that the center of mass of g is

$$\left(\frac{1}{L} \int_g x_1\,ds, \ldots, \frac{1}{L} \int_g x_p\,ds \right),$$

where L is the length of the curve.

5.4 Surface and Surface Area

Determining the surface area of surfaces is a much harder task than finding the area of planar regions or the volume of solids; the definition of surface area itself already causes difficulties. To define surface area, the method used to define area—bounding the value from above and below—does not work. The method of defining arc length (the supremum of the lengths of the inscribed polygonal lines) cannot be applied to define surface area either. This already fails in the simplest cases: one can show that the inscribed polygonal surfaces of a right circular cylinder can have arbitrarily large surface area. In some special cases, such as surfaces of revolution, the definition and computation of the surface area is simpler; see, e.g., Section 16.6 of [7]. To precisely define surface area in the general case, we need the help of multivariable differentiation and integration. Since these are now at our disposal, we may start to define surfaces and to compute their area.

Curves are defined as mappings defined on intervals. Analogously, (although slightly more generally) we define surfaces as mappings from a measurable subset of the plane. More precisely—to avoid conflict with other surface definitions from differential geometry and topology—we will call these maps parametrized surfaces. Therefore, we will say that the mappings $g\colon A \to \mathbb{R}^p$, where $A \subset \mathbb{R}^2$ is measurable, are **parametrized surfaces in** $\mathbb{R}^p$. A parametrized surface g is said to be **continuous** or **differentiable** or **continuously differentiable** if the mapping g has the corresponding property on the set A.

To define surface areas, we first compute the areas of parallelograms. We know the area of the parallelogram $P(a, b)$ spanned by the vectors $a = (a_1, a_2)$ and $b = (b_1, b_2)$ of the plane: it is the absolute value of the determinant $\begin{vmatrix} a_1 & a_2 \\ b_1 & b_2 \end{vmatrix}$, i.e., $|a_1b_2 - a_2b_1|$ (see Theorem 3.31). With some simple algebra, we have

$$(a_1b_2 - a_2b_1)^2 = (a_1^2 + a_2^2)(b_1^2 + b_2^2) - (a_1b_1 + a_2b_2)^2 = |a|^2|b|^2 - \langle a, b\rangle^2,$$

i.e., the area of the parallelogram spanned by the plane vectors a, b is

$$\sqrt{|a|^2|b|^2 - \langle a, b\rangle^2}.$$

Let a and b be arbitrary vectors of $\mathbb{R}^3$. There exists an isometry g such that g maps a and b into the set $\{(x_1, x_2, x_3)\colon x_3 = 0\}$. Identifying this set with $\mathbb{R}^2$, we obtain that the area of the parallelogram spanned by the vectors $g(a)$ and $g(b)$ is $\sqrt{|g(a)|^2|g(b)|^2 - \langle g(a), g(b)\rangle^2}$. Now, isometries change neither the length nor the scalar product of vectors; this latter is true, since $2\langle x, y\rangle = |x + y|^2 - |x|^2 - |y|^2$ for each $x, y \in \mathbb{R}^3$. Assuming that isometries do not change the area of parallelograms either, we can say that *the area of the parallelogram spanned by the vectors* $a, b \in \mathbb{R}^3$ *is* $\sqrt{|a|^2|b|^2 - \langle a, b\rangle^2}$. This area can be defined with the vector multiplication of the vectors a and b in $\mathbb{R}^3$. For a pair of vectors $a = (a_1, a_2, a_3)$ and $b = (b_1, b_2, b_3)$ in $\mathbb{R}^3$, we call the vector

$$(a_2b_3 - b_2a_3, b_1a_3 - a_1b_3, a_1b_2 - b_1a_2)$$

the **vector product** of a and b. We can memorize this with the help of the formula

$$a \times b \overset{\text{def}}{=} \begin{vmatrix} i & j & k \\ a_1 & a_2 & a_3 \\ b_1 & b_2 & b_3 \end{vmatrix},$$

where $i = (1, 0, 0)$, $j = (0, 1, 0)$, $k = (0, 0, 1)$. One can check that the length of $a \times b$ is exactly $\sqrt{|a|^2|b|^2 - \langle a, b\rangle^2}$.

Let $A \subset \mathbb{R}^2$ be measurable, and let $g\colon A \to \mathbb{R}^3$ be a continuously differentiable parametrized surface. Choose a square $N = [a, a+h] \times [b, b+h]$ in A and a point $(c, d) \in N$. Since g is continuously differentiable, g is approximated by $g(c, d) + g'(c, d)(x - c, y - d)$ on N well enough such that—assuming any reasonable definition of the surface area—we expect that the surface area of $g(N)$ is close to the area of the parallelogram $g'(c, d)(N)$ if N is small enough.

Now we compute the area of $g'(c, d)(N)$. Let the components of g be g_1, g_2, g_3 and let us introduce the notation

$$D_1 g = (D_1 g_1, D_1 g_2, D_1 g_3) \quad \text{and} \quad D_2 g = (D_2 g_1, D_2 g_2, D_2 g_3).$$

Then the vectors $D_1 g$ and $D_2 g$ are the same as the column vectors of g's Jacobian matrix. It is easy to see that $g'(c, d)([0, h] \times [0, h])$ is the parallelogram spanned by the vectors $h \cdot D_1 g(c, d)$ and $h \cdot D_2 g(c, d)$; thus $g'(c, d)(N)$ is a translation of this parallelogram. Therefore, the area of $g'(c, d)(N)$ is

$$|h \cdot D_1 g(c, d) \times h \cdot D_2(c, d)| = |D_1 g(c, d) \times D_2 g(c, d)| \cdot h^2.$$

We get that the surface area of $g(N)$ is (supposedly) close to $|D_1 g(c, d) \times D_2 g(c, d)| \cdot t(N)$, for a small enough square N. Thus, for n large enough, this holds for every square $N \in \mathcal{K}_n$ in the interior of A. Let $N_1, \ldots, N_k$ be an enumeration of the squares in the interior of A. Let us choose a point (c_i, d_i) from every square N_i, and take the sum $S = \sum |D_1 g(c_i, d_i) \times D_2 g(c_i, d_i)| \cdot t(N_i)$. This sum will be close to the surface area $g(A)$.

When A is a square, this suggests that the surface area of the parametrized surface g is the number approximated by the above sums S if A is partitioned into small enough squares. Since these sums are also the approximating sums of the integral $\int_A |D_1 g \times D_2 g| \, dx \, dy$, the value of the integral gives the surface area we were looking for.

One can expect the integral to be equal to the surface area even when $A \subset \mathbb{R}^2$ is an arbitrary measurable set. This follows from the fact that the sum S is the same as the approximating sum of the function $|D_1 g \times D_2 g|$ corresponding to the partition $\{N_1, \ldots, N_k, A \setminus \bigcup_{i=1}^{k} N_i\}$, save for a single term. The missing term is small, since $\mu\left(A \setminus \bigcup_{i=1}^{k} N_i\right) = \mu(A) - \underline{\mu}(A, n) \to 0$ as $n \to \infty$. Thus the value of the surface area needs to be $\int_A |D_1 g \times D_2 g| \, dx \, dy$.

From what we have above, it should be more or less clear how to define the surface area with the help of approximating sums. To avoid some technical problems, we define the surface area by the result of the argument; that is, by the integral itself.

Definition 5.41. Let $A \subset \mathbb{R}^2$ be measurable and let $g\colon A \to \mathbb{R}^3$ be a continuously differentiable parametrized surface. If the function $|D_1 g \times D_2 g|$ is integrable on A, then we say that the *surface area of g exists and is equal to*

$$\int_A |D_1 g \times D_2 g| \, dx \, dy. \tag{5.20}$$

Example 5.42. Let us consider the quarter-disk $A = \{(x, y)\colon x, y \geq 0,\ x^2 + y^2 \leq R^2\}$ and find the area of the part of the saddle surface $z = xy$ lying over the set A. Consider the parametrization $g(x,y) = (x,y,xy)$ $((x,y) \in A)$. We have $D_1 g = (1,0,y)$, $D_2 g = (0,1,x)$, which gives

$$|D_1 g \times D_2 g| = \sqrt{(1+y^2)(1+x^2) - x^2 y^2} = \sqrt{1+x^2+y^2},$$

and the surface area is $F = \int_A \sqrt{1+x^2+y^2}\, dx\, dy$. This integral can be computed by substituting with polar coordinates:

$$F = \int_0^{\pi/2} \int_0^R \sqrt{1+r^2} \cdot r\, dr\, d\varphi = \frac{\pi}{2} \cdot \left[\frac{1}{3}(1+r^2)^{3/2}\right]_0^R = \frac{\pi}{6} \cdot \left((1+R^2)^{3/2} - 1\right).$$

Remarks 5.43. **1.** By assumption, $|D_1 g \times D_2 g|$ is continuous on A. Then the integral in (5.20) exists if and only if $|D_1 g \times D_2 g|$ is bounded on the set A (see Theorem 4.14). This is satisfied automatically if A is closed; i.e., the surface area of g exists in this case. The same can be said if g is defined and is continuously differentiable on an open set containing $\operatorname{cl} A$.

2. Let $f\colon [a,b] \to [0,\infty)$ be a continuously differentiable function. The natural parametrization of the surface of revolution we get by rotating graph f is the mapping $g(x,\varphi) = (x, f(x)\cos\varphi, f(x)\sin\varphi)$ $((x,\varphi) \in [a,b] \times [0,2\pi])$. On computing the surface area of the parametrized surface g using Definition 5.41, we obtain

$$2\pi \int_a^b f(x)\sqrt{1+(f'(x))^2}\, dx. \tag{5.21}$$

(See Exercise 5.20); cf. [7, Theorem 16.31].)

3. One can show that for one-to-one continuously differentiable parametrizations defined on bounded and closed sets $H \subset \mathbb{R}^3$, the area of the surface is independent of the parametrization.[12]

The exact meaning of this statement is the following. Let A and B be measurable closed sets in the plane, and let $g\colon A \to \mathbb{R}^3$, $h\colon B \to \mathbb{R}^3$ be injective and continuously differentiable mappings. If $g(A) = h(B)$, then we have

$$\int_A |D_1 g \times D_2 g|\, dx\, dy = \int_B |D_1 h \times D_2 h|\, dx\, dy. \tag{5.22}$$

See Exercise 5.21 for the proof.

Theorem 5.44. *Let $A \subset \mathbb{R}^2$ be a measurable closed set and let $f\colon A \to \mathbb{R}$ be continuously differentiable. Then the surface area of the graph of f is*

$$\int_A \sqrt{1+(D_1 f)^2 + (D_2 f)^2}\, dx\, dy. \tag{5.23}$$

Proof. The mapping $g(x,y) = (x,y,f(x,y))$ $((x,y) \in A)$ is a continuously differentiable parametrization of the graph of f. Since $D_1 g = (1,0,D_1 f)$ and $D_2 g = (0,1,D_2 f)$, we get

[12] For the analogous statement concerning arcs, see [7, Theorem 16.18].

$$|D_1 g \times D_2 g|^2 = (1+(D_1 f)^2)(1+(D_2 f)^2) - (D_1 f)^2 (D_2 f)^2 = 1 + (D_1 f)^2 + (D_2 f)^2,$$

and we obtain the statement by the definition of the surface area. □

Remark 5.45. The area of the parallelogram $P(a, b)$ is

$$\sqrt{|a|^2|b|^2 - \langle a, b\rangle^2} \tag{5.24}$$

not only for vectors a, b belonging to $\mathbb{R}^2$ or $\mathbb{R}^3$, but also for vectors in $\mathbb{R}^p$ for every $p > 3$. The proof is the same as in the case of $p = 3$.

For an arbitrary p, let us denote the value of (5.24) by $|a \times b|$. (When $p \neq 3$, the notation $|a \times b|$ is not the absolute value of the vector $a \times b$, since we defined the vector product $a \times b$ in only three dimensions. For $p \neq 3$, the notation $|a \times b|$ should be considered the abbreviation of (5.24), motivated by the case of the vectors of $\mathbb{R}^3$.)

The ideas of the argument introducing Definition 5.41 can be applied to every parametrized surface mapping into $\mathbb{R}^p$. This justifies the following definition.

Let $A \subset \mathbb{R}^2$ be measurable and let $g\colon A \to \mathbb{R}^p$ be a continuously differentiable parametrized surface. If the function $|D_1 g \times D_2 g|$ is integrable on A, then we say that the **surface area of g exists and its value is**

$$\int_A |D_1 g \times D_2 g|\, dx\, dy.$$

Exercises

5.20. Let $f\colon [a, b] \to [0, \infty)$ be a continuously differentiable function. Show that the surface area of the surface of revolution obtained by rotating graph f about the x-axis and parametrized by $g(x, \varphi) = (x, f(x)\cos\varphi, f(x)\sin\varphi)$ $(x, \varphi) \in [a, b] \times [0, 2\pi]$ is given by (5.21).

5.21. Let A and B be measurable closed sets of the plane and let $g\colon A \to \mathbb{R}^p$, $h\colon B \to \mathbb{R}^p$ be injective and continuously differentiable mappings with $g(A) = h(B)$. Show that

$$\int_A |D_1 g \times D_2 g|\, dx\, dy = \int_B |D_1 h \times D_2 h|\, dx\, dy. \text{ (H)}$$

5.5 Integral Theorems in Three Dimension

Theorem 5.39 can be generalized to every dimension $p > 2$. Unfortunately, even the precise formulation of these generalizations causes difficulties, because the required definition of the necessary notions (e.g., that of the outer normal) is rather complicated. For this reason we restrict ourselves to the case $p = 3$, where we have an intuitive picture of these notions.

First we need to define the surface integral. This integral is the generalization of the integral with respect to arc length to parametrized surfaces.[13]

Let $A \subset \mathbb{R}^2$ be measurable, let $g\colon A \to \mathbb{R}^p$ be a continuously differentiable parametrized surface, and let $f\colon g(A) \to \mathbb{R}$. Let $\{A_1, \ldots, A_n\}$ be a fine partition of the set A, let $(c_i, d_i) \in A_i$ be inner points, and take the approximating sum $\sum_{i=1}^n f(g(c_i, d_i)) \cdot F(g(A_i))$, where F denotes the surface area. The value of the surface integral $\int_A f\, dF$ is the number I that these approximating sums approximate when the partition is fine enough.

We have seen that for a small square $N \subset \operatorname{int} A$ and $(c, d) \in N$, the surface area of $g(N)$ is approximately $|D_1 g(c, d) \times D_2 g(c, d)| \cdot t(N)$. Consider the partition

$$\left\{ N_1, \ldots, N_k, A \setminus \bigcup_{i=1}^{k} N_i \right\},$$

where $N_1, \ldots, N_k$ denote the squares of the square grid $\mathcal{K}_n$ lying in $\operatorname{int} A$. The approximating sum corresponding to this partition differs in only a single term (of small magnitude) from the sum $\sum_{i=1}^k |D_1 g(c_i, d_i) \times D_2 g(c_i, d_i)| \cdot f(g(c_i, d_i)) \cdot \mu(N)$, which is close to the integral $\int_A (f \circ g) \cdot |D_1 g \times D_2 g|\, dx\, dy$ for n large enough.

Therefore, it seems reasonable to define the surface integral not in terms of the approximating sums, but by an integral, similarly to the definition of the surface area.

Definition 5.46. Let $g\colon A \to \mathbb{R}^p$ be a continuously differentiable parametrized surface, where $A \subset \mathbb{R}^2$ is measurable, and let $f\colon g(A) \to \mathbb{R}$. The value of the *surface integral* $\int_A f\, dF$ is, by definition, the value of the integral $\int_A (f \circ g) \cdot |D_1 g \times D_2 g|\, dx\, dy$, assuming that the latter integral exists.

One can show that for a measurable and closed set A and an injective parametrization g, the value of $\int_A f\, dF$ is independent of the parametrization in the sense that if $B \subset \mathbb{R}^2$ is measurable and closed, $h\colon B \to \mathbb{R}^p$ is injective and continuously differentiable, and $g(A) = h(B)$, then

$$\int_A (f \circ g) \cdot |D_1 g \times D_2 g|\, dx\, dy = \int_B (f \circ h) \cdot |D_1 h \times D_2 h|\, dx\, dy.$$

Thus we can talk about surface integrals with respect to surfaces. If $H \subset \mathbb{R}^p$ and $f\colon H \to \mathbb{R}$, then the surface integral $\int_H f\, dF$ is, by definition, the integral $\int_g f\, dF$, where $g\colon A \to \mathbb{R}^p$ is a continuously differentiable and injective mapping on the measurable and closed set with $g(A) = H$. Of course, we need to assume that the set H has such a parametrization. When this holds, then we say that H is a **continuously differentiable surface.**

We need one more notion: the generalization of Definition 5.46 from real-valued functions to vector-valued functions.

[13] We do not cover the generalization of the line integral defined in Definition 5.1 to surfaces. Note that in the two-variable case the line integral can be expressed in terms of the integral with respect to arc length for continuously differentiable curves; see formula (5.18).

For $f\colon H \to \mathbb{R}^q$, by the integral $\int_H f\,dF$ we mean the vector

$$\int_H f\,dF \stackrel{\text{def}}{=} \left(\int_H f_1\,dF, \ldots, \int_H f_q\,dF \right),$$

where $f_1, \ldots, f_q$ denote the coordinate functions of f.

Let the boundary of a bounded set $K \subset \mathbb{R}^3$ be the union of finitely many continuously differentiable surfaces. If ∂K has a tangent plane at the point $x \in \partial K$, we call the unit vector starting from K, perpendicular to the tangent plane, and pointing outward from K the **outer normal** of K. We denote the outer normal at the point $x \in \partial K$ by $n(x)$. The outer normal is not defined on the boundary curves of the surfaces whose union is ∂K; here $n(x)$ is defined as an arbitrary unit vector.

We can now state the three-dimensional variant of the Newton–Leibniz formula.

Theorem 5.47. *Let the bounded set $K \subset \mathbb{R}^3$ be the union of finitely many continuously differentiable surfaces. If the real-valued function f is continuously differentiable on $\mathrm{cl}K$, then we have*

$$\int_{\partial K} fn\,dF = \int_K f'\,dx\,dy\,dz, \tag{5.25}$$

that is,

$$\begin{aligned} &\left(\int_{\partial K} fn_1\,dF, \int_{\partial K} fn_2\,dF, \int_{\partial K} fn_3\,dF \right) = \\ &= \left(\int_K \frac{\partial f}{\partial x}\,dx\,dy\,dz, \int_K \frac{\partial f}{\partial y}\,dx\,dy\,dz, \int_K \frac{\partial f}{\partial z}\,dx\,dy\,dz \right), \end{aligned} \tag{5.26}$$

where $n = (n_1, n_2, n_3)$.

Proof. We may assume that K is a polyhedron. (We obtain the general case by approximating K with a suitable sequence of polyhedra.) We may also assume that K is a convex polyhedron. Indeed, we can represent every polyhedron K as the union of nonoverlapping convex polyhedra by cutting it along the planes of the faces of K. Let $K = K_1 \cup \ldots \cup K_n$ be such a partition, and suppose that

$$\int_{\partial K_i} fn\,dF = \int_{K_i} f'\,dx\,dy\,dz \tag{5.27}$$

holds for every $i = 1, \ldots, n$. Summing these equalities yields (5.25). Indeed, it follows from (4.3) that the sum of the right-hand sides of (5.27) is the same as the right-hand side of (5.25). On summing the left-hand sides, those terms that correspond to the faces of the polyhedra K_i lying in the interior of K cancel each other out.

To prove this, consider a face L of the polyhedron K_i lying in the interior of K, and let S be the plane containing L. There are polyhedra K_j that have a face in S such that K_i and K_j lie on opposite sides of S. For such a polyhedron K_j, the outer normals of the faces of the polyhedra K_i and K_j that lie on S are the opposites of each other. Thus the surface integrals on the intersection of these faces are also the opposites of each other, and consequently, their sum is zero. Therefore, by summing the surface integrals $\int_{\partial K_i} fn\,dF$, the integrals on the faces of ∂K are the only ones that do not cancel out, and the sum of these is exactly the integral $\int_{\partial K} fn\,dF$.

Thus we may assume that K is a convex polyhedron. We prove the equality (5.26) component by component. By symmetry, it is enough to show that the terms corresponding to z are equal to each other on the two sides of (5.26), i.e.,

$$\int\limits_{\partial K} fn_3\,dF = \int\limits_K \frac{\partial f}{\partial z}\,dx\,dy\,dz, \tag{5.28}$$

where $n = (n_1, n_2, n_3)$. Let B be the projection of the polyhedron K on the xy-plane. Then B is a convex polygon, and for every $(x, y) \in B$ the section $K_{(x,y)} = \{z \in \mathbb{R}\colon (x, y, z) \in K\}$ is a segment. Let $K_{(x,y)} = [m(x, y), M(x, y)]$ for every $(x, y) \in B$. Then

$$K = \{(x, y, z)\colon (x, y) \in B,\ m(x, y) \le z \le M(x, y)\},$$

i.e., K is a normal domain in $\mathbb{R}^3$. By Theorem 4.18, the integral on the right-hand side of (5.28) is equal to the integral

$$\int_B \left(\int_{m(x,y)}^{M(x,y)} \frac{\partial f}{\partial z}\,dz \right) dx\,dy.$$

By the Newton–Leibniz formula, the inner integral is $f(M(x, y)) - f(m(x, y))$, and thus

$$\int\limits_K \frac{\partial f}{\partial z}\,dx\,dy\,dz = \int_B f(M(x, y))\,dx\,dy - \int_B f(m(x, y))\,dx\,dy. \tag{5.29}$$

Let the sides of K be $A_1, \ldots, A_n$. Then the integral on the left-hand side of (5.28) equals

$$\sum_{i=1}^{n} \int\limits_{A_i} fn_3\,dF. \tag{5.30}$$

The outer normal is $n(x)$, and its third coordinate restricted to the interior of each side is constant. We form three classes of sides according to whether n_3 is zero, positive, or negative on their interior. The terms of (5.30) corresponding to the first class are zero, and thus their sum is also zero. We show that

$$\sum_{n_3>0} \int\limits_{A_i} fn_3\,dF = \int_B f(M(x, y))\,dx\,dy \tag{5.31}$$

and

$$\sum_{n_3<0} \int\limits_{A_i} fn_3\,dF = -\int_B f(m(x, y))\,dx\,dy. \tag{5.32}$$

From these and from (5.29) the statement of the theorem will follow immediately.

The second class consists of the sides A_i with $n_3 > 0$. In other words, these are the sides whose outer normal points upward (toward the positive direction of the z-axis). It is easy to see that the projections of these sides to the xy-plane are nonoverlapping convex polygons whose union is B. If B_i denotes the projection of the side A_i, then the right-hand side of (5.31) is $\sum_{n_3>0} \int_{B_i} f(M(x,y))\,dx\,dy$. Therefore, in order to prove (5.31), it is enough to show that

$$\int_{A_i} f n_3 \, dF = \int_{B_i} f(M(x,y))\,dx\,dy \tag{5.33}$$

for every side A_i belonging to the second class. Let A_i be such a side. Obviously, if the point (x,y,z) is in A_i, then $z = M(x,y)$. This means that A_i is the graph of the function M restricted to B_i.

We know that the integral on the left-hand side of (5.33) does not depend on the parametrization of A_i, assuming that it is continuous and injective. Hence we may assume that the parametrization of A_i is

$$g(x,y) = (x,y,M(x,y)) \qquad ((x,y) \in B_i),$$

and the integral on the left-hand side of (5.33) is equal to the integral

$$\int_{B_i} f(M(x,y)) \cdot n_3 \cdot |D_1 M \times D_2 M| \, dx\,dy.$$

Since A_i lies in a plane, the function M restricted to B_i is of the form $ax + by + c$, where a, b, c are constants. The vectors $D_1 M = (a,0,0)$ and $D_2 M = (0,b,0)$ are constant, and $|D_1 M \times D_2 M|$ is also constant on the set B_i. Since the area of A_i is $\int_{B_i} |D_1 M \times D_2 M| \, dx\,dy$, the constant $|D_1 M \times D_2 M|$ is the ratio of the areas of A_i and B_i. We show that this ratio is exactly $1/n_3$, which will prove (5.33).

Let the plane containing the side A_i be S. If S is parallel to the xy-plane, then the areas of A_i and B_i are equal to each other. Since $n_3 = 1$ in this case, the claim holds. Now assume that S is not parallel to the xy-plane, and let ℓ be the line where S and the xy-plane intersect.

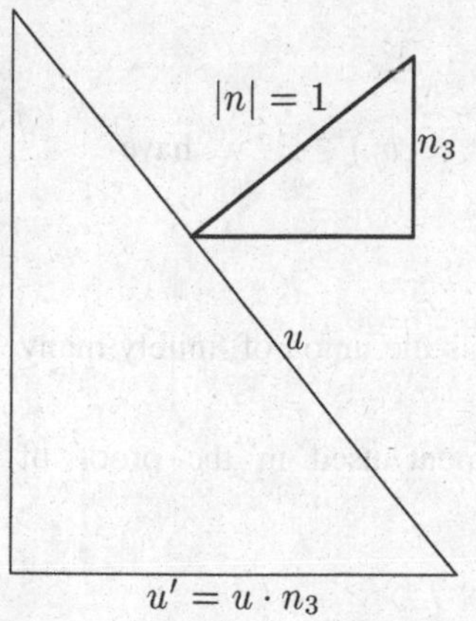

5.5. Figure

If a segment of length u of the plane S is perpendicular to ℓ, then its projection on the xy-plane is a segment of length $u' = u \cdot n_3$. Indeed, take a look at Figure 5.5, whose plane is perpendicular to the line ℓ. The equality $u' = u \cdot n_3$ follows from the similarity of the two right triangles of the figure.

Let H be a triangle in the plane S that has a side perpendicular to ℓ and of length u. The projection of the triangle H to the xy-plane is a triangle H' whose side perpendicular to ℓ is of length $u' = u \cdot n_3$. Since the altitude corresponding to this side of H is not changed by this projection, the ratio of the areas of H and H' is $1/n_3$.

Now, every polygon is the union of nonoverlapping triangles that have a side of fixed direction. Therefore, the ratio of the areas of A_i and B_i is also $1/n_3$. We have proved (5.33) for every i, and as we have seen, (5.31) follows immediately.

Equality (5.32) can be proved entirely similarly, or it can be reduced to (5.31) by reflecting K to the xy-plane. □

Theorem 5.48. *Let the boundary of the bounded set $K \subset \mathbb{R}^3$ be the union of finitely many continuously differentiable surfaces. If $f = (f_1, f_2, f_3)\colon \mathrm{cl}K \to \mathbb{R}^3$ is continuously differentiable, then*

$$\int_{\partial K} \langle f, n\rangle \, dF = \int_K \operatorname{div} f \, dx \, dy \, dz, \tag{5.34}$$

where $\operatorname{div} f \overset{\text{def}}{=} D_1 f_1 + D_2 f_2 + D_3 f_3$, *and*

$$\int_{\partial K} (f \times n) \, dF = -\int_K \operatorname{rot} f \, dx \, dy \, dz, \tag{5.35}$$

where $\operatorname{rot} f \overset{\text{def}}{=} (D_2 f_3 - D_3 f_2, D_3 f_1 - D_1 f_3, D_1 f_2 - D_2 f_1)$.

Proof. Applying (5.25) to f_1, f_2, and f_3, and taking the first component of the first resulting equality, the second component of the second resulting equality, and the third component of the third resulting equality and then summing the equalities obtained, we obtain (5.34). Equality (5.35) can be proved similarly. □

Traditionally, the formula (5.34) is known as the **Gauss**[14]**–Ostrogradsky**[15] **theorem** or **divergence theorem**, , and the formula (5.35) is known as **Stokes's**[16] **theorem**. These formulas are of fundamental importance in physics, e.g., in the theory of the flow of fluids and also in electrodynamics. The physical interpretation of Stokes's theorem is more complicated than that of Green's theorem for the flow of fluids. If $f(x)$ describes the direction and velocity of the flow of a fluid, then $\operatorname{rot} f(x)$ gives the direction of the axis of rotation and the velocity of the rotation at the point x of the flow.

Exercises

5.22. Show that for every $a = (a_1, \ldots, a_p)$, $b = (b_1, \ldots, b_p) \in \mathbb{R}^p$ we have

$$|a \times b|^2 = \sum_{i<j} \begin{vmatrix} a_i & a_j \\ b_i & b_j \end{vmatrix}^2 .$$

5.23. Show that every polyhedron can be expressed as the union of finitely many nonoverlapping tetrahedra. (H)

5.24. Prove Green's theorem following the argument used in the proof of Theorem 5.47.

---

14 Carl Friedrich Gauss (1777–1855), German mathematician.

15 Mikhail Vasilyevich Ostrogradsky (1801–1862), Russian mathematician.

16 George Gabriel Stokes (1819–1903), British mathematician and physicist.

Chapter 6
Infinite Series

6.1 Basics on Infinite Series

If we add infinitely many numbers (that is, if we take the sum of an infinite sequence of numbers), then we get an infinite series. Mathematicians in India investigated infinite series as early as the fifteenth century, while European mathematics caught up with them only in the seventeenth century. Although deep and important discoveries were made both in India and later in Europe, for several centuries the exact notion of convergent series was lacking, and this led to strange or even contradictory results. (For details on the history of infinite series see the "Brief Historical Introduction" of [7], and also the appendix of this chapter.) The debates concerning these contradictions lasted until the nineteenth century, when Augustin-Louis Cauchy defined the sum of an infinite series as the limit of its partial sums. We begin with Cauchy's definition.

Definition 6.1. The *partial sums* of the infinite series $\sum_{n=1}^{\infty} a_n$ are the numbers $s_n = \sum_{i=1}^{n} a_i$ $(n = 1, 2, \ldots)$. If the sequence of partial sums (s_n) is convergent with limit A, then we say that the *infinite series* $\sum_{n=1}^{\infty} a_n$ *is convergent, and its sum is* A. We denote this by $\sum_{n=1}^{\infty} a_n = A$.

If the sequence of partial sums (s_n) is divergent, then we say that the *series* $\sum_{n=1}^{\infty} a_n$ *is divergent.*

If $\lim_{n\to\infty} s_n = \infty$ (or $-\infty$), then we say that *the sum of the series* $\sum_{n=1}^{\infty} a_n$ is ∞ *(or* $-\infty$*)*. We denote this by $\sum_{n=1}^{\infty} a_n = \infty$ (or $-\infty$).

Example 6.2. **1.** The nth partial sum of the series $1 + 1/2 + 1/4 + 1/8 + \ldots$ is $s_n = \sum_{i=0}^{n-1} 2^{-i} = 2 - 2^{-n+1}$. Since $\lim_{n\to\infty} s_n = 2$, the series is convergent, and its sum is 2.
2. The nth partial sum of the series $3/10 + 3/100 + 3/1000 + \ldots$ is

$$s_n = \sum_{i=1}^{n} 3 \cdot 10^{-i} = \frac{3}{10} \cdot \frac{1 - 10^{-n}}{1 - (1/10)}.$$

M. Laczkovich and V.T. Sós, *Real Analysis*, Undergraduate Texts in Mathematics, https://doi.org/10.1007/978-1-4939-7369-9_6

Since $\lim_{n\to\infty} s_n = 3/9 = 1/3$, the series is convergent, and its sum is $1/3$.
3. The nth partial sum of the series $1 + 1 + 1 + \ldots$ is $s_n = n$. Since $\lim_{n\to\infty} s_n = \infty$, the sequence is divergent (and its sum is ∞).
4. The $(2k)$th partial sum of the series $1 - 1 + 1 - \ldots$ is zero, while the $(2k+1)$th partial sum is 1 for all $k \in \mathbb{N}$. Since the sequence (s_n) is oscillating at infinity, the series is divergent (and has no sum).
5. The kth partial sum of the series

$$1 - \frac{1}{2} + \frac{1}{3} - \frac{1}{4} + \ldots \tag{6.1}$$

is

$$s_k = 1 - \frac{1}{2} + \frac{1}{3} - \frac{1}{4} + \ldots + (-1)^{k-1} \cdot \frac{1}{k}.$$

If $n < m$, then we can see that

$$|s_n - s_m| = \left| \frac{1}{n} - \frac{1}{n+1} + \ldots + (-1)^m \cdot \frac{1}{m} \right| < \frac{1}{n}.$$

It follows that the sequence (s_n) satisfies Cauchy's criterion, so it is convergent. This shows that the series (6.1) is convergent. Since $1/2 < s_k < 1$ for every $k > 2$, it follows that the sum s of the series (6.1) satisfies $1/2 \le s \le 1$.

It is well known that in fact, $s = \log 2$. See Exercise 12.92, Remark 13.16, and Example 14.25 of [7].

Remark 6.3. The second example above is a special case of the following fact: *if the infinite decimal expansion of x is $m.a_1a_2\ldots$, then the infinite series*

$$m + \frac{a_1}{10} + \frac{a_2}{10^2} + \ldots \tag{6.2}$$

is convergent, and its sum is x.

In some books on mathematical analysis the decimal fraction $m.a_1a_2\ldots$ is defined as the sum of the infinite series (6.2). In this case, the fact above is just the definition. However, decimal expansions can be defined without the notion of infinite series. For example, we can say that $m.a_1a_2\ldots$ is the decimal expansion of x if

$$m.a_1\ldots a_n \le x \le m.a_1\ldots a_n + \frac{1}{10^n} \tag{6.3}$$

holds for all n. (See, e.g., [7, p. 36].) If we accept this as the definition of decimals, then the statement above becomes a theorem proved as follows. The $(n+1)$st partial sum of the series (6.2) is $m.a_1\ldots a_n$. Now it is clear from (6.3) that $\lim_{n\to\infty} m.a_1\ldots a_n = x$. Therefore, by the definition of the sum of infinite series, we obtain that the series (6.2) is convergent, and its sum is x.

The sum appearing in Example 6.2.1 is a special case of the following theorem.

Theorem 6.4. *The series* $1 + x + x^2 + \ldots$ *is convergent if and only if* $|x| < 1$, *and then its sum is* $1/(1 - x)$.

Proof. We already saw that in the case $x = 1$ the series is divergent, so we may assume that $x \neq 1$. Then the nth partial sum of the series is $s_n = \sum_{i=0}^{n-1} x^i = (1 - x^n)/(1 - x)$. If $|x| < 1$, then $x^n \to 0$ and $s_n \to 1/(1 - x)$. Thus the series is convergent with sum $1/(1 - x)$.

If $x > 1$, then $s_n \to \infty$, so the series is divergent (and its sum is ∞). If, however, $x \leq -1$, then the sequence (s_n) oscillates at infinity, so the series is divergent (with no sum). □

Theorem 6.5. *If the series* $\sum_{n=1}^{\infty} a_n$ *is convergent, then* $\lim_{n\to\infty} a_n = 0$.

Proof. Let the sum of the series be A. Since

$$a_n = (a_1 + \ldots + a_n) - (a_1 + \ldots + a_{n-1}) = s_n - s_{n-1},$$

we have $a_n \to A - A = 0$. □

Remark 6.6. The theorem above states that the condition $a_n \to 0$ is necessary for the convergence of the infinite series $\sum_{n=1}^{\infty} a_n$. It is important to note that this condition is in no way sufficient, since there are many divergent series whose terms tend to zero. A simple example is provided by the following series. The terms of the series $\sum_{i=0}^{\infty} (\sqrt{i+1} - \sqrt{i})$ tend to zero, since $\sqrt{i+1} - \sqrt{i} = 1/(\sqrt{i+1} + \sqrt{i}) < 1/\sqrt{i}$, and $\sqrt{i} \to \infty$.

On the other hand, the nth partial sum is $\sum_{i=0}^{n-1} (\sqrt{i+1} - \sqrt{i}) = \sqrt{n}$, which approaches ∞ as $n \to \infty$, so the series is divergent.

Another well-known example of a divergent series whose terms tend to zero is the series $\sum_{n=1}^{\infty} 1/n$, which is called the **harmonic series**.

Theorem 6.7. *The series* $\sum_{n=1}^{\infty} \frac{1}{n}$ *is divergent.*

Proof. If $n > 2^k$ then

$$\begin{aligned}
s_n &\geq 1 + \frac{1}{2} + + \ldots + \frac{1}{2^k} = \\
&= 1 + \frac{1}{2} + \left(\frac{1}{3} + \frac{1}{4}\right) + \left(\frac{1}{5} + \ldots + \frac{1}{8}\right) + \ldots + \left(\frac{1}{2^{k-1}+1} + \ldots + \frac{1}{2^k}\right) \geq \\
&\geq 1 + \frac{1}{2} + 2 \cdot \frac{1}{4} + 4 \cdot \frac{1}{8} + \ldots + 2^{k-1} \cdot \frac{1}{2^k} = \\
&= 1 + k \cdot \frac{1}{2}.
\end{aligned}$$

Thus $\lim_{n\to\infty} s_n = \infty$, so the series is divergent and its sum is ∞. □

Theorem 6.8.

(i) *A series consisting of nonnegative terms is convergent if and only if the sequence of its partial sums is bounded (from above).*

(ii) *If a series consisting of nonnegative terms is divergent, then its sum is infinite.*

Proof. By the assumption that the terms of the series are nonnegative, we clearly get that the sequence of partial sums of the series is monotonically increasing. If this sequence is bounded from above, then it is convergent, since every bounded and monotone sequence is convergent. (See [7, Theorem 6.2].) Then the series in question is convergent.

If, however, the sequence of partial sums is not bounded from above, then it tends to infinity. Indeed, if a sequence is increasing and is not bounded, then it tends to infinity (see [7, Theorem 6.3]). So the series will be divergent and its sum will be infinity. □

We emphasize that by the above theorem *a series consisting of nonnegative terms always has a sum: either a finite number (if the series converges) or infinity (if the series diverges).*

Example 6.9. The series $\sum_{i=1}^{\infty} 1/i^2$ is convergent, because its nth partial sum is

$$\sum_{i=1}^{n} \frac{1}{i^2} \le 1 + \sum_{i=2}^{n} \frac{1}{(i-1)\cdot i} = 1 + \sum_{i=2}^{n} \left(\frac{1}{i-1} - \frac{1}{i} \right) = 2 - \frac{1}{n} < 2.$$

According to Exercise 4.20, the sum of the series $\sum_{i=1}^{\infty} 1/i^2$ equals $\pi^2/6$. We will give two more proofs of this statement; see Example 7.80 and Theorem 7.92.

In the general case, the following theorem gives an exact condition for the convergence of a series.

Theorem 6.10. (Cauchy's criterion) *The infinite series $\sum_{n=1}^{\infty} a_n$ is convergent if and only if for every $\varepsilon > 0$ there exists an index N such that for every $N \le n < m$,*

$$|a_{n+1} + a_{n+2} + \ldots + a_m| < \varepsilon.$$

Proof. Since $a_{n+1} + a_{n+2} + \ldots + a_m = s_m - s_n$, the statement is clear by Cauchy's criterion for sequences (see Theorem 6.13 of [7]). □

Exercises

6.1. For a fixed $\varepsilon > 0$, give threshold indices above which the partial sums of the following series differ from their actual sums by less than ε.

(a) $\sum_{n=0}^{\infty} 1/2^n$; (b) $\sum_{n=0}^{\infty} (-2/3)^n$;
(c) $\sum_{n=1}^{\infty} (-1)^{n-1}/n$; (d) $\sum_{n=1}^{\infty} 1/(n^2+n)$.

6.2. (a) $\sum_{n=1}^{\infty} 1/(n^2+2n) = ?$ (b) $\sum_{n=1}^{\infty} 1/(n^2+4n+3) = ?$
(c) $\sum_{n=2}^{\infty} 1/(n^3-n) = ?$(H S)

6.3. Show that

$\left(1+\frac{1}{2^c}+\ldots+\frac{1}{n^c}\right)\left(1-\frac{2}{2^c}\right)<1$

for all $n=1,2,\ldots$ and $c>0$. Deduce from this that the series $\sum_{n=1}^{\infty} 1/n^c$ is convergent for all $c>1$. (H)

6.4. Let $a_1, a_2, \ldots$ be an enumeration of the positive integers that do not contain the digit 7 (in decimal representation). Prove that $\sum_{n=1}^{\infty} 1/a_n$ is convergent. (H)

6.5. Prove that if the series $\sum_{n=1}^{\infty} a_n$ is convergent, then

$\lim_{n\to\infty} \frac{a_1+2a_2+\ldots+na_n}{n} = 0$. (S)

6.2 Operations on Infinite Series

The example of the infinite series $1-1+1-1+\ldots$ shows that we cannot manipulate infinite series the way we deal with finite sums. For example, the series $(1-1)+(1-1)+\ldots$ is convergent with sum zero (every term of the sum is zero, and thus every partial sum is zero), but omitting the parentheses results in a divergent series, $1-1+1-1+\ldots$. We will see presently that reordering the terms of an infinite series can change the sum of the series; it can even destroy convergence.

We need to find the "allowed" operations, i.e., the operations that change neither the convergence of a series nor its sum; we should also find out which operations are not (or not always) allowed.

First we consider some simple operations that are allowed in the sense described above.

Theorem 6.11.

(i) *Let $\sum_{n=1}^{\infty} a_n$ be a convergent infinite series with sum A. Then the series $\sum_{n=1}^{\infty} c\cdot a_n$ is also convergent and its sum is $c\cdot A$, for every $c\in\mathbb{R}$.*

(ii) *Let $\sum_{n=1}^{\infty} a_n$ and $\sum_{n=1}^{\infty} b_n$ be a pair of convergent infinite series with sums A and B, respectively. Then the infinite series $\sum_{n=1}^{\infty}(a_n+b_n)$ is also convergent and its sum is $A+B$.*

Proof. (i) If the nth partial sum of the series $\sum_{n=1}^{\infty} a_n$ is s_n, then the nth partial sum of the infinite series $\sum_{n=1}^{\infty} c\cdot a_n$ is $c\cdot s_n$. The statement follows from $\lim_{n\to\infty} c\cdot s_n = c\cdot \lim_{n\to\infty} s_n = c\cdot A$.
(ii) If s_n and t_n are the nth partial sums of the two infinite series, then the nth partial sum of the infinite series $\sum_{n=1}^{\infty}(a_n+b_n)$ is s_n+t_n. The statement follows from $\lim_{n\to\infty}(s_n+t_n) = \lim_{n\to\infty} s_n + \lim_{n\to\infty} t_n = A+B$. □

Remark 6.12. The proof of the following statements are similar to the proofs of (i) and (ii) above.

If $\sum_{n=1}^{\infty} a_n = \pm\infty$ *and* $c > 0$, *then* $\sum_{n=1}^{\infty} c \cdot a_n = \pm\infty$; *if* $\sum_{n=1}^{\infty} a_n = \pm\infty$ *and* $c < 0$, *then* $\sum_{n=1}^{\infty} c \cdot a_n = \mp\infty$.

If $\sum_{n=1}^{\infty} a_n = \pm\infty$ *and* $\sum_{n=1}^{\infty} b_n$ *is convergent, then* $\sum_{n=1}^{\infty}(a_n + b_n) = \pm\infty$.

Theorem 6.13.

(i) *Deleting an arbitrary number of* 0*'s (possibly infinitely many) from, or inserting an arbitrary number of* 0*'s (possibly infinitely many) into a convergent infinite series does not change its convergence or its sum.*

(ii) *Removing finitely many terms from, inserting finitely many terms into, or changing finitely many terms of a convergent infinite series does not change its convergence (but it might change its sum).*

Proof. (i) Let the nth partial sum of the convergent infinite series $\sum_{n=1}^{\infty} a_n$ be s_n. Deleting an arbitrary number of 0's from the series makes the new sequence of partial sums a subsequence of (s_n). (If the series is of the form $a_1 + \ldots + a_N + 0 + 0 + \ldots$, we need to assume that we kept infinitely many 0's.) On the other hand, on inserting an arbitrary number of 0's into the series, we obtain a series whose partial sums have the same elements as the sequence (s_n), with some elements s_n being repeated. (The partial sum s_n will be repeated if we insert a 0 after a_n.) Therefore, (i) follows from the fact that if the sequence (s_n) is convergent, then the new sequence of partial sum is also convergent with the same limit. (See [7, Theorems 5.2, 5.5].)
(ii) Let us assume that we have removed a_k from the terms of the series $\sum_{n=1}^{\infty} a_n$. For $n \geq k$, the nth partial sum of the new series is $s_n - a_k$, i.e., the sequence of the new partial sums converges to $A - a_k$. On the other hand, inserting c between the kth and $(k+1)$st terms changes the nth partial sum of the new series to $s_n + c$, for $n > k$. Thus, the sequence of the partial sums of the new series converges to $A + c$. Neither of these operations changes the convergence of the series. Therefore, repeating these operations finitely many times also results in a convergent series. Changing finitely many terms of the series can be obtained by first removing the terms to be changed and then inserting the new elements into their respective places. By what we proved above, it follows that the operation does not change the convergence of the series. □

We say that the infinite series $\sum_{n=1}^{\infty} c_n$ is obtained by **interleaving** the series $\sum_{n=1}^{\infty} a_n$ and $\sum_{n=1}^{\infty} b_n$ if the sequence (c_n) is the union of the sequences a_n and b_n, and the order of the a_n's and b_n's in the sequence (c_n) is unchanged. (More precisely, the sequence of the indices $(1, 2, \ldots)$ should be divided into two disjoint and strictly monotone subsequences (i_k) and (j_k) such that $a_k = c_{i_k}$ and $b_k = c_{j_k}$ for every k.)

Theorem 6.14. *If the series* $\sum_{n=1}^{\infty} a_n$ *and* $\sum_{n=1}^{\infty} b_n$ *are convergent with the sums* A *and* B, *respectively, then every series obtained by interleaving these series is also convergent with the sum* $A + B$.

Proof. Every series obtained by interleaving the series $\sum_{n=1}^{\infty} a_n$ and $\sum_{n=1}^{\infty} b_n$ can be obtained by inserting 0's into both series at suitable places and then adding the two resulting series. Thus the statement follows from Theorems 6.11 and 6.13. □

Remark 6.15. Similarly, we can show that *if* $\sum_{n=1}^{\infty} a_n = \pm\infty$ *and* $\sum_{n=1}^{\infty} b_n$ *is convergent, then the sum of every series obtained by interleaving* $\sum_{n=1}^{\infty} a_n$ *and* $\sum_{n=1}^{\infty} b_n$ is $\pm\infty$.

By adding parentheses to an infinite series we mean the following: we remove some consecutive terms of the series and replace them by their sum. This operation can be applied to several, even infinitely many, blocks of consecutive terms, assuming that every term of the series belongs to at most one such block. We now give a mathematically precise definition.

Definition 6.16. We call an infinite series of the form $\sum_{i=1}^{\infty}\left(\sum_{n=n_{i-1}}^{n_i-1} a_n\right)$ a *bracketing* of the infinite series $\sum_{n=1}^{\infty} a_n$, where $1 = n_0 < n_1 < \ldots$ is an arbitrary strictly increasing sequence of indices.

Theorem 6.17. *Bracketing a convergent series does not change its convergence, nor does it change its sum.*

Proof. Let (s_n) be the sequence of partial sums of the series. The sequence of partial sums of the bracketed series is a subsequence of (s_n), from which the statement is clear. (See [7, Theorem 5.2].) □

Example 6.18. The series $1 - \frac{1}{2} + \frac{1}{3} - \frac{1}{4} + \ldots$ is convergent and its sum is $\log 2$ (see Example 6.2.5). According to the previous theorem, the series

$$\left(1-\frac{1}{2}\right)+\left(\frac{1}{3}-\frac{1}{4}\right)+\left(\frac{1}{5}-\frac{1}{6}\right)+\ldots$$

is also convergent and its sum is $\log 2$. Since $\frac{1}{n} - \frac{1}{n+1} = 1/(n\cdot(n+1))$ for every n, we have that

$$\frac{1}{1\cdot 2}+\frac{1}{3\cdot 4}+\frac{1}{5\cdot 6}+\ldots=\log 2. \tag{6.4}$$

Exercise 6.13 gives an interesting geometric interpretation of this equality.

Now, we consider the operations that do not preserve the convergence of convergent series. Such operations are, e.g., the deletion of parentheses when some of the terms of the series are bracketed sums. In general, the convergence of the series $\sum_{n=1}^{\infty} a_n$ does not follow from the convergence of the series $\sum_{i=1}^{\infty}\left(\sum_{n=n_{i-1}}^{n_i-1} a_n\right)$. For example, as we mentioned before, the series $(1-1)+(1-1)+\ldots$ is convergent, but the series $1-1+1-1+\ldots$ is divergent. However, if removing the brackets of a convergent series results in another convergent series, then the sums of the two series are the same; this follows from Theorem 6.17.

The following operation we consider is that of reordering the terms of a series.

Definition 6.19. We call a series of the form $\sum_{i=1}^{\infty} a_{\sigma(i)}$ a *reordering* of the infinite series $\sum_{n=1}^{\infty} a_n$ if $\sigma\colon \mathbb{N}^+ \to \mathbb{N}^+$ is an arbitrary one-to-one mapping (i.e., a permutation) of the set of indices to itself.

Example 6.20. **1.** Consider the series

$$1 - \frac{1}{2} - \frac{1}{4} + \frac{1}{3} - \frac{1}{6} - \frac{1}{8} + \frac{1}{5} - \frac{1}{10} - \frac{1}{12} + \ldots. \tag{6.5}$$

Every positive integer appears in exactly one of the denominators, and both the odd and even numbers appear in a monotonically increasing order. An odd integer is followed by two even integers in the denominators, and a positive term is followed by two negative terms. Obviously, (6.5) is a reordering of the infinite series

$$1 - \frac{1}{2} + \frac{1}{3} - \frac{1}{4} + \ldots, \tag{6.6}$$

where the bijection σ is defined by the sequence $(1, 2, 4, 3, 6, 8, 5, 10, 12, \ldots)$. We show that the series (6.5) is convergent with sum $\frac{1}{2} \cdot \log 2$ (which is half the sum of (6.6)). Let s_n be the nth partial sum of the series. Then we have

$$\begin{aligned} s_{3k} &= \left(1 - \frac{1}{2}\right) - \frac{1}{4} + \left(\frac{1}{3} - \frac{1}{6}\right) - \frac{1}{8} + \ldots + \left(\frac{1}{2k-1} - \frac{1}{4k-2}\right) - \frac{1}{4k} = \\ &= \frac{1}{2} - \frac{1}{4} + \frac{1}{6} - \frac{1}{8} + \ldots + \frac{1}{4k-2} - \frac{1}{4k} = \\ &= \frac{1}{2} \cdot \left(1 - \frac{1}{2} + \frac{1}{3} - \frac{1}{4} + \ldots - \frac{1}{2k}\right); \end{aligned}$$

thus $\lim_{k\to\infty} s_{3k} = \frac{1}{2} \cdot \log 2$. Since $s_{3k+1} - s_{3k} = 1/(2k+1) \to 0$ and $s_{3k+2} - s_{3k} = (1/(2k+1)) - (1/(4k+2)) \to 0$ as $k \to \infty$, it follows that $s_n \to \frac{1}{2} \cdot \log 2$ as $n \to \infty$, which is exactly what we wanted to prove.

2. Consider the series

$$1 - \frac{1}{2} + \frac{1}{3} - \frac{1}{4} + \frac{1}{5} - \frac{1}{6} - \frac{1}{8} + \frac{1}{7} - \frac{1}{10} - \frac{1}{12} - \frac{1}{14} - \frac{1}{16} + \frac{1}{9} - \ldots. \tag{6.7}$$

We obtain this series by writing the first two terms of (6.6); then for every $k \geq 2$, we write the term $1/(2k-1)$ followed by the terms $-1/(2i)$ (with $2^{k-1} + 2 \leq 2i \leq 2^k$). The series (6.7) is another reordering of the series (6.6), where the bijection σ is defined by the sequence $(1, 2, 3, 4, 5, 6, 8, 7, 10, 12, 14, 16, \ldots)$. We show that the series (6.7) is divergent. Indeed, the sum of the absolute values of the terms with denominators $2^{k-1} + 2,\ 2^{k-1} + 4, \ldots, 2^k$ is at least $(2^{k-1}/2) \cdot 2^{-k} = 1/4$, for every k. Thus the series does not satisfy the Cauchy criterion.

These examples prove that *we do not have commutativity* in adding infinitely many numbers: the sum may depend on the order of its terms, and what is more, the order of the terms can influence the existence of the sum.

Exercises

6.6. Let $\sum a_n$ be a convergent series of positive terms. Show the existence of a sequence $c_n \to \infty$ such that $\sum c_n \cdot a_n$ is also convergent.

6.7. Let $\sum a_n$ be a divergent series of positive terms. Show the existence of a sequence $c_n \to 0$ such that $\sum c_n \cdot a_n$ is also divergent.

6.8. Show that if $\sum a_n$ is a convergent series of nonnegative terms, then $\sum \sqrt{a_n}/n$ is also convergent. (H)

6.9. Show that if $\sum a_n$ is convergent and (a_n) is monotone, then $n \cdot a_n \to 0$. (H)

6.10. Let $\sum_{n=1}^{\infty} a_n$ be a divergent series of positive terms, and let s_n be its nth partial sum. Show that $\sum_{n=1}^{\infty} a_n/(s_n)^c$ is convergent if and only if $c > 1$. (H)

6.11. Let $\sum_{n=1}^{\infty} a_n$ be a convergent series of positive terms, and let $r_n = a_n + a_{n+1} + \ldots$, for every n. Show that $\sum_{n=1}^{\infty} a_n/(r_n)^c$ is convergent if and only if $c < 1$. (H)

6.12. Let the two series $a_1 + a_2 + \ldots$ and $a_1 - a_2 + a_3 - a_4 + \ldots$ be convergent. Does it follow that the series $a_1 + a_2 - a_3 + a_4 + a_5 - a_6 + a_7 + a_8 - a_9 + \ldots$ is also convergent?

6.13. Let H be the set $\{(x, y)\colon 1 \le x \le 2,\ 0 \le y \le 1/x\}$. Show that H can be tiled by rectangles with areas $1/(1 \cdot 2)$, $1/(3 \cdot 4)$, $1/(5 \cdot 6), \ldots$. More precisely, show that we can find nonoverlapping rectangles in H with areas $\frac{1}{(2n-1) \cdot 2n}$ such that their union covers the set $\{(x, y)\colon 1 \le x \le 2,\ 0 \le y < 1/x\}$. (H)

6.14. Let k and m be given positive integers, and consider the series $\sum_{n=1}^{\infty} \pm 1/n$, where k positive terms are followed by m negative terms. Thus the series corresponding to the case $k = m = 2$ is

$$1 + \frac{1}{2} - \frac{1}{3} - \frac{1}{4} + \frac{1}{5} + \frac{1}{6} - \ldots,$$

and the series corresponding to the case $k = 2$, $m = 1$ is

$$1 + \frac{1}{2} - \frac{1}{3} + \frac{1}{4} + \frac{1}{5} - \frac{1}{6} + \ldots.$$

For what values of k and m will the series be convergent? (H)

6.15. Suppose that $\sum a_n$ is convergent, and its sum is A. Let σ be a permutation of the set $\mathbb{N}^+$ such that $|\sigma(n) - n| \le 100$ for every n. Prove that $\sum a_{\sigma(n)}$ is convergent, and its sum is A.

6.3 Absolute and Conditionally Convergent Series

We will now show that the strange phenomena presented above in connection with reordering the terms do not occur in an important class of infinite series.

Definition 6.21. We say that the infinite series $\sum_{n=1}^{\infty} a_n$ is *absolutely convergent* if the series $\sum_{n=1}^{\infty} |a_n|$ is convergent.

Theorem 6.22.

(i) *Every absolutely convergent series is convergent.*

(ii) *Every reordering of an absolutely convergent series is also absolutely convergent, and its sum equals the sum of the original series.*

Proof. Let $\sum_{n=1}^{\infty} a_n$ be absolutely convergent. Then by the Cauchy criterion, for every $\varepsilon > 0$ there exists N such that $|a_{n+1}| + |a_{n+2}| + \ldots + |a_m| < \varepsilon$ for every $N \le n < m$. By applying the triangle inequality, we get

$$|a_{n+1} + a_{n+2} + \ldots + a_m| \le |a_{n+1}| + |a_{n+2}| + \ldots + |a_m| < \varepsilon,$$

i.e., $\sum_{n=1}^{\infty} a_n$ also satisfies the Cauchy criterion. This proves (i).

Let $\sum_{n=1}^{\infty} b_n$ be a reordering of the series $\sum_{n=1}^{\infty} a_n$. For $\varepsilon > 0$ fixed, choose N such that $|a_{N+1}| + |a_{N+2}| + \ldots + |a_m| < \varepsilon$ for every $m > N$. The terms $a_1, \ldots, a_N$ are present in the series $\sum_{n=1}^{\infty} b_n$ (possibly with different indices). If the maximum of their (new) indices is M, then for $k > M$, the indices of the terms $b_{M+1}, \ldots, b_k$ in the series $\sum_{n=1}^{\infty} a_n$ are greater than N. Thus, for m large enough these terms occur among the terms $a_{N+1}, \ldots, a_m$. This implies

$$|b_{M+1}| + |b_{M+2}| + \ldots + |b_k| \le |a_{N+1}| + |a_{N+2}| + \ldots + |a_m| < \varepsilon,$$

showing that the series $\sum_{n=1}^{\infty} |b_n|$ also satisfies the Cauchy criterion. Hence, it is convergent; that is, the series $\sum_{n=1}^{\infty} b_n$ is also absolutely convergent. Then by (i), it is convergent.

Let $\sum_{n=1}^{\infty} a_n = A$ and $\sum_{n=1}^{\infty} b_n = B$. For $\varepsilon > 0$ fixed, let N and M be the same as above. Let $k > \max(N, M)$ be arbitrary, and let

$$d_k = (a_1 + \ldots + a_k) - (b_1 + \ldots + b_k).$$

Clearly, in the sum d_k the terms $a_1, \ldots, a_N$ are canceled, and thus d_k is a sum of terms of the form $\pm a_n$, where $n > N$. Therefore, with m large enough, we have

$$|d_k| \le |a_N| + |a_{N+1}| + \ldots + |a_m| < \varepsilon.$$

We have proved $\lim_{k\to\infty} d_k = 0$. However, $\lim_{k\to\infty} d_k = A - B$, i.e., $A = B$. □

Remarks 6.23. **1.** The converse of part (i) of Theorem 6.22 is not true: a convergent series is not necessarily absolutely convergent. For example, the series (6.6) is convergent, but the sum of the absolute values of its terms (the harmonic series) is divergent. In other words, the set of absolutely convergent series is a proper subset of the set of convergent series.
2. Reordering the terms of a series of nonnegative terms does not change the sum of the series. Indeed, if a series of nonnegative terms is convergent, then it is also absolutely convergent, and we can apply part (ii) of Theorem 6.22. On the other hand, if a series of nonnegative terms is divergent, then every reordering of the series is also divergent, for otherwise, the reordered series would be convergent, and the original series itself would also be convergent (as a reordered series of a convergent series). Thus, the sum of both series is infinity.

The following theorem gives a simple characterization of absolutely convergent series. In order to state the theorem, we need to introduce some notation.

Notation 6.24. For every number x, let

$$x^+ = \max(x,0) = \begin{cases} x, & \text{if } x \geq 0, \\ 0, & \text{if } x < 0, \end{cases}$$

and

$$x^- = \max(-x,0) = \begin{cases} 0, & \text{if } x \geq 0, \\ -x, & \text{if } x < 0. \end{cases}$$

We call the numbers x^+ and x^- the *positive and negative parts* of x, respectively. It is easy to see that

$$x = x^+ - x^-, \ |x| = x^+ + x^-, \ x^+ = \frac{|x|+x}{2}, \ x^- = \frac{|x|-x}{2}$$

hold for every $x \in \mathbb{R}$.

Theorem 6.25. *The series $\sum_{n=1}^{\infty} a_n$ is absolutely convergent if and only if the series $\sum_{n=1}^{\infty} a_n^+$ and $\sum_{n=1}^{\infty} a_n^-$ are both convergent.*

Proof. If $\sum_{n=1}^{\infty} a_n$ is absolutely convergent, then by Theorem 6.22, it is convergent. The convergence of the series $\sum_{n=1}^{\infty} a_n^+$ and $\sum_{n=1}^{\infty} a_n^-$ follows from the formulas

$$a_n^+ = \frac{|a_n|+a_n}{2} \quad \text{and} \quad a_n^- = \frac{|a_n|-a_n}{2}$$

and from Theorem 6.11.
The proof of the converse is similar, using the fact that $|a_n| = a_n^+ + a_n^-$ for every n. □

Theorem 6.26.

(i) *If* $\sum_{n=1}^{\infty} a_n^+ = \infty$, *then the series* $\sum_{n=1}^{\infty} a_n$ *has a reordering whose sum is positive infinity.*

(ii) *For* $\sum_{n=1}^{\infty} a_n^- = \infty$, *the series* $\sum_{n=1}^{\infty} a_n$ *has a reordering whose sum is negative infinity.*

Proof. (i) The condition implies that the series $\sum_{n=1}^{\infty} a_n$ has infinitely many positive terms. Let $b_1, b_2, \ldots$ be the positive terms of the series, keeping the original order of their indices. Then $\sum_{n=1}^{\infty} b_n = \sum_{n=1}^{\infty} a_n^+ = \infty$. If the series $\sum_{n=1}^{\infty} a_n$ has only finitely many nonpositive terms, then clearly, the sum of the series is positive infinity. Thus, we may assume that the series has infinitely many nonpositive terms; let these terms be $c_1, c_2, \ldots$, keeping the order of their indices. Since $\sum_{n=1}^{\infty} b_n = \infty$, there exist indices $1 = N_1 < N_2 < \ldots$ such that $\sum_{n=N_i}^{N_{i+1}-1} b_n > |c_i| + 1$, for every $i = 1, 2, \ldots$. It is easy to check that the sum of the series

$$b_{N_1} + \ldots + b_{N_2-1} + c_1 + b_{N_2} + \ldots + b_{N_3-1} + c_2 + \ldots$$

is a reordering of $\sum_{n=1}^{\infty} a_n$, and its sum is infinity. Part (ii) can be proved similarly. □

Our next aim is to determine the set of sums of all reorderings of a given series.

Theorem 6.27.

(i) *If the series* $\sum_{n=1}^{\infty} a_n^+$ *and* $\sum_{n=1}^{\infty} a_n^-$ *are convergent, then every reordering of the series* $\sum_{n=1}^{\infty} a_n$ *is convergent with the same sum.*

(ii) *If* $\sum_{n=1}^{\infty} a_n^+ = \infty$ *and* $\sum_{n=1}^{\infty} a_n^-$ *is convergent, then the sum of every reordering of the series* $\sum_{n=1}^{\infty} a_n$ *is positive infinity.*

(iii) *If* $\sum_{n=1}^{\infty} a_n^+$ *is convergent and* $\sum_{n=1}^{\infty} a_n^- = \infty$, *then the sum of every reordering of the series* $\sum_{n=1}^{\infty} a_n$ *is negative infinity.*

(iv) *If* $\sum_{n=1}^{\infty} a_n^+ = \infty$ *and* $\sum_{n=1}^{\infty} a_n^- = \infty$, *then the series* $\sum_{n=1}^{\infty} a_n$ *has a reordering whose sum is infinity, and* $\sum_{n=1}^{\infty} a_n$ *also has a reordering whose sum is negative infinity. Furthermore, assuming also* $a_n \to 0$, *the series* $\sum_{n=1}^{\infty} a_n$ *has a reordering whose sum is* A, *for every given* $A \in \mathbb{R}$.

Proof. Statement (i) follows from Theorems 6.25 and 6.22.
Let σ be a bijection on the positive integers. If the conditions of (ii) hold, then it follows from Theorem 6.22 and Remark 6.23.2 that $\sum_{n=1}^{\infty} (a_{\sigma(n)})^+ = \infty$ and $\sum_{n=1}^{\infty} (a_{\sigma(n)})^-$ is convergent. Since $a_{\sigma(n)} = a_{\sigma(n)}^+ - a_{\sigma(n)}^-$ for every n, Remark 6.12 implies $\sum_{n=1}^{\infty} a_{\sigma(n)} = \infty$.
Statement (iii) can be proved similarly.
By Theorem 6.26, we have only to prove the second half of (iv). Let $\sum_{n=1}^{\infty} a_n^+ = \sum_{n=1}^{\infty} a_n^- = \infty$ and $a_n \to 0$. Then clearly, the series $\sum_{n=1}^{\infty} a_n$ has infinitely many positive and infinitely many negative terms. Let $b_1, b_2, \ldots$ be the sequence of its positive terms (keeping the order of their indices); then we have $\sum_{n=1}^{\infty} b_n = \infty$.

If $c_1, c_2, \ldots$ are the negative terms of $\sum_{n=1}^{\infty} a_n$ (also keeping the order of their indices), then $\sum_{n=1}^{\infty} c_n = -\infty$. Let $A \in \mathbb{R}$ be fixed.
Since $\sum_{n=1}^{\infty} b_n = \infty$, there exists an index N such that $\sum_{n=1}^{N} b_n > A$. Let N_1 be the smallest such N.
Since $\sum_{n=1}^{\infty} c_n = -\infty$, there exists an index M such that $\sum_{n=1}^{N_1} b_n + \sum_{n=1}^{M} c_n < A$. Let M_1 be the smallest such M.
As $\sum_{n=1}^{\infty} b_n = \infty$, there exists an index $N > N_1$ such that $\sum_{n=1}^{N_1} b_n + \sum_{n=1}^{M_1} c_n + \sum_{n=N_1+1}^{N} b_n > A$. Let N_2 be the smallest such N.
As $\sum_{n=1}^{\infty} c_n = -\infty$, there exists an index $M > M_1$ such that $\sum_{n=1}^{N_1} b_n + \sum_{n=1}^{M_1} c_n + \sum_{n=N_1+1}^{N_2} b_n + \sum_{n=M_1+1}^{M} c_n < A$. Let M_2 be the smallest such M.

Repeating the process, we obtain the indices $N_1 < N_2 < \ldots$ and $M_1 < M_2 < \ldots$. Consider the infinite series

$$b_1 + \ldots + b_{N_1} + c_1 + \ldots + c_{M_1} + b_{N_1+1} + \ldots + b_{N_2} + c_{M_1+1} + \ldots + c_{M_2} + \ldots. \tag{6.8}$$

Obviously, this is a reordering of the series $\sum_{n=1}^{\infty} a_n$. We show that the sum of this new series is A.

Let $\varepsilon > 0$ be fixed. Since $a_n \to 0$, there exists K such that $|a_n| < \varepsilon$, when $n > K$. It follows that $b_n < \varepsilon$ and $|c_n| < \varepsilon$ for all $n > K$, since the indices of the terms b_n and c_n in the original series $\sum_{n=1}^{\infty} a_n$ are at least n; thus for $n > K$, these indices are also larger than K.

Let s_n be the nth partial sum of the series (6.8). By our choices of the indices N_i, we have

$$s_{M_{i-1}+N_i-1} \leq A < s_{M_{i-1}+N_i}.$$

Then for $N_i > K$ it follows that

$$A < s_{M_{i-1}+N_i} = s_{M_{i-1}+N_i-1} + b_{N_i} < A + \varepsilon, \tag{6.9}$$

and thus $|s_{M_{i-1}+N_i} - A| < \varepsilon$. Similarly, we obtain $|s_{N_i+M_i} - A| < \varepsilon$, assuming that $M_i > K$. Therefore, $|s_{N_i+M_{i-1}} - A| < \varepsilon$ and $|s_{N_i+M_i} - A| < \varepsilon$ for every i large enough.

For a fixed i, the values of s_n for the indices $N_i + M_{i-1} < n \leq N_i + M_i$ decrease, since we get s_n by adding a negative number (one of the c_j's) to s_{n-1}. Thus, the value of s_n is between the value of $s_{N_i+M_{i-1}}$ and that of $s_{N_i+M_i}$. We have proved that for $\min(N_i, M_i) > K$, we have $|s_n - A| < \varepsilon$. Similarly, for $\min(N_i, M_i) > K$, we have $|s_n - A| < \varepsilon$ for every $N_i + M_i < n \leq N_{i+1} + M_i$. We have proved $|s_n - A| < \varepsilon$ for every n large enough. Since ε was arbitrary, it follows that the sum of the series (6.8) is A. □

We can now rephrase and supplement Theorem 6.22 as follows.

Theorem 6.28. *For every infinite series, the following statements are equivalent:*

(i) *The series is absolutely convergent.*
(ii) *Every reordering of the series is absolutely convergent.*
(iii) *Every reordering of the series is convergent.*
(iv) *Every reordering of the series is convergent, and its sum is the same as the sum of the original series.* □

The infinite series $\sum_{n=1}^{\infty} a_n$ is said to be **conditionally convergent** if it is convergent, but not absolutely convergent. It is clear from Theorems 6.22 and 6.27 that for a conditionally convergent series $\sum_{n=1}^{\infty} a_n$ we have $\sum_{n=1}^{\infty} a_n^+ = \sum_{n=1}^{\infty} a_n^- = \infty$. Since $a_n \to 0$ is also true (based on Theorem 6.5), part (iv) of Theorem 6.27 implies the following theorem, called **Riemann's reordering theorem**.

Theorem 6.29. *If the series $\sum_{n=1}^{\infty} a_n$ is conditionally convergent, then it has a reordering with any prescribed sum (positive infinity, negative infinity, or A for every $A \in \mathbb{R}$), and it also has a divergent reordering with no sum.* □

(For a proof of the last statement, see Exercise 6.18.)

The following theorem says that associativity holds for absolutely convergent series in its most general form (i.e., even for sums of infinitely many terms).

Theorem 6.30. *Let $\sum_{i=1}^{\infty} b_i$ be an absolutely convergent series, where $(b_i)_{i=1}^{\infty}$ is an enumeration of the numbers $a_{k,n}$ $(k, n = 1, 2, \ldots)$. Then the series $\sum_{n=1}^{\infty} a_{k,n}$ is also absolutely convergent for every k, and furthermore, if $\sum_{n=1}^{\infty} a_{k,n} = A_k$ $(k = 1, 2, \ldots)$ and $\sum_{i=1}^{\infty} b_i = A$, then the series $\sum_{k=1}^{\infty} A_k$ is also absolutely convergent, and $\sum_{k=1}^{\infty} A_k = A$.*

Proof. Let $\sum_{i=1}^{\infty} |b_i| = B$. For every k, each partial sum of the series $\sum_{n=1}^{\infty} |a_{k,n}|$ is less than or equal to a suitable partial sum of the series $\sum_{i=1}^{\infty} |b_i|$. Therefore, no partial sum of $\sum_{n=1}^{\infty} |a_{k,n}|$ is larger than B. Thus, the sequence of the partial sums of the series $\sum_{n=1}^{\infty} |a_{k,n}|$ is bounded, and then the series is convergent. For every n, m, we have

$$|a_{1,1} + \ldots + a_{1,n}| + \ldots + |a_{m,1} + \ldots + a_{m,n}| \le \sum_{i=1}^{m} \sum_{j=1}^{n} |a_{i,j}|, \tag{6.10}$$

and the right-hand side is at most B. If $n \to \infty$ on the left-hand side of (6.10), then we obtain $|A_1| + \ldots + |A_m| \le B$. Since this holds for every m, it follows that $\sum_{k=1}^{\infty} A_k$ is absolutely convergent.

Let $\varepsilon > 0$ be fixed. Since $\sum_{i=1}^{\infty} |b_i|$ is convergent, by the Cauchy criterion (Theorem 6.10) it follows that there is an index N such that $\sum_{i=N+1}^{n} |b_i| < \varepsilon$ holds for every $n > N$. Since $(b_i)_{i=1}^{\infty}$ is an enumeration of the numbers $a_{k,n}$ $(k, n = 1, 2, \ldots)$, there exists an index $M \ge N$ such that each of the terms $b_1, \ldots, b_N$ appears among the terms $a_{k,n}$ $(k, n \le M)$. We show that

$$\left|\sum_{k=1}^{m}(a_{k,1}+\ldots+a_{k,n})-(b_1+\ldots+b_n)\right|<\varepsilon \tag{6.11}$$

for every $m,n>M$. Indeed, by the choice of the index M, the terms $a_{k,i}$ ($k\le m$, $i\le n$) include each of $b_1,\ldots,b_N$. Subtracting $b_1+\ldots+b_n$, we get a sum whose terms are of the form $\pm b_i$, where the indices i are distinct and are larger than N. Let p be the largest of these indices. Then the left-hand side of (6.11) is at most $|b_{N+1}|+\ldots+|b_p|<\varepsilon$ by the choice of the index N.

Fixing $m>M$ and letting n approach infinity in (6.11) we obtain

$$\left|\sum_{k=1}^{m}A_k-A\right|\le\varepsilon.$$

Since this is true for every $m>M$, it follows that $\sum_{k=1}^{\infty}A_k=A$. □

In most of the applications of this theorem we have an infinite array of numbers whose terms, in a suitable order, form an absolutely convergent series. Then, adding the array column by column or row by row also gives an absolutely convergent series whose sum is the same as the sum of all the terms of the array.

Example 6.31. **1.** We prove

$$\sum_{n=1}^{\infty}n\cdot x^{n-1}=\frac{1}{(1-x)^2}, \tag{6.12}$$

for every $|x|<1$.

We know that $1+x+x^2+\ldots=1/(1-x)$ for every $|x|<1$. Multiplying by x^i for every $i\ge 0$ and appending i zeros into each row yields

$$\begin{array}{ccccccccccc}
1 & + & x & + & x^2 & + & x^3 & + & \ldots & = & \dfrac{1}{1-x}\\
0 & + & x & + & x^2 & + & x^3 & + & \ldots & = & \dfrac{x}{1-x}\\
0 & + & 0 & + & x^2 & + & x^3 & + & \ldots & = & \dfrac{x^2}{1-x}\\
0 & + & 0 & + & 0 & + & x^3 & + & \ldots & = & \dfrac{x^3}{1-x}\\
\vdots & & \vdots & & \vdots & & \vdots & & \vdots & & \vdots
\end{array} \tag{6.13}$$

These equalities also hold for $|x|$ in place of x. Thus, the sum of finitely many of the absolute values of the terms of this array is at most

$$\sum_{n=0}^{N}\frac{|x|^n}{1-|x|},$$

for an appropriate N. This sum is smaller than $1/(1-|x|)^2$, which proves that the terms on the left-hand side of (6.13) form an absolutely convergent series, regardless of the order in which they are listed. Then, by Theorem 6.30, the sum of this series is

$$\sum_{n=0}^{\infty} \frac{x^n}{1-x} = \frac{1}{(1-x)^2}.$$

Adding the terms of the table column by column and applying Theorem 6.30 implies (6.12). For example, $\sum_{n=1}^{\infty} \frac{n}{2^n} = 2$.

2. Theorem 6.30 supplies us with a new proof of the fact that the harmonic series is divergent. Consider the following equalities:

$$\begin{array}{ccccccccc}
\frac{1}{1\cdot 2} & + & \frac{1}{2\cdot 3} & + & \frac{1}{3\cdot 4} & + & \ldots & = & 1 \\
0 & + & \frac{1}{2\cdot 3} & + & \frac{1}{3\cdot 4} & + & \ldots & = & \frac{1}{2} \\
0 & + & 0 & + & \frac{1}{3\cdot 4} & + & \ldots & = & \frac{1}{3} \\
\vdots & & \vdots & & \vdots & & & & \vdots
\end{array}$$

If the harmonic series were convergent, the terms on the left-hand side would form an absolutely convergent series with the sum $\sum_{n=1}^{\infty}(1/n)$. However, adding the table column by column and applying Theorem 6.30, we would get

$$\frac{1}{2} + \frac{1}{3} + \frac{1}{4} + \ldots = 1 + \frac{1}{2} + \frac{1}{3} + \frac{1}{4} + \ldots,$$

which is impossible.

Exercises

6.16. Show that every convergent series can be bracketed such that the resulting series is absolutely convergent.

6.17. Construct a series that can be bracketed such that the sum of the resulting series is A, for every given real number A. (S)

6.18. Show that every conditionally convergent series has a divergent reordering. (H)

6.19. Let the series $\sum_{n=1}^{\infty} a_n$ be conditionally convergent. Show that for every real number A, there exists a sequence of the indices $n_1 < n_2 < \ldots$ such that the series $\sum_{k=1}^{\infty} a_{n_k}$ is convergent with sum A.

6.20. Find the sums $\sum_{n=1}^{\infty} n^2 x^n$ and $\sum_{n=1}^{\infty} n^3 x^n$, for every $|x| < 1$. Find the sums of the series $\sum_{n=1}^{\infty} n^2/2^n$ and $\sum_{n=1}^{\infty} n^3/2^n$.

6.21. Let $a_n \in \mathbb{R}^p$ for every $n = 1, 2, \ldots$. We say that the infinite series $\sum_{n=1}^{\infty} a_n$ **is convergent and its sum is the vector** $a \in \mathbb{R}^p$ if the sequence of its partial sums converges to a. Show that if $\sum a_n$ is convergent, then $a_n \to 0$.

6.22. Show that if $a_n \in \mathbb{R}^p$ and $\sum |a_n| < \infty$ (i.e., if $\sum a_n$ is **absolutely convergent**), then $\sum a_n$ is convergent.

6.23. Let $a_n \in \mathbb{R}^2$ be an arbitrary sequence of points of $\mathbb{R}^2$, and let S denote the set of the sums of the reorderings of the series $\sum a_n$ that are convergent. Prove that one of the following cases holds: S is empty; S consists of a single element (this holds exactly when the series is absolutely convergent); S consists of a line; S contains the whole plane $\mathbb{R}^2$. $(*)$

6.4 Other Convergence Criteria

Despite the fact that the Cauchy criterion gives an exact condition for convergence, we can rarely use it in practice, since its condition is hard to check. Thus, we need other criteria that are considerably easier to verify, even if it means that in most cases, we get only sufficient, but not necessary, conditions for convergence.

We say that the series $\sum_{n=1}^{\infty} b_n$ is a **majorant** of the series $\sum_{n=1}^{\infty} a_n$, if $|a_n| \le b_n$ holds for every n large enough. The following important convergence criterion is called the **majorant criterion**.

Theorem 6.32. *If the infinite series $\sum_{n=1}^{\infty} a_n$ has a convergent majorant, then $\sum_{n=1}^{\infty} a_n$ is absolutely convergent.*

Proof. Let $\sum_{n=1}^{\infty} b_n$ be a convergent majorant of $\sum_{n=1}^{\infty} a_n$. Then by definition, $|a_n| \le b_n$ holds for every n large enough. Changing finitely many terms does not influence the convergence of the series; thus we may assume that $|a_n| \le b_n$ holds for every n. Then the partial sums of the series $\sum_{n=1}^{\infty} |a_n|$ are not larger than the corresponding partial sums of the series $\sum_{n=1}^{\infty} b_n$. The sequence of the latter is bounded from above, since $\sum_{n=1}^{\infty} b_n$ is convergent. Thus, the sequence of the partial sums of $\sum_{n=1}^{\infty} |a_n|$ is also bounded from above, and then, by Theorem 6.8, $\sum_{n=1}^{\infty} |a_n|$ is convergent. □

Example 6.33. The series $\sum_{n=1}^{\infty} (\sin n)/n^2$ is convergent, since $|(\sin n)/n^2| \le 1/n^2$ for every n, and the series $\sum_{n=1}^{\infty} 1/n^2$ is convergent by Example 6.11.1.

The next corollary is obtained by a simple application of the majorant criterion.

Corollary 6.34. *Every bracketed series of an absolutely convergent series is also absolutely convergent.*

Proof. Every bracketed series obtained from $\sum_{n=1}^{\infty} a_n$ has a majorant that is the appropriately bracketed series obtained from $\sum_{n=1}^{\infty} |a_n|$. If $\sum_{n=1}^{\infty} |a_n|$ is convergent, then by Theorem 6.17, this majorant series is also convergent. Thus the statement follows from Theorem 6.32. □

The next two convergence criteria are used mostly for certain series having terms of a special form.

Theorem 6.35. (Root criterion)

(i) *If there exists a number $q < 1$ such that $\sqrt[n]{|a_n|} < q$ holds for every n large enough, then the series $\sum_{n=1}^{\infty} a_n$ is absolutely convergent.*

(ii) *If $\lim_{n\to\infty} \sqrt[n]{|a_n|} < 1$, then the series $\sum_{n=1}^{\infty} a_n$ is absolutely convergent.*

Proof. (i) We have $|a_n| < q^n$ for every n large enough. Since by Theorem 6.4, the series $\sum_{n=1}^{\infty} q^n$ is convergent, we can apply the majorant criterion.
(ii) Choose a number q such that $\lim_{n\to\infty} \sqrt[n]{|a_n|} < q < 1$. Then $\sqrt[n]{|a_n|} < q$ holds for every n large enough, and thus by (i), the series $\sum_{n=1}^{\infty} a_n$ is absolutely convergent. □

Example 6.36. For $|x| < 1$, the series $\sum_{n=1}^{\infty} n \cdot x^n$ is absolutely convergent. Indeed, $\sqrt[n]{|n \cdot x^n|} = |x| \cdot \sqrt[n]{n} \to |x| < 1$ as $n \to \infty$. (For the sum of the series, see Example 6.31.1.)

Remarks 6.37. **1.** The conditions of the root criterion are not necessary for a series to be convergent: the series $\sum_{n=1}^{\infty} 1/n^2$ is convergent, despite the fact that $\lim_{n\to\infty} \sqrt[n]{1/n^2} = 1$, and thus there exists no $q < 1$ such that $\sqrt[n]{1/n^2} < q$ for every n large enough.
2. For the convergence of the series $\sum_{n=1}^{\infty} a_n$ it is not enough to have $\sqrt[n]{|a_n|} < 1$ for every n large enough (or even for every n). This condition means only that $|a_n| < 1$ for every n large enough, and it does not follow that $a_n \to 0$, which is a necessary condition for convergence.
3. If $\lim_{n\to\infty} \sqrt[n]{|a_n|} > 1$, then the series $\sum_{n=1}^{\infty} a_n$ is divergent, since $|a_n| > 1$ for every n large enough.
4. From the condition $\lim_{n\to\infty} \sqrt[n]{|a_n|} = 1$ we can infer neither the convergence nor the divergence of the series. The series $\sum_{n=1}^{\infty} 1/n^2$ is convergent, while the series $\sum_{n=1}^{\infty} 1/n$ is divergent, even though $\lim_{n\to\infty} \sqrt[n]{1/n^2} = \lim_{n\to\infty} \sqrt[n]{1/n} = 1$.

Theorem 6.38. (Quotient criterion) *Let $a_n \neq 0$ for n large enough.*

(i) *If there exists a number $q < 1$ such that $\left|\frac{a_{n+1}}{a_n}\right| < q$ holds for every n large enough, then the series $\sum_{n=1}^{\infty} a_n$ is absolutely convergent.*

(ii) *If $\lim_{n\to\infty} \left|\frac{a_{n+1}}{a_n}\right| < 1$, then the series $\sum_{n=1}^{\infty} a_n$ is absolutely convergent.*

Proof. (i) If $a_n \neq 0$ and $|a_{n+1}/a_n| < q$ for every $n \geq n_0$, then

$$\begin{aligned} |a_{n_0+1}| &\leq q \cdot |a_{n_0}|, \\ |a_{n_0+2}| &\leq q \cdot |a_{n_0+1}| \leq q^2 \cdot |a_{n_0}|, \\ |a_{n_0+3}| &\leq q \cdot |a_{n_0+2}| \leq q^3 \cdot |a_{n_0}|, \end{aligned} \tag{6.14}$$

and so on. That is, we have $|a_n| \leq q^{n-n_0} \cdot |a_{n_0}|$ for every $n > n_0$. Let $c = q^{-n_0} \cdot |a_{n_0}|$. Then $|a_n| \leq c \cdot q^n$ for every $n > n_0$. Therefore, we can apply the majorant criterion, since by Theorems 6.4 and 6.11, the series $\sum_{n=1}^{\infty} c \cdot q^n$ is convergent.
(ii) Choose a number q such that $\lim_{n\to\infty} |a_{n+1}/a_n| < q < 1$. Then $|a_{n+1}/a_n| < q$ for every n large enough, and thus by (i), the series $\sum_{n=1}^{\infty} a_n$ is absolutely convergent. □

The root criterion is stronger than the quotient criterion in that if a series satisfies the quotient criterion, then it satisfies the root criterion as well; but the converse of this is not true (see Exercise 6.25). However, the quotient criterion is still widely used, because it is often easier to apply than the root criterion.

Example 6.39. **1.** The convergence of the series $\sum_{n=1}^{\infty} nx^n$ can be also verified when $|x| < 1$ by applying the quotient criterion:

$$|(n+1) \cdot x^{n+1}|/|n \cdot x^n| = |x| \cdot (n+1)/n \to |x| < 1$$

as $n \to \infty$.
2. The series $\sum_{n=1}^{\infty} n!/n^n$ is convergent, since

$$\frac{(n+1)!}{(n+1)^{n+1}} : \frac{n!}{n^n} = \frac{n+1}{(n+1) \cdot \left(1+\frac{1}{n}\right)^n} = \frac{1}{\left(1+\frac{1}{n}\right)^n} \to \frac{1}{e} < 1$$

as $n \to \infty$.

Remarks 6.40. **1.** The conditions of the quotient criterion are not necessary for the convergence of a series: the series $\sum_{n=1}^{\infty} 1/n^2$ is convergent, even though $\lim_{n\to\infty} n^2/(n+1)^2 = 1$, and thus there exists no $q < 1$ such that $n^2/(n+1)^2 < q$ for every n large enough.

2. It is not enough for the convergence of the series $\sum_{n=1}^{\infty} a_n$ to have $\left|\frac{a_{n+1}}{a_n}\right| < 1$ for every n large enough (or even for every n). This condition means only that $|a_{n+1}| < |a_n|$ for every n large enough, which does not even imply $a_n \to 0$, which is a necessary condition of the convergence of the series.

3. If $\lim_{n\to\infty} |a_{n+1}/a_n| > 1$, then the series $\sum_{n=1}^{\infty} a_n$ is divergent, since $|a_{n+1}| > |a_n|$ for every n large enough, and then $a_n \not\to 0$.

4. From the condition $\lim_{n\to\infty} |a_{n+1}/a_n| = 1$ we can infer neither the convergence nor the divergence of the series itself. The series $\sum_{n=1}^{\infty} 1/n$ is divergent, while the series $\sum_{n=1}^{\infty} 1/n^2$ is convergent, even though $\lim_{n\to\infty} n/(n+1) = \lim_{n\to\infty} n^2/(n+1)^2 = 1$.

The next two criteria can be applied only if the terms of the series are nonnegative and form a monotonically decreasing sequence. At the same time, these criteria give us not only sufficient, but also necessary, conditions for the convergence of a series.

Theorem 6.41. (Integral criterion) *Let a be an integer, and let f be a monotonically decreasing nonnegative function on the half-line $[a,\infty)$. The infinite series $\sum_{n=a}^{\infty} f(n)$ is convergent if and only if the improper integral $\int_a^{\infty} f(x)\,dx$ is convergent.*

Proof. Let $n > a$ be a fixed integer, and consider the partition of the interval $[a,n]$ by the integers $a, a+1, \ldots, n$. Let s_n and S_n denote the lower and upper sums of the function f corresponding to this partition. Then

$$\sum_{i=a+1}^{n} f(i) = s_n \le \int_a^n f(x)\,dx \le S_n = \sum_{i=a}^{n-1} f(i), \tag{6.15}$$

since—recalling that f is monotonically decreasing—the smallest and the largest values of f on the interval $[i-1,i]$ are $f(i)$ and $f(i-1)$, respectively. Since f is nonnegative, the function $\omega \mapsto \int_a^{\omega} f(x)\,dx$ is monotonically increasing, and thus the improper integral $\int_a^{\infty} f(x)\,dx$ exists: the limit is either finite (when the integral is convergent) or infinite (when the integral is divergent). If the integral is convergent, then the sequence $n \mapsto \int_a^n f(x)\,dx$ is bounded (since it is convergent). The first inequality of (6.15) implies that the sequence (s_n) is also bounded. Then by Theorem 6.10, the series $\sum_{n=a}^{\infty} f(n)$ is convergent.

On the other hand, if the integral is divergent, then the sequence $n \mapsto \int_a^n f(x)\,dx$ converges to infinity. The second inequality of (6.15) implies that the sequence (S_n) also converges to infinity, and then the series $\sum_{n=a}^{\infty} f(n)$ is divergent. □

Example 6.42. For every $c > 0$, the terms of the series $\sum_{n=2}^{\infty} 1/(n \cdot \log^c n)$ are positive and form a monotonically decreasing sequence. Thus, the series is convergent if and only if the integral $\int_2^{\infty} 1/(x \cdot \log^c x)\,dx$ is convergent, i.e., if and only if $c > 1$. (See Example 19.5.1 of [7].)

Theorem 6.43. (Condensation criterion) *Suppose that the sequence (a_n) is nonnegative and monotonically decreasing. Then the series $\sum_{n=1}^{\infty} a_n$ and $\sum_{n=1}^{\infty} 2^n \cdot a_{2^n}$ are either both convergent or both divergent.*

Proof. Let the partial sums of $\sum_{n=1}^{\infty} a_n$ and $\sum_{n=1}^{\infty} 2^n \cdot a_{2^n}$ be denoted by s_n and S_n, respectively. We put $S_0 = 0$. Since $a_{2^n} \ge a_i$ for every $i > 2^n$, we have

$$S_n - S_{n-1} = 2^n \cdot a_{2^n} \ge \sum_{i=2^n+1}^{2^{n+1}} a_i = s_{2^{n+1}} - s_{2^n}$$

for every n, and thus

$$S_n = \sum_{k=1}^{n}(S_k - S_{k-1}) \geq \sum_{k=1}^{n}(s_{2^{k+1}} - s_{2^k}) = s_{2^{n+1}} - s_2.$$

It follows that if the partial sums of $\sum_{n=1}^{\infty} 2^n \cdot a_{2^n}$ are bounded, then the partial sums of the series $\sum_{n=1}^{\infty} a_n$ are also bounded. Similarly, $a_{2^n} \leq a_i$ for every $i \leq 2^n$, and thus

$$S_n - S_{n-1} = 2^n \cdot a_{2^n} \leq 2 \cdot \sum_{i=2^{n-1}+1}^{2^n} a_i = 2 \cdot (s_{2^n} - s_{2^{n-1}}).$$

Then

$$S_n = \sum_{k=1}^{n}(S_k - S_{k-1}) \leq 2 \cdot \sum_{k=1}^{n}(s_{2^k} - s_{2^{k-1}}) = s_{2^n} - s_1.$$

Therefore, if the partial sums of the series $\sum_{n=1}^{\infty} a_n$ are bounded, then the partial sums of $\sum_{n=1}^{\infty} 2^n \cdot a_{2^n}$ are also bounded. Thus, we can apply Theorem 6.10. □

Example 6.44. Consider the series $\sum_{n=2}^{\infty} 1/(n \cdot \log^c n)$, where $c > 0$. The condensation criterion states that the series is convergent if and only if the series $\sum_{n=2}^{\infty} 2^n/2^n \cdot \log^c 2^n = \sum_{n=2}^{\infty} 1/((\log^c 2) \cdot n^c)$ is also convergent, i.e., if and only if $c > 1$.

Our criteria so far have given conditions for the absolute convergence of a series. Another similar criterion can be found in Exercise 6.40. We now introduce a couple of convergence criteria, applicable to series having terms of a certain special form, that do not guarantee the absolute convergence of the series.

Theorem 6.45. (Leibniz criterion) *For every monotone sequence (a_n) converging to zero, the series $\sum_{n=1}^{\infty}(-1)^{n-1}a_n$ is convergent.*

Proof. We may assume that the sequence (a_n) is decreasing. (Otherwise, we turn to the sequence $(-a_n)$.) Let s_n denote the nth partial sum of the series. The conditions imply

$$s_2 \leq s_4 \leq \ldots s_{2n} \leq s_{2n-1} \leq s_{2n-3} \leq \ldots \leq s_3 \leq s_1$$

for every n. Thus the sequence (s_{2n}) is monotonically increasing and bounded from above, and the sequence (s_{2n-1}) is monotonically decreasing and bounded from below. Therefore, both sequences are convergent. Furthermore, $s_{2n} - s_{2n-1} = a_{2n} \to 0$, and thus $\lim_{n\to\infty} s_{2n} = \lim_{n\to\infty} s_{2n-1}$. It follows that the sequence (s_n) is convergent, which is exactly what we wanted to prove. □

Example 6.46. The series $1 - \frac{1}{3} + \frac{1}{5} - \frac{1}{7} + \ldots$ satisfies the conditions of the Leibniz criterion, and thus it is convergent. The sum of the series is $\pi/4$. This is obtained by applying the formula

$$\operatorname{arc\,tg} x = x - \frac{x^3}{3} + \frac{x^5}{5} - \ldots,$$

for $x = 1$ (see Example 7.41.2.).

Theorem 6.47. (Dirichlet[1] criterion) *Suppose that*

(i) *the sequence (a_n) is monotonically decreasing and converges to zero, and*

(ii) *the partial sums of the series $\sum_{n=1}^{\infty} b_n$ are bounded.*

Then the series $\sum_{n=1}^{\infty} a_n b_n$ is convergent.

Note that the Leibniz criterion is a special case of the Dirichlet criterion (let $b_n = (-1)^{n-1}$).

Proof of Theorem 6.47. Let s_n be the nth partial sum of the series $\sum_{n=1}^{\infty} b_n$, and suppose that $|s_n| \le K$ for every n. Let $\varepsilon > 0$ be given. Since $a_n \to 0$, we can choose an index N such that $|a_n| < \varepsilon/K$ holds for every $n \ge N$. If $N \le n < m$, then Abel's[2] inequality[3] gives

$$-\varepsilon < (-K) \cdot a_n \le a_n b_n + \ldots + a_m b_m \le K \cdot a_n < \varepsilon,$$

i.e., $|a_n b_n + \ldots + a_m b_m| < \varepsilon$. Thus the series $\sum_{n=1}^{\infty} a_n b_n$ satisfies the condition of the Cauchy criterion, and so it is convergent. □

Example 6.48. If $x \ne 2k\pi$ $(k \in \mathbb{Z})$ and $c > 0$, then the series $\sum_{n=1}^{\infty} (\cos nx)/n^c$ is convergent. Indeed, the sequence (n^{-c}) is decreasing and tends to zero. On the other hand, for $x \ne 2k\pi$ the partial sums of the series $\sum_{n=1}^{\infty} \cos nx$ are bounded, as seen from the identity

$$\cos x + \cos 2x + \cdots + \cos nx = \frac{\sin(nx/2)}{\sin(x/2)} \cdot \cos \frac{(n+1)x}{2}.$$

Thus the Dirichlet criterion applies.

Theorem 6.49. (Abel's criterion) *Suppose that*

(i) *the sequence (a_n) is monotone and bounded, and*

(ii) *the series $\sum_{n=1}^{\infty} b_n$ is convergent.*

Then the series $\sum_{n=1}^{\infty} a_n b_n$ is also convergent.

---

[1] Lejeune Dirichlet (1805–1859), German mathematician.

[2] Niels Henrik Abel (1802–1829), Norwegian mathematician.

[3] Abel's inequality states that if $c_1 \ge c_2 \ge \cdots \ge c_n \ge 0$ and $m \le d_1 + \cdots + d_k \le M$ for all $k = 1, \ldots, n$, then $c_1 \cdot m \le c_1 d_1 + \cdots + c_n d_n \le c_1 \cdot M$. (See [7, Theorem 14.54].)

Proof. We may assume that the sequence (a_n) is monotonically decreasing, since otherwise we may replace the sequence (a_n) by $(-a_n)$. Let $\lim_{n\to\infty} a_n = a$. Then $(a_n - a)$ is decreasing and converges to zero. Since the series $\sum_{n=1}^{\infty} b_n$ is convergent, the sequence of its partial sums is bounded. Thus, by the Dirichlet criterion, the series $\sum_{n=1}^{\infty}(a_n - a)b_n$ is convergent. The series $\sum_{n=1}^{\infty} a_n b_n$ is obtained by adding the series $\sum_{n=1}^{\infty} a \cdot b_n$ term by term to the series $\sum_{n=1}^{\infty}(a_n - a)b_n$. Therefore, $\sum_{n=1}^{\infty} a_n b_n$ is convergent by Theorem 6.11. □

For example, it follows from Abel's criterion that the convergence of $\sum_{n=1}^{\infty} b_n$ implies the convergence of $\sum_{n=1}^{\infty} \sqrt[n]{2} \cdot b_n$.

Exercises

6.24. Find an example of a convergent series with positive terms that satisfies none of the conditions of the root criterion, quotient criterion, and integral criterion.

6.25. Suppose that $a_n \neq 0$ for every n large enough.
(a) Show that $\lim_{n\to\infty} \left|\frac{a_{n+1}}{a_n}\right| < 1$ implies $\lim_{n\to\infty} \sqrt[n]{|a_n|} < 1$.

(b) Find a sequence (a_n) such that $\lim_{n\to\infty} \sqrt[n]{|a_n|} < 1$, but $\lim_{n\to\infty} \left|\frac{a_{n+1}}{a_n}\right| < 1$ is not true. Can we also find an example such that the limit $\lim_{n\to\infty} \left|\frac{a_{n+1}}{a_n}\right|$ exists?

6.26. Let $a_n > 0,\ b_n > 0$ for every n, and let $a_n/b_n \to 1$. Show that $\sum a_n$ is convergent if and only if $\sum b_n$ is convergent. Find an example to show that this is not true if we omit the condition $a_n > 0,\ b_n > 0$.

6.27. Let $\sum a_n$ and $\sum b_n$ be series of positive terms, and suppose that $a_{n+1}/a_n \leq b_{n+1}/b_n$ for every n large enough. Show that if $\sum b_n$ is convergent, then $\sum a_n$ is also convergent.

6.28. Show that if $a_n \geq 0$ and $\sum a_n = \infty$, then $\sum a_n/(1 + a_n) = \infty$.

6.29. Show that $\sum \operatorname{arc\,tg} a_n$ is convergent whenever $\sum a_n$ is a convergent series of positive terms.

6.30. Show that if $a_n \geq 0$ for every n, and $\sum a_n$ is convergent, then $\sum a_n^2$ is also convergent. Show that the condition $a_n \geq 0$ cannot be omitted.

6.31. Let $\sum a_n$ and $\sum b_n$ be convergent series of positive terms. Does it follow that (a) $\sum a_n b_n$, and (b) $\sum \max(a_n, b_n)$ is also convergent?

6.32. Show that if $\sum a_n$ is convergent and (a_n) is monotone, then $\sum n \cdot a_n^2$ is also convergent. Show that the condition of monotonicity cannot be omitted.

6.33. Decide, for each of the following series, whether it is convergent.

(a) $\sum n^{10}/(3^n - 2^n)$; (b) $\sum 1/\sqrt{n \cdot (n+1)}$;
(c) $\sum n^{100} \cdot q^n$ $(|q| < 1)$; (d) $\sum 1/(\log n + \sqrt{n})$;
(e) $\sum \left(1 - \cos \frac{1}{n}\right)$; (f) $\sum (\log n)/n^2$;
(g) $\sum 1/(n + 1000\sqrt{n})$; (h) $\sum n^2 \cdot e^{-\sqrt{n}}$;
(i) $\sum \frac{1}{\sqrt{n}} \cdot \sin \frac{1}{n}$; (j) $\sum \frac{1}{\sqrt{n}} \cdot \cos \frac{1}{n}$;
(k) $\sum \frac{\log(n!)}{n^3}$; (l) $\sum (\sqrt{n} + \log^7 n)/(n^3 - \log^7 n)$;
(m) $\sum n/\sqrt{n^4 - \log^2 n}$; (n) $\sum \left(\sqrt{n^2+1} - n\right)$;
(o) $\sum \left(\sqrt{n+1} - \sqrt[4]{n^2+1}\right)$; (p) $\sum \left(n^{1/n^2} - 1\right)$;
(q) $\sum \left(\sqrt[n]{2} - 1\right)$; (r) $\sum \frac{\log n}{n} \cdot \left(\sqrt[n]{2} - 1\right)$;
(s) $\sum (\log n)^{\log\log n}/n^2$; (t) $\sum n^{n+\frac{1}{n}} / \left(n + \frac{1}{n}\right)^n$;
(u) $\sum (\log\log n)^{\log n}/(\log n)^{\sqrt{n}}$;
(v) $\sum (\log n)^{\sqrt{n}}/(n^{\log\log n} + (\log\log n)^{\log n})$.

6.34. Decide, for each of the following series, whether it is convergent.

(a) $\sum \left(1 - \frac{1}{n}\right)^{n^2}$; (b) $\sum \left(\frac{n+200}{2n+5}\right)^n$;
(c) $\sum n^7/7^n$; (d) $\sum n^{\log n}/(\log n)^n$;
(e) $\sum \left(\frac{n-1}{n+1}\right)^{n^2-n}$; (f) $\sum n^{n^2+25}/(n+1)^{n^2}$;
(g) $\sum \left(\frac{1}{2} + \frac{1}{n}\right)^n$; (h) $\sum n^{n+\frac{1}{n}} / \left(n + \frac{1}{n}\right)^n$;
(i) $\sum (\log n)^{\log n}/2^n$; (j) $\sum \left(\frac{1+\cos n}{2+\cos n}\right)^{2n}$;
(k) $\sum \left(\frac{\sin n}{e^{\sin n}}\right)^n$.

6.35. Decide, for each of the following series, whether it is convergent.

(a) $\sum 2^n \cdot n!/n^n$; (b) $\sum (n!)^2/2^{n^2}$;
(c) $\sum \frac{1001 \cdot 1002 \cdots (1000+n)}{1 \cdot 3 \cdots (2n-1)}$; (d) $\sum n^{10}/10^n$;
(e) $\sum 1/\binom{2n}{n}$; (f) $\sum n^c/(1+\varepsilon)^n$ $(\varepsilon > 0)$.

6.36. Is the series $\sum 1/(n + \sqrt{n} + \sqrt[3]{n} + \ldots + \sqrt[n]{n})$ convergent? (H)

6.37. Show that if $\sum_{n=1}^{\infty} a_n$ is a convergent series of positive terms, then $\sum_{n=2}^{\infty} (a_n)^{1-(1/\log n)}$ is also convergent. (H)

6.38. Show that the series $\sum_{n=3}^{\infty} 1/(n \cdot \log n \cdot \log\log^c n)$ is convergent if and only if $c > 1$.

6.39. Let $e_0 = 1$, and let $e_{k+1} = e^{e_k}$ for every $k \geq 0$. Furthermore, let $\ell_0(x) \equiv x$ and $\ell_{k+1}(x) = \log(\ell_k(x))$ for every $k \geq 0$ and $x > e_k$.
(a) Show that for every k, the series

$$\sum_{n>e_k} 1/(\ell_0(n) \cdot \ell_1(n) \cdots \ell_k(n) \cdot (\ell_{k+1}(n))^c)$$

is convergent if and only if $c > 1$.
(b) Let $L(x) = k$ if $e_k \leq x < e_{k+1}$ $(k = 1, 2, \ldots)$. Show that the series $\sum_{n=3}^{\infty} 1/(\ell_0(n) \cdot \ell_1(n) \cdots \ell_{L(n)}(n) \cdot (L(n))^c)$ is convergent if and only if $c > 1$.

6.40. Let the terms of the series $\sum_{n=1}^{\infty} a_n$ be positive.
(a) Show that if $\lim_{n\to\infty} n \cdot \left(\frac{a_n}{a_{n+1}} - 1\right) > 1$, then the series $\sum_{n=1}^{\infty} a_n$ is convergent.
(b) Show that if $n \cdot \left(\frac{a_n}{a_{n+1}} - 1\right) \le 1$ for every n large enough, then the series $\sum_{n=1}^{\infty} a_n$ is divergent. **(Raabe's ratio test)**[4] (H)

6.41. Suppose that the sequence (a_n) is monotonically decreasing and tends to zero. Does this imply that the series $a_1 + a_2 - a_3 - a_4 + a_5 + a_6 - \ldots$ is convergent?

6.42. Is the series $\sum_{n=2}^{\infty} \frac{(-1)^n \cdot \sqrt[n]{n}}{\log n}$ convergent? Is it absolutely convergent?

6.43. Is the series $\sum_{n=1}^{\infty} \log\left(1 + \frac{(-1)^{n+1}}{n}\right)$ convergent? Is it absolutely convergent?

6.44. Show that for $c > 0$, the series $\sum_{n=1}^{\infty} \frac{\sin nx}{n^c}$ is convergent for every $x \in \mathbb{R}$.

6.45. We say that the sequence (a_n) **is of bounded variation** if
$\sum_{n=1}^{\infty} |a_{n+1} - a_n| < \infty$.
Show that every sequence of bounded variation is convergent. Find an example of a convergent sequence that is not of bounded variation.

6.46. Show that the sequence (a_n) is of bounded variation if and only if there exist monotone and bounded sequences (b_n) and (c_n) such that $a_n = b_n - c_n$ for every n.

6.47. Let (a_n) be a sequence of bounded variation tending to zero, and suppose that the partial sums of the series $\sum b_n$ are bounded. Show that $\sum a_n b_n$ is convergent.

6.48. Let (a_n) be of bounded variation, and let $\sum b_n$ be convergent. Show that $\sum a_n b_n$ is also convergent.

6.49. Let $a_1 < a_2 < \ldots$ be a sequence of positive integers such that $|\{k \colon a_k \le x\}| \ge cx/\log x$ for every $x \ge 2$ with a constant $c > 0$. Show that $\sum 1/a_k = \infty$. (∗)

6.5 The Product of Infinite Series

Multiplying infinite series is much more complicated than the operations we have explored so far, to such an extent that even defining the operation can be problematic. One way to compute the product of the finite sums $a_1 + \ldots + a_k$ and

[4] Joseph Ludwig Raabe (1801–1859), Swiss mathematician.

$b_1 + \ldots + b_n$ is to add all the products $a_i \cdot b_j$. (This follows from the distributivity of addition and multiplication, combined with the commutativity of addition and multiplication.) By the commutativity of addition, the summation of the $a_i \cdot b_j$ can be performed in an arbitrary order. However, in dealing with infinite series, we have to sum infinitely many products $a_i \cdot b_j$, and (as we saw before) the order of the additions can influence the value of the sum. Thus, when we define the product of two infinite series we need to also define the order of the addition of the products $a_i \cdot b_j$, and then we need to check whether the operation defined in this way satisfies the condition that the sum of the product of two infinite series is the same as the product of the two sums.

Definition 6.50. We call the series

$$a_1b_1 + a_2b_1 + a_2b_2 + a_1b_2 + \ldots + a_nb_1 + a_nb_2 + \ldots + a_nb_n + a_{n-1}b_n + \ldots + a_1b_n + \ldots$$

the *square product* of the series $\sum_{n=1}^{\infty} a_n$ and $\sum_{n=1}^{\infty} b_n$. More precisely, the square product of the series $\sum_{n=1}^{\infty} a_n$ and $\sum_{n=1}^{\infty} b_n$ contains every product a_ib_j in the order $\max(i, j)$, and for $\max(i, j) = n$ we first write the terms a_nb_j for $j = 1, \ldots, n$, followed by the terms a_ib_n for $i = n-1, n-2, \ldots, 1$. (In other words, we list the entries of the following table by going along the sides of its upper left squares, first from left to right, then from bottom to top.)

$$\begin{array}{ccccc} a_1b_1 & a_1b_2 & \ldots & a_1b_n & \ldots \\ & & & \uparrow & \\ a_2b_1 & a_2b_2 & \ldots & a_2b_n & \ldots \\ & & & \uparrow & \\ \ldots & \ldots & \ldots & \ldots & \ldots \\ & & & \uparrow & \\ a_nb_1 \rightarrow & a_nb_2 \rightarrow & \ldots \rightarrow & a_nb_n & \ldots \\ \ldots & \ldots & \ldots & \ldots & \ldots \end{array} \tag{6.16}$$

Theorem 6.51. *If the series $\sum_{n=1}^{\infty} a_n$ and $\sum_{n=1}^{\infty} b_n$ are convergent with sums A and B, respectively, then the square product of $\sum_{n=1}^{\infty} a_n$ and $\sum_{n=1}^{\infty} b_n$ is also convergent, and its sum is $A \cdot B$.*

Proof. Let the nth partial sum of $\sum_{n=1}^{\infty} a_n$, $\sum_{n=1}^{\infty} b_n$ and of their square product be denoted by r_n, s_n, and t_n, respectively. Obviously, $t_{k^2} = r_k \cdot s_k$ holds for every k, implying that $\lim_{k\to\infty} t_{k^2} = A \cdot B$. We need to show that the sequence (t_n) converges to $A \cdot B$.

Let $\varepsilon > 0$ be fixed. We show that $|t_n - A \cdot B| < 2\varepsilon$ for every n large enough.

We know that $r_n \to A$ and $s_n \to B$. In particular, these sequences are bounded. Let $|r_n| \le K$ and $|s_n| \le K$ for every n, where $K > 0$. By Theorem 6.6 we have $a_n \to 0$ and $b_n \to 0$, and thus there exists N_1 such that $|a_n| < \varepsilon/K$ and $|b_n| < \varepsilon/K$ for every $n \ge N_1$.

Since $\lim_{k\to\infty} t_{k^2} = A \cdot B$, there exists N_2 such that $|t_{k^2} - A \cdot B| < \varepsilon$ for every $k \ge N_2$.

Let $n \geq \max(N_1^2, N_2^2)$ be arbitrary. If $k = [\sqrt{n}]$, then we have $k \geq \max(N_1, N_2)$ and $k^2 \leq n < (k+1)^2$. Now, either $n = k^2 + j$, where $0 \leq j \leq k$, or $n = (k+1)^2 - j$, where $0 < j \leq k$.

If $n = k^2 + j$ with $0 \leq j \leq k$, then (by the definition of the square product) we have

$$t_n = t_{k^2} + a_{k+1}b_1 + \ldots + a_{k+1}b_j = t_{k^2} + a_{k+1} \cdot s_j,$$

which implies

$$|t_n - A \cdot B| \leq |t_n - t_{k^2}| + |t_{k^2} - A \cdot B| < |a_{k+1}| \cdot |s_j| + \varepsilon \leq \frac{\varepsilon}{K} \cdot K + \varepsilon = 2\varepsilon.$$

On the other hand, if $n = (k+1)^2 - j$ with $0 < j \leq k$, then we have

$$t_n = t_{(k+1)^2} - (a_j b_{k+1} + \ldots + a_1 b_{k+1}) = t_{(k+1)^2} - b_{k+1} \cdot r_j,$$

which implies

$$|t_n - A \cdot B| \leq |t_n - t_{(k+1)^2}| + |t_{(k+1)^2} - A \cdot B| < |b_{k+1}| \cdot |r_j| + \varepsilon \leq \frac{\varepsilon}{K} \cdot K + \varepsilon = 2\varepsilon.$$

We have proved that $|t_n - A \cdot B| < 2\varepsilon$ for every n large enough. Since ε was arbitrary, we have $t_n \to A \cdot B$. □

We now prove that for absolutely convergent series, the order of the terms of the product series does not affect the sum.

Theorem 6.52. *Let $\sum_{n=1}^{\infty} a_n$ and $\sum_{n=1}^{\infty} b_n$ be absolutely convergent series with sums A and B, respectively. Then adding the terms $a_i b_j$ $(i, j = 1, 2, \ldots)$ in any order, we obtain an absolutely convergent series, whose sum is $A \cdot B$.*

Proof. Let the square sum of the series $\sum_{n=1}^{\infty} a_n$ and $\sum_{n=1}^{\infty} b_n$ be $\sum_{n=1}^{\infty} c_n$. By assumption, the series $\sum_{n=1}^{\infty} |a_n|$ and $\sum_{n=1}^{\infty} |b_n|$ are convergent, and then, by Theorem 6.51, their square product is also convergent. On the other hand, this square product is nothing other than the series $\sum_{n=1}^{\infty} |c_n|$, since $|a_i| \cdot |b_j| = |a_i b_j|$ for every i and j. With this we have proved that $\sum_{n=1}^{\infty} c_n$ is absolutely convergent, and the statement of the theorem follows from Theorem 6.22. □

An important class of infinite series consists of the series of the form $\sum_{n=0}^{\infty} a_n x^n$, whose terms depend on the value of x. These series are called **power series**. Of the series we have seen so far, the series $\sum_{n=0}^{\infty} x^n$ and $\sum_{n=0}^{\infty} (n+1) \cdot x^n$ are power series.

If we want to find the product of the power series $\sum_{n=0}^{\infty} a_n x^n$ and $\sum_{n=0}^{\infty} b_n x^n$, we need to add up the terms $a_i x^i \cdot b_j x^j = a_i b_j \cdot x^{i+j}$. In this case, it is only natural not to follow the order of the square sum, but rather to group the terms based on the value of $i + j$, and then to add these groups, resulting in another power series. This method results in the power series $\sum_{n=0}^{\infty} \left(\sum_{i=0}^{n} a_i b_{n-i}\right) \cdot x^n$.

The previous reordering can be done for every pair of infinite series. To follow the notation of the power series, we will start the indices of the series from zero from now on.

Definition 6.53. We say that the *Cauchy product* of the infinite series $\sum_{n=0}^{\infty} a_n$ and $\sum_{n=0}^{\infty} b_n$ is the infinite series

$$\sum_{n=0}^{\infty}\left(\sum_{i=0}^{n} a_i b_{n-i}\right).$$

(That is, we compute the terms of the Cauchy product by adding the terms along the diagonals in the following table.)

$$\begin{array}{cccccc}
a_0b_0 & a_0b_1 & \dots & \dots & a_0b_n & \dots \\
 & & & & \nearrow & \\
a_1b_0 & a_1b_1 & \dots & a_1b_{n-1} & \dots & \dots \\
 & & \nearrow & & & \\
\vdots & \vdots & \vdots & \vdots & \vdots & \vdots \\
 & & \nearrow & & & \\
\vdots & a_{n-1}b_1 & \dots & \dots & \dots & \dots \\
 & \nearrow & & & & \\
a_nb_0 & a_nb_1 & \dots & \dots & a_nb_n & \dots \\
\vdots & \vdots & \vdots & \vdots & \vdots & \vdots
\end{array} \tag{6.17}$$

Theorem 6.54. *Suppose that the series $\sum_{n=0}^{\infty} a_n$ and $\sum_{n=0}^{\infty} b_n$ are absolutely convergent, and let their sums be A and B, respectively. Then the Cauchy product of these series is also absolutely convergent, and its sum is $A \cdot B$.*

Proof. Since the Cauchy series is the result of reordering and bracketing the square sum, our claim follows from Theorems 6.51, 6.22, 6.17, Corollary 6.34, and Theorem 6.52. □

Example 6.55. If $|x| < 1$, then the series $\sum_{n=0}^{\infty} |x|^n$ is convergent by Theorem 6.5, and thus the series $\sum_{n=0}^{\infty} x^n$ is absolutely convergent. By taking the Cauchy product of this series with itself we get $\sum_{n=0}^{\infty}(n+1) \cdot x^n = \sum_{n=1}^{\infty} n \cdot x^{n-1}$, and our previous theorem implies

$$\sum_{n=1}^{\infty} n \cdot x^{n-1} = \frac{1}{(1-x)^2},$$

for every $|x| < 1$ (as we saw in Example 6.31).

We now show that the Cauchy product of two convergent series can be divergent.

Example 6.56. The series $\sum_{n=0}^{\infty} \frac{(-1)^{n+1}}{\sqrt{n+1}}$ is convergent, since it satisfies the Leibniz criterion. Let $\sum_{i=0}^{\infty} c_n$ be the Cauchy product of this series with itself, i.e., let

$$c_n = (-1)^n \sum_{i=0}^{n} \frac{1}{\sqrt{i+1} \cdot \sqrt{n+1-i}}.$$

By the arithmetic–geometric means inequality, we have

$$\sqrt{(i+1) \cdot (n+1-i)} \leq \frac{(n+2)}{2}$$

for every i, which implies $|c_n| \geq (n+1) \cdot \frac{2}{n+2} \geq 1$ for every n. It follows that the series $\sum_{i=0}^{\infty} c_n$ is divergent.

Thus, the Cauchy product of two convergent series can be divergent if the series are not absolutely convergent. As an improvement on Theorem 6.54 we now show that if at least one of the two series is absolutely convergent, then their Cauchy product will be convergent.

Theorem 6.57. (Mertens's[5] theorem) *Let the series $\sum_{n=0}^{\infty} a_n$ and $\sum_{n=0}^{\infty} b_n$ be convergent with sums A and B, respectively. If at least one of these two series is also absolutely convergent, then their Cauchy product is convergent and its sum is $A \cdot B$.*

Proof. Let the Cauchy product of the series $\sum_{n=0}^{\infty} a_n$ and $\sum_{n=0}^{\infty} b_n$ be $\sum_{k=0}^{\infty} c_k$, and let us denote the nth partial sums of the series $\sum_{n=0}^{\infty} a_n$ and $\sum_{n=0}^{\infty} b_n$ by r_n and s_n, respectively. Furthermore, let $S_n = r_n \cdot s_n - \sum_{k=0}^{n} c_k$. (This is the sum of terms of Table (6.17) forming a triangle.) Since $r_n \to A$ and $s_n \to B$, it is enough to prove that $S_n \to 0$. We may assume that $\sum_{n=1}^{\infty} a_n$ is absolutely convergent, since the roles of the two series are symmetric. Let $\sum_{n=1}^{\infty} |a_n| = M$. It is clear that

$$S_n = \sum_{i=0}^{n} a_i \cdot \sum_{j=0}^{n} b_j - \sum_{i+j \leq n} a_i b_j,$$

i.e., S_n is the sum of the terms $a_i b_j$ satisfying $i \leq n$, $j \leq n$ and $n < i + j$. Thus we have

$$S_n = \sum_{i=1}^{n} \sum_{j=n+1-i}^{n} a_i b_j = \sum_{i=1}^{n} a_i \cdot (s_n - s_{n-i}). \tag{6.18}$$

The idea of the proof is that if n is large, then $s_n - s_{n-i}$ is small for small i (since (s_n) is convergent), while for large i the sum $\sum |a_i|$ is small (since the series $\sum_{n=1}^{\infty} a_n$ is absolutely convergent). That is, S_n is small when n is large. In the sequel we turn this idea into a mathematically precise proof.

---

[5] Franz Carl Joseph Mertens (1840–1927), Polish–Austrian mathematician.

The series (s_n) is bounded, since it is convergent. Let $|s_n| \le K$ for every n. Let $\varepsilon > 0$ be fixed, and apply the Cauchy criterion to both the convergent sequence (s_n) and the convergent series $\sum_{n=1}^{\infty} |a_n|$. We obtain an N such that for every $N \le m < n$, we have both $|s_n - s_m| < \varepsilon$ and $\sum_{i=m}^{n} |a_i| < \varepsilon$.

Let $n > 2N$ be arbitrary. If $i \le N$, then $n - i > N$ and $|s_n - s_{n-i}| < \varepsilon$, which implies

$$\left| \sum_{i=1}^{N} a_i \cdot (s_n - s_{n-i}) \right| \le \varepsilon \cdot \sum_{i=1}^{N} |a_i| \le \varepsilon \cdot M.$$

On the other hand,

$$\left| \sum_{i=N+1}^{n} a_i \cdot (s_n - s_{n-i}) \right| \le \sum_{i=N+1}^{n} |a_i| \cdot 2K \le \varepsilon \cdot 2K.$$

Adding these two estimates, we obtain, by (6.18), that $|S_n| \le (2K + M) \cdot \varepsilon$ for every $n > 2N$. Since $\varepsilon > 0$ was arbitrary, this proves $S_n \to 0$. □

Note that Theorem 6.57 does not claim the absolute convergence of the Cauchy product of two series satisfying the conditions. In fact, this is false in general; see Exercise 6.50.

Exercises

6.50. Show that the Cauchy product of the series $\sum_{n=0}^{\infty} \frac{(-1)^n}{(n+1)^2}$ and $\sum_{n=0}^{\infty} \frac{(-1)^n}{(n+1)}$ is not absolutely convergent.

6.51. Find the square and Cauchy products of the series $1 - \frac{1}{2} - \frac{1}{4} - \frac{1}{8} - \ldots$ and $1 + 1 + \ldots$.

6.52. Find the Cauchy product of the series $\sum_{n=0}^{\infty} x^n/n!$ and $\sum_{n=0}^{\infty} y^n/n!$.

6.53. Show that the series $\sum_{n=1}^{\infty} n^2 \cdot x^n$ is absolutely convergent for every $|x| < 1$, and find its sum. Do the same with the series $\sum_{n=1}^{\infty} n^3 \cdot x^n$.

6.54. In the series $1 - \frac{1}{2} + \frac{1}{4} - \frac{1}{8} - \frac{1}{16} + \frac{1}{32} - \ldots$, the nth positive term is followed by n negative terms. Find the sum of the series.

6.6 Summable Series

It was realized already by Euler that the investigation of certain divergent series can lead to useful results for convergent series as well (see the appendix of this chapter). Therefore, it would not be wise to banish all divergent series from mathematical analysis, and it would be useful to assign sum-like values to certain divergent series in a noncontradictory manner. The nineteenth century saw the emergence of several such methods. Here we discuss only the simplest of these, which requires only the convergence of the averages of the partial sums of a series, instead of the convergence of the partial sums themselves. Later, while exploring the topic of power series we will discuss another, much more general, method.

Definition 6.58. We say that the infinite series $\sum_{n=1}^{\infty} a_n$ is *summable* with sum A if the partial sums $s_n = \sum_{i=1}^{n} a_i$ satisfy

$$\lim_{n\to\infty} \frac{s_1 + \ldots + s_n}{n} = A. \tag{6.19}$$

Example 6.59. The series $1 - 1 + 1 - 1 - \ldots$ is summable with sum $1/2$. Indeed, the partial sums satisfy

$$s_n = \begin{cases} 1 & \text{if } n \text{ is odd,} \\ 0 & \text{if } n \text{ is even.} \end{cases}$$

If $n \to \infty$, then we have

$$\left| \frac{s_1 + \ldots + s_n}{n} - \frac{1}{2} \right| \le \frac{1}{n} \to 0.$$

Consistency requires that we not assign different numbers to the same infinite series. In other words, it is a natural requirement that if a series is convergent and summable at the same time, then the two sums should be equal. We show that this is true; furthermore, we show that the convergence of a series automatically implies its summability.

Theorem 6.60. *If the infinite series $\sum_{n=1}^{\infty} a_n$ is convergent and its sum is A, then the series is also summable with sum A.*

Proof. Let s_n denote the nth partial sum of the series. We have to prove that if $s_n \to A$, then $(s_1 + \ldots + s_n)/n \to A$. For a given $\varepsilon > 0$ there exists an N such that if $n \ge N$, then $|s_n - A| < \varepsilon$. Let $|s_1 - A| + \ldots + |s_N - A| = K$. If $n \ge N$, then

$$|s_n - A| = \left| \frac{(s_1 - A) + \ldots + (s_n - A)}{n} \right| \le \frac{|s_1 - A| + \ldots + |s_n - A|}{n} \le$$
$$\le \frac{K + n\varepsilon}{n} < 2\varepsilon,$$

given that $n > K/\varepsilon$. Thus $s_n \to A$. □

By the previous theorem, the set of summable series is larger than the set of convergent series: if a series is convergent, it is also summable, while the converse is not necessarily true, as we can see from the example of $1 - 1 + 1 - 1 - \ldots$. Of course, the summable series constitute but a very small subset of the full set of all infinite series. One can show that if the infinite series $\sum_{n=1}^{\infty} a_n$ is summable, then $a_n/n \to 0$ (see Exercise 6.56).

The following theorem of Tauber[6] describes the exact relationship between summable and convergent series.

Theorem 6.61. *The infinite series $\sum_{n=1}^{\infty} a_n$ is convergent if and only if it is summable and satisfies*

$$\lim_{n\to\infty} \frac{a_1 + 2a_2 + \ldots + na_n}{n} = 0. \tag{6.20}$$

Proof. If a series is convergent, then it is also summable by Theorem 6.60, while (6.20) follows from Exercise 6.5.

Now let $\sum_{n=1}^{\infty} a_n$ be summable with sum A, and let s_n denote the nth partial sum of the series. Then

$$\frac{s_1 + \ldots + s_{n-1}}{n} = \frac{n-1}{n} \cdot \frac{s_1 + \ldots + s_{n-1}}{n-1} \to 1 \cdot A = A$$

as $n \to \infty$. Since

$$\frac{a_1 + 2a_2 + \ldots + na_n}{n} = \frac{s_n + (s_n - s_1) + \ldots + (s_n - s_{n-1})}{n} =$$
$$= s_n - \frac{s_1 + \ldots + s_{n-1}}{n},$$

it follows that (6.20) implies $s_n \to A$, i.e., the series is convergent and its sum is A. □

The following corollary is also due to Tauber.

Corollary 6.62. *If the series $\sum_{n=1}^{\infty} a_n$ is summable and $n \cdot a_n \to 0$, then it is convergent.*

Proof. It is easy to see that if $n \cdot a_n \to 0$, then (6.20) holds. (See the proof of Theorem 6.60.) Therefore, we can apply Theorem 6.61. □

Hardy[7] and Landau[8] recognized that it is enough to assume the boundedness of $n \cdot a_n$ in the previous theorem (see Exercise 6.60).

It is not hard to prove that the series $\sum_{n=1}^{\infty} \sin nx$ is divergent for every $x \neq k\pi$, while the series $\sum_{n=1}^{\infty} \cos nx$ is divergent for every x (see Exercise 6.61). We now show that for $x \neq 2k\pi$ both series are summable, and we find their sums.

---

[6] Alfred Tauber (1866–1942), Austrian mathematician.

[7] Godfrey Harold Hardy (1877–1947), British mathematician.

[8] Edmund Georg Hermann Landau (1877–1938), German mathematician.

Theorem 6.63.

(i) *The series* $\sum_{n=1}^{\infty} \sin nx$ *is summable for every* $x \in \mathbb{R}$, *and its sum is zero if* $x = 2k\pi$ $(k \in \mathbb{Z})$, *and* $(1/2) \cdot \operatorname{ctg}(x/2)$ *if* $x \neq 2k\pi$ $(k \in \mathbb{Z})$.

(ii) *The series* $\sum_{n=1}^{\infty} \cos nx$ *is summable for every* $x \neq 2k\pi$ $(k \in \mathbb{Z})$ *and its sum is* $-1/2$.

Lemma 6.64. *For* $x \neq 2k\pi$ $(k \in \mathbb{Z})$ *we have*

$$|\sin x + \ldots + \sin nx| \leq \frac{1}{|\sin(x/2)|} \quad \textit{and} \quad |\cos x + \ldots + \cos nx| \leq \frac{1}{|\sin(x/2)|}$$

for every $n = 1, 2, \ldots$.

Proof. Adding the identities

$$2 \sin \frac{x}{2} \sin jx = \cos\left(jx - \frac{x}{2}\right) - \cos\left(jx + \frac{x}{2}\right)$$

and

$$2 \sin \frac{x}{2} \cos jx = \sin\left(jx + \frac{x}{2}\right) - \sin\left(jx - \frac{x}{2}\right)$$

for $j = 1, \ldots, n$ and dividing them by $2 \sin(x/2)$ yields

$$\sin x + \ldots + \sin nx = \frac{\cos \frac{x}{2} - \cos\left(nx + \frac{x}{2}\right)}{2 \sin \frac{x}{2}} \tag{6.21}$$

and

$$\cos x + \ldots + \cos nx = \frac{\sin\left(nx + \frac{x}{2}\right) - \sin \frac{x}{2}}{2 \sin \frac{x}{2}}, \tag{6.22}$$

from which the statements of the lemma follow immediately. □

Proof of Theorem 6.63. For $x = 2k\pi$, every term of the series $\sum_{n=1}^{\infty} \sin nx$ is zero, and thus each of its partial sums and its sum are also zero. Thus we may assume that $x \neq 2k\pi$ $(k \in \mathbb{Z})$. Let us use the notation $s_n = s_n(x) = \sum_{j=1}^{n} \sin jx$ and $c_n = c_n(x) = \sum_{j=1}^{n} \cos jx$. Slightly rewriting (6.21) and (6.22) gives

$$s_n(x) = \frac{1}{2} \operatorname{ctg} \frac{x}{2} - \frac{1}{2} \operatorname{ctg} \frac{x}{2} \cdot \cos nx + \frac{1}{2} \cdot \sin nx$$

and

$$c_n(x) = -\frac{1}{2} + \frac{1}{2} \operatorname{ctg} \frac{x}{2} \cdot \sin nx + \frac{1}{2} \cdot \cos nx.$$

Therefore, we have

$$\frac{s_1 + \ldots + s_n}{n} = \frac{1}{2}\operatorname{ctg}\frac{x}{2} - \frac{1}{2}\operatorname{ctg}\frac{x}{2} \cdot \frac{c_n(x)}{n} + \frac{1}{2} \cdot \frac{s_n(x)}{n} \tag{6.23}$$

and

$$\frac{c_1 + \ldots + c_n}{n} = -\frac{1}{2} + \frac{1}{2}\operatorname{ctg}\frac{x}{2} \cdot \frac{s_n(x)}{n} + \frac{1}{2} \cdot \frac{c_n(x)}{n}. \tag{6.24}$$

Since the sequences (s_n) and (c_n) are bounded by Lemma 6.64, we obtain

$$\frac{s_1 + \ldots + s_n}{n} \to \frac{1}{2}\operatorname{ctg}\frac{x}{2}$$

and

$$\frac{c_1 + \ldots + c_n}{n} \to -\frac{1}{2}$$

as $n \to \infty$. This is what we wanted to prove. □

Exercises

6.55. Is the series $1 - 2 + 3 - 4 + 5 - \ldots$ summable?

6.56. Show that if the series $\sum_{n=1}^{\infty} a_n$ is summable, then $a_n/n \to 0$. (S)

6.57. Show that if a series of nonnegative terms is summable, then it is also convergent.

6.58. Show that if $a_1 \geq a_2 \geq \ldots \geq 0$, then the series $a_1 - a_2 + a_3 - a_4 + \ldots$ is summable.

6.59. Show that if $x, y \in [0, 2\pi]$ and $x \neq y$, then the series $\sum_{n=1}^{\infty} \sin nx \sin ny$ is summable with sum zero. (H)

6.60. Show that if the series $\sum_{n=1}^{\infty} a_n$ is summable and the sequence $(n \cdot a_n)$ is bounded from below or bounded from above, then the series is convergent. (∗ H S)

6.61. (a) Show that the series $\sum_{n=1}^{\infty} \sin nx$ is convergent if and only if $x = k\pi$ $(k \in \mathbb{Z})$.
(b) Show that the series $\sum_{n=1}^{\infty} \cos nx$ is divergent for every x.
(c) Show that the series $\sum_{n=1}^{\infty} \sin n^2 x$ is convergent if and only if $x = k\pi$ $(k \in \mathbb{Z})$. (H S)

6.7 Appendix: On the History of Infinite Series

The mathematicians of the eighteens and nineteenth centuries believed that infinite series had fixed "predestined" sums, and that the arithmetic of infinite series was more or less the same as that of finite series. (See the historical introduction of [7].) The resulting oddities (such as the formula $1+2+4+\ldots=-1$) and contradictions (such as the question of the sum of the series $1-1+1-1+\ldots$) were widely known, and they caused heated debates. Leonhard Euler explores these problems in detail in his book *Introduction to Differentiation,* published in 1755 (see p. 61 of [4]). Euler adds the examples

$$1-2+4-8+16-\ldots=\frac{1}{3} \tag{6.25}$$

and

$$1-3+9-27+81-\ldots=\frac{1}{4} \tag{6.26}$$

to $1+2+4+\ldots=-1$. (All these equations follow from the formula $1+x+x^2+\ldots=1/(1-x)$ by plugging in $x=-2$, $x=-3$, and $x=-1$, respectively.) Then he says:

"It is clear that the sum of the series (6.25) *cannot be equal to* $1/3$*, since the more terms we actually sum, the farther away the result gets from* $1/3$*. But the sum of any series ought to be a limit the closer to which the partial sums should approach, the more terms are added.*

From this we conclude that series of this kind, which are called divergent, have no fixed sums, since the partial sums do not approach any limit that would be the sum for the infinite series."

Yet Euler does not reject the idea of working with divergent series. Moreover, he argues that the sums attributed to the divergent series can help us find true and important results, and so, in a certain sense, the sums similar to Examples (6.25) and (6.26) are also correct. Euler writes ([4, p. 61]):

"[T]hese sums, even though they seem not to be true, never lead to error. Indeed, if we allow them, then we can discover many excellent results that we would not have if we rejected them out of hand. Furthermore, if these sums were really false, they would not consistently lead to true results ...."

Of course, Euler's statement claiming that these sums never lead to errors can be disproved easily: when we "derived" from the formula $1+x+x^2+\ldots=1/(1-x)$ that the sum of the series $1-1+1-1-\ldots$ is on the one hand $1/2$, while on the other hand the sum is either zero or 1, we could only be right at most once, and we made errors at least twice. Euler based his claims on his own experience: his brilliant mathematical intuition led to great results based on his operations on divergent series. However, we cannot be convinced of the truth of these results until we support Euler's claims with precise definitions and flawless proofs.

Euler thought that the root of the problem was that the sum of a series is understood as one of two different things: the result of a formal operation (which also led to the formula $1 + x + x^2 + \ldots = 1/(1 - x)$), and, in the case of convergent series, the value to which the sequence of the partial sums converges. In the end, he suggests that we work with divergent series as well, but always keeping in mind whether the actual series is convergent (i.e., whether the sequence of its partial sums converges to the sum of the series) or divergent.

Euler's attitude did not dissolve the uncertainties surrounding infinite series at all. In 1826, Abel still wrote: "*In the field of mathematics we can hardly find a single infinite series whose sum is strictly defined.*"

The solution of the problem can be credited to Cauchy, who, in his book *Algebraic Analysis* turned Euler's distinction between convergent and divergent series into a strict mathematical definition and rejected the idea that a divergent series can have a finite sum. This definition weeded out the anomalies regarding infinite series, and Cauchy's definition became generally accepted.

In the end, however, Euler was also right: with the help of the theory of summable series and its generalizations, divergent series became legitimate objects of mathematical analysis.

Chapter 7
Sequences and Series of Functions

We have seen several sequences and series whose terms depended on a parameter or variable. Such sequences are a^n, $\sqrt[n]{a}$, $a^n/n!$, $\left(1+\frac{a}{n}\right)^n$ and every power series. Now we turn to the systematic investigation of those sequences and series that depend on a variable.

7.1 The Convergence of Sequences of Functions

Definition 7.1. Let $f_1, f_2, \ldots$ be real valued functions defined on the set H. (We do not assume that $H \subset \mathbb{R}$.) We say that the sequence of functions (f_n) *converges pointwise to the function* $f\colon H \to \mathbb{R}$, if $\lim_{n\to\infty} f_n(x) = f(x)$ for every $x \in H$. We use the notation $f_n \to f$.

Examples 7.2. **1.** Let $H = [0,1]$ and $f_n(x) = x^n$ for every $x \in [0,1]$ and $n = 1, 2, \ldots$. Then the sequence of functions (f_n) converges pointwise to the function

$$f(x) = \begin{cases} 0 & \text{if} \quad 0 \le x < 1, \\ 1 & \text{if} \quad x = 1 \end{cases} \tag{7.1}$$

(see [7, Theorem 4.16]).

2. Let $H = \mathbb{R}$ and $f_n(x) = \operatorname{arc\,tg}(nx)$ for every $x \in \mathbb{R}$ and $n = 1, 2, \ldots$. Since $\lim_{x\to\pm\infty} \operatorname{arc\,tg} x = \pm\pi/2$ and $\operatorname{arc\,tg} 0 = 0$, the sequence of functions (f_n) converges pointwise to the function

$$f(x) = \begin{cases} -\pi/2 & \text{if} \quad x < 0, \\ 0 & \text{if} \quad x = 0, \\ \pi/2 & \text{if} \quad x > 0 \end{cases} \tag{7.2}$$

M. Laczkovich and V.T. Sós, *Real Analysis*, Undergraduate Texts in Mathematics, https://doi.org/10.1007/978-1-4939-7369-9_7

These examples show that the *pointwise limit of continuous functions is not necessarily continuous.* This may be surprising, since a seemingly convincing argument claims the exact opposite. Indeed, let the sequence of functions (f_n) be convergent pointwise to the function f, and let every function f_n be continuous at the point a. If x is close to a, then $f_n(x)$ is close to $f_n(a)$ since the functions f_n are continuous at a. Furthermore, for n large enough $f_n(x)$ is close to $f(x)$, and $f_n(a)$ is close to $f(a)$ (since $f_n(x) \to f(x)$ and $f_n(a) \to f(a)$). Since

$$|f(x) - f(a)| \le |f(x) - f_n(x)| + |f_n(x) - f_n(a)| + |f_n(a) - f(a)|,$$

we might believe that $f(x)$ has to be close to $f(a)$ if x is close to a; i.e., that f is continuous at a.

Examples 7.2 show that this argument cannot be correct. The problem is the fact that while for a fixed n and ε there exists some δ such that $|f_n(x) - f_n(a)| < \varepsilon$ for $|x - a| < \delta$ is true, but this δ might depend not only on ε, but also on n. For x fixed, $|f(x) - f_n(x)| < \varepsilon$ holds for some $n > n_0$, but this n_0 may depend on x itself, and it is also possible that the $\delta = \delta_n$ depending on the indices $n > n_0$ is so small that x is not in the interval $(a - \delta_n, a + \delta_n)$, and $|f_n(x) - f_n(a)| < \varepsilon$ does not necessarily hold.

Our previous examples illustrate the fallacies of this incorrect argument. Let $f_n(x) = x^n$, $a = 1$, and let ε be $1/2$. If $|f_n(x) - f_n(a)| = |x^n - 1| < 1/2$, then we have $\sqrt[n]{1/2} < x$. However, for $x < 1$ fixed $|f(x) - f_n(x)| = |0 - x^n| < 1/2$ holds if and only if $x < \sqrt[n]{1/2}$, and $|f_n(x) - f_n(a)| < 1/2$ does not hold for these values of n.

If we want to ensure the continuity of the limit function, it suffices to assume that the numbers δ_n have a common, positive lower bound (i.e., there is some δ that works for every n), or the indices n_0 have a common upper bound (i.e., there is some n_0 that works for all x). We give a definition for both conditions below.

Definition 7.3. Let $f_1, f_2, \ldots$ be real functions defined on $H \subset \mathbb{R}^p$. We say that the sequence of functions (f_n) is *uniformly equicontinuous on* H, if for every $\varepsilon > 0$ there exists a $\delta > 0$ such that whenever $x, y \in H$ and $|x - y| < \delta$, then $|f_n(x) - f_n(y)| < \varepsilon$ for all n.

Theorem 7.4. *Let the sequence of functions (f_n) be uniformly equicontinuous on $H \subset \mathbb{R}^p$. If $f_n \to f$ pointwise on H, then f is continuous on H.*

Proof. Let $\varepsilon > 0$ be fixed, and let $\delta > 0$ be chosen according to the definition of the uniform equicontinuity of (f_n). If $x, y \in H$ and $|x - y| < \delta$, then we have $|f_n(x) - f_n(y)| < \varepsilon$ for all n. Since $f_n(x) \to f(x)$ and $f_n(y) \to f(y)$, it follows that $|f(x) - f(y)| \le \varepsilon$. Obviously, this implies the (uniform) continuity of f on H. □

Unfortunately, the condition of uniform equicontinuity is very hard to verify and check. (For an exception, see Exercise 7.18.) In general, the condition requiring the existence of a common n_0, for every x, is more useful.

Definition 7.5. Let $f_1, f_2, \ldots$ be real functions defined on H. (We do not assume $H \subset \mathbb{R}^p$.) We say that the sequence of functions (f_n) *is uniformly convergent to the function* $f\colon H \to \mathbb{R}$, if for every $\varepsilon > 0$ there exists an n_0 such that $|f_n(x) - f(x)| < \varepsilon$ for every $x \in H$ and every $n \geq n_0$.

(Note that, while the conditions of Definitions 7.3 and 7.5 are different, these definitions coincide in some cases; see Exercise 7.19.)

Examples 7.6. **1.** The sequence of functions x^n *converges uniformly* to the constant zero function on the interval $[0, a]$, for every $0 < a < 1$. Indeed, for $\varepsilon > 0$ fixed, $a^n \to 0$ implies the existence of some n_0 such that $a^{n_0} < \varepsilon$. We have

$$x^n \leq a^n \leq a^{n_0} < \varepsilon$$

for all $x \in [0, a]$ and $n \geq n_0$, which implies our claim.

2. The sequence of functions x^n *does not converge uniformly* to the constant zero function on $[0, 1)$. Indeed, for every n there is an $x \in [0, 1)$ such that $x^n > 1/2$ (every number works between $\sqrt[n]{1/2}$ and 1), thus there is no n_0 for $\varepsilon = 1/2$ such that $|x^n - 0| < \varepsilon$ holds for every $x \in [0, 1)$ and $n \geq n_0$.

3. The sequence of functions $\operatorname{arc\,tg}(nx)$ *converges uniformly* to the constant $\pi/2$ function on the interval $[a, \infty)$, for every $a > 0$. Indeed, let $\varepsilon > 0$ be given. Since $\lim_{x\to\infty} \operatorname{arc\,tg} x = \pi/2$, there is an n_0 such that $\pi/2 - \varepsilon < \operatorname{arc\,tg}(na) < \pi/2$ for all $n \geq n_0$. If $x \geq a$ and $n > n_0$, then we have $\pi/2 - \varepsilon < \operatorname{arc\,tg}(na) \leq \operatorname{arc\,tg}(nx) < \pi/2$, proving the uniform convergence on $[a, \infty)$.

4. The sequence of functions $\operatorname{arc\,tg}(nx)$ *does not converge uniformly* to the constant $\pi/2$ function on the interval $(0, \infty)$. Indeed, $\lim_{x\to 0} \operatorname{arc\,tg} x = 0$ implies the existence of $c > 0$ such that $\operatorname{arc\,tg} c < 1$. Thus $\operatorname{arc\,tg}(nx) < 1$ for every n at the point $x = c/n > 0$, and there is no n_0 for $\varepsilon = 1/2$ such that $|\operatorname{arc\,tg} nx - (\pi/2)| < \varepsilon$ holds for every $x \in (0, \infty)$ and $n \geq n_0$.

Remark 7.7. Obviously, if the sequence of functions (f_n) is uniformly convergent on the set H, then (f_n) is also uniformly convergent on every subset of H.

On the other hand, Examples 7.6 show that a sequence of functions can be uniformly convergent on every closed subinterval of an interval I, without being uniformly convergent on I itself.

Weierstrass' approximation theorem states that for every continuous function $f\colon B \to \mathbb{R}$ defined on a box $B \subset \mathbb{R}^p$ and for every $\varepsilon > 0$ there exists a polynomial p such that $|f(x) - p(x)| < \varepsilon$ for every $x \in B$. (See Theorem 1.54 and Exercises 1.59–1.63. See also Remark 7.85, where we give another proof of the one dimensional case.) We can reformulate this theorem as follows.

Theorem 7.8. *Let $B \subset \mathbb{R}^p$ be a box. Then, for every continuous function $f\colon B \to \mathbb{R}$ there exists a sequence of polynomials (p_n) such that (p_n) converges uniformly to f on B.*

Proof. By Weierstrass' approximation theorem, for every n there exists a polynomial p_n such that $|f(x) - p_n(x)| < 1/n$ for every $x \in B$. Obviously, the sequence of polynomials (p_n) converges uniformly to f on B. □

The following theorem gives the precise condition for the uniform convergence – without even knowing the limit function.

Theorem 7.9. (Cauchy criterion) *The sequence of functions (f_n) is uniformly convergent on the set H if and only if for every $\varepsilon > 0$ there exists an N such that*

$$|f_n(x) - f_m(x)| < \varepsilon \tag{7.3}$$

for every $x \in H$ and $n, m \geq N$.

Proof. Let $f_n \to f$ be uniformly convergent on the set H. For $\varepsilon > 0$ fixed, let N be an index such that $|f_n(x) - f(x)| < \varepsilon/2$ for every $x \in H$ and $n \geq N$. Obviously, (7.3) holds for every $x \in H$ and $n, m \geq N$.

Now let (f_n) satisfy the condition of the theorem. Then for every $x \in H$ fixed, $(f_n(x))$ is a Cauchy sequence of real numbers, thus it is convergent. Let $f(x) = \lim_{n\to\infty} f_n(x)$ for every $x \in H$.

In this way we defined the function $f\colon H \to \mathbb{R}$. We now prove that $f_n \to f$ uniformly on the set H. Let $\varepsilon > 0$ be fixed, and let N be an index such that (7.3) holds for every $x \in H$ and $n, m \geq N$. For $n \geq N$ fixed, (7.3) implies

$$|f_n(x) - f(x)| = \lim_{m\to\infty} |f_n(x) - f_m(x)| \leq \varepsilon$$

for every $x \in H$, proving our statement. □

Theorem 7.10. *Let the sequence of functions (f_n) converge uniformly to the function f on the set $H \subset \mathbb{R}^p$, and let α be a limit point of H. (For $p = 1$, i.e., when $H \subset \mathbb{R}$, we allow the cases $\alpha = \infty$ and $\alpha = -\infty$ as well.) If $\lim_{x\to\alpha,\, x\in H} f_n(x) = b_n$ exists and it is finite for every n, then the finite limit $\lim_{x\to\alpha,\, x\in H} f(x) = b$ also exists and $\lim_{n\to\infty} b_n = b$.*

Proof. Let $\varepsilon > 0$ be fixed, and choose the index N such that $|f_n(x) - f_m(x)| < \varepsilon$ holds for every $x \in H$ and $n, m \geq N$. Then

$$|b_n - b_m| = \lim_{\substack{x\to\alpha \\ x\in H}} |f_n(x) - f_m(x)| \leq \varepsilon$$

for every $n, m \geq N$. This shows that the sequence (b_n) satisfies the Cauchy condition, and hence it is convergent. Let $b = \lim_{n\to\infty} b_n$.

Let $\varepsilon > 0$ be given, and choose an index n such that $|f_n(x) - f(x)| < \varepsilon$ for every $x \in H$ and, furthermore, $|b_n - b| < \varepsilon$ also holds. (Every large enough n works.) Since

$$\lim_{\substack{x\to\alpha\\ x\in H}} f_n(x) = b_n,$$

α has a punctured neighborhood $\dot{U}$ such that $|f_n(x) - b_n| < \varepsilon$ for every $x \in H \cap \dot{U}$. Then

$$|f(x) - b| \le |f(x) - f_n(x)| + |f_n(x) - b_n| + |b_n - b| < 3\varepsilon$$

for every $x \in H \cap \dot{U}$, which proves $\lim_{x\to\alpha,\, x\in H} f(x) = b$. □

Remarks 7.11. **1.** The previous theorem can be formalized as follows. When we have uniform convergence, then

$$\lim_{n\to\infty} \lim_{\substack{x\to\alpha\\ x\in H}} f_n(x) = \lim_{\substack{x\to\alpha\\ x\in H}} \lim_{n\to\infty} f_n(x),$$

i.e., the limits $x \to \alpha$ and $n \to \infty$ are interchangeable.

2. Examples 7.2 show that the statement of the theorem is not necessarily true when we assume pointwise convergence only. For example,

$$\lim_{n\to\infty} \lim_{x\to 1-0} x^n = 1 \neq 0 = \lim_{x\to 1-0} \lim_{n\to\infty} x^n,$$

and

$$\lim_{n\to\infty} \lim_{x\to 0+0} \operatorname{arc\,tg}(nx) = 0 \neq \pi/2 = \lim_{x\to 0+0} \lim_{n\to\infty} \operatorname{arc\,tg}(nx).$$

Theorem 7.12. *Let the sequence of functions (f_n) converge uniformly to the function f on the set $H \subset \mathbb{R}^p$. If each f_n is continuous at a point $a \in H$ restricted to H, then f is also continuous at a restricted to H.*

Proof. If a is an isolated point of H, then there is nothing to prove, since every function is continuous at a restricted to H (see Remark 1.43). On the other hand, when a is a limit point of H, then we have $f_n(a) = \lim_{x\to a,\, x\in H} f_n(x)$ for every n. By Theorem 7.10, this gives

$$\lim_{\substack{x\to a\\ x\in H}} f(x) = \lim_{n\to\infty} f_n(a) = f(a),$$

and thus f is continuous at a. □

Remark 7.13. Our previous theorem states that the uniform limit of continuous functions is also continuous. In general, uniform convergence of a sequence of functions is not necessary for the continuity of the limit function to hold. Consider the functions $f_n(x) = x^n - x^{2n}$. It is easy to see that the sequence of functions (f_n)

is pointwise convergent to the constant zero function on the interval $[0,1]$. Thus the limit function is continuous. However, the convergence is not uniform, since

$$f_n\left(\sqrt[n]{1/2}\right)=\frac{1}{2}-\frac{1}{4}=\frac{1}{4}$$

for every n, and the condition of uniform convergence does not hold for $\varepsilon<1/4$.

The following theorem shows that, in some special cases, the continuity of the limit function implies uniform convergence.

Theorem 7.14. (Dini's[1] theorem) *Let the functions f_n be continuous on a bounded and closed set $K\subset\mathbb{R}^p$, and suppose that $f_1(x)\le f_2(x)\le\ldots$ for every $x\in K$. If the sequence of functions (f_n) is pointwise convergent to a continuous function on K, then the convergence is uniform.*

Proof. Let the sequence of functions (f_n) be pointwise convergent to the function f, and let f be continuous on K. We may assume that f is the constant zero function on K, since otherwise we could switch to the sequence of functions (f_n-f).

Obviously, $f_n(x)\le 0$ for every $x\in K$ and every n. Suppose that the convergence is not uniform. Then there exists an $\varepsilon>0$ such that, for infinitely many n, there is a point $x_n\in K$ with $f_n(x_n)<-\varepsilon$. We may assume that there is such an x_n for every n, because we can remove the functions f_n from the sequence for which such an x_n does not exist. Since K is bounded, so is (x_n), and thus (x_n) has a convergent subsequence by the Bolzano-Weierstrass theorem. We may assume that the sequence (x_n) is convergent itself, since may remove those functions f_n for which n does not belong to the subsequence (n_k).

Let $\lim_{n\to\infty}x_n=c$. Then, as K is closed, we have $c\in K$ and, by assumption, the functions f_n are continuous at c. Since $f_n(c)\to 0$, there exists an index N such that $|f_N(c)|<\varepsilon$. As the function f_N is continuous at c, there is a neighborhood U of c such that $|f_N(x)|<\varepsilon$ for every $x\in U\cap K$. Since $x_n\to c$, we have $x_n\in U$ for every n large enough. Choose n such that $n\ge N$ and $x_n\in U$. Then $-\varepsilon<f_N(x_n)\le f_n(x_n)$. On the other hand, $f_n(x_n)<-\varepsilon$ by the choice of x_n. We reached a contradiction, which proves the theorem. □

Now we consider the possibility of interchanging the limit operation and integration.

Examples 7.15. **1.** Let $f_n(x)=(n+1)\cdot x^n$ for $0\le x<1$, and $f_n(1)=0$ for every n. We have $f_n(x)\to 0$ for every $x\in[0,1]$, i.e., the sequence of functions (f_n) is pointwise convergent to the constant zero function on $[0,1]$. However, $\int_0^1 f_n(x)\,dx=1$ for every n, thus

$$\lim_{n\to\infty}\int_0^1 f_n(x)\,dx=1\neq 0=\int_0^1\left(\lim_{n\to\infty}f_n(x)\right)\,dx.$$

[1] Ulisse Dini (1845–1918), Italian mathematician.

2. Let $f_n(0) = f_n(1/n) = f(1) = 0$, $f_n(1/(2n)) = n$, and let f_n be linear on each of the intervals $[0, 1/(2n)]$, $[1/(2n), 1/n]$, $[1/n, 1]$ for every $n = 1, 2, \ldots$. We have $f_n(x) \to 0$ for every $x \in [0, 1]$. Indeed, this is obvious for $x = 0$. For $0 < x \le 1$, however, we have $f_n(x) = 0$ for every $n > 2/x$. On the other hand, $\int_0^1 f_n(x)\,dx = 1/2$ for every n, and thus

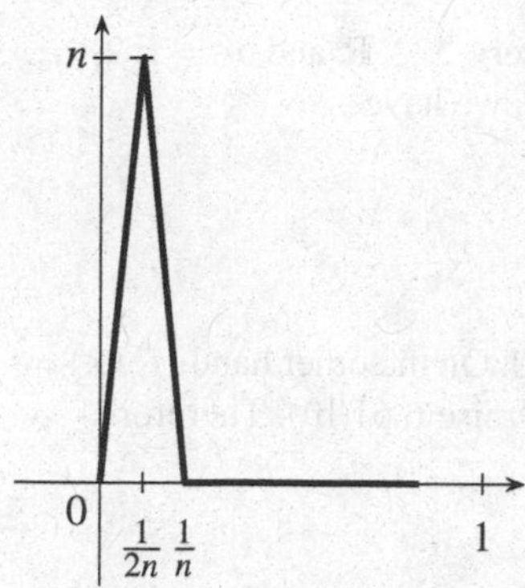

7.1. Figure

$$\lim_{n\to\infty} \int_0^1 f_n(x)\,dx = \frac{1}{2} \ne 0 = \int_0^1 \left(\lim_{n\to\infty} f_n(x)\right) dx.$$

By the following theorem, uniform convergence is a remedy for this problem as well.

Theorem 7.16. *Let the sequence of functions (f_n) be uniformly convergent to the function f on the nonempty and Jordan measurable set $A \subset \mathbb{R}^p$. If f_n is integrable on A for every n, then f is also integrable on A, and*

$$\int_A f(x)\,dx = \lim_{n\to\infty} \int_A f_n(x)\,dx. \tag{7.4}$$

Proof. By uniform convergence, there exists an index $N(\varepsilon)$ for every $\varepsilon > 0$ such that $|f_n(x) - f(x)| < \varepsilon$ for every $x \in A$ and $n \ge N(\varepsilon)$. Since the functions f_n are bounded, f is also bounded. Let $\varepsilon > 0$ and $n \ge N(\varepsilon)$ be fixed. Let $\omega(g; B)$ denote the oscillation of the function g on the nonempty set B, and let $\Omega_F(g)$ denote the oscillatory sum of the function g corresponding to the partition F. Since f_n is integrable on A, there is a partition $F = \{A_1, \ldots, A_n\}$ such that $\Omega_F(f_n) < \varepsilon$. Since $|f(x) - f_n(x)| < \varepsilon$ for every $x \in A$, we have $\omega(f; A_i) \le \omega(f_n; A_i) + 2\varepsilon$ for every i. Thus we have

$$\Omega_F(f) \le \Omega_F(f_n) + 2\varepsilon\mu(A) < \varepsilon \cdot (1 + 2\mu(A)).$$

Because this is true for every $\varepsilon > 0$, it follows that f is integrable on A.

Let $\int_A f_n(x)\,dx = I_n$ and $\int_A f(x)\,dx = I$. If $n \ge N(\varepsilon)$, then

$$|I_n - I| = \left| \int_A (f_n - f)\,dx \right| \le \int_A |f_n - f|\,dx \le \varepsilon \cdot \mu(A).$$

This is true for every $\varepsilon > 0$ and for every $n \ge N(\varepsilon)$, and thus we have $I_n \to I$. $\square$

Finally, we consider the possibility of interchanging the limit operation and differentiation.

Examples 7.17. **1.** Let $f_n(x) = \big(\sin(nx)\big)/n$ for every $x \in \mathbb{R}$ and $n = 1, 2, \ldots$. Then $f_n \to 0$ uniformly on $\mathbb{R}$. Indeed, for $\varepsilon > 0$ fixed, we have

$$\left|\frac{\sin(nx)}{n}\right| \leq \frac{1}{n} < \varepsilon$$

for every $x \in \mathbb{R}$ and $n > 1/\varepsilon$. Thus $(\lim_{n\to\infty} f_n)' \equiv 0$. On the other hand, $f_n'(x) = \cos(nx)$ does not converges to zero for any x (see Exercise 6.61(b)). Therefore,

$$\lim_{n\to\infty} f_n' \neq \left(\lim_{n\to\infty} f_n\right)'.$$

Moreover, for $x = 0$ we have $\lim_{n\to\infty} f_n'(x) = 1$, while $(\lim_{n\to\infty} f_n)'(x) = 0$.

2. By theorem 7.8, there exists a sequence of polynomials that converges uniformly to the function $f(x) = |x|$ on the interval $[-1, 1]$. (This can be proved directly; see Exercise 7.3.) This example shows that the limit of a sequence of differentiable functions is not necessarily differentiable. Moreover, since there are continuous and nowhere differentiable functions (see Theorem 7.38), Theorem 7.8 implies the existence of a uniformly convergent sequence of differentiable functions (and even polynomials) whose limit is nowhere differentiable.

As we saw above, taking the derivative and taking the limit are not interchangeable operations, not even for uniformly convergent sequences. We now prove that assuming the uniform convergence of *the sequence of the derivatives* implies the interchangeability of these operations.

Theorem 7.18. *Let the functions f_n be continuously differentiable on a bounded interval I, and suppose that*

(i) *the sequence of functions (f_n') is uniformly convergent to the function g on I, and*

(ii) *there exists at least one point $x_0 \in I$ such that the sequence $(f_n(x_0))$ is convergent.*

Then the sequence of functions (f_n) converges uniformly to a function f on I, the function f is differentiable, and $f'(x) = g(x)$ for every $x \in I$.

Proof. By Theorem 7.12, g is continuous on I. Let $f(x) = \int_{x_0}^{x} g(t)\,dt$ for every $x \in I$. The uniform convergence of the sequence of functions (f_n') implies that for every $\varepsilon > 0$ there exists an index N such that $|f_n'(x) - g(x)| < \varepsilon$ for every $x \in I$ and $n \geq N$. Let $|I|$ denote the length of the interval I. Then, for every $x \in I$ and $n \geq N$ we have

$$|f_n(x) - f_n(x_0) - f(x)| = \left| \int_{x_0}^{x} (f_n'(t) - g(t))\, dt \right| \le \varepsilon \cdot |x - x_0| \le \varepsilon \cdot |I|.$$

Therefore, we have $f_n \to f + b$ uniformly on I, where $b = \lim_{n\to\infty} f_n(x_0)$. Since g is continuous and f is the integral function of g, we have $f' = g$ (see [7, part (iii) of Theorem 15.5]), and the theorem is proved. □

Remark 7.19. The conditions of the theorem above can be relaxed: instead of the continuous differentiability of the functions f_n it is enough to assume that f_n is differentiable for every n (see Exercise 7.17).

Exercises

7.1. Show that the sequence of functions $(\cos nx)$ is only convergent at the points $x = 2k\pi$ $(k \in \mathbb{Z})$, and that the sequence of functions $(\sin nx)$ is only convergent at the points $x = k\pi$ $(k \in \mathbb{Z})$.

7.2. Find the points of convergence of the following sequences of functions. Find the intervals where these sequences of functions are uniformly convergent.

(a) $\sqrt[n]{|x|}$, (b) $x^n/n!$, (c) $x^n - x^{n+1}$,
(d) $x^n/(1 + x^{2n})$, (e) $\left(1 + \frac{x}{n}\right)^n$, (f) $\sqrt[n]{1 + x^{2n}}$,
(g) $\sqrt{x^2 + n^{-2}}$.

7.3. Let $p_0 \equiv 0$ and let $p_{n+1}(x) = p_n(x) + (x^2 - p_n^2(x))/2$ for every $x \in \mathbb{R}$ and $n = 0, 1, \ldots$. Show that $0 \le |x| - p_n(x) \le 2/(n+1)$ for every n and $x \in [-1, 1]$. (H)

7.4. Let $f_n\colon [a, b] \to \mathbb{R}$ be continuous for every $n = 1, 2, \ldots$. Show that if (f_n) is uniformly convergent on (a, b), then it is uniformly convergent on $[a, b]$.

7.5. Let $f_n\colon \mathbb{R} \to \mathbb{R}$ be continuous for every $n = 1, 2, \ldots$. Show that if (f_n) is uniformly convergent on $\mathbb{Q}$, then it is uniformly convergent on $\mathbb{R}$.

7.6. Construct a pointwise convergent sequence of continuous functions $f_n\colon [0, 1] \to \mathbb{R}$ such that the limit function has a point of discontinuity at every rational point of $[0, 1]$.

7.7. Construct a sequence of continuous functions $f_n\colon [0, 1] \to \mathbb{R}$ such that $f_n \to 0$ pointwise on $[0, 1]$, but no subsequence of (f_n) converges uniformly on any subinterval of $[0, 1]$. (∗ H)

7.8. Show that if the sequence of functions $f_n\colon H \to \mathbb{R}$ is uniformly convergent on every countable subset of H, then (f_n) is also uniformly convergent on H.

7.9. Let $f_n \to 0$ uniformly on H. Show that the sequence of functions $g_n = \max(f_1, \ldots, f_n)$ converges uniformly on H.

7.10. Find an example of a sequence of continuous functions $f_n \colon [0,1] \to [0,1]$ such that (f_n) does not have a pointwise convergent subsequence (on any interval). (H)

7.11. Let $f_n \colon [a,b] \to \mathbb{R}$ be monotone for every $n = 1, 2 \ldots$. Show that if the sequences $(f_n(a))$ and $(f_n(b))$ are bounded, then (f_n) has a pointwise convergent subsequence. (H)

7.12. Let $f_n \colon [a,b] \to \mathbb{R}$ be monotone for every n. Show that if (f_n) is pointwise convergent to a continuous function on $[a,b]$, then (f_n) is uniformly convergent on $[a,b]$. (H)

7.13. Let $f_n(x) = n^2(x^{n-1} - x^n)$ $(x \in [0,1])$. Show that $f_n \to 0$ pointwise, but not uniformly on $[0,1]$. Check that $\int_0^1 f_n \, dx$ does not converge to zero.

7.14. Let $f \colon [0,1] \to \mathbb{R}$ be continuous. Show that the sequence of functions

$$f_n(x) = \frac{1}{n} \cdot \left(f\left(\frac{x}{n}\right) + f\left(\frac{x+1}{n}\right) + \ldots + f\left(\frac{x+n-1}{n}\right) \right)$$

is uniformly convergent on $[0,1]$. What is the limit of this sequence of functions?

7.15. Does there exist a sequence of continuously differentiable functions $f_n \colon \mathbb{R} \to \mathbb{R}$ such that $f_n(x) \to x$ and $f_n'(x) \to 0$ for every x? (∗)

7.16. Let $f_n \colon \mathbb{R} \to \mathbb{R}$ be continuously differentiable functions, and let $f_n \to f$ pointwise, where f is an everywhere differentiable function. Show that there exists a point x and a sequence of indices $n_1 < n_2 < \ldots$ such that $\lim_{k\to\infty} f_{n_k}'(x) = f'(x)$. (H)

7.17. Let the functions f_n be differentiable on a bounded interval I. Suppose that the sequence of functions (f_n') converges to the function g uniformly on I, and there is a point $x_0 \in I$ such that the sequence $(f_n(x_0))$ is convergent. Show that the sequence of functions (f_n) converges to a function f uniformly on I, where f is differentiable, and $f'(x) = g(x)$ for every $x \in I$. (S)

7.18. Show that if $H \subset \mathbb{R}$, $K > 0$, and the functions $f_n \colon H \to \mathbb{R}$ have the property $|f_n(x) - f_n(y)| \le K \cdot |x - y|$, for every $x, y \in H$ and $n = 1, 2 \ldots$, then the sequence of functions (f_n) is uniformly equicontinuous.

7.19. Let $f_n \colon [a,b] \to \mathbb{R}$ be continuous functions, and let the sequence of functions (f_n) be pointwise convergent to the function f on $[a,b]$. Show that the $f_n \to f$ convergence is uniform on $[a,b]$ if and only if the sequence of functions (f_n) is uniformly equicontinuous. (I.e., the conditions of Definitions 7.3 and 7.5 are the same in this special case.) (H)

7.2 The Convergence of Series of Functions

So far we considered sequences of functions defined on a set H. Now we turn to infinite series of functions defined on a set; we call them **series of functions.**

Definition 7.20. Let $f_1, f_2, \ldots$ be real valued functions defined on the set H. We say that *the series of functions* $\sum_{n=1}^{\infty} f_n$ *is pointwise convergent, and its sum is the function* $f: H \to \mathbb{R}$, if the infinite series $\sum_{n=1}^{\infty} f_n(x)$ is convergent, and its sum is $f(x)$ for every $x \in H$. We use the notation $\sum_{n=1}^{\infty} f_n = f$.

Obviously, $\sum_{n=1}^{\infty} f_n = f$ if and only if the sequence of functions $s_n = \sum_{i=1}^{n} f_i$ is pointwise convergent to the function f on H.

Definition 7.21. Let $\sum_{n=1}^{\infty} f_n = f$ on the set H. We say that the series of functions $\sum_{n=1}^{\infty} f_n$ *is uniformly convergent on* H, if the sequence of functions $s_n = \sum_{i=1}^{n} f_i$ is uniformly convergent to f on H.

Examples 7.22. **1.** We show that the power series $\sum_{n=0}^{\infty} x^n$ is uniformly convergent on the interval $[-a, a]$, for every $0 < a < 1$. We know that $\sum_{n=0}^{\infty} x^n = 1/(1-x)$ for every $x \in (-1, 1)$. Since $s_n(x) = \sum_{i=0}^{n-1} x^i = (1-x^n)/(1-x)$, we have

$$\left| s_n(x) - \frac{1}{1-x} \right| = \frac{|x|^n}{1-x} \le \frac{a^n}{1-a}$$

for every $|x| \le a$. Since $a^n \to 0$, it follows that for every $\varepsilon > 0$ there exists some n_0 such that $a^{n_0}/(1-a) < \varepsilon$. Therefore, $|s_n(x) - (1/(1-x))| < \varepsilon$ for every $|x| \le a$ and $n > n_0$.

2. The convergence of the power series $\sum_{n=0}^{\infty} x^n$ is *not* uniform on the interval $(-1, 1)$. Indeed, the function s_n is bounded on $(-1, 1)$ (because s_n is a polynomial). On the other hand, $1/(1-x)$ is not bounded on $(-1, 1)$, and thus, for any ε, there does not exists n such that $|s_n(x) - (1/(1-x))| < \varepsilon$ holds for every $x \in (-1, 1)$.

We know that every reordering of an infinite series is convergent if and only if the series is absolutely convergent, and its sum is the same as the sum of the original series. (See 6.28.) It follows that every reordering of a series of functions $\sum_{n=1}^{\infty} f_n$ is pointwise convergent on the set H if and only if the series of functions $\sum_{n=1}^{\infty} |f_n|$ is pointwise convergent on H.

Definition 7.23. We say that the series of functions $\sum_{n=1}^{\infty} f_n$ is *absolutely convergent* on the set H if $\sum_{n=1}^{\infty} |f_n|$ is pointwise convergent on H.

Remark 7.24. The absolute and uniform convergence of series of functions are two independent properties, i.e., neither follows from the other. E.g., the power series $\sum_{n=0}^{\infty} x^n$ is absolutely convergent, but it is not uniformly convergent on the interval $(-1, 1)$ (see Example 7.22.2.). On the other hand, it is easy to see, that if $\sum_{n=0}^{\infty} a_n$ is a convergent infinite series, where f_n is the constant a_n function for every n on a non-empty set H, then the series of functions $\sum_{n=1}^{\infty} f_n$ is uniformly convergent on H. Choosing $a_n = (-1)^{n-1}/n$ yields a uniformly convergent series of functions that is not absolutely convergent.

The exact condition of the uniform convergence of a series of functions is given by the following theorem.

Theorem 7.25. (Cauchy criterion) *The series of functions $\sum_{n=1}^{\infty} f_n$ is uniformly convergent on the set H if and only if, for every $\varepsilon > 0$, there exists an N such that*

$$\left|\sum_{i=n+1}^{m} f_i(x)\right| < \varepsilon$$

holds for every $x \in H$ and $N \leq n < m$.

Proof. This is immediate from Theorem 7.9. □

Theorem 7.25 implies that if $\sum_{n=1}^{\infty} |f_n|$ is uniformly convergent on the set H, then $\sum_{n=1}^{\infty} f_n$ is also uniformly convergent on H. The following theorem goes one step further.

Theorem 7.26. *If $\sum_{n=1}^{\infty} |f_n|$ is uniformly convergent on the set H, then every reordering of the series of functions $\sum_{n=1}^{\infty} f_n$ is also uniformly convergent on H.*

Proof. Let $\varepsilon > 0$ be fixed. By the Cauchy criterion, there exists an N such that $\sum_{i=N+1}^{m} |f_i(x)| < \varepsilon$ for every $x \in H$ and $N < m$.

Let $\sum_{k=1}^{\infty} f_{n_k}$ be a reordering of our original series of functions. For K large enough, the indices $n_1, \ldots, n_K$ contain each of the numbers $1, \ldots, N$, and thus $n_k > N$ for every $k \geq K$. If $K \leq p < q$ and $m = \max_{p<k\leq q} n_k$, then $N < n_k \leq m$ for every $p < k \leq q$. Thus

$$\left|\sum_{k=p+1}^{q} f_{n_k}(x)\right| \leq \sum_{k=p+1}^{q} |f_{n_k}(x)| \leq \sum_{i=N+1}^{m} |f_i(x)| < \varepsilon$$

for every $x \in H$. Hence, the series of functions $\sum_{n=1}^{\infty} f_{n_k}$ is uniformly convergent on H by the Cauchy criterion. □

The converse of this theorem is also true: if every reordering of a series of functions $\sum_{n=1}^{\infty} f_n$ is uniformly convergent, then $\sum_{n=1}^{\infty} |f_n|$ is also uniformly convergent (see Exercise 7.24). Using this statement it is not difficult to show that *there exists a series of functions $\sum_{n=1}^{\infty} f_n$ such that $\sum_{n=1}^{\infty} f_n$ is absolutely and uniformly convergent on H, but it has a non-uniformly convergent reordering.* E.g., the series of functions $\sum_{n=1}^{\infty} (-x)^n (1-x)$ on the interval $[0, 1]$ has this property (see Exercise 7.25).

Since the conditions of the Cauchy criterion are (generally) hard to verify, it is important to have some sufficient criteria for the uniform convergence that are easy to check. The following theorem is the most important of these.

Theorem 7.27. (Weierstrass criterion) *Suppose we have real numbers a_n and an index n_0 such that the infinite series $\sum_{n=1}^{\infty} a_n$ is convergent, and $|f_n(x)| \le a_n$ for every $x \in H$ and $n \ge n_0$. Then the series of functions $\sum_{n=1}^{\infty} f_n$ is uniformly convergent on the set H.*

Proof. Let $\varepsilon > 0$ be fixed. By the Cauchy criterion for infinite series (Theorem 6.10) there exists an index N such that

$$|a_{n+1} + a_{n+2} + \ldots + a_m| < \varepsilon$$

for every $N \le n < m$. Obviously, for every $x \in H$ and $n \ge \max(n_0, N)$, we have

$$|f_{n+1}(x) + f_{n+2}(x) + \ldots + f_m(x)| < \varepsilon,$$

and thus the uniform convergence of the series of functions $\sum_{n=1}^{\infty} f_n$ follows from Theorem 7.25. □

In Example 7.22.1. we saw that the power series $\sum_{n=0}^{\infty} x^n$ is uniformly convergent of on the interval $[-a, a]$, for every $0 < a < 1$. We now prove a similar statement for every power series with the help of the Weierstrass criterion.

Theorem 7.28. *If $x_0 \neq 0$ and the infinite series $\sum_{n=0}^{\infty} a_n x_0^n$ is convergent, then the power series $\sum_{n=0}^{\infty} a_n x^n$ is absolutely and uniformly convergent on the interval $[-q|x_0|, q|x_0|]$, for every $0 < q < 1$.*

Proof. Since $\sum_{n=0}^{\infty} a_n x_0^n$ is convergent, the sequence $(a_n x_0^n)$ converges to zero, and thus there exists an n_0 such that $|a_n x_0^n| \le 1$ for every $n \ge n_0$. If $x \in [-q|x_0|, q|x_0|]$, then

$$|a_n x^n| = |a_n x_0^n| \cdot \left| \frac{x}{x_0} \right|^n \le 1 \cdot q^n$$

for every $n \ge n_0$. Since the series $\sum_{n=0}^{\infty} q^n$ is convergent, it follows from the Weierstrass criterion that $\sum_{n=0}^{\infty} |a_n x^n|$ is uniformly convergent on $[-q|x_0|, q|x_0|]$. □

We say that the sequence of functions (f_n) is **uniformly bounded on** H, if there exists a number K such that $|f_n(x)| \le K$ for every n and $x \in H$. The following two criteria can also be used in many cases.

Theorem 7.29. *Let f_n and g_n $(n = 1, 2, \ldots)$ be real valued functions on the set H, and suppose that*

(i) *the sequence* $(f_n(x))$ *is monotone decreasing for every* $x \in H$,

(ii) $f_n \to 0$ *uniformly on* H, *and*

(iii) *the partial sums of the series* $\sum_{n=1}^{\infty} g_n$ *are uniformly bounded on* H.

Then the series of functions $\sum_{n=1}^{\infty} f_n g_n$ *is uniformly convergent on* H.

Proof. Let the nth partial sum of the series $\sum_{n=1}^{\infty} g_n$ be s_n, and let $|s_n(x)| \le K$ for every n and every $x \in H$ with $K > 0$. Let $\varepsilon > 0$ be fixed. By condition (ii), we can pick an index N such that $|f_n(x)| < \varepsilon/K$ for every $n \ge N$ and $x \in H$. If $N \le n < m$ then, by the Abel inequality (see p. 214), we have

$$-\varepsilon < f_n(x) \cdot (-K) \le f_n(x)g_n(x) + \ldots + f_m(x)g_m(x) \le f_n(x) \cdot K < \varepsilon,$$

that is, $|f_n(x)g_n(x) + \ldots + f_m(x)g_m(x)| < \varepsilon$ for every $x \in H$. Therefore, the series $\sum_{n=1}^{\infty} f_n g_n$ satisfies the conditions of the Cauchy criterion, hence it is uniformly convergent. □

Corollary 7.30. (Dirichlet criterion) *Let* (λ_n) *be a monotone decreasing sequence of real numbers that converges to zero. If the series of the partial sums of the series of functions* $\sum_{n=1}^{\infty} g_n$ *is uniformly bounded on* H, *then* $\sum_{n=1}^{\infty} \lambda_n g_n$ *is uniformly convergent on* H. □

Example 7.31. For $c > 0$, the series of functions $\sum_{n=1}^{\infty} (\sin nx)/n^c$ and $\sum_{n=1}^{\infty} (\cos nx)/n^c$ are uniformly convergent on the interval $[\delta, 2\pi - \delta]$, for every $0<\delta<\pi$.

Indeed, on one hand, the sequence (n^{-c}) is monotone decreasing and converges to zero, on the other hand, by Lemma 6.64, the partial sums of the series $\sum_{n=1}^{\infty} \sin nx$ and $\sum_{n=1}^{\infty} \cos nx$ are uniformly bounded on $[\delta, 2\pi - \delta]$.

Theorem 7.32. *Let* f_n *and* g_n $(n = 1, 2, \ldots)$ *be real valued functions on the set* H, *and suppose that*

(i) *the sequence of functions* (f_n) *is uniformly bounded on* H,

(ii) *the sequence* $(f_n(x))$ *is monotone for every* $x \in H$, *and*

(iii) *the series of functions* $\sum_{n=1}^{\infty} g_n$ *is uniformly convergent on* H.

Then the series of functions $\sum_{n=1}^{\infty} f_n g_n$ *is also uniformly convergent on* H.

Proof. We use the following variant of Abel's inequality: if $c_1, \ldots, c_n$ is a monotone sequence of real numbers and $|d_1 + \ldots + d_k| \le M$ for every $k = 1, \ldots, n$, then

$$\left| \sum_{i=1}^{n} c_i d_i \right| \le (|c_1| + 2|c_n|) \cdot M.$$

In order to prove this we may assume that $c_1 \ge \ldots \ge c_n$, since otherwise we could switch to the numbers $-c_i$. We have $c_1 - c_n \ge \ldots \ge c_{n-1} - c_n \ge 0$. Thus, by Abel's inequality,

$$\left|\sum_{i=1}^{n}(c_i - c_n)d_i\right| \le (c_1 - c_n)\cdot M,$$

and

$$\left|\sum_{i=1}^{n} c_i d_i\right| \le \left|\sum_{i=1}^{n}(c_i - c_n)d_i\right| + |c_n|\cdot\left|\sum_{i=1}^{n} d_i\right| \le$$
$$\le (c_1 - c_n + |c_n|)\cdot M \le (|c_1| + 2|c_n|)\cdot M.$$

Now we turn to the proof of our theorem. Let $|f_n(x)| < K$ for every n, and let $\varepsilon > 0$ be fixed. The uniform convergence of the series of functions $\sum_{n=1}^{\infty} g_n$ implies the existence of an N such that $|\sum_{i=n}^{m} g_i(x)| < \varepsilon/(3K)$ for every $N \le n < m$ and $x \in H$. Since the sequence $(f_i(x))$ is monotone, the inequality above implies

$$\left|\sum_{i=n}^{m} f_i(x)g_i(x)\right| \le (|f_n(x)| + 2|f_m(x)|)\cdot(\varepsilon/(3K)) \le \varepsilon$$

for every $N \le n < m$ and $x \in H$. Applying the Cauchy criterion yields the uniform convergence of $\sum_{n=1}^{\infty} f_n g_n$ on H. □

Corollary 7.33. (Abel's criterion) *Suppose that the sequence of functions (f_n) is uniformly bounded on H, and the sequence $(f_n(x))$ is monotone for every $x \in H$. If the infinite series $\sum_{n=1}^{\infty} \mu_n$ is convergent, then the series of functions $\sum_{n=1}^{\infty} \mu_n f_n$ is uniformly convergent on H.* □

The following application of Abel's criterion is a supplement to Theorem 7.28.

Theorem 7.34. *Let $x_0 \ne 0$, and let the infinite series $\sum_{n=0}^{\infty} a_n x_0^n$ be convergent. Then the power series $\sum_{n=0}^{\infty} a_n x^n$ is uniformly convergent on the interval $[0, x_0]$.*

Proof. Apply Abel's criterion with $\mu_n = a_n x_0^n$ and $f_n(x) = (x/x_0)^n$. □

Similarly to the case of the sequences of functions, we need to know whether certain properties of the terms of a series of functions can be transmitted to the sum. We know that the properties of continuity, integrability and differentiability are inherited by finite sums. As for infinite sums, Theorems 7.10 and 7.12 imply the following.

Theorem 7.35. *Suppose that $\sum_{n=1}^{\infty} f_n = f$ uniformly on the set $H \subset \mathbb{R}^p$, and let α be a limit point of the set H. (For $H \subset \mathbb{R}$, we also allow $\alpha = \infty$ and $\alpha = -\infty$.) If the limit $\lim_{x\to\alpha,\, x\in H} f_n(x) = b_n$ exists and it is finite for every n, then the infinite series $\sum_{n=1}^{\infty} b_n$ is convergent, and $\lim_{x\to\alpha,\, x\in H} f(x) = \sum_{n=1}^{\infty} b_n$.* □

Theorem 7.36. *Suppose that $\sum_{n=1}^{\infty} f_n = f$ uniformly on the set $H \subset \mathbb{R}^p$. If the functions f_n are continuous at a point $a \in H$ restricted to H, then f is also continuous at a restricted to H.* □

Examples 7.37. **1.** Consider the series $\sum_{k=0}^{\infty} b^k \cos(a^k x)$ and $\sum_{k=0}^{\infty} b^k \sin(a^k x)$, where $|b| < 1$ and $a \in \mathbb{R}$ is arbitrary. Both series of functions are uniformly convergent on $\mathbb{R}$, since they satisfy the conditions of Weierstrass' criterion. Thus, by Theorem 7.36, both sums are continuous.

2. Let $\langle x \rangle$ denote the distance of the real number x to the closest integer. The function $\langle x \rangle$ is continuous and periodic with the period 1 on the real line. The function

$$T(x) = \sum_{n=0}^{\infty} \frac{\langle 2^n x \rangle}{2^n}$$

is called the **Takagi function**[2] (named after its first "inventor"). (Figure 7.2 shows the restriction of the Takagi function to $[0,1]$. Note that the figure is only an approximation. If fact, the function has infinitely many local extrema in every interval.) Since the series of functions above also satisfies the conditions of the Weierstrass criterion and its terms are continuous, it follows by Theorem 7.36, that the Takagi function is continuous on the real line.

We hinted at the existence of everywhere continuous but nowhere differentiable functions several times so far. Example 7.37.2. presents one such function.

Theorem 7.38. *The Takagi function is continuous everywhere, but it is nowhere differentiable.*

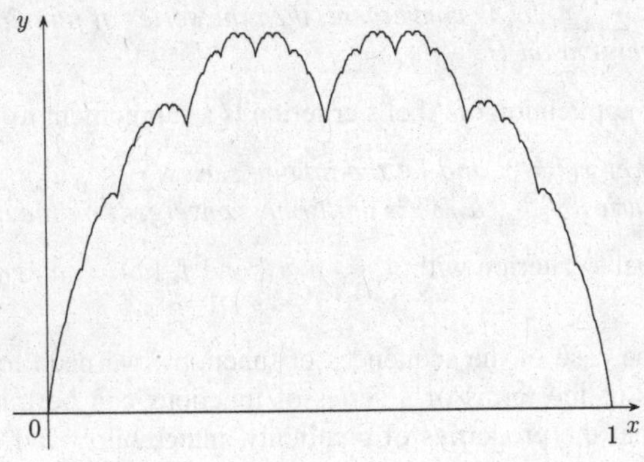

7.2. Figure

Proof. We only have to prove that T is not differentiable at any point $a \in \mathbb{R}$. Let $a \in \mathbb{R}$ be arbitrary, and let x_i denote the largest number $k/2^i$ such that $k \in \mathbb{Z}$ and $k/2^i \le a$. Also, let y_i denote the smallest number $k/2^i$ such that $k \in \mathbb{Z}$ and $k/2^i > a$. Then we have $x_i \le a < y_i$ and $y_i - x_i = 1/2^i$ for every i. Suppose that T is differentiable at a. Then

---

[2] Takagi Teiji (1875–1960), Japanese mathematician.

$$\lim_{i\to\infty} \frac{T(y_i) - T(x_i)}{y_i - x_i} = T'(a). \tag{7.5}$$

Indeed if $x_i < a$, for every i, then this follows from Exercise 7.33. On the other hand, if $x_i = a$ for some i, then $x_n = a$ for every $n \geq i$, and (7.5) follows from the definition of differentiability. Now we have

$$\frac{T(y_i) - T(x_i)}{y_i - x_i} = \sum_{n=0}^{\infty} \frac{\langle 2^n y_i \rangle - \langle 2^n x_i \rangle}{2^n (y_i - x_i)} = \sum_{n=0}^{i-1} \frac{\langle 2^n y_i \rangle - \langle 2^n x_i \rangle}{2^{n-i}} \tag{7.6}$$

for every i, since $2^n y_i$ and $2^n x_i$ are integers for $n \geq i$, and $\langle 2^n y_i \rangle = \langle 2^n x_i \rangle = 0$. If $n < i$, then $2^n y_i$ and $2^n x_i$ are rational numbers with denominator 2^{i-n} such that their numerators are adjacent integers. Thus $\langle 2^n y_i \rangle - \langle 2^n x_i \rangle = \pm 1/2^{i-n}$, and each of the terms of the sum on the right-hand side of (7.6) is either 1 or -1. It follows that $(T(y_i) - T(x_i))/(y_i - x_i)$ is an integer. Furthermore, this integer is even if i is even, and odd if i is odd. A sequence with these properties cannot be convergent, which proves that (7.5) cannot hold. □

The combination of Theorems 7.36 and 7.34 imply the following.

Theorem 7.39. (Abel's continuity theorem) *Let $x_0 \neq 0$, and let the infinite series $\sum_{n=0}^{\infty} a_n x_0^n$ be convergent. Then the sum of the power series $\sum_{n=0}^{\infty} a_n x^n$ is continuous on the interval $[0, x_0]$.* □

The following theorem is an immediate corollary of Theorem 7.16.

Theorem 7.40. (Term by term integrability) *Suppose that $\sum_{n=1}^{\infty} f_n = f$ uniformly on the Jordan measurable set $A \subset \mathbb{R}^p$. If f_n is integrable on A for every n, then f is also integrable on A, and*

$$\int_A f(x)\, dx = \sum_{n=1}^{\infty} \int_A f_n(x)\, dx.$$

□

Examples 7.41. **1.** We know that $\sum_{n=0}^{\infty} x^n = 1/(1-x)$ for every $|x| < 1$, and the convergence is uniform on every closed subinterval of $(-1, 1)$ (see Example 7.22.1. and Theorem 7.28). Thus we can integrate the series term by term on the interval $[0, x]$ for every $|x| < 1$. We obtain

$$-\log(1-x) = \sum_{n=1}^{\infty} \frac{1}{n} \cdot x^n \tag{7.7}$$

for every $|x| < 1$. Since the series is also convergent at $x = -1$, Abel's continuity theorem (Theorem 7.39) implies that (7.7) holds for every $x \in [-1, 1)$. For $x = -1$ we get the well-known identity $\sum_{n=1}^{\infty} (-1)^{n-1}/n = \log 2$.

2. Since $\sum_{n=0}^{\infty}(-1)^n x^{2n} = 1/(1+x^2)$, for every $|x| < 1$ and the convergence is uniform on every closed subinterval of $(-1, 1)$ (by Theorem 7.28), we can integrate this series term by term on the interval $[0, x]$ for every $|x| < 1$. We get

$$\operatorname{arc\,tg} x = \sum_{n=1}^{\infty} \frac{(-1)^{n-1}}{2n-1} \cdot x^{2n-1} \tag{7.8}$$

for every $|x| < 1$. Since the series is also convergent at $x = \pm 1$, Abel's continuity theorem (Theorem 7.39) implies that (7.8) holds for every $|x| \le 1$. Putting $x = 1$ we get

$$\frac{\pi}{4} = 1 - \frac{1}{3} + \frac{1}{5} - \frac{1}{7} + \ldots .$$

The following theorem is an immediate corollary of Theorem 7.18.

Theorem 7.42. (Term by term differentiation) *Let the functions f_n be continuously differentiable on a bounded interval I. Suppose that*

(i) *$\sum_{n=1}^{\infty} f_n' = g$ uniformly on the interval I, and*

(ii) *there exist a point $x_0 \in I$ such that the infinite series $\sum_{n=1}^{\infty} f_n(x_0)$ is convergent.*

Then the series of functions $\sum_{n=1}^{\infty} f_n$ is uniformly convergent on I. If $\sum_{n=1}^{\infty} f_n = f$, then f is differentiable, and $f'(x) = g(x)$, for every $x \in I$. That is, we have

$$\left(\sum_{n=1}^{\infty} f_n\right)'(x) = \sum_{n=1}^{\infty} f_n'(x),$$

for every $x \in I$. □

Note that we can relax the condition of continuous differentiability of the functions f_n: it is enough to assume that the functions f_n are differentiable (see Remark 7.19).

Example 7.43. For $0 < b < 1$ and $|ab| < 1$, the functions $\sum_{k=0}^{\infty} b^k \cos(a^k x)$ and $\sum_{k=0}^{\infty} b^k \sin(a^k x)$ are differentiable everywhere. Indeed, the series we get from differentiating term by term are uniformly convergent, and we can apply Theorem 7.42.

One can show (though the proof is rather complicated) that, for every $0 < b < 1$ and $|ab| \ge 1$, the functions $\sum_{k=0}^{\infty} b^k \cos(a^k x)$ and $\sum_{k=0}^{\infty} b^k \sin(a^k x)$ are nowhere differentiable.

If we assume that a is an odd integer and $ab > 2\pi + 1$, then the proof of the nowhere differentiability of $\sum_{k=0}^{\infty} b^k \cos(a^k x)$ is simpler; see Exercise 7.35.

Examples 7.44. Let f be infinitely differentiable at 0. How fast can the sequence of numbers $|f^{(n)}(0)|$ grow? If $f(x) = e^{ax}$ with $a > 1$, then we have $f^{(n)}(0) = a^n$, and so the growth of the sequence is exponential. With the help of Theorem 7.42

we can construct functions with the property that $|f^{(n)}(0)|$ converges to infinity at a rate faster than exponential.

1. Let

$$f(x) = \sum_{k=1}^{\infty} \frac{1}{2^k} \cdot \cos\left(kx + \frac{\pi}{4}\right)$$

for every x. Obviously, this series is everywhere convergent. By taking the term by term derivative of the series n times, we get one of the series

$$\pm \sum_{k=1}^{\infty} \frac{k^n}{2^k} \cdot \cos\left(kx + \frac{\pi}{4}\right), \quad \pm \sum_{k=1}^{\infty} \frac{k^n}{2^k} \cdot \sin\left(kx + \frac{\pi}{4}\right), \tag{7.9}$$

depending on the remainder of n when divided by 4. The resulting series is uniformly convergent on $\mathbb{R}$ in each case. Indeed, for n fixed we have $2^k > k^{n+2}$ for every k large enough. Thus, for each n, the absolute values of the kth term of the nth series is less than $1/k^2$ for every k large enough, and we may apply Weierstrass' criterion. Applying Theorem 7.42 n times implies that f is n times differentiable on $\mathbb{R}$, and its nth derivative is the sum of one of the series of (7.9). Since $\cos(\pi/4) = \sin(\pi/4) = \sqrt{2}/2$, we obtain

$$|f^{(n)}(0)| = \sum_{k=1}^{\infty} \frac{k^n}{2^k} \cdot \frac{\sqrt{2}}{2} > \frac{1}{2} \cdot \frac{n^n}{2^n} = \frac{1}{2} \cdot \left(\frac{n}{2}\right)^n,$$

which grows faster than any exponential sequence.

2. Consider the function

$$g(x) = \sum_{k=1}^{\infty} \frac{2}{2^k} \cdot \cos\left(k^2 x + \frac{\pi}{4}\right).$$

Repeating the argument above, we get that g is infinitely differentiable, and $|g^{(n)}(0)| > n^{2n}/2^n$ for every n.

We can easily modify the construction to yield a functions f such that $|f^{(n)}(0)|$ converges to infinity faster than an arbitrary given sequence. The function $\sum_{k=1}^{\infty} a_k \cdot \cos\left(b_k x + \frac{\pi}{4}\right)$ will have the desired properties, for an appropriate choice of the sequences $a_k \to 0$ and $b_k \to \infty$.

In fact, *for every sequence* (a_k), *there exists an infinitely differentiable function* f *with* $f^{(k)}(0) = a_k$ *for every* k. See Exercises 7.36 and 7.37.

Exercises

7.20. At which points x are the following series of functions convergent? On which intervals are they uniformly convergent?

(a) $\sum x^n/(1+x^n)$, (b) $\sum n/x^n$, (c) $\sum n^x$,
(d) $\sum \left(\frac{1+x}{1-x}\right)^n$, (e) $\sum (\log n)^x$, (f) $\sum ne^{-nx}$,
(g) $\sum \sin(x^n)/n^2$, (h) $\sum \sqrt[n]{x^{2n}+1/2^n}$, (i) $\sum x^2 e^{-nx}$,
(j) $\sum \sin(x/n^2)$, (k) $\sum(\operatorname{arc\,tg}(nx))/(n^2+x^2)$,
(l) $\sum(-1)^n/(x+2^n)$, (m) $\sum x/(x^2+n^3)$.

7.21. Show that the uniform convergence of the series $\sum_{n=1}^{\infty} f_n$ on the set H implies that $f_n \to 0$ uniformly on H. Show that the converse of this statement is not true.

7.22. Let $\sum_{n=1}^{\infty} f_n$ be uniformly convergent on $[a,b]$. Is it true that the series $\sum_{n=1}^{\infty} \sup\{|f_n(x)|\colon x \in [a,b]\}$ is necessarily convergent? What happens if we assume the continuity of the functions f_n? (H)

7.23. Find an example of a uniformly convergent series of functions which does not satisfy the conditions of the Weierstrass criterion.

7.24. Show that if every reordering of the series of functions $\sum_{n=1}^{\infty} f_n$ is uniformly convergent on H, then $\sum_{n=1}^{\infty} |f_n|$ is also uniformly convergent on H.

7.25. Show that the series of functions $\sum_{n=1}^{\infty}(-x)^n(1-x)$ is both absolutely and uniformly convergent on $[0,1]$, but it has a reordering that is not uniformly convergent on $[0,1]$.

7.26. Show that we cannot omit any of the three conditions of Theorem 7.29. Show the same for Theorem 7.32.

7.27. Prove that the series of functions $\sum_{n=1}^{\infty}(\sin nx)/n$ is not uniformly convergent on $\mathbb{R}$. (H)

7.28. Let the functions $f_n\colon [a,b] \to \mathbb{R}$ be continuous, and suppose that the series $\sum_{n=1}^{\infty} f_n(x)$ satisfy the Leibniz criterion for series of real numbers with alternating sign for every $x \in [a,b]$. (That is, $|f_1(x)| \geq |f_2(x)| \geq \ldots$, $f_n(x) \to 0$ as $n \to \infty$, $f_n(x)$ is nonnegative for every n even, and $f_n(x)$ is nonpositive for every n odd for every $x \in [a,b]$.) Show that the series of functions $\sum_{n=1}^{\infty} f_n$ is uniformly convergent on $[a,b]$.

7.29. Show that the function $f(x) = \sum_{n=1}^{\infty} e^{-n^2 x}$ is infinitely differentiable on $(0,\infty)$.

7.30. Show that the partial sums of $\sum_{k=1}^{\infty} \sin kx$ are not bounded on $(0, 2\pi)$.

7.31. Show that there exist continuous functions $f_n\colon [a,b] \to \mathbb{R}$ such that, for every continuous function $f\colon [a,b] \to \mathbb{R}$, there is a bracketing of the series $\sum_{n=1}^{\infty} f_n$ whose sum equals f on $[a,b]$.

7.32. Show that the statement of the previous exercise is not true if we omit the condition of continuity of both f and f_n.

7.33. Let f differentiable at the point a. Prove that if $x_n < a < y_n$ for all n, and if $y_n - x_n \to 0$, then

$$\lim_{n\to\infty} \frac{f(y_n) - f(x_n)}{y_n - x_n} = f'(a). \text{ (H S)}$$

7.34. Let T be the Takagi-function (Example 7.37.2.). Show that if $a = k/2^n$ with $k \in \mathbb{Z}$ and $n \in \mathbb{N}$, then $T'_+(a) = \infty$ and $T'_-(a) = -\infty$.

7.35. Show that if $0 < b < 1$ and a is an odd integer such that $ab > 2\pi + 1$, then the function $f(x) = \sum_{k=0}^{\infty} b^k \cos(a^k x)$ $(x \in \mathbb{R})$ is continuous everywhere, but it is nowhere differentiable. (H S)

7.36. Show that for every positive integer n, $a \in \mathbb{R}$ and $\varepsilon,\ K > 0$, there exists an infinitely differentiable function g such that $g^{(n)}(0) = a$, $g^{(i)}(0) = 0$ for every $0 \le i < n$, and $|g^{(i)}(x)| < \varepsilon$ for every $0 \le i < n$ and $|x| < K$. (H S)

7.37. Show that for every sequence of numbers (a_k) there exists an infinitely differentiable function f such that $f^{(k)}(0) = a_k$ for every k. (∗ H S)

7.3 Taylor Series and Power Series

Recall the definitions of the Taylor polynomials and Taylor series of functions of a single variable: if f is n times differentiable at the point x_0, we call the polynomial

$$\sum_{k=0}^{n} \frac{f^{(k)}(x_0)}{k!}(x - x_0)^k$$

the nth Taylor polynomial of the function f corresponding to the point x_0. If f is infinitely differentiable at point x_0, then we call the infinite series

$$\sum_{k=0}^{\infty} \frac{f^{(k)}(x_0)}{k!}(x - x_0)^k$$

the Taylor series of the function f corresponding to the the point x_0. (See [7, Definitions 13.6 and 13.8]).

If a Taylor series is convergent at a point x and its sum is $f(x)$, we say that **the Taylor series represents f at the point x.**

Examples 7.45. **1.** Every polynomial p is represented everywhere by its Taylor series corresponding to every point x_0. Indeed, if the degree of the polynomial p is n, then $p^{(k)} \equiv 0$, for every $k > n$, and by Taylor's formula[3] we have

$$p(x) = \sum_{k=0}^{n} \frac{p^{(k)}(x_0)}{k!}(x - x_0)^k \tag{7.10}$$

for every x.

2. It is easy to check that the Taylor series of the function $1/(1-x)$ at the point $x_0 = 0$ is exactly the geometric series $\sum_{n=0}^{\infty} x^n$. Thus the function $1/(1-x)$ is represented on the interval $(-1, 1)$ by its Taylor series corresponding to the point $x_0 = 0$.

3. An easy consequence of Taylor's formula is the following theorem. *If f is infinitely differentiable on the interval I and there exists a K such that $|f^{(n)}(x)| \le K$ for every $x \in I$ and $n \ge 1$, then the Taylor series of f corresponding to every point $x_0 \in I$ represents f everywhere on I.* (See [7, Theorem 13.9].)

It follows that the functions e^x, $\sin x$, $\cos x$, $\operatorname{sh} x$, $\operatorname{ch} x$ are represented everywhere by their Taylor series corresponding to every point $x_0 \in \mathbb{R}$. For $x_0 = 0$ we obtain the following formulas.

$$\sin x = x - \frac{x^3}{3!} + \frac{x^5}{5!} - \ldots + \frac{x^{4n+1}}{(4n+1)!} - \ldots,$$

$$\cos x = 1 - \frac{x^2}{2!} + \frac{x^4}{4!} - \ldots + \frac{x^{4n}}{(4n)!} - \ldots,$$

$$e^x = 1 + x + \frac{x^2}{2!} + \ldots + \frac{x^n}{n!} + \ldots,$$

$$\operatorname{sh} x = x + \frac{x^3}{3!} + \frac{x^5}{5!} + \ldots + \frac{x^{2n+1}}{(2n+1)!} + \ldots,$$

$$\operatorname{ch} x = 1 + \frac{x^2}{2!} + \frac{x^4}{4!} + \ldots + \frac{x^{2n}}{(2n)!} + \ldots.$$

(See Example 13.14 of [7].)

---

[3] Taylor's formula (with the Lagrange remainder) states the following. *Let the function f be $(n+1)$ times differentiable on the interval $[a, x]$ (or, on $[x, a]$ if $x < a$). Then there exists a number $c \in (a, x)$ (or $c \in (x, a)$) such that*

$$f(x) = \sum_{k=0}^{n} \frac{f^{(k)}(a)}{k!}(x-a)^k + \frac{f^{(n+1)}(c)}{(n+1)!}(x-a)^{n+1}.$$

See Theorem 13.7 of [7].

4. According to Example 7.41.1., the function $\log(1-x)$ is represented everywhere on the interval $[-1,1)$ by its Taylor series corresponding to $x_0 = 0$.

However, not every function is represented by its Taylor series on an (arbitrarily small) neighborhood of the point x_0.

Examples 7.46. **1.** By Example 7.44.2., there exists an infinitely differentiable function g such that $|g^{(n)}(0)| > n^{2n}/2^n$ for every n. Then the Taylor series of the function g at 0 is not convergent at any $x \neq 0$, since $x \neq 0$ implies

$$\left|\frac{g^{(n)}(0)}{n!} \cdot x^n\right| > \frac{n^{2n}}{2^n \cdot n^n} \cdot |x|^n = \left(\frac{n|x|}{2}\right)^n \to \infty$$

as $n \to \infty$.

One can prove that there exists a function f such that f is infinitely differentiable everywhere, and the Taylor series of f corresponding to any point x_0 is divergent at every $x \neq x_0$.

2. It is also possible that the Taylor series is convergent, but it does not represents the function. One can show that the function $f(x) = e^{-1/x^2}$, $f(0) = 0$ is infinitely differentiable on $\mathbb{R}$, and $f^{(n)}(0) = 0$ for every n. (See [7, Remark 13.17].) Then the Taylor series of f corresponding to the point 0 is the $\sum_{n=0}^{\infty} 0$ series, which is convergent everywhere, but it does not represent f at any $x \neq 0$. This also shows that different functions can have the same Taylor series.

Definition 7.47. We say that the function f is *analytic at the point* x_0, if f is infinitely differentiable at x_0, and its Taylor series at the point x_0 represents f in a neighborhood of x_0.

For example, the functions e^x, $\sin x$, $\cos x$, $\operatorname{sh} x$, $\operatorname{ch} x$ are everywhere analytic functions, and the functions $\log(1+x)$ and $1/(1-x)$ are analytic at the point $x_0 = 0$. (In fact, the function $\log(1+x)$ is analytic at every point $x_0 > -1$, and the function $1/(1-x)$ is analytic at every point $x_0 \neq 1$; see Examples 7.56).

The following theorem gives a sufficient condition for the analiticity of a function.

Theorem 7.48. *Let f be infinitely differentiable on the open interval I, and suppose that there is a positive number c such that $|f^{(n)}(x)| \le (cn)^n$ for every $x \in I$ and $n > 0$. Then f is analytic at every point of the interval I.*

Proof. Let $x_0 \in I$ be arbitrary. Applying Taylor's formula with the Lagrange remainder we get that, for every n and $x \in I \setminus \{x_0\}$, there exists a $d \in I$ such that

$$\left| f(x) - \sum_{k=0}^{n-1} \frac{f^{(k)}(x_0)}{k!}(x-x_0)^k \right| = \left|\frac{f^{(n)}(d)}{n!}\right| \cdot |x-x_0|^n. \tag{7.11}$$

Since $n! > (n/e)^n$ and $|f^{(n)}(d)| \le (cn)^n$ by assumption, the right-hand side of (7.11) is at most $(ec \cdot |x-x_0|)^n$. Put $\eta = 1/(ec)$. If $|x-x_0| < \eta$ and $x \in I$, then the right-hand side of (7.11) converges to zero as $n \to \infty$. This means that the Taylor series of f at the point x_0 represents f on the interval $(x_0 - \eta, x_0 + \eta) \cap I$. □

We will later see that the converse of the Theorem above holds as well. That is, if f is analytic at x_0, then the conditions of Theorem 7.48 are satisfied in a neighborhood of x_0. In order to prove this, however, we first need to get acquainted with the theory of power series. We have already seen several results about power series (see Theorems 7.28, 7.34, and 7.39). We will now systematically explore the topic in more details.

We say that the **domain of convergence** of the power series $\sum_{n=0}^{\infty} a_n x^n$ is the set of numbers $x \in \mathbb{R}$ such that the series is convergent at x. Let T denote the domain of convergence of the series. Note that $T \neq \emptyset$, since every power series is convergent at the point $x = 0$, i.e., $0 \in T$. We call the number $R = \sup T$ (which can be infinite, when T is not bounded from above) the **radius of convergence** of the power series.

By Theorem 7.28, if the power series is convergent at a point x_0, then it is convergent at every point $x \in (-|x_0|, |x_0|)$. From this it is clear that $\inf T = -R$. The statements (i)-(iii) of the following theorem are also easy consequences of Theorem 7.28.

Theorem 7.49. *Let R be the radius of convergence of the power series $\sum_{n=0}^{\infty} a_n x^n$. Then the following are true.*

(i) *If $R = 0$, then the domain of convergence of the series is the single-element set $\{0\}$.*

(ii) *If $0 < R < \infty$, then the domain of convergence of the series is one of the intervals $[-R, R]$, $[-R, R)$, $(-R, R]$, or $(-R, R)$.*

(iii) *If $R = \infty$, then the domain of convergence of the series is the whole real line.* □

Examples 7.50. **1.** The domain of convergence of the power series $\sum_{n=0}^{\infty} n! \cdot x^n$ is the single-element set $\{0\}$. Indeed, for $x \neq 0$, the terms of the series do not converge to zero, and thus the series is divergent.

2. The domain of convergence of the power series $\sum_{n=0}^{\infty} x^n$ is the interval $(-1, 1)$ (see Theorem 6.4).

3. The domain of convergence of the power series $\sum_{n=1}^{\infty} ((-1)^{n-1}/n) x^n$ is the interval $(-1, 1]$ by Example 7.41.

4. The domain of convergence of the series $\sum_{n=1}^{\infty} (1/n) x^n$ is the interval $[-1, 1)$. This follows trivially from the previous example.

5. The domain of convergence of the power series $\sum_{n=1}^{\infty} (1/n^2) x^n$ is the interval $[-1, 1]$. It is clear that the series is convergent at every point $x \in [-1, 1]$. However, for $|x| > 1$, the terms of the series do not converge to zero, thus the series is divergent there.

6. The domain of convergence of the power series $\sum_{n=0}^{\infty} (1/n!) x^n$ is the whole real line (see Example 7.45.3).

These examples suggest that the radius of convergence of the power series $\sum_{n=1}^{\infty} a_n x^n$ depends on the order of magnitude of the sequence $|a_n|$. The famous Cauchy–Hadamard[4] formula gives a precise mathematical formulation of this statement. The formula is discussed in the first appendix of this chapter.

The statement of the following theorem is an immediate consequence of Theorems 7.28 and 7.39.

Theorem 7.51. *Every power series is uniformly convergent on every bounded and closed subinterval of its domain of convergence. The sum of every power series is continuous on its whole domain of convergence.* □

Much more is true in the interior of the domain of convergence of a power series.

Theorem 7.52. *Let the radius of convergence R of the power series $\sum_{n=0}^{\infty} a_n x^n$ be positive (or the infinity), and let $f(x) = \sum_{n=0}^{\infty} a_n x^n$ for every $|x| < R$. Then f is infinitely differentiable on the interval $(-R, R)$ and*

$$f^{(k)}(x) = \sum_{n=k}^{\infty} n(n-1)\cdots(n-k+1)\cdot a_n \cdot x^{n-k} \tag{7.12}$$

for every $|x| < R$ and $k \geq 1$.

Proof. We first prove that the power series $\sum_{n=1}^{\infty} n \cdot a_n \cdot x^{n-1}$ is uniformly convergent on the interval $[-q, q]$ for every $0 < q < R$. Let $q < r < R$ be fixed. Since the series $\sum_{n=0}^{\infty} a_n r^n$ is convergent, it follows that $\lim_{n\to\infty} a_n r^n = 0$. Thus there exists an n_0 such that $|a_n| < r^{-n}$, for every $n > n_0$. Therefore, if $|x| \leq q$, then

$$|n \cdot a_n \cdot x^{n-1}| \leq q^{-1} n(q/r)^n$$

for every $n > n_0$. Since the series $\sum_{n=1}^{\infty} n(q/r)^n$ is convergent by Example 6.36, applying the Weierstrass criterion we obtain that the power series $\sum_{n=1}^{\infty} n \cdot a_n x^{n-1}$ is uniformly convergent on $[-q, q]$. Since this is true for every $0 < q < R$, Theorem 7.42 implies that the sum of the power series $\sum_{n=0}^{\infty} a_n x^n$ (i.e., the function f) is differentiable on $(-R, R)$, and its derivative is $\sum_{n=1}^{\infty} n \cdot a_n x^{n-1}$ there. Repeating this argument for this latter power series we get that f' is also differentiable on $(-R, R)$, and its derivative is $\sum_{n=2}^{\infty} n(n-1)a_n x^{n-2}$ there. Applying induction on k we get (7.12) for every k. □

Example 7.53. Applying Theorem 7.52 for the power series $\sum_{n=0}^{\infty} x^n$ we obtain that

$$\sum_{n=k}^{\infty} n(n-1)\cdots(n-k+1)\cdot x^{n-k} = \frac{k!}{(1-x)^{k+1}} \tag{7.13}$$

[4] Jacques Hadamard (1865–1963), French mathematician.

for every $|x| < 1$ and $k \geq 0$. (We have seen the special case $k = 1$ in Examples 6.31.1. and 6.55.)

From now on, we will also call the series of form $\sum_{n=0}^{\infty} a_n(x - x_0)^n$ power series (more precisely, **power series around the point** x_0). It follows immediately from Theorem 7.52 that if the power series $\sum_{n=0}^{\infty} a_n(x - x_0)^n$ is convergent on the interval $(x_0 - R, x_0 + R)$ and its sum is $f(x)$ there, then f is infinitely differentiable on $(x_0 - R, x_0 + R)$, and

$$f^{(k)}(x) = \sum_{n=k}^{\infty} n(n-1)\cdots(n-k+1) \cdot a_n \cdot (x - x_0)^{n-k} \tag{7.14}$$

for every $x \in (x_0 - R, x_0 + R)$ and $k \geq 0$. The following theorem is a simple, but important corollary of this fact.

Theorem 7.54. *Let the power series $\sum_{n=0}^{\infty} a_n(x - x_0)^n$ be convergent on $(x_0 - R, x_0 + R)$, where $R > 0$, and let its sum be $f(x)$ there. Then $a_n = f^{(n)}(x_0)/n!$ for every n. In other words, the power series is equal to the Taylor series of its sum corresponding to the point x_0.*

Proof. Apply (7.14) to $x = x_0$. □

Corollary 7.55. *A function f is analytic at the point x_0 if and only if there exists a power series around x_0 which represents f on a neighborhood of x_0.* □

We say that the function $f\colon I \to \mathbb{R}$ is **analytic on the open interval** I, if f is analytic at every point of I. Examples 7.45 showed that the polynomials and the functions e^x, $\sin x$, $\cos x$, $\operatorname{sh} x$, and $\operatorname{ch} x$ are each analytic everywhere. We will now prove that several other elementary functions are analytic on their respective domain.

Examples 7.56. **1.** We show that the function $1/x$ is analytic at every point $a \neq 0$. Let $|x - a| < |a|$. We have

$$\frac{1}{x} = \frac{1}{a + (x - a)} = \frac{1}{a} \cdot \frac{1}{1 + (x - a)/a} = \frac{1}{a} \cdot \sum_{n=0}^{\infty} \frac{(-1)^n}{a^n} \cdot (x - a)^n.$$

Thus $1/x$ is represented by a power series on the interval $(0, 2a)$ (or on the interval $(-2a, 0)$, if $a < 0$). Thus, by Corollary 7.55, $1/x$ is analytic at a.

2. Generalizing the previous example we show that every rational function is analytic everywhere on its domain. Let $S = p/q$, where $p = \sum_{i=0}^{n} a_i x^i$ and $q = \sum_{j=0}^{m} b_j x^j$ are polynomials. First we prove that if $q(0) \neq 0$, then S is analytic at 0.

We may assume that $q(0) = 1$; then $q(x) = 1 - r(x)$, where $r(x) = -\sum_{j=1}^{m} b_j x^j$. The function $\sum_{j=1}^{m} |b_j| \cdot |x|^j$ is continuous and vanishes at the point 0. Therefore, we can choose a $\delta > 0$ such that $\sum_{j=1}^{m} |b_j| \cdot |x|^j < 1$ holds for every $|x| < \delta$. Then $|r(x)| < 1$ for every $x \in (-\delta, \delta)$, and thus

$$\frac{p(x)}{q(x)} = \left(\sum_{i=0}^{n} a_i x^i\right) \cdot \frac{1}{1-r(x)} = \left(\sum_{i=0}^{n} a_i x^i\right) \cdot \sum_{k=0}^{\infty} r(x)^k =$$
$$= \left(\sum_{i=0}^{n} a_i x^i\right) \cdot \sum_{k=0}^{\infty} \left(\sum_{j=1}^{m} c_j x^j\right)^k.$$

By performing the multiplications and reordering the resulting terms according to the exponents of x we get a power series. This operation does not change the sum of the series since, by $\sum_{j=1}^{m} |b_j| \cdot |x|^j < 1$ it follows that each of the series appearing in the argument is absolute convergent, and we can apply Theorem 6.30. We leave the details to the reader. In this way we represented the function S on the interval $(-\delta, \delta)$ by the sum of a power series. Thus, by Corollary 7.55, the function S is analytic at 0.

Now, let x_0 be an arbitrary point where q is non-zero. The function $S_1(x) = S(x + x_0)$ is also a rational function which does not disappear at 0. If $S_1(x) = \sum_{n=0}^{\infty} c_n x^n$ for every $|x| < \delta$, then $S(x) = \sum_{n=0}^{\infty} c_n (x - x_0)^n$ for every $|x - x_0| < \delta$. Thus S is analytic at x_0.

3. With the help of Theorem 7.54 we can give a new proof of the fact that every exponential function is analytic on $\mathbb{R}$. Indeed, for $a > 0$ and $x_0 \in \mathbb{R}$ we have

$$a^x = a^{x_0} \cdot e^{\log a \cdot (x - x_0)} = a^{x_0} \cdot \sum_{n=0}^{\infty} \frac{(\log a)^n}{n!} (x - x_0)^n.$$

Thus a^x is represented by a power series around x_0, i.e., a^x is analytic at x_0.

4. We now show that the function $(1 + x)^c$ is analytic at 0 for every $c \in \mathbb{R}$. The nth derivative of the function at 0 is $c(c-1) \cdots (c-n+1)$, and the Taylor series at 0 is

$$\sum_{n=0}^{\infty} \frac{c(c-1) \cdots (c-n+1)}{n!} x^n. \tag{7.15}$$

We prove that *the series represents the function on the interval* $(-1, 1)$.

If c is a non-negative integer, then by the binomial theorem the sum of the series (7.15) is $(1 + x)^c$ for every x. Thus we may assume that $c \notin \mathbb{N}$.

We first show that the series (7.15) is convergent for every $|x| < 1$. Indeed, for $x \neq 0$, the ratio of the $n + 1$st and nth terms of the series is $(x \cdot (c - n))/(n + 1)$. Since this converges to $-x$ as $n \to \infty$, it follows from the ratio test that the series is convergent for $|x| < 1$. Let the sum of the series be $f(x)$. By Theorem 7.52, f is differentiable on $(-1, 1)$, and

$$f'(x) = \sum_{n=1}^{\infty} n \cdot \frac{c(c-1) \cdots (c-n+1)}{n!} x^{n-1} \tag{7.16}$$

there. The power series on the right-hand side is convergent on $(-1, 1)$, thus it is also absolutely convergent there. Multiplying the series by $(1 + x)$, then reordering it according to the exponents of x yields, by doing some algebra, c-times the series (7.15). The absolute convergence ensures that reordering does not change the sum of the series, thus we proved that $(1 + x)f'(x) = cf(x)$, for every $|x| < 1$. Then we have

$$\begin{aligned}\left(f(x)\cdot(1+x)^{-c}\right)' &= f'(x)\cdot(1+x)^{-c} - c\cdot f(x)\cdot(1+x)^{-c-1} = \\ &= (1+x)^{-c-1}\cdot\left((1+x)f'(x) - cf(x)\right) = 0,\end{aligned}$$

and thus $f(x)/(1 + x)^c$ is constant on $(-1, 1)$. Since $f(0) = 1$, we necessarily have $f(x)/(1 + x)^c = 1$ and $f(x) = (1 + x)^c$, which is what we wanted to prove. □

To emphasize the analogy with the binomial theorem, let us use the notation

$$\frac{c(c-1)\cdots(c-n+1)}{n!} = \binom{c}{n}$$

for every $c \in \mathbb{R}$ and $n \in \mathbb{N}$. (If $n = 0$, let $\binom{c}{0} = 1$ for every c.) We call the numbers $\binom{c}{n}$ **generalized binomial coefficients**. Using this notation, the previously proved statement has the form

$$(1+x)^c = \sum_{n=0}^{\infty}\binom{c}{n}x^n \tag{7.17}$$

for every $c \in \mathbb{R}$ and $|x| < 1$. The series on the right-hand side is called **binomial series.**

5. We prove that the power function x^c is analytic on $(0, \infty)$ for every c. Indeed, for $a > 0$ and $|x - a| < a$, we have

$$x^c = a^c\cdot\left(1 + \frac{x-a}{a}\right)^c = a^c\cdot\sum_{n=0}^{\infty}\binom{c}{n}\left(\frac{x-a}{a}\right)^n,$$

i.e., x^c is represented on the interval $(0, 2a)$ by a power series around a. Thus, by Corollary 7.55, x^c is analytic at a.

6. We prove that the function $\log x$ is analytic on $(0, \infty)$. Let $a > 0$ and $|x - a| < a$. We have

$$\begin{aligned}\log x &= \log a + \log\frac{x}{a} = \log a + \log\left(1 + \frac{x-a}{a}\right) = \\ &= \log a + \sum_{n=1}^{\infty}\frac{(-1)^{n-1}}{n\cdot a^n}\cdot(x-a)^n.\end{aligned}$$

Thus $\log x$ is represented on the interval $(0, 2a)$ by a power series, i.e., $\log x$ is analytic at a.

Remark 7.57. Let f be analytic on the open interval I. For every $x_0 \in I$ let $r(x_0)$ denote the largest positive number (or the infinity) such that the Taylor series of f at x_0 represents f on the neighborhood of x_0 of radius $r(x_0)$.

Let us determine the value of $r(x_0)$ for the functions of Examples 7.45 and 7.56. For polynomials, and for the functions e^x, $\sin x$, $\cos x$, $\operatorname{sh} x$ and $\operatorname{ch} x$, we have $r(x_0) = \infty$ for every x_0. For the functions $1/x$, $(1+x)^c$, $\log(1+x)$, we proved that $r(x_0) = x_0$ for every $x_0 > 0$, and thus $r(x_0)$ is the largest number such that f is analytic on the neighborhood of x_0 with radius $r(x_0)$. We might believe that this is always true, that is, if f is analytic on the interval (a, b), then $r(x_0) \geq \min(x_0 - a, b - x_0)$ for every $x_0 \in (a, b)$.

However, this conjecture is false. Consider the function $f(x) = 1/(1+x^2)$ on $\mathbb{R}$. By Example 7.56.2, f is analytic everywhere. On the other hand, the Taylor series of f at 0 is the series $\sum_{n=0}^{\infty}(-1)^n x^{2n}$. Indeed, this power series represents f on the interval $(-1, 1)$ and thus, by Theorem 7.54, this is the Taylor series of the function corresponding to the point zero. However, this series is divergent for $|x| \geq 1$, therefore, $r(x_0) = 1$. We find that the function f is analytic everywhere, but its Taylor series corresponding to 0 represents f only on the interval $(-1, 1)$.

This phenomenon might be very surprising at first. What could determine the value of the radius $r(x_0)$ if not the largest interval where f is analytic? In order to answer this question, we have to step out to the complex plane. See the second appendix of this chapter for the details.

We now return to the converse of Theorem 7.48.

Lemma 7.58. *Let the power series $\sum_{n=0}^{\infty} a_n(x - x_0)^n$ be convergent on $(x_0 - R, x_0 + R)$ $(R > 0)$, and let its sum be $f(x)$. Then, for every $0 < q < R$, there exists a $c > 0$ such that $|f^{(k)}(x)| \leq (ck)^k$ for every $|x - x_0| \leq q$ and $k > 0$.*

Proof. Let $q < r < R$ be fixed. Since the series $\sum_{n=0}^{\infty} a_n r^n$ is convergent, we have $a_n r^n \to 0$. Then there exists a $K > 1$ such that $|a_n| \leq K/r^n$ for every $n > 0$. Thus, if $|x - x_0| \leq q$ then, using (7.14) and (7.13) we get

$$\begin{aligned}
\left|f^{(k)}(x)\right| &\leq \sum_{n=k}^{\infty} n(n-1)\cdots(n-k+1)\cdot|a_n|\cdot|x-x_0|^{n-k} \leq \\
&\leq \sum_{n=k}^{\infty} n(n-1)\cdots(n-k+1)\cdot\frac{K}{r^n}\cdot q^{n-k} = \\
&= \frac{K}{r^k}\cdot\sum_{n=k}^{\infty} n(n-1)\cdots(n-k+1)\cdot(q/r)^{n-k} = \\
&= \frac{K\cdot k!}{r^k\cdot(1-(q/r))^{k+1}} < \frac{rK}{r-q}\cdot\left(\frac{k}{r-q}\right)^k
\end{aligned}$$

for every $k > 0$.

Thus $\left|f^{(k)}(x)\right| \le (ck)^k$, where $c = rK/(r-q)^2$. Since this is true for every $k > 0$, the lemma is proved. □

The previous lemma has important consequences.

Theorem 7.59. *The function f is analytic at the point x_0 if and only if there exist positive numbers δ and c such that f is infinitely differentiable on $(x_0 - \delta, x_0 + \delta)$, and $|f^{(n)}(x)| \le (cn)^n$ for every $x \in (x_0 - \delta, x_0 + \delta)$ and $n > 0$.*

Proof. We proved the "if" part of the theorem in Theorem 7.48. The "only if part" follows from Lemma 7.58. □

Theorem 7.60. *If the power series $\sum_{n=0}^{\infty} a_n(x-x_0)^n$ is convergent on $(x_0 - R, x_0 + R)$, then its sum is analytic at every point of the interval $(x_0 - R, x_0 + R)$.*

Proof. This is clear by Lemma 7.58 and Theorem 7.48. □

We note that the theorem can also be proved directly (see Exercise 7.58).

Example 7.61. By applying the previous theorem, we show that the function $\arcsin x$ is analytic on $(-1, 1)$. Apply (7.17) with $c = -1/2$ and with $-x^2$ in place of x. We get

$$\frac{1}{\sqrt{1-x^2}} = \sum_{n=0}^{\infty} (-1)^n \cdot \binom{-1/2}{n} x^{2n}$$

for every $|x| < 1$. If $|x| < 1$, then the series on the right-hand side is uniformly convergent on the interval $[0, x]$, and thus, by Theorem 7.40, we may integrate the series term-by-term there. We obtain

$$\arcsin x = \int_0^x \frac{dt}{\sqrt{1-t^2}} = \sum_{n=0}^{\infty} (-1)^n \cdot \binom{-1/2}{n} \cdot \frac{1}{2n+1} x^{2n+1}$$

for every $x \in (-1, 1)$. Therefore, by Theorem 7.60, $\arcsin x$ is analytic on $(-1, 1)$.

Note that $\binom{-1/2}{n} = (-1)^n \binom{2n}{n}/4^n$, which gives

$$\arcsin x = \sum_{n=0}^{\infty} \frac{\binom{2n}{n}}{4^n(2n+1)} \cdot x^{2n+1} \tag{7.18}$$

for every $|x| < 1$.

Remark 7.62. Let f be infinitely differentiable on the open interval I. For every $x_0 \in I$, let $R(x_0)$ denote the radius of convergence of the Taylor series $\sum_{n=0}^{\infty} \frac{f^{(n)}(x_0)}{n!}(x-x_0)^n$. If f is analytic on I, then $R(x_0) > 0$ for every $x_0 \in I$, since the Taylor series has to represent f on a neighborhood of x_0. The converse of this is not true: $R(x_0) > 0$ $(x_0 \in I)$ does not imply that f is analytic at every point of the interval I. E.g., let $f(x) = e^{-1/x^2}$ and $f(0) = 0$. As we saw in Examples 7.46.2, f is not analytic at 0, but we have $R(x_0) > 0$ for every x_0. Indeed, $R(0) = \infty$, since the Taylor series at 0 is convergent everywhere. On the other hand, f is analytic on both of the half lines $(-\infty, 0)$ and $(0, \infty)$; this easily follows from Exercise 7.60. Thus $R(x_0) > 0$ for every $x_0 \neq 0$.

If we want to ensure that f is analytic on I, we need to assume more than $R(x_0) > 0$ $(x_0 \in I)$. With the help of Theorem 7.59, it easy to see that if for every bounded and closed interval $J \subset I$, we have

$$\inf\{R(x_0) \colon x_0 \in J\} > 0, \tag{7.19}$$

then f is analytic on I (see Exercise 7.64). The converse of this claim is also true: if f is analytic on I, then (7.19) holds for every bounded and closed interval $J \subset I$ (see Exercise 7.65).

The following theorem presents an important property of analytic functions.

Theorem 7.63. *Let $f \colon I \to \mathbb{R}$ be analytic on the open interval I. If there exists a sequence (x_n) of roots of f converging to a point $x_0 \in I$ such that $x_0 \neq x_n$ for every n, then f is the constant zero function.*

Proof. We know that f is infinitely differentiable on I. Thus f is continuous, which implies $f(x_0) = 0$. Applying Rolle's theorem (see [7, Theorem 12.49]) successively, we get that for every $k > 0$ there exists a sequence converging to x_0, and such that its terms are different from x_0, and $f^{(k)}$ is zero at each term of the sequence. Since $f^{(k)}$ is continuous, it follows that $f^{(k)}(x_0) = 0$ for every k.

This means that the Taylor series of f at x_0 is constant zero. Since f is analytic at x_0, we have that f is constant zero on a neighborhood of x_0. Let b be the supremum of the set of those points x for which $x > x_0$, $x \in I$, and f is constant zero on the interval $[x_0, x]$. Put $b = \sup I$. Then $b = \sup I$. Indeed, suppose this is not true; that is, $b < \sup I$. Then $b \in I$ and f is constant zero on the interval $[x_0, b)$. Then it follows that $f^{(k)}(b) = 0$ for every k, and thus the sum of the Taylor series of the function f corresponding to the point b is constant zero. Since f is analytic at b, f is constant zero on a neighborhood of b. However, this contradicts the definition of b.

We proved that $b = \sup I$, i.e., $f(x) = 0$ for every point $x > x_0$, $x \in I$. We can prove $f(x) = 0$ for $x < x_0$, $x \in I$ similarly. □

Remark 7.64. Theorem 7.63 can be rephrased as follows. If f is analytic on the open interval I and I has a bounded and closed subinterval where f has infinitely many roots, then f is the constant zero function.

It is easy to see that if the functions f and g are analytic on the interval I, then $f - g$ is also analytic on I. Applying Theorem 7.63 to $f - g$ yields the following.

Theorem 7.65. *Let f and g be analytic on the open interval I, and let $f(x_n) = g(x_n)$ for every n, where $x_n \to x_0 \in I$, and $x_0 \neq x_n$ for every n. Then $f(x) = g(x)$ for every $x \in I$.* □

Remarks 7.66. **1.** The previous theorem states that if I is an open interval, $x_n \to x_0 \in I$, and $x_0 \neq x_n$ for every n, then every function that is analytic on I is determined by its values at the points x_n. For this reason, we call Theorem 7.65 the **unicity theorem**. (The statement of the unicity theorem is the analogue of the fact that if two kth order polynomials are equal at $(k+1)$ points, then they are equal to each other.)

2. It is important to note that in Theorems 7.63 and 7.65 the condition requiring that the limit of the sequence (x_n) is in I is essential. One can prove that the function $\sin(1/x)$ is analytic on the half line $(0, \infty)$ (see Exercise 7.61). This function is zero at the points $x_n = 1/(n\pi)$, where $x_n \to 0$. Still, the function $\sin(1/x)$ is not the constant zero function.

3. The properties described by Theorems 7.63 and 7.65 are shared neither by the class of continuous, nor the class of differentiable functions. Moreover, these properties are not shared even by the functions that are infinitely differentiable. E.g., let $f(x) = 0$ if $x \leq 0$ and $f(x) = e^{-1/x^2}$ if $x > 0$. One can show that f is infinitely differentiable on the whole real line (and $f^{(n)}(0) = 0$ for every n). (See [7, Remark 13.17].) Now, if $x_n < 0$, $x_n \to 0$, then $f(x_n) = 0$, since f is zero on $(-\infty, 0]$, but f is not constant on $\mathbb{R}$.

Exercises

7.38. Find the radius of convergence of the following power series.

(a) $\sum n^c x^n$, (b) $\sum x^n/(a^n + b^n)$,

(c) $\sum n! x^n$, (d) $\sum n! x^{n^2}$,

(e) $\sum x^n / \binom{2n}{n}$, (f) $\sum ((\log n)^{\log n}/2^n) x^n$,

(g) $\sum (1 + \frac{1}{n})^{n^2} x^n$, (h) $\sum (n^n/n!) x^n$,

(i) $\sum 2^{-n^2} x^n$, (j) $\sum 2^{-n^n} x^{n!}$.

7.39. Find power series that represent the following functions on a neighborhood of the given points:

(a) $1/x^2$, $x_0 = 3$; (b) $(2x-5)/(x^2-5x+6)$, $x_0=0$;

(c) 3^x, $x_0 = 0$; (d) 3^x, $x_0 = 2$;

(e) $\log x$, $x_0 = 10$; (f) e^{x^3}, $x_0 = 0$;

(g) $\log(1+x^2)$, $x_0 = 0$; (h) $\log(1+x+x^2)$, $x_0 = 0$;

(i) $\sin x^2$, $x_0 = 0$; (j) $\operatorname{sh}(1+x^3)$, $x_0 = 0$;

(k) $f(x) = \operatorname{ch}\sqrt{x}$ if $x \geq 0$, and $f(x) = \cos\sqrt{-x}$ if $x < 0$, $x_0 = 0$;

(l) $1/(1+x^2)$, $x_0 = 1$; (m) $1/(1+x^2)^2$, $x_0 = 0$;

(n) $x \cdot \operatorname{arc\,tg} x - \log\sqrt{1+x^2}$, $x_0 = 0$;

(o) $1/\sqrt{x}$, $x_0 = 2$.

7.40. Find the 357th derivative of $\operatorname{arc\,tg} x$ at 0. Find the 42nd derivative of e^{x^2} at 0. Find the 78th derivative of $\log(1+x+x^2)$ at 0. Find the 80th derivative of $(\operatorname{arc\,tg} x)^2$ at 0.

7.41. Find the sum of the following infinite series:

(a) $\sum_{n=1}^{\infty} n/3^n$, (b) $\sum_{n=1}^{\infty} 1/(n \cdot 2^n)$,
(c) $\sum_{n=1}^{\infty} n^2/5^n$, (d) $\sum_{n=0}^{\infty} 1/(2n+1)!$,
(e) $\sum_{n=1}^{\infty} 4^n/(2n)!$, (f) $\sum_{n=0}^{\infty} 1/((2n+1)2^n)$,
(g) $\sum_{n=1}^{\infty} (1-\sqrt{e})^n/n$.

7.42. Find the value of the limit

$$\lim_{n\to\infty}\left(1+\frac{1}{n}\right)^{n^2} \cdot e^{-n}.$$

7.43. Find those pairs of numbers (a, b) for which the sequence

$$n^a\left(e-\left(1+\frac{1}{n}\right)^{n+b}\right)$$

is convergent. (H)

7.44. Let $f(x) = \sum_{n=0}^{\infty} a_n x^n$, for every $|x| < r$, where the coefficients a_n are non-negative. Show that the power series is convergent at r if and only if f is bounded on $(-r, r)$.

7.45. Evaluate the sum $\sum_{n=0}^{\infty}\left(\frac{1}{3n+1}-\frac{1}{3n+2}\right)$. (H)

7.46. Find the sums of the following power series in $(-1, 1)$:

(a) $\sum_{n=0}^{\infty} \frac{1}{3n+1} x^n$, (b) $\sum_{n=1}^{\infty} \frac{x^n}{n\cdot(n+1)}$.

7.47. Show that if $c > -1$, then

$$2^c = 1 + \binom{c}{1} + \binom{c}{2} + \ldots.$$

Prove also that the series on the right-hand side is divergent for $c \leq -1$. (H)

7.48. Show that if $c \geq 0$, then

$$1 - \binom{c}{1} + \binom{c}{2} - \ldots = 0.$$

Prove also that the series on the left-hand side is divergent for $c < 0$. (H)

7.49. Show that the power series of $\arcsin x$ is convergent at $x = 1$. Use this to prove

$$\frac{\pi}{2} = \sum_{n=0}^{\infty} \frac{\binom{2n}{n}}{4^n(2n+1)}. \text{ (H)}$$

7.50. Prove the converse of Abel's theorem: if $\sum_{n=0}^{\infty} a_n x^n$ is uniformly convergent on $[0, x_0)$, then it is convergent at x_0.

7.51. Construct a function f such that f is infinitely differentiable everywhere, the Taylor series of f at 0 is convergent everywhere, and the Taylor series represents f on $[-1, 1]$, but does not represent f anywhere else.

7.52. True or false? If a power series is convergent at the point $x_0 > 0$, then its sum is differentiable from the left at x_0. (H)

7.53. True or false? If a power series is convergent at the point $x_0 > 0$, then the (finite or infinite) left-hand side derivative of its sum exists at x_0. (∗ H)

7.54. Show that if f is analytic on the open interval I, then the primitive function of f is also analytic on I.

7.55. Show that the function $\operatorname{arc\,tg} x$ is analytic on $\mathbb{R}$.

7.56. Show that if f is analytic on $\mathbb{R}$, then its graph consists of finitely many monotone segments over every bounded and closed interval.

7.57. Show that if f is analytic on $\mathbb{R}$, then its graph consists of finitely many convex or concave segments over every bounded and closed interval.

7.58. Let $f(x)=\sum_{n=0}^{\infty} a_n x^n$ for every $x \in (-r,r)$. Show that for every $x_0 \in (-r,r)$, the Taylor series of f corresponding to x_0 represents f on the interval $(x_0-\delta, x_0+\delta)$, where $\delta = \min(x_0+r, r-x_0)$. (H)

7.59. Show that if f and g are analytic on (a,b), then $f+g$, $f\cdot g$, and (whenever $g \neq 0$) f/g are also analytic there.

7.60. (a) Suppose that the function f is represented by the power series $\sum_{k=0}^{\infty} a_k(x-a)^k$ on the interval $(a-\delta, a+\delta)$, $f(a)=b$, and the function g is represented by the power series $\sum_{n=0}^{\infty} b_n(x-b)^n$ on the interval $(b-\varepsilon, b+\varepsilon)$. Suppose further that $\sum_{k=1}^{\infty} |a_k|\cdot|x-a|^k < \varepsilon$ for every $x \in (a-\delta, a+\delta)$. Show that the function $g \circ f$ is represented by its Taylor series corresponding to the point a on the interval $(a-\delta, a+\delta)$. (H)
(b) Let f be analytic on the open interval I, and let g be analytic on the open interval J, where $J^f(I)$. Show that $g \circ f$ is analytic on I.

7.61. Show that the function $\sin(1/x)$ is analytic on the half-line $(0,\infty)$.

7.62. Show that if f is infinitely differentiable on (a,b) and $f^{(n)}(x) \geq 0$ for every $x \in (a,b)$ and every n, then f is analytic on (a,b). (∗)

7.63. Show that if f is infinitely differentiable on (a,b) and the sign of $f^{(n)}$ is the same everywhere on (a,b) for every n, then f is analytic on (a,b). (∗)

7.64. Let f be infinitely differentiable on the open interval I, and suppose that for every bounded and closed interval $J \subset I$ there is a $\delta > 0$ such that, for every $x_0 \in J$, the radius of convergence of the Taylor series of the function f corresponding to the point x_0 is at least δ. Show that f is analytic on I. (H)

7.65. Show that if f is analytic on an open interval I, then for every bounded and closed interval $J \subset I$ there exists a $\delta > 0$ such that for every $x_0 \in J$, the radius of convergence of the Taylor series of f corresponding to the point x_0 is at least δ. (H)

7.66. Show that for every continuous function $f\colon \mathbb{R} \to \mathbb{R}$ there exists an everywhere analytic function g such that $g(x) > f(x)$ for every x. (∗ H)

In the following exercises we consider sequences with the property that every term of the sequence equals a linear combination of the previous k terms with coefficients independent of the term. E.g., the Fibonacci[5] sequence[6] is such a sequence. The precise definition is as follows. We say that the sequence $(a_n)_{n=0}^{\infty}$ satisfies a **linear recursion**, if there exist real numbers $c_1, \ldots, c_k$ such that

$$a_n = c_1 a_{n-1} + \ldots + c_k a_{n-k} \tag{7.20}$$

holds for every $n \geq k$.

---

[5] Fibonacci (Leonardo Pisano) (about 1170–1240), Italian mathematician.

[6] The sequence (u_n) of the Fibonacci numbers is defined by $u_0 = 0$, $u_1 = 1$ and $u_n = u_{n-1} + u_{n-2}$ $(n \geq 2)$.

7.67. Let u_n denote the nth Fibonacci number. Show that the radius of convergence of the power series $\sum_{n=0}^{\infty} u_n x^n$ is positive. Find the sum of the power series, find the Taylor series of this sum corresponding to 0, and use these to give a closed formula for the Fibonacci number u_n.

7.68. Show that the sequence $(a_n)_{n=0}^{\infty}$ satisfies a linear recursion if and only if the radius of convergence of the power series $\sum_{n=0}^{\infty} a_n x^n$ is positive, and its sum is a rational function.

7.69. Let $0.a_1a_2\ldots$ be the number $t \in [0,1]$, written as a decimal. Show that the sum of $\sum_{n=0}^{\infty} a_n x^n$ is a rational function if and only if t is a rational number.

7.70. Let $c_1, \ldots, c_k \geq 0$ and $c_1 + \ldots + c_k = 1$. Show that if the sequence $(a_n)_{n=0}^{\infty}$ satisfies the recursion (7.20), then (a_n) is convergent.

7.71. Let $C_0 = C_1 = 1$, and for $n \geq 2$, let C_n denote the number of triangulations of a convex $n+2$-vertex polygon.

(i) Show that $C_{n+1} = \sum_{i=0}^{n} C_i C_{n-i}$ for every $n \geq 0$.
(ii) Show that the power series $\sum_{n=0}^{\infty} C_n x^n$ is convergent in a neighborhood of 0.
(iii) Show that if $\sum_{n=0}^{\infty} C_n x^n = f(x)$, then we have $f(x) - xf^2(x) - 1 = 0$ and $f(x) = (1 - \sqrt{1-4x})/(2x)$.
(iv) Show that $C_n = \frac{1}{n+1}\binom{2n}{n}$ for every $n \geq 0$. (H)

The numbers C_n are called the **Catalan numbers**[7].

7.4 Abel Summation

When discussing summable series in the previous chapter, we defined a class of divergent infinite series (namely, the set of summable series) to which we assigned sum-like values. This was done as follows: instead of the sequence of partial sums s_n, we considered the sequence of their arithmetic means (t_n), and if this was convergent, we said that the limit of the sequence (t_n) was the sum of the series.

In fact, this method is only one of several methods assigning a sum-like value to divergent series. An infinite system of such methods was presented by Hölder[8].

Hölder's idea was the following. The series $1 - 2 + 3 - 4 + \ldots$ is not summable, since the sequence of its partial sums is

$$(s_n) = (1, -1, 2, -2, 3, -3, \ldots),$$

[7] Eugène Charles Catalan (1814–1894), Belgian mathematician.

[8] Otto Ludwig Hölder (1859–1937), German mathematician.

and the sequence of their arithmetic means is

$$(t_n) = (1, 0, 2/3, 0, 3/5, 0, \ldots, n/(2n-1), 0, \ldots).$$

The series (t_n) is divergent, since the subsequence of its even terms converges to 0, while the subsequence of its odd terms converges to $1/2$. This means that the series $1-2+3-4+\ldots$ is not summable. (This also follows from the fact that a_n/n does not tend to zero; see Exercise 6.56).)

But if we take the arithmetic means of the sequence (t_n), the resulting sequence is convergent and converges to $1/4$. This motivated the following definition. Let s_n denote the partial sums of the infinite series $\sum_{n=1}^{\infty} a_n$, furthermore, let $t_n = (s_1 + \ldots + s_n)/n$ and $u_n = (t_1 + \ldots + t_n)/n$ for every n. If the sequence (u_n) is convergent with limit A, then we say that the infinite series $\sum_{n=1}^{\infty} a_n$ is $(H, 2)$ **summable,** and its $(H, 2)$ sum is A. The series $1-2+3-4+\ldots$ is $(H, 2)$ summable and its $(H, 2)$ sum is $1/4$.

The process can be continued: take the arithmetic mean of the sequence of partial sums, then take the arithmetic mean of the resulting sequence, and continue this method for k steps. If the sequence we get after the kth step is convergent with limit A, then we say that the infinite series $\sum_{n=1}^{\infty} a_n$ is (H, k) **summable,** and its (H, k) sum is A.

Since the sequence of the arithmetic means of a convergent sequence also converges to the same limit, it is obvious that if an infinite series is (H, k) summable, then it is also (H, m) summable for every $m > k$, and its (H, m) sum is the same as its (H, k) sum. Therefore, these summation methods are more and more efficient in the sense that they assign a sum-like value to a wider and wider set of infinite series. The previous example shows that the set of $(H, 2)$ summable series is strictly larger than the set of $(H, 1)$ summable (i.e., the summable in the original sense) series. One can show that in general the set of $(H, k+1)$ summable series is strictly larger than the set of (H, k) summable sets (see Exercise 7.77).

As another application of the theory of power series we now introduce another, even more effective, summing method. It follows from Theorem 7.39 that if the series $\sum_{n=0}^{\infty} a_n$ is convergent with sum A, then

$$\lim_{x \to 1-0} \sum_{n=0}^{\infty} a_n x^n = A. \tag{7.21}$$

This observation motivates the following definition.

Definition 7.67. We say that the infinite series $\sum_{n=0}^{\infty} a_n$ is *Abel summable and its Abel sum is* A, if the power series $\sum_{n=0}^{\infty} a_n x^n$ is convergent on $(-1, 1)$ and (7.21) holds.

Thus it follows from Theorem 7.39 that if a series is convergent and its sum is A, then the series is Abel summable and its Abel sum is also A. Even more is true.

Theorem 7.68. *If an infinite series is summable and its sum is A, then the series is Abel summable and its Abel sum is also A.*

Proof. Let the infinite series $\sum_{n=0}^{\infty} a_n$ be summable. By Exercise 6.56 we have $a_n/n \to 0$, and thus the power series $\sum_{n=0}^{\infty} a_n x^n$ is convergent on $(-1, 1)$, since it can be majorized by the series $\sum n \cdot |x|^n$. Let the sum of the series $\sum_{n=0}^{\infty} a_n x^n$ be $f(x)$. We need to show that $\lim_{x \to 1-0} f(x) = A$.

Let $s_n = \sum_{i=0}^{n} a_i$. It is easy to check that the Cauchy product of the series $\sum_{n=0}^{\infty} x^n$ and $\sum_{n=0}^{\infty} a_n x^n$ is the series $\sum_{n=0}^{\infty} s_n x^n$. Since the series are absolutely convergent on $(-1, 1)$, it follows that their Cauchy product is also absolutely convergent, and its sum is the product of the sums of the two original series. (See Theorem 6.54.) Therefore,

$$\frac{f(x)}{1-x} = \sum_{n=0}^{\infty} s_n x^n \tag{7.22}$$

for every $x \in (-1, 1)$, where the series on the right-hand side is absolutely convergent. Taking the Cauchy product of the right-hand side of (7.22) and $\sum_{n=0}^{\infty} x^n$, we obtain

$$\frac{f(x)}{(1-x)^2} = \sum_{n=0}^{\infty} (s_0 + \ldots + s_n) x^n \tag{7.23}$$

for every $x \in (-1, 1)$. By assumption, the series $\sum_{n=0}^{\infty} a_n$ is summable. If its sum is A, then $(s_0 + \ldots + s_n)/(n+1) \to A$, i.e., the sequence

$$c_n = \frac{s_0 + \ldots + s_n}{n+1} - A$$

converges to zero. Writing the terms of the series on the right-hand side of (7.23) in terms of the numbers c_n, we get

$$\begin{aligned}\sum_{n=0}^{\infty} ((n+1)A + (n+1)c_n) x^n &= A \cdot \sum_{m=1}^{\infty} m x^{m-1} + \sum_{n=0}^{\infty} (n+1) c_n x^n = \\ &= \frac{A}{(1-x)^2} + \sum_{n=0}^{\infty} (n+1) c_n x^n.\end{aligned}$$

Multiplying (7.23) by $(1-x)^2$ we get

$$f(x) = A + (1-x)^2 \cdot \sum_{n=0}^{\infty} (n+1) c_n x^n.$$

To finish our proof, we need to show that

$$\lim_{x\to 1-0}(1-x)^2\cdot\sum_{n=0}^{\infty}(n+1)c_nx^n=0.$$

Let $\varepsilon>0$ be fixed. Since $c_n\to 0$, there exists an index N such that $|c_n|<\varepsilon$ holds for every $n\geq N$. Now we have

$$\left|(1-x)^2\cdot\sum_{n=N}^{\infty}(n+1)c_nx^n\right|<\varepsilon\cdot(1-x)^2\cdot\sum_{n=N}^{\infty}(n+1)x^n<$$

$$<\varepsilon\cdot(1-x)^2\cdot\sum_{n=0}^{\infty}(n+1)x^n=\varepsilon. \qquad (7.24)$$

Since $\lim_{x\to 1-0}(1-x)^2\cdot\sum_{n=0}^{N-1}(n+1)c_nx^n=0$, hence

$$\left|(1-x)^2\cdot\sum_{n=0}^{N-1}(n+1)c_nx^n\right|<\varepsilon \qquad (7.25)$$

for $1-\delta<x<1$. Finally, comparing (7.24) and (7.25) we obtain

$$\left|(1-x)^2\cdot\sum_{n=0}^{\infty}(n+1)c_nx^n\right|<2\varepsilon$$

if $1-\delta<x<1$. This is what we wanted to prove. □

Remark 7.69. The previous theorem (along with its proof) can be easily generalized to show that if an infinite series is (H,k) summable for a k, then the series is necessarily Abel summable and its Abel sum is the same as its (H,k) sum.

For example, the series $1-2+3-4+\ldots$ is Abel summable, since

$$\sum_{n=0}^{\infty}(-1)^n\cdot(n+1)x^n=\frac{1}{(1+x)^2}$$

for every $x\in(-1,1)$, and the right-hand side converges to $1/4$ as $x\to 1-0$. This agrees with the fact that the $(H,2)$ sum of the series is $1/4$.

Thus the Abel summation is more "efficient" than all (H,k) summation. Furthermore, as the following example shows, there are Abel summable series that are not (H,k) summable for any k.

Example 7.70. The function $e^{1/(1+x)}$ is analytic on $(-1,1)$, and its Taylor series at 0 represents the function there (see Exercise 7.60 (a)). Let this series be $\sum_{n=0}^{\infty}a_nx^n$. Then

$$\lim_{x \to 1-0} \sum_{n=0}^{\infty} a_n x^n = \lim_{x \to 1-0} e^{1/(1+x)} = e^{1/2},$$

i.e., the series $\sum_{n=0}^{\infty} a_n$ is Abel summable (and its Abel sum is $e^{1/2}$).

On the other hand, the series $\sum_{n=0}^{\infty} a_n$ is not (H, k) summable for any k. This follows from the fact that every (H, k) summable series satisfies $\lim_{n \to \infty} a_n/n^k = 0$ (see Exercise 7.76); however, this is not true for the series above (see Exercise 7.78).

Exercises

7.72. Show that if a series of non-negative terms is (H, k) summable, then it is convergent.

7.73. Show that if a series of non-negative terms is Abel summable, then it is convergent.

7.74. Show that the series $1^2 - 2^2 + 3^2 - 4^2 + 5^2 - \ldots$ is $(H, 3)$ summable and find its $(H, 3)$ sum.

7.75. Check that the series $1^2 - 2^2 + 3^2 - 4^2 + 5^2 - \ldots$ is Abel summable and its Abel sum is the same as its $(H, 3)$ sum.

7.76. Show that if the series $\sum_{n=0}^{\infty} a_n$ is (H, k) summable then

$$\lim_{n \to \infty} a_n/n^k = 0.$$

(H)

7.77. Show that the series $1^k - 2^k + 3^k - 4^k + 5^k - \ldots$ is Abel summable, but it is not (H, k) summable.

7.78. Let $\sum_{n=0}^{\infty} a_n x^n$ be the Taylor series of the function $e^{1/(1+x)}$ corresponding to the point 0. Show that the sequence (a_n/n^k) $(n = 1, 2, \ldots)$ is not bounded for any k. (H)

7.5 Fourier Series

The development of the theory of differentiation and integration in the seventeenth century was mostly motivated by physics-based problems. Physics retained its key role in the motivation of analysis: in the eighteenth and nineteenth centuries several physics-based problem arose that elucidated some basic notions of analysis (such as functions, limits and infinite series), and also helped the emergence of new mathematical theories. Such problems were to find the equation of a vibrating string or the equation of heat conduction[9], which led to the following question. The series of the form

$$a_0 + \sum_{n=1}^{\infty} (a_n \cos nx + b_n \sin nx) \tag{7.26}$$

(where a_n and b_n are constants) are called **trigonometric series**. Since the functions $\cos nx$ and $\sin nx$ are 2π-periodic for every $n \in \mathbb{N}$, it follows that if the series (7.26) is convergent everywhere, then its sum is also 2π-periodic. Now the question is: can we write every 2π-periodic function in this form? If not, which 2π-periodic function can be written as the sum of a trigonometric series? These question led to the emergence of the theory of the Fourier series. The answers were found as late as in the twentieth century.

Proving the uniqueness of the representation was easier than showing its existence. Georg Cantor proved in 1870 that if a function can be written as the sum of a series of the form (7.26), then the representation is unique, i.e., the coefficients a_n and b_n are uniquely determined. We only prove Cantor's theorem for the case when the series (7.26) is uniformly convergent (for the general case, see [16, (3.1) Theorem].

Theorem 7.71. *Let the series* (7.26) *be uniformly convergent on* $\mathbb{R}$, *and let* $f(x)$ *be the sum of the series. Then* f *is continuous, and we have*

$$a_0 = \frac{1}{2\pi} \int_0^{2\pi} f(x)\,dx \tag{7.27}$$

and

$$a_n = \frac{1}{\pi} \int_0^{2\pi} f(x) \cos nx\,dx, \qquad b_n = \frac{1}{\pi} \int_0^{2\pi} f(x) \sin nx\,dx, \qquad (n \geq 1). \tag{7.28}$$

Lemma 7.72. *For every integer* $n \geq 1$ *we have*

$$\int_0^{2\pi} \sin^2 nx\,dx = \int_0^{2\pi} \cos^2 nx\,dx = \pi. \tag{7.29}$$

---

[9] For the details, see the third appendix of this chapter.

For every pair of integers n and m we have

$$\int_0^{2\pi} \sin nx \cos mx \, dx = 0. \tag{7.30}$$

For every pair of distinct integers n and m we have

$$\int_0^{2\pi} \cos nx \cos mx \, dx = \int_0^{2\pi} \sin nx \sin mx \, dx = 0. \tag{7.31}$$

Proof. The statements follow from the identities

$$\cos^2 x = (1 + \cos 2x)/2, \quad \sin^2 x = (1 - \cos 2x)/2,$$

$$\cos x \cos y = \tfrac{1}{2}\left(\cos(x+y) + \cos(x-y)\right),$$
$$\sin x \sin y = \tfrac{1}{2}\left(\cos(x-y) - \cos(x+y)\right),$$
$$\sin x \cos y = \tfrac{1}{2}\left(\sin(x+y) + \sin(x-y)\right),$$

using the fact that $\int_0^{2\pi} \cos kx \, dx = \int_0^{2\pi} \sin kx \, dx = 0$ for every integer $k \neq 0$. □

Proof of Theorem 7.71. Since the terms of the series (7.26) are continuous, the continuity of the sum f follows from Theorem 7.36. By Theorem 7.40, the series (7.26) is integrable term by term on the interval $[0, 2\pi]$, which gives (7.27) immediately.

We now prove that the series

$$a_0 \cos mx + \sum_{n=1}^{\infty} (a_n \cos nx \cos mx + b_n \sin nx \cos mx) \tag{7.32}$$

is also uniformly convergent on $\mathbb{R}$ for every $m > 0$. Let $\varepsilon > 0$ be fixed. By the uniform convergence of the series (7.26), there exists an N such that the nth partial sum of the series is closer to its sum than ε for every $n > N$ and for every $x \in \mathbb{R}$. Since $|\cos mx| \leq 1$, the nth partial sum of the series (7.32) is also closer to its sum (i.e., to $f(x) \cos mx$) than ε for every $n > N$ and for every $x \in \mathbb{R}$. Thus the series (7.32) is indeed uniformly convergent, and then it is term by term integrable on $[0, 2\pi]$. Therefore, by Lemma 7.72 we obtain $\int_0^{2\pi} f(x) \cos mx \, dx = \pi \cdot a_m$. We get $\int_0^{2\pi} f(x) \sin mx \, dx = \pi \cdot b_m$ in the same way, which proves (7.28). □

Remark 7.73. In the formulas (7.27) and (7.28) we could integrate f on any interval of length 2π and not just on the interval $[0, 2\pi]$. This follows from the fact that if f is p-periodic for some $p > 0$, and integrable on $[0, p]$, then it is integrable on every interval $[a, a + p]$, and

$$\int_{a}^{a+p} f\,dx = \int_{0}^{p} f\,dx. \tag{7.33}$$

This can be proved as follows. Let $(k-1)p \le a < kp$, where k is an integer. The periodicity of f implies $\int_a^{kp} f\,dx = \int_{a-(k-1)p}^{p} f\,dx$ and $\int_{kp}^{a+p} f\,dx = \int_0^{a-(k-1)p} f\,dx$, and thus (7.33) follows.

Formulas (7.27) and (7.28) were already known by Euler. However, their systematic investigation started with the work of Fourier[10], who used (7.27) and (7.28) and the series of the form (7.26) to solve the equation of the heat conduction.

Definition 7.74. Let $f\colon \mathbb{R} \to \mathbb{R}$ be 2π-periodic and integrable on $[0, 2\pi]$. The numbers defined by (7.27) and (7.28) are called the *Fourier coefficients* of f, and the series (7.26) written with these coefficients is called the *Fourier series* of f.

If the Fourier series of f is convergent at a point x and its sum is $f(x)$, then we say that the Fourier series *represents f at the point x*.

Examples 7.75. **1.** Let f be the 2π-periodic function such that $f(x) = x^2$ for every $x \in [-\pi, \pi]$. Obviously, f is an even function.

It follows that the coefficients b_n of the Fourier series of the function (i.e., the coefficients of the terms $\sin nx$) are zero. Indeed, by Remark 7.73,

$$b_n = \frac{1}{\pi}\int_{-\pi}^{\pi} f(x)\sin nx\,dx.$$

Now, the value of the integral is zero, since on the right-hand side we integrate an odd function on the interval $[-\pi, \pi]$.

Let us find the coefficients a_n. First, we have

$$a_0 = \frac{1}{2\pi}\int_{-\pi}^{\pi} x^2\,dx = \frac{\pi^2}{3}.$$

If $n > 0$, then using integration by parts yields

$$\begin{aligned} a_n &= \frac{1}{\pi}\int_{-\pi}^{\pi} x^2 \cos nx\,dx = \frac{1}{\pi}\cdot\left[x^2\cdot\frac{\sin nx}{n}\right]_{-\pi}^{\pi} - \frac{2}{\pi n}\int_{-\pi}^{\pi} x\sin nx\,dx = \\ &= 0 + \frac{2}{\pi n^2}\left[x\cos nx\right]_{-\pi}^{\pi} - \frac{2}{\pi n^2}\int_{-\pi}^{\pi}\cos nx\,dx = \\ &= (-1)^n\frac{4}{n^2}. \end{aligned}$$

[10] Jean Baptiste Fourier (1768–1830), French mathematician.

Thus the Fourier series of the function is

$$\frac{\pi^2}{3} - 4 \cdot \left(\cos x - \frac{\cos 2x}{2^2} + \frac{\cos 3x}{3^2} - \ldots \right). \tag{7.34}$$

Later we will see that this series represents the function everywhere (see Example 7.80).

2. Let f be the 2π-periodic function such that $f(x) = (\pi - x)/2$ for every $x \in (0, 2\pi)$, and $f(k\pi) = 0$ for every integer k. Integration by parts gives $a_n = 0$ for every n and $b_n = 1/n$ for every $n > 0$. I.e., the Fourier series of f is the series $\sum_{n=1}^{\infty} (\sin nx)/n$. This series also represents the function everywhere (see Theorem 7.86).

3. Let f be the 2π-periodic function such that $f(x) = (x^2/4) - (\pi x/2) + (\pi^2/6)$ for every $x \in [0, 2\pi]$. Performing integration by parts twice, we obtain $b_n = 0$ for every n and $a_n = 1/n^2$ for every $n > 0$. Thus the Fourier series of f is the series $\sum_{n=1}^{\infty} (\cos nx)/n^2$. This series also represents the function everywhere (see Theorem 7.79).

Theorem 7.71 can be rephrased as follows. *If a trigonometric series is uniformly convergent, then the series is necessarily the Fourier series of its sum.*

In general, we cannot expect that the Fourier series of every function represents the function. Indeed, if we change the value of the function f at a single point, then the integrals of (7.27) and (7.28) do not change, thus the Fourier series and its sum are also unchanged. Thus we can always force a function *not* to be represented by its Fourier series at a fixed point. The real question is, whether a continuous and 2π-periodic function is represented by its Fourier series, or not?

In order to understand the nature of this question better, let us recall some results on power series and Taylor series. In this context, we can draw a parallel between trigonometric series and power series on one hand, and the Fourier series of continuous functions and the Taylor series of infinitely differentiable functions on the other hand. According to this analogy, the statement corresponding to Theorem 7.71 is that a convergent power series is always equal to the Taylor series of its sum (Theorem 7.54). On the other hand, we know that the function $f(x) = e^{-1/x^2}$, $f(0) = 0$ is infinitely differentiable, its Taylor series is convergent everywhere, but the sum of its Taylor series is equal to the function at no point (except for the origin). The analogous statement would be that there exists a continuous function whose Fourier series is convergent, but its sum is not equal to the function anywhere. Is this true? In Example 7.46.1. we defined a function which is infinitely differentiable, and whose Taylor series is divergent everywhere (except for the origin). Is there a continuous function whose Fourier series is divergent everywhere?

These questions were not answered until the end of the nineteenth century and in the twentieth century. Lipót Fejér[11] proved the following theorem in 1900 (the proof of which goes beyond the limits of this book.)

[11] Lipót Fejér (1880–1959), Hungarian mathematician.

Theorem 7.76. (Fejér's theorem) *The Fourier series of a continuous and 2π-periodic function is summable everywhere, and its sum is equal to the function itself. Consequently, if the Fourier series of a continuous function is convergent at a point, then its sum must be the value of the function at that point.*

The first examples of continuous functions whose Fourier series were divergent at some points were constructed in the nineteenth century. It was unsolved until 1966 whether the Fourier series of a continuous function can be divergent everywhere or not. Carleson[12] proved that this is impossible: the Fourier series of every continuous (further, every integrable) function converges to the value of the function almost everywhere (in a certain, well-defined way; see Chapter 8). I.e., the Fourier series behave better than the Taylor series in both sense.

Several sufficient conditions are known that guarantee the representation of a function by its Fourier series. E.g., it follows from the results of Dirichlet and Riemann that if f is continuous and monotone on an interval (a, b), then its Fourier series represents f there. It follows immediately that the functions of Examples 7.75 are represented by their Fourier series everywhere. (We will prove this presently without using the results of Dirichlet and Riemann.)

In lack of space, we cannot delve into the theory of the Fourier series. The topic is explored by several books; see [15], [2], or [16]. We will prove, however, that if the Fourier series of a continuous function is uniformly convergent, then the Fourier series represents the function everywhere. We prove this in two steps. First, we show that if the Fourier series of a continuous function is the constant zero function, then the function is also identically zero. (The analogue of this statement for Taylor series is false: the Taylor series of the function $f(x) = e^{-1/x^2}$, $f(0) = 0$ is the constant zero function.)

Theorem 7.77. *Let $f\colon \mathbb{R} \to \mathbb{R}$ be continuous and 2π-periodic. If every Fourier coefficient of f is zero, then f is the constant zero function.*

Proof. I. The function $\cos^n x$ can be written as $\sum_{k=0}^{n} c_k \cos kx$ with suitable constants $c_0, \ldots, c_n$. We prove this by induction on n. The claim is true for $n = 1$. If it is true for n and

$$\cos^n x = \sum_{k=0}^{n} c_k \cos kx,$$

then

$$\cos^{n+1} x = \sum_{k=0}^{n} c_k \cos kx \cos x.$$

Using the identity $\cos kx \cos x = \frac{1}{2} \cdot (\cos(k+1)x + \cos(k-1)x)$ for every k and reordering the resulting terms yield the statement for $n + 1$.

From this observation it follows that if all the Fourier coefficients of the function f are zero, then $\int_0^{2\pi} f(x) \cos^n x \, dx = 0$ for every non-negative integer n. Which,

[12] Lennart Carleson (1928–), Swedish mathematician.

in turn, implies that

$$\int_0^{2\pi} f(x) \cdot p(\cos x)\, dx = 0,$$

for every polynomial p.

II. We now show that if f is continuous, 2π-periodic, and every Fourier coefficient of f is zero, then $f(0) = 0$. Suppose that $f(0) \neq 0$. Since the Fourier coefficients of the function $\frac{1}{f(0)} \cdot f(x)$ are also zero, we may assume that $f(0) = 1$. By the continuity of f, there exists a $0 < \delta < \pi/2$ such that $f(x) > 1/2$ for every $|x| < \delta$. The idea of the proof is to find a polynomial p such that $p(\cos x)$ is greater than 1 around the point 0, positive on $(-\delta, \delta)$, and small enough on $[-\pi, \pi] \setminus (-\delta, \delta)$ to make the integral $\int_{-\pi}^{\pi} f(x) \cdot p(\cos x)\, dx$ positive. (By Remark 7.73, we may integrate on any interval of length 2π.) Since the integral has to be zero, this will lead to a contradiction.

We show that the polynomial $p(x) = (x + \varepsilon)^N$ satisfies the conditions if ε is small enough and N is large enough. Let $0 < \varepsilon < 1$ small enough for $(\cos\delta) + \varepsilon < 1$ to hold. Let $\varphi(x) = (\cos x) + \varepsilon$. Since $\varphi(0) > 1$, there exists some $\eta > 0$ such that $\varphi(x) > 1$ for every $|x| < \eta$. Obviously, $\varphi(x) > 0$ is also true for every $|x| < \delta$. Note that ε and η only depend on δ.

For $x \in [\delta, \pi]$ we have $-1 + \varepsilon \le \varphi(x) \le (\cos\delta) + \varepsilon$, and thus $|\varphi(x)| \le q$, where $q = \max(1 - \varepsilon, (\cos\delta) + \varepsilon) < 1$. Therefore, if $x \in [\delta, \pi]$, then $|\varphi(x)^N| \le q^N$. Since φ is even, this is also true on $[-\pi, -\delta]$. Let $|f(x)| \le K$ for every $x \in [-\pi, \pi]$. Then we have

$$\left| \int_\delta^\pi f(x) \cdot \varphi(x)^N\, dx \right| \le \pi \cdot K \cdot q^N \quad \text{and} \quad \left| \int_{-\pi}^{-\delta} f(x) \cdot \varphi(x)^N\, dx \right| \le \pi \cdot K \cdot q^N.$$

On the other hand,

$$\int_{-\delta}^{\delta} f(x) \cdot \varphi(x)^N\, dx > \int_{-\eta}^{\eta} f(x) \cdot \varphi(x)^N\, dx > \int_{-\eta}^{\eta} f(x)\, dx > \frac{1}{2} \cdot 2\eta = \eta.$$

Thus $\int_{-\pi}^{\pi} f(x) \cdot \varphi(x)^N\, dx > \eta - 2\pi K q^N > 0$ for N large enough, which is a contradiction.

III. Let f be continuous and 2π-periodic, and let every Fourier coefficients of f be zero. Then, for every fixed $a \in \mathbb{R}$, the function $f(x + a)$ is also continuous and 2π-periodic We show that the Fourier coefficients of $f(x + a)$ are also zero. Indeed,

$$\int_0^{2\pi} f(x+a)\cos nx\,dx = \int_a^{2\pi+a} f(x)\cos(n(x-a))\,dx =$$

$$= \int_0^{2\pi} f(x)(\cos nx\cos a + \sin nx \sin a)\,dx =$$

$$= \cos a \cdot \int_0^{2\pi} f(x)\cos nx\,dx + \sin a \cdot \int_0^{2\pi} f(x)\sin nx\,dx =$$

$$= (\cos a)\cdot 0 + (\sin a)\cdot 0 = 0,$$

and we can prove $\int_0^{2\pi} f(x+a)\sin nx\,dx = 0$ similarly. Part II. of our proof implies that $f(0+a) = 0$. Since this is true for every a, the function f is identically zero. □

Remark 7.78. We say that the integrable functions $f, g\colon [a,b] \to \mathbb{R}$ are **orthogonal** if $\int_a^b f \cdot g\,dx = 0$. (As for the motivation of this notion, see [7, Remark 14.57].) Using this terminology, formulas (7.30) and (7.31) state that on the interval $[0, 2\pi]$ any two of the functions $\cos nx$ $(n = 0, 1, \ldots)$ and $\sin nx$ $(n = 1, 2, \ldots)$ are orthogonal. Now Theorem 7.77 states that if a continuous and 2π-periodic function f is orthogonal to every one of these functions, then f is identically zero. Thus Theorem 7.77 says that the system of trigonometric functions cannot be extended with respect to their orthogonality, i.e, *the trigonometric function form a complete system.* For this reason we also call Theorem 7.77 the **completeness theorem**.

We can now easily prove that if the Fourier series of a function is uniformly convergent, then it represents the function.

Theorem 7.79. *Let $f\colon \mathbb{R} \to \mathbb{R}$ be continuous and 2π-periodic. If the Fourier series of f is uniformly convergent on $\mathbb{R}$, then its sum is equal to $f(x)$ everywhere.*

Proof. Let the sum of the Fourier series of f be g. By Theorem 7.71, g is continuous, and the Fourier coefficient of f and g are the same. It follows easily that the Fourier coefficients of the continuous and 2π-periodic function $(f - g)$ are zero. Then, by Theorem 7.77, $f - g = 0$, i.e., $f = g$. □

Example 7.80. From Theorem 7.79 it follows that the Fourier series of the function of Example 7.75.1. represents the function everywhere, since the function is continuous, and its Fourier series is uniformly convergent by the Weierstrass criterion. We get that

$$\frac{\pi^2}{3} - 4\cdot\left(\cos x - \frac{\cos 2x}{2^2} + \frac{\cos 3x}{3^2} - \ldots\right) = x^2$$

for every $|x| \le \pi$. Plugging $x = \pi$ we obtain that $\sum_{n=1}^{\infty} 1/n^2 = \pi^2/6$.

We now prove that every smooth enough function is represented by its Fourier series.

Lemma 7.81. *If the function $f\colon \mathbb{R} \to \mathbb{R}$ is 2π-periodic and k times continuously differentiable on $\mathbb{R}$, then there exists an $M > 0$ such that*

$$|a_n| \le M/n^k, \quad \textit{and} \quad |b_n| \le M/n^k \tag{7.35}$$

hold for the Fourier coefficients of f for every $n \ge 1$.

Proof. We prove by induction on k. For $k = 0$ the condition simply means (by definition) that f is continuous on $[0, 2\pi]$. In this case, f is bounded. Suppose that $|f(x)| \le K$ for every $x \in [0, 2\pi]$ with an appropriate positive number K. Then it follows from (7.28) that $|a_n| \le 2K$ and $|b_n| \le 2K$ for every $n \ge 1$. (Actually, it is also true that the coefficients a_n and b_n converge to zero as $n \to \infty$, but we will not need this. See Exercise 7.96.)

Suppose that the statement of the theorem is true for k, and let $f\colon \mathbb{R} \to \mathbb{R}$ be 2π-periodic and $(k+1)$-times continuously differentiable. Then f' is k-times continuously differentiable and, by the induction hypothesis, the Fourier coefficients of f' satisfy the inequalities (7.35) with an appropriate $M > 0$. Integration by parts gives

$$\int_0^{2\pi} f(x) \cos nx \, dx = \left[f(x) \cdot \frac{\sin nx}{n} \right]_0^{2\pi} - \int_0^{2\pi} f'(x) \cdot \frac{\sin nx}{n} \, dx =$$

$$= 0 - \frac{1}{n} \cdot \int_0^{2\pi} f'(x) \cdot \sin nx \, dx,$$

and

$$\left| \int_0^{2\pi} f(x) \cos nx \, dx \right| \le M/n^{k+1}.$$

Similarly (by using $f(2\pi) = f(0)$), we get that

$$\left| \int_0^{2\pi} f(x) \sin nx \, dx \right| \le M/n^{k+1}$$

for every $n \ge 1$. □

Theorem 7.82. *If $f\colon \mathbb{R} \to \mathbb{R}$ is 2π-periodic and twice continuously differentiable, then it is represented by its Fourier series everywhere.*

Proof. By Lemma 7.81, the Fourier coefficients of f satisfy $|a_n| \le M/n^2$ and $|b_n| \le M/n^2$ with an appropriate M. Thus, by the Weierstrass criterion, the Fourier series of f is uniformly convergent, and we can apply Theorem 7.79. □

Remark 7.83. The conditions of the theorem can be significantly relaxed. For example, it follows from Dirichlet's convergence theorems that every differentiable and 2π-periodic function is represented by its Fourier series.

We call the finite sums of the form $a_0 + \sum_{k=1}^{n}(a_k \cos kx + b_k \sin kx)$ (where a_k and b_k are constants) **trigonometric polynomials**. Every trigonometric polynomial is continuous and 2π-periodic. Therefore, if a function $f\colon \mathbb{R} \to \mathbb{R}$ can be written as the uniform limit of a sequence of trigonometric polynomials, then f is continuous (and of course 2π-periodic). We now show that the converse is also true.

Theorem 7.84. (Weierstrass' 2nd approximation theorem) *If $f\colon \mathbb{R} \to \mathbb{R}$ is continuous and 2π-periodic, then for every $\varepsilon > 0$ there exists a trigonometric polynomial t such that $|f(x) - t(x)| < \varepsilon$ holds for every x.*

Proof. If f is twice continuously differentiable, then f is represented by its Fourier series everywhere by Theorem 7.82. In the proof of Theorem 7.82 we also showed that the Fourier series of f is uniformly convergent. Thus, for every $\varepsilon > 0$, the nth partial sum of the Fourier series satisfies the conditions of the theorem, if n is large enough.

Therefore, it is enough to prove that if f is continuous and 2π-periodic, then f can be uniformly approximated by twice continuously differentiable and 2π-periodic functions.

Let $\varepsilon > 0$ be fixed. Since f is uniformly continuous on $[0, 2\pi]$, there exists a $\delta > 0$ such that $|f(x) - f(y)| < \varepsilon$ whenever $x, y \in [0, 2\pi]$ and $|x - y| < \delta$. Fix a partition $0 = x_0 < x_1 < \ldots < x_n = 2\pi$ of the interval $[0, 2\pi]$ finer than δ. For every $i = 1, \ldots, n$ choose a function $\varphi_i\colon [x_{i-1}, x_i] \to \mathbb{R}$ with the following properties:

(i) φ_i is continuously differentiable on $[x_{i-1}, x_i]$,
(ii) $\varphi_i(x_{i-1}) = \varphi_i'(x_{i-1}) = \varphi_i(x_i) = \varphi_i'(x_i) = 0$,
(iii) the sign of φ_i does not change on (x_{i-1}, x_i), and
(iv) $\int_{x_{i-1}}^{x_i} \varphi_i(x)\,dx = f(x_i) - f(x_{i-1})$.

It is easy to see that the function $c \cdot (x - x_{i-1})^2 \cdot (x_i - x)^2$ has these properties with a suitable constant c.

Let $\varphi(x) = \varphi_i(x)$ for every $x \in [x_{i-1}, x_i]$ and every $i = 1, \ldots, n$, and let us extend φ periodically to $\mathbb{R}$. Clearly, φ is continuously differentiable and 2π-periodic. Let $g(x) = f(0) + \int_0^x \varphi(t)\,dt$ for every $x \in \mathbb{R}$. It is easy to see that g is twice continuously differentiable. We show that g is 2π-periodic and $|f(x) - g(x)| < \varepsilon$ for every x. The 2π-periodicity of the function g follows from

$$\int_0^{2\pi} \varphi(t)\,dt = \sum_{i=1}^{n} \int_{x_{i-1}}^{x_i} \varphi_i(x)\,dx = \sum_{i=1}^{n} (f(x_i) - f(x_{i-1})) = f(2\pi) - f(0) = 0.$$

The values of the functions f and g are equal at the points x_i. Indeed,

$$g(x_i) = f(0) + \int_0^{x_i} \varphi(t)\,dt = f(0) + \sum_{j=1}^{i} \big(f(x_j) - f(x_{j-1})\big) = f(x_i).$$

If $x \in [x_{i-1}, x_i]$, then $|f(x) - f(x_{i-1})| < \varepsilon$ and $|f(x) - f(x_i)| < \varepsilon$, since $|x_i - x_{i-1}| < \delta$ implies $|x - x_{i-1}| < \delta$ and $|x - x_i| < \delta$. On the other hand $g(x)$ lies between the numbers $g(x_{i-1}) = f(x_{i-1})$ and $g(x_i) = f(x_i)$, since the sign of φ_i does not change, and thus g is monotone on $[x_{i-1}, x_i]$. Thus we have $|g(x) - f(x)| < \varepsilon$. Since f and g are 2π-periodic, this is true for every x. □

Remark 7.85. Using Theorem 7.84 we can give a new proof of Weierstrass' (first) approximation theorem, i.e., Theorem 7.8, for functions of one variable. Let $f\colon [0,1] \to \mathbb{R}$ be continuous, and let $\varepsilon > 0$ be fixed. We will construct a polynomial p such that $|f - p| < \varepsilon$ on the interval $[0,1]$.

Extend f continuously to the interval $[0, 2\pi]$ in such a way that $f(2\pi) = f(0)$ also holds. (E.g., let f be linear on the interval $[1, 2\pi]$.) Extend the resulting function to $\mathbb{R}$ 2π-periodically, and denote the resulting function also by f. Since f is continuous on $\mathbb{R}$, it follows from Theorem 7.84 that there exists a trigonometric polynomial t such that $|f(x) - t(x)| < \varepsilon/2$ for every x. We know that the functions $\cos x$ and $\sin x$ are represented everywhere by their Taylor series corresponding to the point zero, and that these Taylor series are uniformly convergent on every bounded interval. From this it follows easily that the trigonometric polynomial t is also represented everywhere by its Taylor series corresponding to the point zero, and this Taylor series is uniformly convergent on $[0,1]$. Thus, if p_n denotes the nth partial sum of the Taylor series of t, then for n large enough, $|t(x) - p_n(x)| < \varepsilon/2$ holds for every $x \in [0,1]$. Then $|f(x) - p_n(x)| < 2\varepsilon$ for every $x \in [0,1]$, which proves the statement.

Finally, we find the sum of two important trigonometric series.

Theorem 7.86.

(i) *The series $\sum_{n=1}^{\infty} (\sin nx)/n$ is convergent for every $x \in \mathbb{R}$, and its sum is 2π-periodic. The sum of the series is zero for $x = k\pi$ ($k \in \mathbb{Z}$), and*

$$\sum_{n=1}^{\infty} \frac{\sin nx}{n} = \frac{\pi - x}{2} \tag{7.36}$$

for every $0 < x < 2\pi$.

(ii) *The series $\sum_{n=1}^{\infty} (\cos nx)/n$ is divergent for every $x = 2k\pi$ ($k \in \mathbb{Z}$), and convergent with sum*

$$\sum_{n=1}^{\infty} \frac{\cos nx}{n} = -\log\left|2\sin\frac{x}{2}\right| \tag{7.37}$$

for every $x \neq 2k\pi$ ($k \in \mathbb{Z}$).

Proof. Let the derivative of the nth partial sums of the two series be $s_n(x) = \sum_{j=1}^{n} \sin jx$ and $c_n(x) = \sum_{j=1}^{n} \cos jx$, respectively. By Lemma 6.64, the sequences $(|s_n(x)|)$ and $(|c_n(x)|)$ are bounded if $x \neq 2k\pi$ (their upper bound is $1/|\sin(x/2)|$). Now the convergence of the two series follow from Dirichlet's criterion (Theorem 6.47).

It is clear that the sum of the series $\sum_{n=1}^{\infty} (\sin nx)/n$ at the points $x = 2k\pi$ is zero. Since the function on the right-hand side of (7.37) is 2π-periodic, it is enough to show that (7.36) and (7.37) hold for every $0 < x < 2\pi$.

Let $0 < \delta < \pi$ be fixed. If $\delta \le x \le 2\pi - \delta$, then $\sin(x/2) \ge \sin(\delta/2)$ and $|\operatorname{ctg}(x/2)| \le 1/\sin(\delta/2)$. It follows from (6.23) and (6.24) that

$$\left| \frac{s_1(x) + \ldots + s_n(x)}{n} - \frac{1}{2} \operatorname{ctg} \frac{x}{2} \right| \le \frac{1}{2\sin^2(\delta/2)} \cdot \frac{1}{n} + \frac{1}{2\sin(\delta/2)} \cdot \frac{1}{n}$$

and

$$\left| \frac{c_1(x) + \ldots + c_n(x)}{n} + \frac{1}{2} \right| \le \frac{1}{2\sin^2(\delta/2)} \cdot \frac{1}{n} + \frac{1}{2\sin(\delta/2)} \cdot \frac{1}{n},$$

for every $x \in [\delta, 2\pi - \delta]$. This means that the sequence of functions $(s_1 + \ldots + s_n)/n$ converges uniformly to the function $\frac{1}{2}\operatorname{ctg}(x/2)$, and the sequence of functions $(c_1 + \ldots + c_n)/n$ converges uniformly to the constant $-1/2$ function on the interval $[\delta, 2\pi - \delta]$. By Theorem 7.16 this implies

$$\int_{\pi}^{x} \frac{s_1(t) + \ldots + s_n(t)}{n}\, dt \to \int_{\pi}^{x} \frac{1}{2} \operatorname{ctg} \frac{t}{2}\, dt = \log \sin \frac{x}{2} \tag{7.38}$$

and

$$\int_{\pi}^{x} \frac{c_1(t) + \ldots + c_n(t)}{n}\, dt \to \frac{\pi - x}{2} \tag{7.39}$$

for every $x \in [\delta, 2\pi - \delta]$ as $n \to \infty$.

Let $S_n(x) = \sum_{j=1}^{n} (\sin jx)/j$ and $C_n(x) = \sum_{j=1}^{n} (\cos jx)/j$. Then

$$\int_{\pi}^{x} c_n(t)\, dt = \int_{\pi}^{x} (\cos t + \ldots + \cos nt)\, dt = \left[\sin t + \ldots + \frac{\sin nt}{n} \right]_{\pi}^{x} = S_n(x)$$

and

$$\int_{\pi}^{x} s_n(t)\, dt = \int_{\pi}^{x} (\sin t + \ldots + \sin nt)\, dt = \left[-\cos t - \ldots - \frac{\cos nt}{n} \right]_{\pi}^{x} = -C_n(x) - \alpha_n,$$

where $\alpha_n = 1 - (1/2) + \ldots + (-1)^{n-1}(1/n)$.

Thus $S_n(x) = \int_\pi^x c_n(t)\,dt$ and $C_n(x) = -\int_\pi^x s_n(t)\,dt - \alpha_n$, and we have

$$\frac{S_1(x)+\ldots+S_n(x)}{n} = \int_\pi^x \frac{c_1(t)+\ldots+c_n(t)}{n}\,dt$$

and

$$\frac{C_1(x)+\ldots+C_n(x)}{n} = -\int_\pi^x \frac{s_1(t)+\ldots+s_n(t)}{n}\,dt - \frac{\alpha_1+\ldots+\alpha_n}{n}.$$

Taking (7.38) and (7.39) into account we find that $(S_1(x)+\ldots+S_n(x))/n \to (\pi - x)/2$ and $(C_1(x)+\ldots+C_n(x))/n \to -\log\sin(x/2) - \log 2$ for every $x \in [\delta, 2\pi - \delta]$, since $\alpha_n \to \log 2$, and thus $(\alpha_1+\ldots+\alpha_n)/n \to \log 2$. Therefore, the series $\sum_{n=1}^\infty (\sin nx)/n$ and $\sum_{n=1}^\infty (\cos nx)/n$ are summable, and their sum is $(\pi - x)/2$ and $-\log\big(2\sin(x/2)\big)$, respectively. We have already proved that these series are convergent. Thus, by Theorem 6.60, their sum coincides with their sum as summable series. Since this holds for every $x \in [\delta, 2\pi - \delta]$ and every $0 < \delta < \pi$, the theorem is proved. □

Remark 7.87. Statement (i) of the theorem says that the function of Example 7.75.2 is represented everywhere by its Fourier series.

Exercises

7.79. Each of the following functions is defined on an interval of length 2π. Extend the functions to $\mathbb{R}$ as 2π-periodic functions, and find their Fourier series.
(a) $f(x) = x$ $(x \in [-\pi, \pi))$;
(b) $f(x) = |x|$ $(x \in [-\pi, \pi))$;
(c) $f(x) = x(\pi - |x|)$ $(x \in [-\pi, \pi))$;
(d) $f(x) = x$ $(x \in [0, 2\pi))$;
(e) $f(x) = x^2$ $(x \in [0, 2\pi))$
(f) $f(x) = 1$ $(x \in [0, \pi))$ and $f(x) = -1$ $(x \in [-\pi, 0))$;
(g) $f(x) = (x-\pi)^2$ $(x \in [0, \pi))$ and $f(x) = (x+\pi)^2$ $(x \in [-\pi, 0))$;
(h) $f(x) = |\sin x|$ $(x \in [0, 2\pi))$;
(i) $f(x) = \cos x$ $(|x| \le \pi/2)$ and $f(x) = 0$ $(\pi/2 < |x| \le \pi)$.

7.80. Find the points where the functions of the previous exercise are represented by their respective Fourier series. (H)

7.81. Show that if the 2π-periodic function f is even (odd), then its Fourier series has $b_n = 0$ ($a_n = 0$) for every $n \geq 1$.

7.82. Find the sum of the following series:

(a) $\sum_{n=1}^{\infty} q^n \sin nx$ ($|q| < 1$); (b) $\sum_{n=1}^{\infty} q^n \cos nx$ ($|q| < 1$);

(c) $\sum_{n=2}^{\infty} \frac{\sin nx}{n(n-1)}$; (d) $\sum_{n=2}^{\infty} \frac{\cos nx}{n(n-1)}$;

(e) $\sin x + \frac{\sin 2x}{2!} + \frac{\sin 3x}{3!} + \ldots$; (f) $\sin x + \frac{\sin 3x}{3} + \frac{\sin 5x}{5} + \ldots$;

(g) $\sin x - \frac{\sin 3x}{3} + \frac{\sin 5x}{5} - \ldots$; (h) $\frac{\cos 2x}{1\cdot 3} + \frac{\cos 4x}{3\cdot 5} + \ldots$;

(i) $\sin x + \frac{\sin 3x}{3^3} + \frac{\sin 5x}{5^3} + \ldots$.

7.83. Find the functions whose Fourier series is of the form

(a) $\sum_{n=1}^{\infty} a_n \cos 2nx$; (b) $\sum_{n=1}^{\infty} a_n \cos(2n+1)x$;

(c) $\sum_{n=1}^{\infty} a_n \sin 2nx$; (d) $\sum_{n=1}^{\infty} a_n \sin(2n+1)x$.

7.84. Show that the trigonometric polynomial $\sum_{n=1}^{N}(a_n \cos nx + b_n \sin nx)$ has a root.

7.85. Show that if $f(x) = 1 + \sum_{n=1}^{N}(a_n \cos nx + b_n \sin nx) \geq 0$ for every x, then $f(x) \leq N + 1$ for every x. (∗ H)

7.86. Show that if $\sum_{i=0}^{n} |a_i|^2 = 1$, then

$$\int_0^1 |a_0 + a_1 x + \ldots + a_n x^n| \, dx \leq \pi/2.$$

7.87. Show that if $n_1, \ldots, n_k$ are distinct integers, then

$$\frac{1}{2\pi}\int_0^{2\pi} |\cos n_1 x + \ldots + \cos n_k x| \, dx \leq \sqrt{\frac{k}{2}}.$$

7.88. Let the function f be 2π-periodic, and suppose that the Fourier series of the functions $f(x)$ and $f(x+1)$ are the same. Show that the Fourier series of f has $a_n = b_n = 0$ for every $n \geq 1$.

7.89. Construct a 2π-periodic and continuous function f such that the sequence $(n \cdot a_n)$ is not bounded, where a_n is the coefficient of $\cos nx$ in the Fourier series of f. (H)

7.90. Let (a_n) be a sequence such that $\sum_{n=1}^{\infty} |a_n \cos nx| \leq 1$ at every point of a non-degenerated interval. Show that $\sum_{n=1}^{\infty} |a_n| < \infty$. (∗ H)

7.91. Let (a_n) and (b_n) be sequences such that

$$\sum_{n=1}^{\infty} |a_n \cos nx + b_n \sin nx| \leq 1$$

at every point of a non-degenerated interval. Show that $\sum_{n=1}^{\infty}(|a_n| + |b_n|) < \infty$. (∗ H)

7.92. Let (a_n) be a monotone decreasing sequence that converges to zero. Show that the partial sums of $\sum_{n=1}^{\infty} a_n \sin nx$ are bounded if and only if the sequence $(n \cdot a_n)$ is bounded. (∗ H)

7.93. Let (a_n) be a monotone decreasing sequence that converges to zero. Show that $\sum_{n=1}^{\infty} a_n \sin nx$ is uniformly convergent on $\mathbb{R}$ if and only if $n \cdot a_n \to 0$. (∗ H)

7.94. Let f be continuously differentiable on $[0, \pi]$. Show that if $\int_0^\pi f\, dx = 0$ or $f(0) = f(\pi) = 0$, then $\int_0^\pi f^2\, dx \leq \int_0^\pi (f')^2\, dx$.

7.95. Let $f\colon \mathbb{R} \to \mathbb{R}$ be 2π-periodic and suppose that f is a piecewise constant function on $[0, 2\pi]$. Show that the sequence of the Fourier coefficients of f converges to zero.

7.96. Let $f\colon \mathbb{R} \to \mathbb{R}$ be 2π-periodic and integrable on $[0, 2\pi]$. Show that the sequence of the Fourier coefficients of f converges to zero. (Riemann's lemma) (H)

7.97. Let $f\colon \mathbb{R} \to \mathbb{R}$ be 2π-periodic and integrable on $[0, 2\pi]$. Show that if every Fourier coefficient of f is zero, then $f(x) = 0$ at every point x, where f is continuous.

7.98. Let $f\colon \mathbb{R} \to \mathbb{R}$ be 2π-periodic and integrable on $[0, 2\pi]$. Show that if the Fourier series of f is uniformly convergent, then the sum of this series is $f(x)$ at every point x, where f is continuous.

7.6 Further Applications

The infinite series $\sum_{i=1}^{\infty} 1/i^s$ is convergent for every $s > 1$; we denote its sum by $\zeta(s)$ (see Exercise 6.3). The above defined ζ function (defined on the half line $(1, \infty)$) is one of the most investigated functions of mathematics. The motive behind this interest is the discovery, made by Riemann in 1859, that the function ζ can be extended to the complex plane (except for the point 1) as a differentiable function, and that the properties of this extended function are in close relation with the distribution of the prime numbers. A famous (and after 150 years still unsolved) conjecture of Riemann claims that the real part of every non-real root of the complex extension of the ζ function is $1/2$ [19].

We know that $\zeta(2) = \pi^2/6$ (see Exercise 4.20 and Example 7.80.) Our next goal it to find the value of $\zeta(2k)$ for every positive integer k (see Theorem 7.92). To do this, we need to define a sequence of polynomials which also appears in other fields of mathematics (e.g., in combinatorics and probability theory).

Theorem 7.88. *There exists a uniquely defined sequence of polynomials $B_0(x)$, $B_1(x)$, ... with the following properties: $B_0(x) \equiv 1$, furthermore, $B_n'(x) = B_{n-1}(x)$ and $\int_0^1 B_n(x)\,dx = 0$, for every $n > 0$.*

Proof. We prove the existence and uniqueness of the polynomials $B_n(x)$ by induction. Let $n > 0$ and let the polynomial $B_{n-1}(x)$ be given. Let $F(x) = \int_0^x B_{n-1}(t)\,dt$. Clearly, there is a unique function f such that $f'(x) = B_{n-1}(x)$ and $\int_0^1 f\,dx = 0$, namely, the function $f(x) = F(x) - c$, where $c = \int_0^1 F(x)\,dx$. It is also obvious that f is a polynomial. □

We call the polynomials given by the previous theorem the **Bernoulli**[13] **polynomials**. The first five Bernoulli polynomials are the following:

$$B_0(x) = 1, \qquad B_1(x) = x - \frac{1}{2}, \qquad B_2(x) = \frac{1}{2}x^2 - \frac{1}{2}x + \frac{1}{12},$$

$$B_3(x) = \frac{1}{6}x^3 - \frac{1}{4}x^2 + \frac{1}{12}x, \qquad B_4(x) = \frac{1}{24}x^4 - \frac{1}{12}x^3 + \frac{1}{24}x^2 - \frac{1}{720}.$$

For $n > 1$, we have

$$B_n(1) - B_n(0) = \int_0^1 B_n'(x)\,dx = \int_0^1 B_{n-1}(x)\,dx = 0,$$

thus

$$B_n(0) = B_n(1) \qquad (n = 2, 3, \ldots). \tag{7.40}$$

[13] Jacob Bernoulli (1654–1705), Swiss mathematician.

It is clear from the construction of the Bernoulli polynomials that $B_n(x)$ is an nth-order polynomial with rational coefficients. We call the numbers $B_n = n! \cdot B_n(0)$ **Bernoulli numbers**. Below, we list the first few Bernoulli numbers:

$$\begin{aligned} &B_0 = 1, && B_1 = -1/2, && B_2 = 1/6, && B_3 = 0,\\ &B_4 = -1/30, && B_5 = 0, && B_6 = 1/42, && B_7 = 0,\\ &B_8 = -1/30, && B_9 = 0, && B_{10} = 5/66, && B_{11} = 0,\\ &B_{12} = -691/2730, && B_{13} = 0, && B_{14} = 7/6, && B_{15} = 0,\\ &B_{16} = -3617/510, && B_{17} = 0 && B_{18} = 43\,867/798. && \end{aligned} \tag{7.41}$$

Let us return to the topic of Fourier series. By Theorem 7.79, the function in Example 7.75.3 is represented everywhere by its Fourier series, since the function is continuous and its Fourier series is uniformly convergent by the Weierstrass criterion. We get that

$$\frac{x^2}{4} - \frac{\pi}{2}x + \frac{\pi^2}{6} = \sum_{n=1}^{\infty} \frac{\cos nx}{n^2},$$

for every $x \in [0, 2\pi]$. The left-hand side of the formula can be simplified if we replace x by $2\pi x$:

$$x^2 - x + \frac{1}{6} = \sum_{n=1}^{\infty} \frac{\cos 2n\pi x}{\pi^2 n^2} \qquad (x \in [0,1]). \tag{7.42}$$

This formula can be generalized the following way.

Theorem 7.89. *For every $x \in [0,1]$ and every positive integer k, we have*

$$B_{2k+1}(x) = (-1)^{k-1} \sum_{n=1}^{\infty} \frac{2\sin 2n\pi x}{(2n\pi)^{2k+1}} \tag{7.43}$$

and

$$B_{2k}(x) = (-1)^{k-1} \sum_{n=1}^{\infty} \frac{2\cos 2n\pi x}{(2n\pi)^{2k}}. \tag{7.44}$$

Proof. Since the left-hand side of (7.42) is $2 \cdot B_2(x)$, thus (7.44) holds for $k = 1$. Now, consider the series $\sum_{n=1}^{\infty} (2\sin 2n\pi x)/(2n\pi)^3$. By Weierstrass' criterion, the series is uniformly convergent on $\mathbb{R}$, and then its sum – which we denote by f – is everywhere continuous. By differentiating term by term we get a series which is $1/2$-times the right-hand side of (7.42), which is a uniformly convergent series. Thus, applying Theorem 7.42, we obtain that f is everywhere differentiable and its derivative is $B_2(x)$ for every $x \in [0,1]$. By the definition of the polynomial $B_3(x)$, we have $B_3'(x) = B_2(x)$, and thus $f(x) = B_3(x) + c$ on the interval $[0,1]$. The series defining f is term by term integrable on every interval (because it is uniformly convergent), thus we have $\int_0^1 f\,dx = 0$, since the integral of each term of the series is zero on $[0,1]$. It follows that

$$0 = \int\limits_0^1 f\,dx = \int\limits_0^1 (B_3(x) + c)\,dx = c,$$

because $\int_0^1 B_3(x)\,dx = 0$ by the definition of $B_3(x)$. We get that $c = 0$ and $f(x) = B_3(x)$ on the interval $[0,1]$, which proves (7.43), for $k = 1$.

Now, consider the series $\sum_{n=1}^{\infty} (2\cos 2n\pi x)/(2n\pi)^4$. Repeating the argument above (nearly word by word), we get that the sum of the series is $B_4(x)$, for every $x \in [0,1]$. Continuing the process yields (7.43) and (7.44), for every positive integer k. □

Remarks 7.90. **1.** Replacing x by $2\pi x$ in the equality (7.36), we get

$$x - \frac{1}{2} = -\sum_{n=1}^{\infty} \frac{\sin 2n\pi x}{n\pi}, \tag{7.45}$$

for every $x \in (0,1)$. As $B_1(x) = x - \frac{1}{2}$, this means that the equality (7.43) is also true for $k = 0$, at least when $x \in (0,1)$. The equality does not hold for $x = 0$ and $x = 1$, since the right-hand side of (7.45) is zero at these points.

2. According to Example 7.31, the right-hand side of (7.45) is uniformly convergent on the interval $[\delta, 2\pi - \delta]$ for every $0 < \delta < \pi$. Using this fact, we can give a new proof of Theorem 7.89. Consider the series $\sum_{n=1}^{\infty} (2\cos 2n\pi x)/(2n\pi)^2$, and let its sum be denoted by f. By taking the term by term derivative of the series, we get the right-hand side of (7.45), and thus $f'(x) = B_1(x)$ for every $x \in (0,1)$. We get that $f(x) = B_2(x) + c$ on the interval $(0,1)$. Since f and $B_2(x)$ are continuous everywhere, we obtain that $f(x) = B_2(x) + c$ for every $x \in [0,1]$. Now, integrating the series defining f term by term, we get $\int_0^1 f\,dx = 0$, i.e., $f(x) = B_2(x)$ for every $x \in [0,1]$. The rest of the proof is the same as the original proof of Theorem 7.89.

Theorem 7.89 has several interesting corollaries. First, plugging $x = 0$ into (7.43) implies the following theorem.

Theorem 7.91. *We have $B_{2k+1} = 0$, for every integer $k \geq 1$.* □

On the other hand, by plugging $x = 0$ into (7.44), the right-hand side becomes $(-1)^{k-1}\zeta(2k) \cdot 2/(2\pi)^{2k}$, while the left-hand side is $B_{2k}/(2k)!$. This gives the formula proved by Euler:

Theorem 7.92. *We have*

$$\zeta(2k) = (-1)^{k-1} \frac{(2\pi)^{2k}}{2(2k)!} B_{2k}$$

for every $k \geq 1$. □

E.g., $\zeta(2) = \pi^2/6$, $\zeta(4) = \pi^4/90$ and $\zeta(6) = \pi^6/945$. Theorem 7.92 also implies that B_{2k} is positive when k is odd, and is negative when k is even.

Our next goal is to prove the following formula.

Theorem 7.93.

$$\pi \operatorname{ctg} \pi x = \lim_{N\to\infty} \sum_{n=-N}^{N} \frac{1}{x-n}, \tag{7.46}$$

for every $x \in \mathbb{R} \setminus \mathbb{Z}$.

An important ingredient of the proof is the fact that the function $f(x) = \operatorname{ctg} \pi x$ satisfies the functional equation

$$f\left(\frac{x}{2}\right) + f\left(\frac{x+1}{2}\right) = 2f(x) \tag{7.47}$$

for every $x \in \mathbb{R} \setminus \mathbb{Z}$ (check this fact). We note that there exists several elementary functions that satisfy similar equations. We say that the function f is **k-replicative,** if there exists a constant a_k such that

$$f\left(\frac{x}{k}\right) + f\left(\frac{x+1}{k}\right) + \ldots + f\left(\frac{x+k-1}{k}\right) = a_k f(x) \tag{7.48}$$

for every $x \in D(f)$. (E.g., the function $\operatorname{ctg} \pi x$ is 2-replicative with $a_2 = 2$.) It is easy to see that the function $x - \frac{1}{2}$ is k-replicative for every $k > 1$ with the constant $a_k = 1$. By induction on n it is not too hard to see that the Bernoulli polynomial $f(x) = B_n(x)$ is also k-replicative for every $k > 1$ with the constant $a_k = k^{1-n}$ (see Exercise 7.103). One can prove that this property characterizes the Bernoulli polynomials in the following sense: if a polynomial p is k-replicative for any $k > 1$, then p is a constant multiple of a Bernoulli polynomial (see Exercise 7.104).

The function $f(x) = \operatorname{ctg} \pi x$ is also k-replicative for every $k > 1$; this follows from either Theorem 6.63 or Theorem 7.93 (see Exercise 7.105). If k is a power of 2 then this follows immediately from (7.47) (see Exercise 7.107). For the proof of Theorem 7.93 we only need the $k = 2$ case, that is, (7.47).

Lemma 7.94. *If the function $f\colon [0,1] \to \mathbb{R}$ is continuous and satisfies* (7.47) *for every $x \in [0,1]$, then f is constant.*

Proof. Let the greatest value of f be M, and let $M = f(x)$ for some $x \in [0,1]$. Each term on the left-hand side of (7.47) is at most M, and thus (7.47) can hold only if $f(x/2) = M$. Repeating this argument for $x/2$, we get that $f(x/4) = M$ and, in general $f(x/2^n) = M$ for every n. Now f is continuous from the right at 0, and hence $f(0) = M$. By a similar argument we get that $f(0) = \min\{f(x) : x \in [0,1]\}$, and thus $f \equiv f(0)$. □

Proof of Theorem 7.93. We show that the limit on the right-hand side of (7.46) exists for every $x \notin \mathbb{Z}$. By doing some simple algebra, we get that

$$\sum_{n=-N}^{N} \frac{1}{x-n} = \frac{1}{x} + \sum_{n=1}^{N} \frac{2x}{x^2-n^2}. \tag{7.49}$$

If $K > 0$, $|x| \le K$ and $n > 2K$, then $x^2 < n^2/2$, $|x^2 - n^2| > n^2/2$, and thus

$$\left| \frac{2x}{x^2-n^2} \right| < \frac{2K}{n^2/2} = \frac{4K}{n^2}.$$

This implies that the series of functions $\sum_{n=1}^{\infty} 2x/(x^2-n^2)$ satisfies the Weierstrass criterion on $([-K,K] \setminus \mathbb{Z}) \cup \{0\}$. Thus the series is uniformly convergent on this set. This is true for every $K > 0$, which implies that the series is convergent for every $x \in (\mathbb{R} \setminus \mathbb{Z}) \cup \{0\}$, and its sum is continuous there. Thus the limit on the right-hand side of (7.46) exists if $x \notin \mathbb{Z}$. Denote the limit by $g(x)$. We can see that g is continuous on $\mathbb{R} \setminus \mathbb{Z}$, and

$$\lim_{x \to 0} \left(g(x) - \frac{1}{x} \right) = 0. \tag{7.50}$$

Now we show that the function g is 1-periodic. Indeed, for $x \notin \mathbb{Z}$ we have

$$\begin{aligned} g(x+1) - g(x) &= \lim_{N \to \infty} \left(\sum_{n=-N}^{N} \frac{1}{x+1-n} - \sum_{n=-N}^{N} \frac{1}{x-n} \right) = \\ &= \lim_{N \to \infty} \left(\frac{1}{x+1+N} - \frac{1}{x-N} \right) = 0. \end{aligned}$$

We now prove that the function g satisfies the functional equation (7.47). For a fixed N we have

$$\begin{aligned} \sum_{n=-N}^{N} \frac{1}{(x/2)-n} &+ \sum_{n=-N}^{N} \frac{1}{((x+1)/2)-n} - 2 \cdot \sum_{n=-2N}^{2N} \frac{1}{x-n} = \\ &= \sum_{n=-N}^{N} \frac{2}{x-2n} + \sum_{n=-N}^{N} \frac{2}{x+1-2n} - \sum_{n=-2N}^{2N} \frac{2}{x-n} = \\ &= \frac{2}{x+1+2N}. \end{aligned}$$

Since this converges to $g\left(\frac{x}{2}\right) + g\left(\frac{x+1}{2}\right) - 2g(x)$ on the one hand and to zero on the other hand as $N \to \infty$, we can see that g satisfies (7.47) as we stated. Thus the same is true for the function $h(x) = \pi \operatorname{ctg} \pi x - g(x)$.

Applying the L'Hospital rule, we get that

$$\lim_{x\to 0}\left(\pi\operatorname{ctg}\pi x-\frac{1}{x}\right)=0. \tag{7.51}$$

Thus (7.50) implies $\lim_{x\to 0} h(x)=0$. Since h is 1-periodic (because $\pi\operatorname{ctg}\pi x$ and g are also 1-periodic), hence $\lim_{x\to 1} h(x)=0$ is also true. Therefore, if we define the function h to be 0 at the points 0 and 1, then this extended function (which we also denote by h) is continuous on $[0,1]$. We know that $h\left(\frac{x}{2}\right)+h\left(\frac{x+1}{2}\right)=2h(x)$, for every $x\in(0,1)$. The continuity of h implies that this holds for $x=0$ and $x=1$ as well. Then, by Lemma 7.94, $h=0$ on the interval $[0,1]$. Since h is 1-periodic, it follows that $h(x)=0$ for every $x\notin\mathbb{Z}$, and the theorem is proved. □

Comparing the statement of the previous theorem with (7.49), we get that

$$\pi\operatorname{ctg}\pi x=\frac{1}{x}+\sum_{n=1}^{\infty}\frac{2x}{x^2-n^2} \tag{7.52}$$

for every $x\in\mathbb{R}\setminus\mathbb{Z}$. If $|x|<1$ and $n\in\mathbb{N}^+$, then

$$\frac{1}{n^2-x^2}=\frac{1}{n^2}\cdot\frac{1}{1-\frac{x}{n}^2}=\sum_{i=1}^{\infty}\frac{1}{n^{2i}}\cdot x^{2i-2}.$$

Since the series $\sum_{n=1}^{\infty}1/(n^2-x^2)$ is convergent, it follows from Theorem 6.30 that the terms x^{2i-2}/n^{2i} written in any order form an (absolutely) convergent series of sum $\sum_{n=1}^{\infty}1/(n^2-x^2)$. Now, for i fixed, we have $\sum_{n=1}^{\infty}x^{2i-2}/n^{2i}=\zeta(2i)\cdot x^{2i-2}$. Then using Theorem 6.30 again we obtain

$$\sum_{n=1}^{\infty}\frac{1}{n^2-x^2}=\sum_{i=1}^{\infty}\zeta(2i)\cdot x^{2i-2}.$$

By comparing this with (7.52) we have the following theorem.

Theorem 7.95. *For every $|x|<1$, we have*

$$\pi\operatorname{ctg}\pi x=\frac{1}{x}-2\cdot\sum_{i=1}^{\infty}\zeta(2i)\cdot x^{2i-1}.\ \square \tag{7.53}$$

We can use this to find the power series of the tangent function with respect to the point 0. It is easy to see that

$$\operatorname{tg}x=\operatorname{ctg}x-2\cdot\operatorname{ctg}(2x)$$

for every $x \neq k\pi/2$. Thus, applying (7.53) to x and $2x$ and taking the difference of the two series term by term we obtain

$$\pi \text{tg}\, \pi x = \sum_{i=1}^{\infty} 2\zeta(2i)(2^{2i} - 1) \cdot x^{2i-1}$$

for every $|x| < 1/2$. If we express $\zeta(2i)$ with the help of the Bernoulli numbers using Theorem 7.92 and replace πx by x, then we obtain the power series of the function $\text{tg}\, x$ around 0.

Theorem 7.96. *For every $|x| < \pi/2$, we have*

$$\text{tg}\, x = \sum_{i=1}^{\infty} (-1)^{i-1} \frac{(2^{2i} - 1)2^{2i} B_{2i}}{(2i)!} \cdot x^{2i-1}. \square \tag{7.54}$$

Next, we prove Euler's celebrated product formula.

Theorem 7.97. *For every $x \in \mathbb{R}$, we have*

$$\sin \pi x = \pi x \cdot \lim_{N \to \infty} \prod_{n=1}^{N} \left(1 - \frac{x^2}{n^2}\right). \tag{7.55}$$

Proof. Using (7.49), we can reformulate the statement of Theorem 7.93 as follows:

$$\sum_{n=1}^{\infty} \frac{2x}{x^2 - n^2} = \pi \text{ctg}\, \pi x - \frac{1}{x} \tag{7.56}$$

for every $x \notin \mathbb{Z}$. In the proof of Theorem 7.93 we showed that the series on the left-hand side is uniformly convergent on the interval $(-1, 1)$. Thus the equation (7.56) can be integrated term by term on the interval $[0, x]$ for every $0 < x < 1$. Since

$$\int_0^x \frac{2t}{t^2 - n^2}\, dt = \left[\log(n^2 - t^2)\right]_0^x = \log(n^2 - x^2) - \log(n^2) = \log\left(1 - \frac{x^2}{n^2}\right),$$

we find that the integral of the left-hand side of (7.56) equals

$$\lim_{N \to \infty} \sum_{n=1}^{N} \log\left(1 - \frac{x^2}{n^2}\right) = \lim_{N \to \infty} \log \prod_{n=1}^{N} \left(1 - \frac{x^2}{n^2}\right). \tag{7.57}$$

At the same time, the integral of the right-hand side of (7.56) is

$$\int_0^x \left(\pi \operatorname{ctg} \pi t - \frac{1}{t}\right) dt = [\log \sin \pi t - \log t]_0^x = \tag{7.58}$$

$$= \left[\log \frac{\sin \pi t}{t}\right]_0^x = \log \frac{\sin \pi x}{\pi x}.$$

If we compare (7.57) with (7.58), then raise e to the power of the two equal sides of the resulting equality and rearrange the result, then we get (7.55).

The proof is still not complete yet, since we only proved the equality for $x \in (0,1)$. Let $P_N(x)$ denote the product $x \cdot \prod_{n=1}^{N} \left(1 - \frac{x^2}{n^2}\right)$. Then

$$P_N(x) = \frac{(-1)^N}{(n!)^2} \cdot \prod_{n=-N}^{N} (x-n), \tag{7.59}$$

and thus

$$\frac{P_N(x+1)}{P_N(x)} = \frac{x+N+1}{x-N} \to -1$$

as $N \to \infty$, for every $x \notin \mathbb{Z}$. Since $\sin \pi(x+1) = -\sin \pi x$, this implies that if (7.55) holds for some $x \notin \mathbb{Z}$, then it holds for $x+1$ as well, and the converse is also true. It is clear from this that (7.55) holds for every $x \notin \mathbb{Z}$. Since both sides of (7.55) are zero when $x \in \mathbb{Z}$, the theorem is proved. □

Exercises

7.99. Show that the function $\zeta(x)$ is infinitely differentiable on $(1, \infty)$.

7.100. Show that

$$B_n(x) = \frac{x^n}{n!} + \frac{B_1}{1!} \cdot \frac{x^{n-1}}{(n-1)!} + \ldots + \frac{B_n}{n!} \qquad (n = 1, 2, \ldots).$$

7.101. Show that

$$1 + \binom{n}{1} B_1 + \binom{n}{2} B_2 + \ldots + \binom{n}{n-1} B_{n-1} = 0 \qquad (n = 2, 3, \ldots).$$

7.102. Formulate a conjecture for the value of the denominator of B_{2n} with the help of table (7.41) (and, if necessary, using also the values of some other Bernoulli numbers). (S)

7.103. Show that the Bernoulli polynomial $B_n(x)$ is k-replicative with the constant $a_k = k^{1-n}$ for every $k > 1$.

7.104. Show that if a polynomial p is k-replicative (for any $k > 1$), then p is a constant multiple of a Bernoulli polynomial.

7.105. Show that the function $\operatorname{ctg} \pi x$ is k-replicative for every $k > 1$. (H)

7.106. Show that if a function f is k_1-replicative and k_2-replicative, then it is $k_2 \cdot k_2$-replicative.

7.107. (i) Let (a_n) be a sequence defined by the following recursion. Let $a_0 = a_1 = 1$, and let

$$2(n+1)a_{n+1} = \sum_{i=0}^{n} a_i a_{n-i} \quad (n \geq 1). \tag{7.60}$$

Show that (a) $0 < a_n \leq 1/2$ for every $n \geq 2$, and (b) $n! \cdot a_n$ is an integer for every n. (H)

(ii) It follows from (i) that the power series $\sum_{n=0}^{\infty} a_n x^n$ is convergent on $(-1, 1)$. Let the sum of this power series be $f(x)$.
Show that $1 + f(x)^2 = 2f'(x)$ for every $x \in (-1, 1)$.

(iii) Solve the differential equation $1 + y^2 = 2y'$, and show that

$$f(x) = \operatorname{tg}\left(\frac{x}{2} + \frac{\pi}{4}\right) = \operatorname{tg} x + \frac{1}{\cos x}$$

for every $x \in (-1, 1)$.

(iv) Show that

$$\operatorname{tg} x = \sum_{n=1}^{\infty} a_{2n-1} x^{2n-1} \quad \text{and} \quad 1/\cos x = \sum_{n=0}^{\infty} a_{2n} x^{2n}$$

for every $x \in (-1, 1)$.

7.108. With the help of the previous exercise and of (7.54), show that $(2^{2n} - 1)2^{2n} B_{2n}/(2n)$ is an integer for every n.

7.109. Show that the denominator of B_{2n} divides $2^{2n}(2^{2n} - 1)$.

7.110. Show that if $p > 3$ is prime, then p divides the numerator of B_{2p}.

7.111. Check that (7.55) yields Wallis' formula[14]:

$$\pi = \lim_{n\to\infty} \left(\frac{2 \cdot 4 \cdots 2n}{1 \cdot 3 \cdots (2n-1)}\right)^2 \cdot \frac{1}{n},$$

when applied with $x = 1/2$.

---

[14] John Wallis (1616–1703), English mathematician.

7.7 First Appendix: The Cauchy–Hadamard Formula

In this appendix our aim is to prove the Cauchy–Hadamard formula (see Theorem 7.101 below) that gives the radius of convergence of a power series $\sum_{n=0}^{\infty} a_n x^n$ in terms of the coefficients a_n.

We say that α is a **cluster point** of the sequence (a_n) if (a_n) has a subsequence that converges to α. (Here α can be a real number, or either one of ∞ and $-\infty$.) From now on, it is useful to extend the ordering of the real numbers to ∞ and $-\infty$. We define the extended ordering by putting $-\infty < a < \infty$ for every real a.

Theorem 7.98. *Every sequence has a largest cluster point. α is the largest cluster point of the sequence (a_n) if and only if, for every $b < \alpha$ infinitely many terms of the sequence are larger than b, and for every $b > \alpha$ only finitely many terms of the sequence are larger than b.*

Proof. First suppose that the sequence (a_n) is not bounded from above. Then (a_n) has a subsequence converging to the infinity, thus its largest cluster point is ∞. Clearly, the statement of the theorem is true in this case.

Now let (a_n) be bounded from above. Let S be the set of numbers $b \in \mathbb{R}$ such that infinitely many terms a_n are larger than b. If $S = \emptyset$, then $a_n \to -\infty$, and the only cluster point of the series is $-\infty$. In this case the statement of the theorem holds again.

Therefore, we may assume that S is non-empty and is bounded from above. Let $\alpha = \sup S$. If $b < \alpha$, then there exists a $b' \in S$ such that $b < b' < \alpha$. Then infinitely many terms a_n are larger than b', thus infinitely many terms are larger than b as well. It is clear that α is the only value with the property that the sequence has infinitely many terms larger than b, when $b < \alpha$, while the sequence only has finitely many terms larger than b, when $b > \alpha$.

We still need to prove that α is the largest cluster point of the sequence (a_n). For every k, infinitely many a_n satisfies $a_n > \alpha - (1/k)$, and only finitely many of these satisfy $a_n > \alpha + (1/k)$. Thus we can choose the terms a_{n_k} such that $\alpha - (1/k) < a_{n_k} \le \alpha + (1/k)$. Since we can choose from infinitely many n at every step, may can also assume that $n_1 < n_2 < \ldots$. The resulting subsequence (a_{n_k}) converges to α, thus α is a cluster point of the sequence.

On the other hand, if β is a cluster point and $a_{n_k} \to \beta$, then $a_{n_k} > b$ holds for every $b < \beta$ when k is large enough. Thus infinitely many terms of the sequence are larger than b, and it follows that $b \in S$ and $\alpha \ge b$. This is true for every $b < \beta$, hence $\alpha \ge \beta$. Therefore, α is the largest cluster point of the sequence. □

By modifying Theorem 7.98 in the obvious way, we get that every sequence has a smallest cluster point, and this smallest cluster point α has the following property: the sequence has infinitely many terms smaller than b for every $b > \alpha$, while the sequence has only finitely many terms smaller than b for every $b < \alpha$.

Definition 7.99. The largest cluster point of the sequence (a_n) is called the *limit superior* of the sequence, denoted by $\limsup_{n\to\infty} a_n$.

The smallest cluster point of the sequence (a_n) is called the *limit inferior* of the sequence, denoted by $\liminf_{n\to\infty} a_n$.

It is easy to see that for every sequence (a_n) we have

$$\limsup_{n\to\infty} a_n = \lim_{n\to\infty} (\sup\{a_n, a_{n+1}, \ldots\}). \tag{7.61}$$

In the case when $\sup\{a_n, a_{n+1}, \ldots\} = \infty$ for every n, (7.61) means that the left-hand side equals ∞. Similarly, for every sequence (a_n), we have

$$\liminf_{n\to\infty} a_n = \lim_{n\to\infty} (\inf\{a_n, a_{n+1}, \ldots\}) \tag{7.62}$$

(see Exercise 7.114).

Theorem 7.100. *For every (a_n) we have $\liminf_{n\to\infty} a_n \le \limsup_{n\to\infty} a_n$. The equality $\liminf_{n\to\infty} a_n = \limsup_{n\to\infty} a_n$ holds if and only if the sequence has a (finite or infinite) limit, and then*

$$\lim_{n\to\infty} a_n = \liminf_{n\to\infty} a_n = \limsup_{n\to\infty} a_n. \tag{7.63}$$

Proof. It is clear from the definition that $\liminf_{n\to\infty} a_n \le \limsup_{n\to\infty} a_n$, and if $\lim_{n\to\infty} a_n$ exists, then (7.63) holds.

On the other hand, if $\liminf_{n\to\infty} a_n = \limsup_{n\to\infty} a_n = \alpha$, then α is the only cluster point of the sequence. In this case it is clear from Theorem 7.98 that $\lim_{n\to\infty} a_n = \alpha$. □

The following theorem gives the radius of convergence of power series in terms of the coefficients.

Theorem 7.101. (Cauchy–Hadamard formula) *The radius of convergence of the power series $\sum_{n=0}^{\infty} a_n x^n$ is*

$$R = \frac{1}{\limsup_{n\to\infty} \sqrt[n]{|a_n|}}.$$

In the case when $\limsup_{n\to\infty} \sqrt[n]{|a_n|} = 0$ the formula should be interpreted as $R = \infty$, and if $\limsup_{n\to\infty} \sqrt[n]{|a_n|} = \infty$, then the formula means $R = 0$.

Proof. Let $R_0 = 1/\limsup_{n\to\infty} \sqrt[n]{|a_n|}$ and let R denote the radius of convergence of the power series. We need to prove that $R = R_0$.

If $R_0 < x$, then $\limsup_{n\to\infty} \sqrt[n]{|a_n|} > 1/x$, and thus $\sqrt[n]{|a_n|} > 1/x$ for infinitely many n. For every such n, we have $|a_n x^n| > 1$, which means that the absolute value of infinitely many terms of the power series is larger than 1, hence the series is divergent. Thus the power series is divergent at every point $x > R_0$, i.e., $R \le R_0$. For $R_0 = 0$, this implies $R = R_0$.

If $R_0 > 0$ and $0 < x < R_0$, then $\limsup_{n\to\infty} \sqrt[n]{|a_n|} < 1/x$. Fix a number q such that

$$\limsup_{n\to\infty} \sqrt[n]{|a_n|} < q < \frac{1}{x}.$$

Then $\sqrt[n]{|a_n|} \le q$ holds for every n large enough. Thus $|a_n x^n| \le (qx)^n$ for n large enough, and then, by the majorant criterion, we find that the power series is convergent at x. Since this is true for every $0 < x < R_0$, we have $R \ge R_0$. Since $R \le R_0$ is also true, we have $R = R_0$. □

Exercises

7.112. Find the lim sup and lim inf of the following sequences.

(a) $(-1)^n$,
(b) $(n + (-2)^n)/(n + 2^n)$,
(c) $\sin n$,
(d) $a_n = 2\sqrt{n^2+2}$ if n is even, and $a_n = 3\sqrt[3]{n^3+3}$ if n is odd.

7.113. Let $\{x\}$ denote the fractional part of x. Show that the value of the Riemann function at the point x is equal to $1 - \limsup\{nx\}$.

7.114. Show that (7.61) and (7.62) are true for every sequence.

7.115. Show that for arbitrarily sequences (a_n) and (b_n) the following is true:

$$\liminf a_n + \liminf b_n \le \liminf(a_n + b_n) \le \liminf a_n + \limsup b_n \le$$
$$\le \limsup(a_n + b_n) \le \limsup a_n + \limsup b_n.$$

7.116. Show that if $a_n > 0$ for every n, then

$$\limsup(a_n)^{1/n} \le \limsup(a_{n+1}/a_n).$$

Use this to prove (again) that the root criterion is stronger than the ratio criterion.

7.117. Let a_n denote the nth decimal digit of the number $\sqrt{2}$. Find (i) $\limsup \sqrt[n]{a_n}$ and (ii) $\liminf \sqrt[n]{a_n}$. (H)

7.118. Suppose that $(a_{n+1} - a_n) \to 0$. Show that for every number

$$s \in [\liminf a_n, \limsup a_n]$$

there is a subsequence (a_{n_k}) such that $a_{n_k} \to s$.

7.119. Let (a_n) be a sequence such that $a_{n+m} \le a_n + a_m$ for every n, m. Show that the (finite or infinite) limit of the sequence (a_n/n) exists. (H)

7.120. Show that the function $f\colon \mathbb{R} \to \mathbb{R}$ satisfies $f(\limsup a_n) = \limsup f(a_n)$ for every bounded sequence (a_n) if and only if f is continuous and monotone increasing.

7.121. Let $A_1, A_2, \ldots$ be subsets of X, and let χ_H denote the characteristic function of the set $H \subset X$, i.e., let $\chi_H(x) = 1$ for $x \in H$, and $\chi_H(x) = 0$ for $x \notin H$. Find the sets B and C that satisfy $\limsup \chi_{A_n}(x) = \chi_B(x)$ and $\liminf \chi_{A_n}(x) = \chi_C(x)$ for every $x \in X$.

7.8 Second Appendix: Complex Series

We define the absolute value of the complex number $a + bi$ as $|a + bi| = \sqrt{a^2 + b^2}$. One can prove that the usual properties of the absolute value function are satisfied, thus $|u + v| \le |u| + |v|$ and $|uv| = |u| \cdot |v|$ holds for every pair of complex numbers u and v.

Recall that we can define convergence for sequences of complex numbers: we say that the sequence of complex numbers $z_n = a_n + b_n i$ converges to the complex number $z = a + bi$, if $a_n \to a$ and $b_n \to b$ both hold. (See the second appendix of Chapter 11 of [7].) This condition is equivalent to $|z_n - z| \to 0$.

The theory of infinite series can also be extended to series of complex terms. We say that the complex infinite series $\sum_{n=1}^{\infty} a_n$ is convergent and its sum is A, if the sequence of its partial sums $s_n = a_1 + \ldots + a_n$ converges to A. The series $\sum_{n=1}^{\infty} a_n$ is called divergent, it it is not convergent. It is easy to check that most of the theorems we proved about infinite series in Chapter 6 are also true for complex series as well, without any changes.

However, the complex case of Riemann's reordeing theorem needs some modifications. It is still true that a series is absolutely convergent if and only if each of its reordered series is convergent with the same sum. But if a complex series is conditionally convergent, we cannot claim that for every complex number A, there is a reordering of the series with the sum A. (E.g., if every term of the series is real, then the sum of every reordering is also real.) The correct statement is that if a series is conditionally convergent, then the sums of its convergent reorderings constitute a line, or cover the whole complex plane.

The theory of the power series can also be extended to complex series. The theorem corresponding to Theorem 7.49 states that for every power series $\sum_{n=0}^{\infty} a_n z^n$ there exists a $0 \le R \le \infty$ such that the power series is convergent at every point z with $|z| < R$, and the power series is divergent at every point z with $|z| > R$. Thus the domain of convergence of a power series can be either the origin (if $R = 0$), an open disk, possibly with some of its boundary points (if $0 < R < \infty$), or the whole convex plane (if $R = \infty$). One can show (with an identical proof) that the Cauchy–Hadamard formula is also true for complex power series.

It is also true that the sum f of a power series is continuous on the interior of its domain of convergence. This means that if $z_n \to z$ and $|z| < R$, then $f(z_n) \to f(z)$.[15].

This is what explains why the radius of convergence of the Taylor series of the function $1/(1+x^2)$ corresponding to the point 0 is 1 (and not larger), despite the fact that the function is analytic on the whole real line. The Taylor series in question is the series $\sum_{n=0}^{\infty}(-1)^n \cdot x^{2n}$. It is easy to see that this series is also convergent at every *complex* number whose absolute value is less than 1, with the sum $1/(1+x^2)$. If the radius of convergence of the series was larger than 1, the sum would be continuous in the open disc $B(0, R)$ with some $R > 1$. Thus the sum would be continuous at the complex number i, which is impossible, since the sum of the series at $x = ti$ is

$$\frac{1}{1+(ti)^2} = \frac{1}{1-t^2}$$

for every $t \in (0, 1)$, and this can be arbitrarily large when t is close enough to 1.

The definition of the differentiability of a complex function is the same as it was in the real case. I.e., the differential quotient $f'(a)$ is nothing else than the limit of the difference quotient $(f(z) - f(a))/(z - a)$ as $z \to a$. Complex differentiability is a much stronger assumption than real differentiability. Cauchy discovered that if the function f is differentiable on the interior of a disk, then f is *infinitely* differentiable, furthermore, f is analytic there.

Let the radius of convergence of the power series $\sum_{n=0}^{\infty} a_n z^n$ be positive, and let the sum of the power series be $f(z)$ on a small (real) neighborhood of the origin. It follows from Cauchy's theorem, that the radius of convergence of the power series is the largest number R (or the infinity) such that f can be extended analytically to the interior of the origin-centered disk with radius R.

With the help of complex power series we can also extend numerous elementary functions to the complex plane. E.g., the power series $\sum_{n=0}^{\infty} z^n/n!$ is convergent for every complex z. The sum of this series is the extension of the e^x function to the complex plane. One can show that the sum of the series is the same as the limit of the sequence $\left(1 + \frac{z}{n}\right)^n$ for every complex number z, i.e., e^z (see the second appendix of Chapter 11 of [7]).

The power series of $\sin x$, $\cos x$, $\operatorname{sh} x$, and $\operatorname{ch} x$ are also convergent on the whole complex plane, and their corresponding sums define extensions of these functions to the complex plane. (These extensions coincide with the extensions introduced in the second appendix of Chapter 11 of [7]).)

Since these results fall into the field of complex function theory, they are outside the scope of this book. However, we can see that the topic of power series is an important common point of real and complex analysis.

---

[15] In the general case, as opposed to Theorem 7.51, it is not true that the sum is also continuous at the boundary points of the convergence domain.

7.9 Third Appendix: On the History of the Fourier Series

As we mentioned before, part of the motivating factors behind the emergence of the theory of Fourier series were two problems from physics: the equation of a vibrating string and the equation of heat conduction.

Suppose a string is spanned between the points 0 and L of the real line. How can we describe the motion of the vibrating string? As early as in 1715, Taylor already found the equation of the motion a vibrating string has to satisfy. He argued as follows. We may assume that every particle of the string only makes small, vertical motions. By Newton's law, the acceleration of a particle is proportional to the force acting on it. Now, the force acting on the particle comes from the fact that the string "wants to straighten out", thus it acts with a force proportional to the string's curvature[16].

Suppose that the function $c(t) \cdot f(x)$ describes the motion of the string in the sense that at time t the shape of the string is described by the graph of the function $x \mapsto c(t) \cdot f(x)$ $(x \in [0, L])$. The acceleration of the particle above the point $(x, 0)$ is $c''(t) \cdot f(x)$, while the curvature of the string is $c(t) \cdot f''(x)$. By Newton's law, we have

$$c''(t) \cdot f(x) = \rho \cdot c(t) \cdot f''(x), \tag{7.64}$$

where $\rho \neq 0$ is a constant.

Since the endpoints of the string are fixed, we have $c(t)f(0) = c(t)f(L) = 0$ for every t. We may assume that c is not constant zero, and thus

$$f(0) = f(L) = 0. \tag{7.65}$$

Fix a t such that $c(t) \neq 0$. Then, by (7.64), f satisfies the differential equation $f''(x) = b \cdot f(x)$ for every $x \in [0, L]$, where $b = c''(t)/(\rho \cdot c(t))$. If $b = 0$, then f is linear and, according to (7.65), f is constant zero. We can exclude this case (which describes the situation when the string is not moving).

If $b > 0$ and $b = a^2$, then we have $f(x) = \alpha e^{ax} + \beta e^{-ax}$. (See [7, Chapter 13, p. 280].) This solution only satisfies the conditions of (7.65) when $\alpha = \beta = 0$, so it can also be excluded.

Thus, necessarily, we have $b < 0$ and $b = -a^2$. Then we have $f(x) = \alpha \sin (ax) + \beta \cos(ax)$ (see also [7, Chapter 13, p. 280]). As $f(0) = f(L) = 0$, we have $\beta = 0$ and $aL = n\pi$, where n is an integer.

---

[16] The curvature of the string is the change of the steepness of its tangent, i.e., the second derivative of the function describing the shape of the string. In fact, when calculating the curvature, we need to take the arc length of the graph (of the string) into consideration as well. If $s(g; [a, b])$ denotes the arc length of graph g on the interval $[a, b]$, then the curvature of the graph of g at the point $(a, g(a))$ is $\lim_{x\to a}(g'(x) - g'(a))/s(g; [a, x])$. Since $\lim_{x\to a} s(g; [a, x])/(x - a) = \sqrt{1 + (g'(a))^2}$ (see [7, Theorem 13.41]), it follows that if g is twice-differentiable on a neighborhood of a, then the curvature is equal to $g''(a)/\sqrt{1 + (g'(a))^2}$. However, in our case the function describes the shape of the string, thus its derivative is small, and we can neglect the term $(g'(a))^2$ of the denominator.

We get that if the motion of the string is described by a function of the form $c(t) \cdot f(x)$, then $f(x) = \sin \frac{n\pi}{L} x$ (and c satisfies the differential equation $c'' = (-\rho a^2) \cdot c$). These solutions correspond to what are called the standing waves of the string.

In 1747 d'Alembert[17] discovered that if φ is a $2L$-periodic function, then the function $\varphi(at + x) - \varphi(at - x)$ is zero at the points $x = 0$ and $x = L$, furthermore, φ satisfies the motion equation given by Taylor in the sense that its second derivative with respect to t is equal to a constant times its second derivative with respect to x. d'Alembert concluded that the formulas $\varphi(at + x) - \varphi(at - x)$ describe the motion of a vibrating string, where φ is an "arbitrary" $2L$-periodic function.

In 1753 Daniel Bernoulli[18], following the ideas of Taylor and taking into consideration the physical phenomenon that any sound is made of its harmonics, claimed that the motion of a vibrating string is described by formulas of the form $\sum_{n=1}^{\infty} c_n(t) \sin \frac{n\pi}{L} x$, for appropriate coefficient functions $c_n(t)$. However, d'Alembert did not accept this statement on the basis that, combined with his results, this would imply that "every" periodical function can be written as a sum of trigonometric functions, which is "obviously" impossible. Several well-known mathematicians joined the discussion (including Euler and Lagrange).

In 1822 Fourier published his book on heat conduction, which investigated the following problem – among many others. Given a homogeneous rod, insulated from its surroundings in such a way that heat only flows in the inside of the rod and not between the rod and its surroundings. The problem is to describe the change of temperature in the interior of the rod.

It is known that the quantity of heat stored by a solid depends on the mass and the temperature of the solid: the larger the temperature, the larger the stored amount of heat. More precisely, the quantity of heat is proportional to the mass and the temperature (using an appropriate scaling for the temperature). In other words, the quantity of heat of a solid of mass m is $\alpha \cdot m \cdot T$, where α is a constant (called specific heat), and T is the temperature.

Returning to Fourier's problem, the homogeneousness of the rod means that the mass of a segment $[a, b]$ of the rod is $\gamma \cdot (b - a)$, where γ is a constant (density). Thus the quantity of heat stored by the segment $[a, b]$ is $\delta \cdot (b - a) \cdot T$, where δ is the product of the specific heat and the density, and T is the temperature of the segment. If the temperature changes along the segment, namely the temperature of the rod is $T(x)$ at the point x, then following a familiar argument we can see that the amount of heat of the segment is $\int_a^b \delta \cdot T(x)\, dx$. (See, e.g., the computation of the work and pressure in Chapter 14 of [7].)

For simplicity, let us assume that the temperature of the rod is given by the function $c(t) \cdot f(x)$ in the sense that the temperature at the time t of the point x of the rod is $c(t) \cdot f(x)$. Then, at time t, the quantity of heat of the segment $[a, b]$ is

---

[17] Jean d'Alembert (1717–1783), French natural scientist.

[18] Daniel Bernoulli (1700–1782), Swiss mathematician.

$$H = \int_a^b \delta \cdot c(t) f(x)\, dx = \delta \cdot c(t) \cdot \int_a^b f(x)\, dx.$$

The quantity of heat of a segment $[a, b]$ changes in time because the segment loses or gains heat through its endpoints. If, at a given moment, the temperature as the function of the place is decreasing in a small neighborhood of the point b, then the segment $[a, b]$ loses heat at that point; if the temperature is an increasing function of the place, then the segment gains heat through b. Fourier assumed (rightly so) that the speed of heat conduction at the point b is proportional with the derivative of the temperature (as a function of the place) at the point b, in other words, the speed of the heat conduction is $\kappa \cdot c(t) \cdot f'(b)$, where κ is a positive constant (the conductivity constant). The situation is reversed at the point a: if the temperature is decreasing in a neighborhood of a then the segment $[a, b]$ of the rod gains heat, and if the temperature is increasing, the segment loses heat. Thus the rate of heat conduction at the point a is $-\kappa \cdot c(t) \cdot f'(a)$. In the end, we get that the rate of change of the quantity of heat H is $H'(t) = \kappa \cdot c(t) \cdot f'(b) - \kappa \cdot c(t) \cdot f'(a)$, i.e.,

$$\delta \cdot c'(t) \cdot \int_a^b f(x)\, dx = \kappa \cdot c(t) \cdot (f'(b) - f'(a)).$$

Dividing by $b - a$ and taking the limit as $b \to a$, we get that $\delta \cdot c'(t) \cdot f(a) = \kappa \cdot c(t) \cdot f''(a)$. This is true for every a, thus we have

$$c'(t) \cdot f(x) = \rho \cdot c(t) \cdot f''(x) \tag{7.66}$$

for every x, where $\rho > 0$ is a constant.

Following Fourier's argument, let us consider another simplification of our problem: let the temperature at the endpoints of the rod be 0 degrees. (Imagine that the endpoints of the rod are glued to a large tank of temperature 0, which guarantees the constant 0 degree temperature at the endpoints.) Let L be the length of the rod; then $c(t)f(0) = c(t)f(L) = 0$ for every t. We may assume that c is not constant zero, thus we have

$$f(0) = f(L) = 0. \tag{7.67}$$

We can solve the equation (7.66) similarly to (7.64). Fix a t such that $c(t) \neq 0$. By (7.66), f satisfies the differential equation $f''(x) = b \cdot f(x)$ for every $x \in [0, L]$, where $b = c'(t)/(\rho \cdot c(t))$. If $b = 0$, then f is linear, and then, by (7.67), f is the constant zero function. We may exclude this case (which corresponds to the case when the temperature of the rod is zero everywhere all the time). If $b > 0$ and $b = a^2$, then $f(x) = \alpha e^{ax} + \beta e^{-ax}$. This solution satisfies the condition (7.67) only when $\alpha = \beta = 0$, thus it can also be excluded. Necessarily, we have $b < 0$ and $b = -a^2$. Then $f(x) = \alpha \sin(ax) + \beta \cos(ax)$. Since $f(0) = 0$, we have $\beta = 0$, and, by

$f(L) = 0$, $aL = n\pi$, where n is an integer. We found that if the temperature of the rod is described by a function of the form $c(t) \cdot f(x)$, then $f(x) = \sin \frac{n\pi}{L} x$ (and c satisfies the differential equation $c' = (-\rho a^2) \cdot c$).

Based on this argument Fourier concluded that the temperature of the rod in the general case is described by functions of the form $\sum_{n=1}^{\infty} c_n(t) \sin \frac{n\pi}{L} x$, similarly to the case of the equation of a vibrating string. Then Fourier claimed – sharing the opinion of Bernoulli – that "every" 2π-periodic function can be written in the form $\sum_{n=0}^{\infty}(a_n \cos nx + b_n \sin nx)$.

The debate concerning trigonometric series – considered to be the most famous dispute of great consequence in the history of mathematics – involved several topics. First of all, the notion of a function was not cleared up. In Fourier's work one can detect the enormous conceptual difficulties that had to be overcome in order to arrive from the concept of function as a formula to the concept of function as correspondence. At one place, Fourier wrote: *"It is remarkable that we can express by convergent series ... the ordinates of lines and surfaces which are not subject to a continuous law.* Or elsewhere: *"A function [can be] completely arbitrary, that is to say a succession of given values, subject or not to a common law, and answering to all the values of x.... The function represents a succession of values or ordinates each of which is arbitrary ... They succeed each other in any manner whatever, and each of them is given as if it were a single quantity."*

Still, from other details of the proofs given by Fourier (which, in general, are devoid of any precision) it is obvious, that the most general functions imaginable by Fourier are defined by not a single, but by several formulas on some consecutive intervals. This is, and not more, what he meant that "the values do not follow a single, common rule."

Another unclear notion was that of continuity. For Euler, the continuous functions were the ones defined by an "analytic formula". He called the functions whose graph can be written by a "freely moving hand" connected functions. In the end, Cauchy cleared up the definition of continuity (in the modern sense). However, Cauchy thought – and also proved (albeit incorrectly) – that an infinite sum of continuous functions is also necessarily continuous. This meant further complications, since it was realized by several people that the sum of the trigonometric series

$$\sin x + \frac{\sin 3x}{3} + \frac{\sin 5x}{5} + \dots$$

is the constant $\pi/4$ function on the interval $(0, \pi)$, while it is the constant $-\pi/4$ function on the interval $(\pi, 2\pi)$. Fourier tried to help this by claiming that the graph of this function consists of vertical segments (!) at the points $k\pi$, saving continuity.

In the end it was Dirichlet who cleared things up. In a seminal paper written in 1829, he introduced the modern definition of a function (= correspondence) (and illustrated it with a function now bearing his name), and he also gave precise proof of the first convergence theorems of Fourier series. The mathematical theory of the Fourier series starts with Dirichlet.

In fact, set theory also owes to the theory of the Fourier series for its existence. When Cantor proved in 1870 that every function can be written as the sum of a trigonometric function in at most one way, he started to think about what happens if this is only true save for a few particular points. Investigating these exceptional sets led Cantor to invent ordinal numbers, countable sets, and finally set theory.

Chapter 8
Miscellaneous Topics

8.1 Approximation of Sums

As we saw in Theorem 7.92, the sum of the series $\sum_{n=1}^{\infty} 1/n^{2k}$ can be given explicitly for every positive integer k. These series, however, are exceptions: in general, the sum of an arbitrary series cannot be expressed in closed form. In fact, no closed expression is known for the sum $\sum_{n=1}^{\infty} 1/n^3$.

The same is true for finite sums. Finding the sum of many terms can be, in general, a very hard problem. It can happen that we find a "closed formula" for a finite sum (as in the cases of the arithmetic and the geometric series), but in most cases it is impossible.

Thus, it is imperative to determine the sums at least approximately. In fact, this is useful even when there is a closed formula available, since a good approximation can sometimes be more valuable in practice than a formula.

In this section we introduce some methods of approximating sums.

Theorem 8.1. *Let $a < b$ be integers, and let $f\colon [a, b] \to \mathbb{R}$ be monotonically increasing. Then*

$$\int_a^b f(x)\,dx + f(a) \le \sum_{i=a}^{b} f(i) \le \int_a^b f(x)\,dx + f(b). \tag{8.1}$$

If f is monotonically decreasing, then the reverse inequalities hold.

Proof. If f is monotonically increasing, then $f(i) \le f(x) \le f(i+1)$ for every $x \in [i, i+1]$, and thus $f(i) \le \int_i^{i+1} f\,dx \le f(i+1)$. Adding these inequalities for $i = a, a+1, \ldots, b-1$ gives

$$\sum_{i=a}^{b} f(i) - f(b) \le \int_a^b f(x)\,dx \le \sum_{i=a}^{b} f(i) - f(a),$$

M. Laczkovich and V.T. Sós, *Real Analysis*, Undergraduate Texts in Mathematics, https://doi.org/10.1007/978-1-4939-7369-9_8

from which (8.1) is clear. The case of a decreasing f is similar, or it can be reduced to the case of increasing functions by applying it to $-f$. □

Example 8.2. **1.** Let us apply Theorem 8.1 to $[a, b] = [0, n]$ and $f(x) = x^k$ ($k = 1, 2, 3$). We get

$$\frac{n^2}{2} \le \sum_{i=1}^{n} i \le \frac{n^2}{2} + n; \qquad \frac{n^3}{3} \le \sum_{i=1}^{n} i^2 \le \frac{n^3}{3} + n^2; \qquad \frac{n^4}{4} \le \sum_{i=1}^{n} i^3 \le \frac{n^4}{4} + n^3. \tag{8.2}$$

2. Applying the theorem to the same interval and $f(x) = \sqrt{x}$ gives

$$\frac{2}{3} \cdot n^{3/2} \le \sum_{i=1}^{n} \sqrt{i} \le \frac{2}{3} \cdot n^{3/2} + \sqrt{n}. \tag{8.3}$$

3. Let $[a, b] = [1, n]$ and $f(x) = 1/x$. We get

$$\log n < \sum_{i=1}^{n} \frac{1}{i} \le \log n + 1. \tag{8.4}$$

(We applied the theorem to the monotonically decreasing function f and omitted the $1/n$ term on the left-hand side.)

4. Finally, let $[a, b] = [1, n]$ and $f(x) = \log x$. We have $\int_1^n \log x \, dx = n \log n - n + 1$, and thus

$$n \log n - n + 1 \le \sum_{i=1}^{n} \log i \le n \log n - n + \log n + 1.$$

Exponentiating of all three "sides" of the inequalities results in

$$e \cdot \left(\frac{n}{e}\right)^n \le n! \le e \cdot n \cdot \left(\frac{n}{e}\right)^n. \tag{8.5}$$

This approximation is rather crude; we will see much finer estimates later. (Note also that the inequalities $(n/e)^n \le n!$ and $n! \le n \cdot (n/e)^n$ ($n \ge 7$) can be proved by induction on n.)

In order to improve Theorem 8.1 we will now prove an often used formula.

Theorem 8.3. (Euler's summation formula) *Let $a < b$ be integers, let f be differentiable on $[a, b]$, and let f' be integrable on $[a, b]$. Then*

$$\sum_{i=a}^{b} f(i) = \int_a^b f(x) \, dx + \frac{f(a) + f(b)}{2} + \int_a^b \left(\{x\} - \frac{1}{2}\right) \cdot f'(x) \, dx. \tag{8.6}$$

Proof. If i is an integer and $x \in [i, i+1)$, then $\{x\} = x - i$, and thus

$$\int_i^{i+1} \left(\{x\} - \frac{1}{2}\right) \cdot f'(x)\,dx = \int_i^{i+1} \left(x - i - \frac{1}{2}\right) \cdot f'(x)\,dx =$$
$$= \left[\left(x - i - \frac{1}{2}\right) \cdot f(x)\right]_i^{i+1} - \int_i^{i+1} f(x)\,dx = \tag{8.7}$$
$$= \frac{f(i) + f(i+1)}{2} - \int_i^{i+1} f(x)\,dx.$$

Adding the equations above for $i = a, a+1, \ldots, b-1$, we get (8.6). □

Now we turn to improving Theorem 8.1. Naturally, if we want a sharper bound than that of (8.1), we need to have stricter assumptions than just the monotonicity of f.

Theorem 8.4. *Let $a < b$ be integers and let $f\colon [a, b] \to \mathbb{R}$ be differentiable, monotonically decreasing, and convex. Then*

$$\int_a^b f(x)\,dx + \frac{f(a) + f(b)}{2} \le \sum_{i=a}^{b} f(i) \le \int_a^b f(x)\,dx + \frac{f(a) + f(b)}{2} - f'(a)/8. \tag{8.8}$$

The reverse inequalities hold if f is a differentiable, monotonically increasing, and concave function.

Proof. If f is differentiable, monotonically decreasing, and convex, then f' is monotonically increasing and nonpositive. Thus, applying the second mean value theorem of integration (see [7, Theorem 15.8]), we obtain

$$\int_a^b \left(\{x\} - \frac{1}{2}\right) \cdot f'(x)\,dx = f'(a) \cdot \int_a^c \left(\{x\} - \frac{1}{2}\right)\,dx \tag{8.9}$$

with a suitable real number $c \in [a, b]$. Let $d = \{c\}$. Then

$$0 \ge \int_0^d \left(x - \frac{1}{2}\right)\,dx = (d^2 - d)/2 \ge -1/8.$$

Since $\int_{n-1}^{n} \left(\{x\} - \frac{1}{2}\right)\,dx = 0$ for every integer n, we have

$$0 \ge \int_a^c \left(\{x\} - \frac{1}{2}\right)\,dx \ge -\tfrac{1}{8}. \tag{8.10}$$

Comparing this to (8.9) and (8.6), we obtain (8.8). □

Example 8.5. **1.** Let us apply Theorem 8.4 with $[a, b] = [1, n]$ and $f(x) = \sqrt{x}$. Since f is concave and increasing, we obtain

$$\frac{2}{3} \cdot n^{3/2} + \frac{\sqrt{n}}{2} - \frac{1}{6} - \frac{1}{16} \le \sum_{i=1}^{n} \sqrt{i} \le \frac{2}{3} \cdot n^{3/2} + \frac{\sqrt{n}}{2} - \frac{1}{6}, \tag{8.11}$$

which is much sharper than (8.3).
2. Applying (8.8) to the function $1/x$ on the interval $[1, n]$, we get

$$\log n + \frac{1}{2} < \sum_{i=1}^{n} \frac{1}{i} \le \log n + \frac{5}{8} + \frac{1}{2n}. \tag{8.12}$$

3. Let $[a, b] = [1, n]$ and $f(x) = \log x$. Again, we get

$$n \log n - n + 1 + \frac{\log n}{2} - \frac{1}{8} \le \log n! \le n \log n - n + 1 + \frac{\log n}{2}.$$

Putting all three "sides" into the exponent of e yields

$$c_1 \cdot \left(\frac{n}{e}\right)^n \cdot \sqrt{n} \le n! \le c_2 \cdot \left(\frac{n}{e}\right)^n \cdot \sqrt{n}, \tag{8.13}$$

for some positive constants c_1 and c_2. We will make this inequality more precise in Theorem 8.10.

According to Stirling's formula, the sequence $(n/e)^n \sqrt{n}/n!$ is convergent (and converges to $1/\sqrt{2\pi}$). The following theorem is a generalization of this statement.

Theorem 8.6.

(i) *Let f be differentiable, monotonically decreasing, and convex on $[a, \infty)$. Then the sequence*

$$a_n = \sum_{i=a}^{n} f(i) - \int_a^n f(x)\,dx - \frac{f(a) + f(n)}{2} \quad (n = a, a+1, \ldots)$$

is monotonically increasing and convergent.

(ii) *Let f be differentiable, monotonically decreasing, and convex, and let $\lim_{x \to \infty} f'(x) = 0$. Then the improper integral $\int_a^\infty \left(\{x\} - \frac{1}{2}\right) \cdot f'(x)\,dx$ is convergent, and*

$$\lim_{n \to \infty} a_n = \int_a^\infty \left(\{x\} - \frac{1}{2}\right) \cdot f'(x)\,dx. \tag{8.14}$$

Proof. (i) Let f be differentiable, monotonically decreasing, and convex. Then f' is monotonically increasing and nonpositive on the half-line $[a, \infty)$. By (8.6) we have

$$a_{n+1} - a_n = \int_n^{n+1} \left(\{x\} - \frac{1}{2}\right) \cdot f'(x)\, dx.$$

Thus, by the second mean value theorem of integration and by (8.10) we obtain

$$a_{n+1} - a_n = f'(n) \cdot \int_n^c \left(\{x\} - \frac{1}{2}\right) dx \geq 0,$$

since the right-hand side is the product of two nonpositive numbers. Thus the sequence (a_n) is monotonically increasing. On the other hand, (a_n) is bounded from above by (8.8). Therefore, (a_n) is convergent.

(ii) Suppose that $\lim_{x\to\infty} f'(x) = 0$ in addition to the conditions of (i). We claim that the improper integral on the right-hand side of (8.14) is convergent. Indeed, it is known that the improper integral $\int_a^\infty g(x)h(x)\, dx$ is convergent, provided that f is monotone, $\lim_{x\to\infty} g(x) = 0$, and the function $x \mapsto \int_a^x h(t)\, dt$ is bounded. (See [7, Theorem 19.22].) These conditions are satisfied for $f' = g$ and $\{x\} - 1/2 = h$; see (8.10). Therefore, (8.14) follows from (8.6). □

Remark 8.7. **1.** If the function f is monotonically increasing and concave, then the sequence a_n is decreasing and convergent. The proof is similar or can be reduced to part (i) of the theorem. It is also obvious that part (ii) also holds for differentiable monotonically increasing concave functions whose derivative converges to zero at infinity.

2. Applying the theorem above to the function $\log x$, we obtain that the sequence

$$\log n! - \log\left(\left(\frac{n}{e}\right)^n \sqrt{n}\right) = \log\left(n!/\left(\frac{n}{e}\right)^n \sqrt{n}\right)$$

is convergent. Therefore, the sequence $(n/e)^n \sqrt{n}/n!$ is also convergent. Its limit, according to Stirling's formula, is $\sqrt{2\pi}$.

Example 8.8. **1.** Theorem 8.6 implies that the sequence

$$\left(\sum_{i=1}^n \sqrt{i}\right) - \frac{2}{3} \cdot n^{3/2} - \frac{\sqrt{n}}{2}$$

is decreasing and convergent. Its limit lies between $-1/6$ and $-1/4$ by (8.11).

2. By Theorem 8.6, the sequence $(\sum_{i=1}^{n} 1/i) - \log n - (1 + (1/n))/2$ is convergent. Thus the sequence $(\sum_{i=1}^{n} 1/i) - \log n$ is also convergent. We denote the limit of this sequence by γ (**Euler's constant**). It follows from (8.14) that

$$\gamma = \frac{1}{2} - \int_1^\infty \frac{\{x\} - (1/2)}{x^2}\,dx = 1 - \int_1^\infty \frac{\{x\}}{x^2}\,dx. \tag{8.15}$$

By the inequality (8.12), $0.5 \le \gamma \le 0.625$. This constant appears in analysis and also in number theory. It is known that $\gamma = 0.5772157\ldots$; in fact, millions of decimal digits of Euler's constant are known. However, it is a longstanding open problem whether the number γ is rational or irrational.

We are now going to improve our approximations of (8.2). Since for $k > 1$, the function x^k is increasing and convex on $[1, n]$, our current methods are not applicable. However, by improving Euler's summation formula, we can handle these sums as well.

At the core of the proof of Euler's summation formula lies a partial integration. It is a natural idea to make the formula more precise by further partial integrations. The function $B_m(x)$ appearing in the next theorem is the mth Bernoulli polynomial (see Theorem 7.88).

Theorem 8.9. (Euler's summation formula, general form) *Let $a < b$ and $m \ge 1$ be integers. If the function f is m times differentiable, and $f^{(m)}$ is integrable on $[a, b]$, then*

$$\begin{aligned}\sum_{i=a}^{b} f(i) = \int_a^b f(x)\,dx + \frac{f(a)+f(b)}{2} + \frac{B_2}{2!}\cdot(f'(b) - f'(a)) + \ldots +\\ + \frac{B_m}{m!}\cdot\left(f^{(m-1)}(b) - f^{(m-1)}(a)\right) + \int_a^b B_m(\{x\})\cdot f^{(m)}(x)\,dx.\end{aligned} \tag{8.16}$$

Proof. We prove the result by induction on m. If $m = 1$, then Theorem 8.3 implies the statement, since $B_1(\{x\}) = \{x\} - \frac{1}{2}$. In the induction step it is enough to prove that if f is $(m+1)$ times differentiable and $f^{(m+1)}$ is integrable on $[a, b]$, then

$$\begin{aligned}\int_a^b B_m(\{x\})\cdot f^{(m)}(x)\,dx = &\frac{B_{m+1}}{(m+1)!}\cdot\left(f^{(m)}(b) - f^{(m)}(a)\right) + \\ &+ \int_a^b B_{m+1}(\{x\})\cdot f^{(m+1)}(x)\,dx.\end{aligned} \tag{8.17}$$

Let $a \le i < b$ be an integer. Using partial integration, we get

$$\int_i^{i+1} B_m(\{x\}) \cdot f^{(m)}(x)\,dx = \int_i^{i+1} B'_{m+1}(x-i) \cdot f^{(m)}(x)\,dx =$$
$$= \Big[B_{m+1}(x-i) \cdot f^{(m)}(x)\Big]_i^{i+1} - \int_i^{i+1} B_{m+1}(x-i) \cdot f^{(m+1)}(x)\,dx =$$
$$= \frac{B_{m+1}}{(m+1)!} \cdot \left(f^{(m)}(i+1) - f^{(m)}(i)\right) -$$
$$- \int_i^{i+1} B_{m+1}(\{x\}) \cdot f^{(m+1)}(x)\,dx, \tag{8.18}$$

since by (7.40), $B_{m+1}(1) = B_{m+1}(0) = B_{m+1}/(m+1)!$. Adding equations (8.18) for $i = a, a+1, \ldots, b-1$, we obtain (8.17). □

Using Theorem 8.9, we can give sharper estimates of the sums $1^k + 2^k + \ldots + n^k$ in (8.2); moreover, we can write these sums as polynomials in n. Applying the theorem with the choices of $a = 0,\ b = n,\ f(x) = x^k,\ m = k+1$ results in

$$1^k + 2^k + \ldots + n^k = \frac{1}{k+1} \cdot n^{k+1} + \frac{1}{2} \cdot n^k + \frac{k}{2!} B_2 \cdot n^{k-1} + \ldots + \frac{k!}{k!} B_k \cdot n.$$

This is Bernoulli's famous formula[1]:

$$1^k + 2^k + \ldots + n^k = \frac{1}{k+1} \cdot n^{k+1} + \frac{1}{2} \cdot n^k +$$
$$+ \frac{1}{k+1} \cdot \left(\binom{k+1}{2} B_2 \cdot n^{k-1} + \ldots + \binom{k+1}{k} B_k \cdot n\right). \tag{8.19}$$

Notice that the right-hand side is a $(k+1)$th-degree polynomial in n.

We will now apply Theorem 8.9 to estimate $n!$, by improving (8.13). The next result also gives a quantitative version of Stirling's formula.

Theorem 8.10. *For every positive integer n, we have*

$$\left(\frac{n}{e}\right)^n \cdot \sqrt{2\pi n} \le n! \le \left(\frac{n}{e}\right)^n \cdot \sqrt{2\pi n} \cdot e^{1/(12n)}. \tag{8.20}$$

Proof. Let $a_n = (n/e)^n \sqrt{2\pi n}/n!$ for every n. It follows from Theorem 8.6 and Remark 8.8 that the sequence (a_n) is strictly monotonically increasing. By Stirling's formula, it converges to 1. Hence $a_n < 1$ for every n, and thus the first inequality of (8.20) holds.

[1] The formula was introduced by Jacob Bernoulli in his book on probability theory (1713).

The second inequality is equivalent to $\log a_n \geq -1/(12n)$ for every n. In order to prove this, apply Theorem 8.9 to the function $f(x) = \log x$ on the interval $[1, n]$ for $m = 3$. Since $B_2 = 1/6$ and $B_3 = 0$, we obtain

$$\log n! = \int_1^n \log x\,dx + \frac{\log n}{2} + \frac{1}{12}\left(\frac{1}{n} - 1\right) + 2 \cdot \int_1^n \frac{B_3(\{x\})}{x^3}\,dx. \quad (8.21)$$

By $\int_1^n \log x\,dx = n\log n - n + 1 = \log(n/e)^n + 1$, (8.21) gives

$$\log a_n = a - \frac{1}{12n} - 2 \cdot \int_1^n \frac{B_3(\{x\})}{x^3}\,dx \quad (8.22)$$

for some constant a. Since $a_n \to 1$, (8.22) implies

$$0 = a - 2 \cdot \int_1^\infty \frac{B_3(\{x\})}{x^3}\,dx.$$

Subtracting this from (8.22), we obtain

$$\log a_n = -\frac{1}{12n} + 2 \cdot \int_n^\infty \frac{B_3(\{x\})}{x^3}\,dx. \quad (8.23)$$

In order to complete the proof, we have to show that

$$\int_n^\infty \frac{B_3(\{x\})}{x^3}\,dx \geq 0$$

for every n. It is enough to prove that

$$\int_k^{k+1} \frac{B_3(\{x\})}{x^3}\,dx \geq 0$$

for every $k \geq 1$. Since the function $1/x^3$ is monotonically decreasing and nonnegative on the interval $[k, k+1]$, the second mean value theorem of integration gives

$$\int_k^{k+1} \frac{B_3(\{x\})}{x^3}\,dx = k^{-3} \cdot \int_k^c B_3(\{x\})\,dx = k^{-3} \cdot \int_0^d B_3(x)\,dx,$$

where $c \in [k, k+1]$ and $d = c - k \in [0, 1]$. It is enough to show that $\int_0^d B_3(x)dx \geq 0$ for every $d \in [0, 1]$.

A simple computation shows that 0, 1/2, and 1 are all roots of the polynomial $B_3(x) = (x^3/6) - (x^2/4) + (x/12)$. Since $B_3(x)$ cannot have more roots, the sign of the polynomial $B_3(x)$ does not change on the intervals $(0, 1/2)$ and $(1/2, 1)$. The polynomial $B_3(x)$ is locally decreasing at 1/2, since $B_3'(1/2) =$

$B_2(1/2) < 0$. Therefore, $B_3(x) > 0$ for every $x \in (0, 1/2)$, and $B_3(x) < 0$ for every $x \in (1/2, 1)$. Thus $\int_0^d B_3(x)\, dx \geq 0$ for every $d \in [0, 1/2]$. On the other hand, if $d \in [1/2, 1]$, then

$$\int_0^d B_3(x)\, dx = -\int_d^1 B_3(x)\, dx \geq 0,$$

and the theorem is proved. □

Exercises

8.1. Show that

$$1 + \frac{1}{2} + \ldots + \frac{1}{n} - \log n = \gamma + \frac{1}{2n} - \frac{1}{12n^2} - 6 \cdot \int_n^\infty \frac{B_3(\{x\})}{x^4}\, dx$$

for every positive integer n. Show that the absolute value of the integral is at most c/n^4.

8.2. Double check that the right-hand sides of (8.19), for $k = 1,\ 2,\ 3$, are $n(n+1)/2$, $n(n+1)(2n+1)/6$, and $(n(n+1)/2)^2$, respectively.

8.3. Show that

$$1^k + 2^k + \ldots + (n-1)^k = k! \cdot (B_{k+1}(n) - B_{k+1}(0)).$$

8.2 Approximation of Definite Integrals

It is well known that the indefinite integral of an elementary function is not necessarily elementary (see, e.g., [7, Section 15.5]). The same is true for definite integrals: it is not guaranteed that we can express the definite integral of an elementary function in "closed form," i.e., by a formula containing only known functions and constants (and no limits). In fact, it is the exception when we can express the value of a definite integral in closed form[2].

So far, the simplest method for finding definite integrals has proven to be the Newton–Leibniz formula. However, as we saw above, we cannot always use this method. Furthermore, even when the Newton–Leibniz formula is applicable, the computation is not always feasible.

---

[2] It might happen that the indefinite integral of an elementary function is not elementary, yet we can express its definite integral in closed form on some intervals. For example, the indefinite integrals of the functions $x \cdot \operatorname{ctg} x$ and $x^2/\sin^2 x$ are not elementary, but the value of their integrals on the interval $[0, \pi/2]$ are $\pi \cdot \log 2$ and $2\pi \cdot \log 2$, respectively. See Problems 19.20 and 19.22 in [7].

Another method for finding definite integrals is using the definition itself. However, working with upper and lower sums can be quite messy (except for finding the integrals of monotone functions), and thus it is natural to approximate the integral with the help of approximating sums. We know that if F_n is a sequence of partitions of $[a, b]$ such that the mesh of F_n tends to zero, then

$$\int_a^b f(x)\,dx = \lim_{n\to\infty} \sigma_{F_n},$$

with arbitrary approximating sums σ_{F_n} (see [7, (iii) of Theorem 14.23]). However, we still need to estimate the precision of the approximation, i.e., the error $\left|\int_a^b f(x)\,dx - \sigma_{F_n}\right|$. In the sequel we give methods of approximation of integrals with estimates of the error depending on some conditions on the function. Recall that the mesh of a partition $F\colon a = x_0 < x_1 < \ldots < x_n = b$ is $\delta(F) = \max_{1\le i\le n}(x_i - x_{i-1})$. The approximating sums corresponding to the partition F are the sums

$$\sigma_F = \sigma_F(f;(c_i)) = \sum_{i=1}^{n} f(c_i)(x_i - x_{i-1}),$$

where $c_i \in [x_{i-1}, x_i]$ for every $i = 1, \ldots, n$.

Theorem 8.11. *If f is monotone on $[a, b]$, then*

$$\left|\int_a^b f(x)\,dx - \sigma_F\right| \le \delta(F)\cdot(f(b) - f(a))$$

for every partition F of $[a, b]$, and for every approximating sum σ_F corresponding to F.

Proof. We may assume that f is increasing. Then the lower and upper sums corresponding to F are $s_F = \sum_{i=1}^n f(x_{i-1})(x_i - x_{i-1})$ and $S_F = \sum_{i=1}^n f(x_i)(x_i - x_{i-1})$. Let $I = \int_a^b f\,dx$. Then $s_F \le I \le S_F$ and $s_F \le \sigma_F \le S_F$, and thus

$$|I - \sigma_F| \le S_F - s_F = \sum_{i=1}^{n}(f(x_i) - f(x_{i-1}))(x_i - x_{i-1}) \le$$

$$\le \delta(F)\cdot\sum_{i=1}^{n}(f(x_i) - f(x_{i-1})) = \delta(F)\cdot(f(b) - f(a)).$$

□

Our next two theorems concern the approximation of the integrals of Lipschitz functions. By the mean value theorem, if f is differentiable on $[a,b]$ and $|f'(x)| \le K$ at every point $x \in [a,b]$, then $|f(x) - f(y)| \le K \cdot |x-y|$ for every $x, y \in [a,b]$. This means that the Lipschitz condition is satisfied whenever $|f'(x)| \le K$ for every $x \in [a,b]$.

Theorem 8.12. *Let f be Lipschitz on $[a,b]$ and let $|f(x) - f(y)| \le K \cdot |x-y|$ for every $x, y \in [a,b]$. Then for every partition F and every corresponding approximating sum σ_F we have*

$$\left| \int_a^b f(x)\,dx - \sigma_F \right| \le \frac{K}{2} \cdot \delta(F) \cdot (b-a). \tag{8.24}$$

Proof. Let $a \le \alpha < \beta \le b$ and $c \in [\alpha, \beta]$. Since $|f(x) - f(c)| \le K \cdot |x-c|$ for every x, we have

$$\begin{aligned} \left| \int_\alpha^\beta f(x)\,dx - f(c)\cdot(\beta-\alpha) \right| &= \left| \int_\alpha^\beta (f(x)-f(c))\,dx \right| \le K \cdot \int_\alpha^\beta |x-c|\,dx = \\ &= K \cdot \int_\alpha^c (c-x)\,dx + K \cdot \int_c^\beta (x-c)\,dx = \qquad (8.25) \\ &= K \cdot \left(\frac{1}{2}(c-\alpha)^2 + \frac{1}{2}(\beta-c)^2 \right) \le \frac{K}{2} \cdot (\beta-\alpha)^2. \end{aligned}$$

Applying this to the subintervals $[x_{i-1}, x_i]$ and using that

$$\int_a^b f(x)\,dx = \sum_{i=1}^n \int_{x_{i-1}}^{x_i} f(x)\,dx,$$

we get

$$\begin{aligned} \left| \int_a^b f(x)dx - \sum_{i=1}^n f(c_i)(x_i - x_{i-1}) \right| &\le \sum_{i=1}^n \frac{K}{2} \cdot (x_i - x_{i-1})^2 \le \\ &\le \frac{K}{2} \cdot \sum_{i=1}^n \delta(F) \cdot (x_i - x_{i-1}) = \\ &= \frac{K}{2} \cdot \delta(F) \cdot (b-a). \qquad \square \end{aligned}$$

By choosing $c = (\alpha + \beta)/2$ we can improve the upper estimate of (8.25) to $(K/4) \cdot (\beta - \alpha)^2$. Therefore, taking $c_i = (x_{i-1} + x_i)/2$ for every i, we obtain the following.

Theorem 8.13. *Let f be Lipschitz on $[a, b]$, and let $|f(x) - f(y)| \le K \cdot |x - y|$ for every $x, y \in [a, b]$. Then for every partition $F: a = x_0 < \ldots < x_n = b$ we have*

$$\left| \int_a^b f(x)\,dx - \sum_{i=1}^{n} f\left(\frac{x_{i-1} + x_i}{2}\right) \cdot (x_i - x_{i-1}) \right| \le \frac{K}{4} \cdot \delta(F) \cdot (b - a). \qquad \square$$

Notice that the sum

$$\sum_{i=1}^{n} f((x_{i-1} + x_i)/2) \cdot (x_i - x_{i-1})$$

is the approximating sum corresponding to the midpoints of the subintervals $[x_{i-1}, x_i]$. If $f \ge 0$, this is the sum of the areas of the rectangles with the appropriate heights (see Figure 8.1).

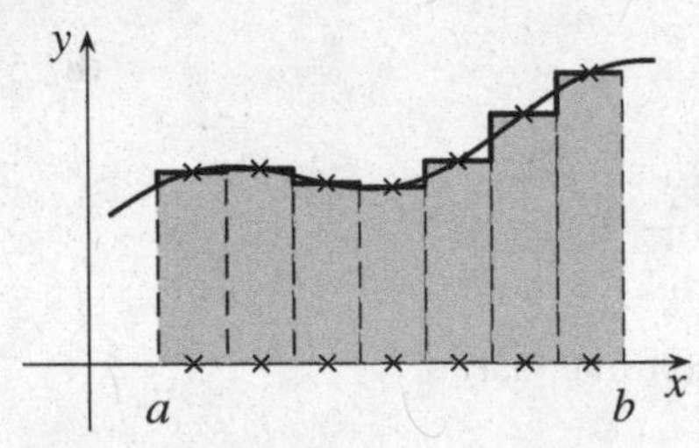

8.1. Figure

We now show that we can achieve similar precision using the sum of the areas of trapezoids defined by consecutive base points and the corresponding points of the graph. In the following theorem we assume a little more than the Lipschitz property. In fact, the Lipschitz property would be sufficient (see Exercise 8.4).

Theorem 8.14. *Let f be differentiable and suppose that f' is integrable on the interval $[a, b]$, and let $|f'(x)| \le K$ for every $x \in [a, b]$. Then for every partition $F: a = x_0 < \ldots < x_n = b$ we have*

$$\left| \int_a^b f(x)\,dx - \sum_{i=1}^{n} \left(\frac{f(x_{i-1}) + f(x_i)}{2}\right) \cdot (x_i - x_{i-1}) \right| \le \frac{K}{4} \cdot \delta(F) \cdot (b - a). \tag{8.26}$$

Proof. Let $a \le \alpha < \beta \le b$ and $\gamma = (\alpha + \beta)/2$. We have

$$\int_{\alpha}^{\beta} f(x)\,dx = \int_{\alpha}^{\beta} f(x) \cdot (x-\gamma)'\,dx =$$
$$= [f(x) \cdot (x-\gamma)]_{\alpha}^{\beta} - \int_{\alpha}^{\beta} f'(x)(x-\gamma)\,dx = \tag{8.27}$$
$$= \frac{f(\alpha)+f(\beta)}{2} \cdot (\beta-\alpha) - \int_{\alpha}^{\beta} f'(x)(x-\gamma)\,dx,$$

and thus

$$\left| \int_{\alpha}^{\beta} f(x)\,dx - \frac{f(\alpha)+f(\beta)}{2} \cdot (\beta-\alpha) \right| \le \int_{\alpha}^{\beta} K \cdot |x-\gamma|\,dx = \frac{K}{4} \cdot (\beta-\alpha)^2.$$

Applying this inequality to the subintervals $[x_{i-1}, x_i]$, we obtain (8.26) in the same way as in the proof of Theorem 8.12. □

Intuitively, the smoother the function, the better it is approximated by its inscribed polygons, and the sum of the areas of the corresponding trapezoids approximates the integral better. For similar reasons, we also expect that the areas of the rectangles corresponding to the midpoints of the partition intervals approximate the integral better, with the same precision. We now prove both statements. In particular, we prove that the error is *quadratic* in $\delta(F)$ for every twice-differentiable function.

Theorem 8.15. *Let f be twice differentiable on $[a, b]$ and let $|f''(x)| \le K$ for every $x \in [a, b]$. Then for every partition $F\colon a = x_0 < \ldots < x_n = b$ we have*

$$\left| \int_{a}^{b} f(x)\,dx - \sum_{i=1}^{n} f\left(\frac{x_{i-1}+x_i}{2}\right) \cdot (x_i - x_{i-1}) \right| \le \frac{K}{24} \cdot \delta(F)^2 \cdot (b-a). \tag{8.28}$$

Proof. Let $a \le \alpha < \beta \le b$ and $\gamma = (\alpha + \beta)/2$. Applying Taylor's formula yields

$$f(x) = f(\gamma) + f'(\gamma) \cdot (x-\gamma) + \frac{f''(c)}{2} \cdot (x-\gamma)^2,$$

and thus

$$|f(x) - f(\gamma) - f'(\gamma) \cdot (x-\gamma)| \le \frac{K}{2} \cdot (x-\gamma)^2$$

for every x. Since $\int_\alpha^\beta (x-\gamma)\,dx = 0$, we have

$$\left|\int_\alpha^\beta f(x)\,dx - f(\gamma)\cdot(\beta-\alpha)\right| = \left|\int_\alpha^\beta (f(x)-f(\gamma)-f'(\gamma)(x-\gamma))\,dx\right| \le$$

$$\le \frac{K}{2}\cdot\int_\alpha^\beta (x-\gamma)^2 dx = \frac{K}{24}\cdot(\beta-\alpha)^3.$$

Applying this to the subintervals $[x_{i-1}, x_i]$ and summing the results gives (8.28) as before. □

In order to get some further estimates, we need the following theorem.

Theorem 8.16. (Taylor's formula with integral remainder) *Let the function f be $(n+1)$ times differentiable, and let $f^{(n+1)}$ be integrable on the interval $[a, x]$. Then*

$$\begin{aligned} f(x) = & f(a) + \frac{f'(a)}{1!}(x-a) + \ldots + \frac{f^{(n)}(a)}{n!}(x-a)^n + \\ & + \frac{1}{n!}\cdot\int_a^x f^{(n+1)}(t)\cdot(x-t)^n\,dt. \end{aligned} \tag{8.29}$$

We allow both $x < a$ and $x > a$.

Proof. We prove the result only for $a < x$; the proof of the $x < a$ case is similar. Let

$$R(t) = \left[f(t) + \frac{f'(t)}{1!}(x-t) + \cdots + \frac{f^{(n)}(t)}{n!}(x-t)^n\right] - f(x)$$

for every $t \in [a, x]$. An easy computation gives $R'(t) = \frac{f^{(n+1)}(t)}{n!}(x-t)^n$. Thus, by the Newton–Leibniz formula we get

$$R(a) = R(a) - R(x) = \frac{1}{n!}\cdot\int_x^a f^{(n+1)}(t)\cdot(x-t)^n\,dt,$$

which is exactly (8.29). □

Theorem 8.17. *Let f be twice differentiable, and let f'' be integrable on $[a, b]$. If $|f''(x)| \le K$ for every $x \in [a, b]$, then for every partition $F\colon a = x_0 < \ldots < x_n = b$ we have*

$$\left|\int_a^b f(x)\,dx - \sum_{i=1}^n \left(\frac{f(x_{i-1}) + f(x_i)}{2}\right)\cdot(x_i - x_{i-1})\right| \le \frac{K}{12}\cdot\delta(F)^2\cdot(b-a). \tag{8.30}$$

Proof. Let $a \le \alpha < \beta \le b$. We show that

$$\left| \int_\alpha^\beta f(x)\,dx - \frac{f(\alpha)+f(\beta)}{2} \cdot (\beta-\alpha) \right| \le \frac{K}{12} \cdot (\beta-\alpha)^3. \tag{8.31}$$

Let $F(x) = \int_\alpha^x f(t)\,dt$ for every $x \in [\alpha, \beta]$. Then $F' = f$, and thus F is three times differentiable on $[\alpha, \beta]$. Taylor's formula (8.29) gives

$$\int_\alpha^\beta f(x)\,dx = F(\beta) = f(\alpha)(\beta-\alpha) + \frac{1}{2} f'(\alpha)(\beta-\alpha)^2 + \frac{1}{2} \cdot \int_\alpha^\beta f''(t)(\beta-t)^2\,dt. \tag{8.32}$$

Applying formula (8.29) to f yields

$$f(\beta) = f(\alpha) + f'(\alpha)(\beta-\alpha) + \int_\alpha^\beta f''(t)(\beta-t)\,dt. \tag{8.33}$$

If we multiply (8.33) by $(\beta-\alpha)/2$ and subtract the result from (8.32), we obtain

$$\begin{aligned} \int_\alpha^\beta f(x)\,dx - &\frac{f(\alpha)+f(\beta)}{2} \cdot (\beta-\alpha) = \\ &= \frac{1}{2} \cdot \int_\alpha^\beta f''(t)(\beta-t)^2\,dt - \frac{1}{2} \cdot \int_\alpha^\beta f''(t)(\beta-t)(\beta-\alpha)\,dt = \\ &= -\int_\alpha^\beta f''(t) \cdot \varphi(t)\,dt, \end{aligned} \tag{8.34}$$

where $\varphi(t) = \frac{1}{2}(\beta-t)(t-\alpha)$. The function φ is continuous and nonnegative on $[\alpha, \beta]$. Applying the general equation (8.34) to the function $f_0(x) = (x-\alpha)^2$ gives $-(\beta-\alpha)^3/6 = -2 \cdot \int_\alpha^\beta \varphi(t)\,dt$, i.e., $\int_\alpha^\beta \varphi(t)\,dt = (\beta-\alpha)^3/12$.

Now, if $|f''| \le K$, then (8.34) implies

$$\left| \int_\alpha^\beta f(x)\,dx - \frac{f(\alpha)+f(\beta)}{2} \cdot (\beta-\alpha) \right| \le \int_\alpha^\beta K \cdot \varphi(t)\,dt = K \cdot (\beta-\alpha)^3/12,$$

which proves (8.31). We get (8.30) following the usual argument. □

Remark 8.18. **1.** We can omit the integrability condition on f'' of the previous theorem. Moreover, it is enough to assume that f is differentiable and f' is Lipschitz, i.e., $|f'(x) - f'(y)| \le K \cdot |x-y|$ for every $x, y \in [a, b]$ (see Exercise 8.5).

2. Applying Theorem 8.17, we can give a new proof of the convergence of $(n/e)^n \cdot \sqrt{n}/n!$. Let

$$a_n = \int_n^{n+1} \log x \, dx - \frac{\log n + \log(n+1)}{2}$$

for every $n > 0$. Since $\log x$ is concave, we have $a_n \geq 0$. On the other hand, the absolute value of the second derivative of $\log x$ is $|-1/x^2| \leq 1/n^2$ for every $x \in [n, n+1]$, and thus (8.31) implies $0 \leq a_n \leq 1/(12n^2)$. It follows that the sequence $\sum_{k=1}^{n-1} a_k$ is monotonically increasing and bounded, hence convergent. Since

$$\begin{aligned}\sum_{k=1}^{n-1} a_k &= \int_1^n \log x \, dx - \sum_{k=1}^{n} \log k + (\log n)/2 = \\ &= n \log n - n + 1 + (\log n)/2 - \log n! = \\ &= 1 + \log\left(\frac{(n/e)^n \sqrt{n}}{n!}\right),\end{aligned}$$

it follows that the sequence $(n/e)^n \sqrt{n}/n!$ is convergent. □

The core idea of approximating with trapezoids (Theorem 8.17) is to replace the function f on every interval $[x_{i-1}, x_i]$ by a linear function whose value is the same as that of f at x_{i-1} and x_i. We can expect to get a better approximation using polynomials of higher degree instead of linear functions. Let us see what happens if we approximate the function f on the interval $[x_{i-1}, x_i]$ by the (at most) quadratic polynomial whose value at the points $\alpha = x_{i-1}$, $\beta = x_i$, and $\gamma = (x_{i-1} + x_i)/2$ is the same as that of f. It is easy to see that if p is an (at most) quadratic polynomial, then

$$\int_\alpha^\beta p(x) \, dx = \frac{p(\alpha) + 4p(\gamma) + p(\beta)}{6} \cdot (\beta - \alpha), \tag{8.35}$$

where $\gamma = (\alpha + \beta)/2$ (see Exercise 8.6). The question is how accurately the quantity $(f(\alpha) + 4f(\gamma) + f(\beta)) \cdot (\beta - \alpha)/6$ approximates the integral of f on $[\alpha, \beta]$.

Lemma 8.19. (Simpson's[3] approximation) *Let f be four times differentiable and let $f^{(4)}$ be integrable on $[\alpha, \beta]$. If $|f^{(4)}(x)| \leq K$ for every $x \in [\alpha, \beta]$, then*

$$\left| \int_\alpha^\beta f(x) \, dx - \frac{f(\alpha) + 4f(\gamma) + f(\beta)}{6} \cdot (\beta - \alpha) \right| \leq \frac{K}{2880} \cdot (\beta - \alpha)^5, \tag{8.36}$$

where $\gamma = (\alpha + \beta)/2$.

[3] Thomas Simpson (1710–1761), British mathematician.

Proof. The proof is analogous to the proof of Theorem 8.17. To simplify the computation, let us consider the $\alpha = -1$, $\beta = 1$ case first. Let $F(x) = \int_0^x f(t)\,dt$ for every $x \in [-1, 1]$. Then $F' = f$, and F is five times differentiable on $[-1, 1]$. Thus, formula (8.29) gives

$$F(x) = f(0)x + \frac{1}{2}f'(0)x^2 + \frac{1}{6}f''(0)x^3 + \frac{1}{24}f'''(0)x^4 + \frac{1}{24} \cdot \int_0^x f^{(4)}(t)(x-t)^4\,dt,$$

and

$$\int_{-1}^{1} f(x)\,dx = F(1) - F(-1) =$$

$$= 2f(0) + \frac{1}{3}f''(0) + \frac{1}{24} \cdot \int_0^1 f^{(4)}(t)(1-t)^4\,dt - \frac{1}{24} \cdot \int_0^{-1} f^{(4)}(t)(-1-t)^4\,dt.$$

Applying formula (8.29) to f, we get

$$f(x) = f(0) + f'(0)x + \frac{1}{2}f''(0)x^2 + \frac{1}{6}f'''(0)x^3 + \frac{1}{6} \cdot \int_0^x f^{(4)}(t)(x-t)^3\,dt,$$

and thus

$$\frac{f(-1) + 4f(0) + f(1)}{3} =$$

$$= 2f(0) + \frac{1}{3}f''(0) + \frac{1}{18} \cdot \int_0^1 f^{(4)}(t)(1-t)^3\,dt + \frac{1}{18} \cdot \int_0^{-1} f^{(4)}(t)(-1-t)^3\,dt.$$

Therefore,

$$\begin{aligned} &\int_{-1}^{1} f(x)\,dx - \frac{f(-1) + 4f(0) + f(1)}{3} = \\ &= \int_0^1 f^{(4)}(t)\left[\frac{1}{24}(1-t)^4 - \frac{1}{18}(1-t)^3\right]\,dt + \\ &\quad + \int_{-1}^{0} f^{(4)}(t)\left[\frac{1}{24}(1+t)^4 - \frac{1}{18}(1+t)^3\right]\,dt = \\ &= -\int_{-1}^{1} f^{(4)}(t) \cdot \psi(t)\,dt, \end{aligned} \tag{8.37}$$

where $\psi(t) = \frac{1}{72}(1 - |t|)^3(1 + 3|t|)$. The function ψ is continuous and nonnegative on $[-1, 1]$. Applying (8.37) to the function $f(x) = x^4$, we find that $(2/5) - (2/3) = -24 \cdot \int_{-1}^{1} \psi(t)\,dt$, i.e., $\int_{-1}^{1} \psi(t)\,dt = 1/90$. If $|f''| \le K$, then (8.37) implies

$$\left| \int_{-1}^{1} f(x)\,dx - \frac{f(-1)+4f(0)+f(1)}{3} \right| \le \int_{-1}^{1} K \cdot \psi(t)\,dt = K/90. \tag{8.38}$$

Now consider the general case, and let $f\colon [\alpha, \beta] \to \mathbb{R}$ satisfy the conditions of the lemma. Applying (8.38) to the function

$$\overline{f}(x) = f\left(\frac{\alpha+\beta}{2} + x \cdot \frac{\beta-\alpha}{2}\right) \qquad (x \in [-1,1]),$$

we get (8.36). □

The following theorem is obtained from the previous lemma by applying the argument above.

Theorem 8.20. *Let f be four-times differentiable on $[a,b]$, and let $|f^{(4)}(x)| \le K$, for every $x \in [a,b]$. Then for every partition $F\colon a = x_0 < \ldots < x_n = b$, we have*

$$\left| \int_a^b f(x)\,dx - \sum_{i=1}^{n} \frac{f(x_{i-1}) + 4f\left((x_{i-1}+x_i)/2\right) + f(x_i)}{6} \cdot (x_i - x_{i-1}) \right| \le$$

$$\le \frac{K}{2880} \cdot \delta(F)^4 \cdot (b-a).$$

In numerical computations it is simplest to work with the uniform partition of the interval $[a,b]$. The base points of this partition are $x_i = a + (b-a) \cdot i/n$ $(i = 0, \ldots, n)$, and the mesh of the partition is $(b-a)/n$. Therefore, applying Theorems 8.11, 8.14, 8.17, and 8.20, we obtain the following theorem.

Theorem 8.21.

(i) *If f is monotone on $[a,b]$, then*

$$\left| \int_a^b f(x)\,dx - \frac{b-a}{n} \cdot \sum_{i=1}^{n} f(x_i) \right| \le \frac{f(b)-f(a)}{n} \cdot (b-a).$$

(ii) *Let f be differentiable and f' integrable on $[a,b]$. If $|f'(x)| \le K$ for every $x \in [a,b]$, then*

$$\left| \int_a^b f(x)\,dx - \left(\frac{f(a)+f(b)}{2} + \sum_{i=1}^{n-1} f(x_i) \right) \cdot \frac{b-a}{n} \right| \le \frac{K}{4n} \cdot (b-a)^2.$$

(iii) *Let f be twice differentiable and f'' integrable on $[a,b]$. If $|f''(x)| \le K$ for every $x \in [a,b]$, then*

$$\left| \int_a^b f(x)\,dx - \left(\frac{f(a)+f(b)}{2} + \sum_{i=1}^{n-1} f(x_i) \right) \cdot \frac{b-a}{n} \right| \le \frac{K}{12n^2} \cdot (b-a)^3.$$

(iv) **(Simpson's formula).** *Let f be four times differentiable and $f^{(4)}$ integrable on $[a,b]$. If $|f^{(4)}(x)| \le K$ for every $x \in [a,b]$, then*

$$\left| \int_a^b f(x)\,dx - \left(f(a) + f(b) + 2 \cdot \sum_{i=1}^{k-1} f(x_{2i}) + 4 \cdot \sum_{i=0}^{k-1} f(x_{2i+1}) \right) \cdot \frac{b-a}{6n} \right| \le$$

$$\le \frac{K}{2880n^4} \cdot (b-a)^5, \qquad \square$$

for $n = 2k$.

Exercises

8.4. Show that if $|f(x) - f(y)| \le K \cdot |x-y|$ for every $x, y \in [a,b]$, g is differentiable, and g' is integrable on $[a,b]$, then

$$\left| \int_a^b fg'\,dx - [fg]_a^b \right| \le K \cdot \int_a^b |g'|\,dx. \qquad \text{(H)}$$

8.5. Show that we can relax the conditions of Theorem 8.14 to $|f(x) - f(y)| \le K \cdot |x-y|$ for every $x, y \in [a,b]$.

8.6. Show that we can relax the conditions of Theorem 8.17 to f being differentiable and $|f'(x) - f'(y)| \le K \cdot |x-y|$ for every $x, y \in [a,b]$.

8.7. Show that (8.35) holds for every at most *cubic* polynomial p.

8.3 Parametric Integrals

When we say that $\int_0^1 x^c\,dx = 1/(c+1)$ for every $c \ne -1$, we consider an integral depending on the parameter c. Similarly, when we state that the improper integral $\int_1^\infty x^c\,dx$ is convergent if $c < -1$ and is divergent if $c \ge -1$, we consider the integral $\int_1^\infty x^c\,dx$ depending on the different values of the parameter c. In general, by a **parametric integral** we mean an integral of the form $\int_H f(t,x)\,dx$, where T is an interval, f is defined on the set $T \times H$, and we integrate the section function f_t on H. Here, H can be a Jordan measurable set on which the section function $x \mapsto f_t(x) = f(t,x)$ is integrable, or an interval on which the improper integral of the section f_t is convergent for every $t \in T$.[4] In the first case we are talking about **parametric Riemann integrals,** while in the second case, we are talking about **parametric improper integrals**[5].

The parametric integrals play an important role in both the applications of analysis and the solutions of certain theoretical problems. In some cases, considering a definite integral as a special case of some parametric integral might be very useful for calculating its value. Consider, for example, the integral

$$\int_0^1 \frac{x-1}{\log x}\,dx. \tag{8.39}$$

The function $(x-1)/\log x$ is integrable on $[0,1]$, since it is both bounded and continuous on $(0,1)$. Indeed, we have $\log x < x-1 < 0$ for $0 < x < 1$, and thus $0 < (x-1)/\log x < 1$. Since the primitive function of $(x-1)/\log x$ is not an elementary function (see Exercise 8.8), calculating the exact value of the integral might look hopeless at first. However, if our goal is calculating the value of the parametric integral

$$F(t) = \int_0^1 \frac{x^t-1}{\log x}\,dx, \tag{8.40}$$

with parameter $t \geq 0$, then it is only natural to calculate the derivative of $F(t)$. We get

$$F'(t) = \int_0^1 \frac{\partial}{\partial t}\left(\frac{x^t-1}{\log x}\right)dx = \int_0^1 x^t\,dx = \frac{1}{t+1}. \tag{8.41}$$

It remains to be seen whether this step is justified, i.e., whether we can take the derivative "behind" the integral. We will have our answer presently. If, however, (8.41) is correct, then $F(t) = \log(t+1) + c$ for some constant c. Since $F(0) = \int_0^1 0\,dx = 0$, we have $c = 0$, i.e., $F(t) = \log(t+1)$ for every $t \geq 0$. Thus, the value of the integral (8.39) is $F(1) = \log 2$.

Let us now consider the integral

$$\int_0^\infty \frac{e^{-x} - e^{-2x}}{x}\,dx, \tag{8.42}$$

---

[4] Obviously, the parameter can be denoted by any letter. In choosing the letter t (instead of the letter c) we want to indicate that we consider the parameter to be a variable, i.e., we want to think of the value of the integral as a function of t.

[5] The two cases do not exclude each other: if H is a bounded interval and f_t is Riemann integrable on H for every $t \in T$, then the integral $\int f(t,x)\,dx$ is both a parametric Riemann integral and a parametric improper integral. See [7, Remark 19.4.2].

whose convergence is easy to see. We can calculate the integral by reducing it to (8.39) using the substitution $e^{-x} = y$, but we can also use the method applied above. Consider the parametric improper integral

$$G(t) = \int_0^\infty \frac{e^{-x} - e^{-tx}}{x}\, dx \tag{8.43}$$

with parameter $t > 0$, and calculate its derivative:

$$\begin{aligned} G'(t) &= \int_0^\infty \frac{\partial}{\partial t}\left(\frac{e^{-x} - e^{-tx}}{x}\right) dx = \int_0^\infty \frac{-e^{-tx} \cdot (-x)}{x}\, dx = \\ &= \int_0^\infty e^{-tx}\, dx = \left[-\frac{e^{-tx}}{t}\right]_{x=0}^\infty = \frac{1}{t}. \end{aligned} \tag{8.44}$$

We should check whether this step is justifiable, similarly to the case of our previous example. But if (8.44) holds, then $G(t) = \log t + c$, and by $G(1) = 0$, we have $G(t) = \log t$ for every $t > 0$. Thus, the value of the integral (8.42) is $G(2) = \log 2$.

In the arguments above we need to justify why the derivatives of F and G equal the integrals of the derivatives of the respective integrands. In other words, the question is whether the results of the integration with respect to x and the derivation with respect to t of the functions $(x^t - 1)/\log x$ and $(e^{-x} - e^{-tx})/x$ are independent of the order of these operations. Similar questions can be asked about any functions $f(t, x)$, and not only about derivation but also about taking limits or integration.

First, we want to find conditions that ensure the continuity, integrability, and differentiability of a parametric Riemann integral. That is, we consider the case that H is a Jordan measurable set, possibly in $\mathbb{R}^p$.

In the following three theorems we make the following assumptions: $H \subset \mathbb{R}^p$ is a Jordan measurable set, f is defined on the set $[a, b] \times H$, and the section function $x \mapsto f(t, x)$ is integrable on H for every $t \in [a, b]$. Let

$$F(t) = \int_H f(t, x)\, dx \qquad (t \in [a, b]).$$

Theorem 8.22. *If f is continuous and bounded on $[a, b] \times H$, then F is continuous on $[a, b]$.*

Proof. First, let us assume that H is closed. Then $[a, b] \times H$ is also closed. Since H is bounded (because it is Jordan measurable), $[a, b] \times H$ is bounded as well. Thus, by Heine's theorem (Theorem 1.53), f is uniformly continuous on $[a, b] \times H$.

Let $t_0 \in [a, b]$ and $\varepsilon > 0$ be fixed. By the uniform continuity, there exists $\delta > 0$ such that $|f(u, x) - f(v, y)| < \varepsilon$ whenever $(u, x), (v, y) \in [a, b] \times H$ and $|(u, x) - (v, y)| < \delta$. Thus, if $t \in [a, b]$ and $|t - t_0| < \delta$, then $|f(t, x) - f(t_0, x)| < \varepsilon$ for every $x \in H$. Therefore,

$$|F(t) - F(t_0)| = \left| \int_H (f(t,x) - f(t_0,x))\,dx \right| \leq \int_H |f(t,x) - f(t_0,x)|\,dx \leq$$
$$\leq \varepsilon \cdot \mu(H),$$

where $\mu(H)$ denotes the Jordan measure of H. This proves that F is continuous at t_0.

We now consider the general case, in which H is not necessarily closed (but Jordan measurable and consequently bounded). Let f be bounded and continuous. Suppose that $|f(t,x)| \leq K$ for every $(t,x) \in [a,b] \times H$, and let $\varepsilon > 0$ be fixed. Since H is Jordan measurable, there exists a set $A \subset H$ such that $\mu(A) > \mu(H) - \varepsilon$, and A is the union of finitely many closed boxes. Then A is bounded and closed, and thus the function $F_1(t) = \int_A f(t,x)\,dx$ is continuous on $[a,b]$ (by what we proved above). Furthermore, applying the formula (4.3), we get

$$|F(t) - F_1(t)| = \left| \int_H f(t,x)\,dx - \int_A f(t,x)\,dx \right| = \left| \int_{H \setminus A} f(t,x)\,dx \right| \leq$$
$$\leq K \cdot \mu(H \setminus A) < K \cdot \varepsilon$$

for every $t \in [a,b]$. It follows that F can be obtained as the uniform limit of a sequence of continuous functions. Then, by Theorem 7.12, F itself is continuous. □

The following theorem is a straightforward application of Theorem 4.17 on calculating the integral of a multivariable function. We know that the section functions of a two-variable integrable function are not necessarily integrable, whence the upper and lower integrals in the statement of the following theorem (see Remark 4.21.2).

Theorem 8.23. *If f is integrable on $[a,b] \times H$, then F is also integrable on $[a,b]$, and*

$$\int_a^b F(t)\,dt = \int_H \left(\overline{\int_a^b} f(t,x)\,dt \right) dx = \int_H \left(\underline{\int_a^b} f(t,x)\,dt \right) dx.$$

Proof. Let $B \subset \mathbb{R}^p$ be a box containing the set H. Define f to be zero at the points of the set $[a,b] \times (B \setminus H)$. The section function $x \mapsto f(t,x)$ of this extended function f is integrable on B, with $\int_B f(t,x)\,dx = F(t)$ for every $t \in [a,b]$ (see (4.3)). Now the statement of this theorem follows from Theorem 4.17. □

Recall that if the section function $t \mapsto f(t,x)$ $(t \in [a,b])$ is differentiable for a fixed $x \in H$, then its derivative is called the partial derivative of the function f with respect to t, and it is denoted by either $\frac{\partial}{\partial t} f$ or $D_1 f$.

Theorem 8.24. *If the partial derivative $D_1 f$ exists and is continuous and bounded on $[a,b] \times H$, then F is differentiable on $[a,b]$, and $F'(t) = \int_H D_1 f(t,x)\,dx$ for every $t \in [a,b]$.*

Proof. Let $I(t) = \int_H D_1 f(t,x)\,dx$ for every $t \in [a,b]$. By Theorem 8.22, I is continuous on $[a,b]$.

Let $u \in [a,b]$ be arbitrary. Applying Theorem 8.23 to the interval $[a,u]$ and the function $D_1 f$ yields[6]

$$\int_a^u I(t)\,dt = \int_H \left(\int_a^u D_1 f\,dt \right) dx = \int_H (f(u,x) - f(a,x))\,dx = F(u) - F(a).$$

Thus, the function $F - F(a)$ is the integral function of the function I. Since I is continuous, it follows that F is differentiable, and $F'(t) = I(t)$ for every $t \in [a,b]$. (See [7, Theorem 15.5].) □

Remark 8.25. One can show that the condition on the continuity of the partial derivative function $D_1 f$ can be omitted from Theorem 8.24.

Example 8.26. Let us take another look at the integral (8.39). For every $t \geq 0$ the function $x \mapsto f(t,x) = (x^t - 1)/\log x$ is integrable on $(0,1)$, since it is continuous and bounded. The function $D_1 f(t,x) = x^t$ is continuous and bounded on the set $[0,b] \times (0,1)$ for every $b > 0$. Thus, by Theorem 8.24, the function $F(t) = \int_0^1 ((x^t - 1)/\log x)\,dx$ is differentiable, and (8.41) holds. Then, as we saw, $F(t) = \log(t+1)$ follows, and thus $F(1) = \log 2$.

Now we turn to the investigation of parametric improper integrals. These integrals are more important for applications than the parametric Riemann integrals. Unfortunately, the results on these integrals are more complicated, in that they require stricter conditions on the integrands. This is unavoidable: if, for example, we want to generalize Theorem 8.22 to parametric improper integrals, we need to assume more than continuity and boundedness of the function f. Consider the following simple example.

Example 8.27. The two-variable function $f(t,x) = t \cdot e^{-tx}$ is continuous and bounded on the set $[0,1] \times [0,\infty)$, and the improper integral $F(t) = \int_0^\infty f(t,x)\,dx$ is convergent for every $t \geq 0$. If $t > 0$, then

$$F(t) = \int_0^\infty t \cdot e^{-tx}\,dx = \left[-e^{-tx}\right]_0^\infty = 1.$$

On the other hand, $F(0) = 0$, and thus the function F is not continuous at the point $t = 0$.

Now consider the analogue of Theorem 8.23 for improper integrals. Let f be defined on the set $[a,b] \times [c,\infty)$, and let the improper integral $F(t) = \int_c^\infty f(t,x)\,dx$ be convergent for every $t \in [a,b]$. We need conditions that guarantee that

---

[6] The function $D_1 f$ is integrable on $[a,b] \times H$, since it is bounded and continuous there. See Theorem 4.14.

$$\int_a^b F(t)\,dt = \int_c^\infty \left(\overline{\int_a^b} f(t,x)\,dt \right) dx \tag{8.45}$$

holds. The condition of the integrability of f on the set $[a,b] \times [c,\infty)$ is out of the question, since we defined the integral of multivariable functions only on Jordan measurable (and hence bounded) sets. However, (8.45) does not necessarily hold without adding extra conditions—not even in the case that f is continuous and bounded, and every integral appearing in the theorem exists. This is illustrated by the following example.

Example 8.28.
Consider the open boxes $P_k = (2^{-k}, 2^{-k+1}) \times (2^{k-1}, 2^k)$ and $N_k = (2^{-k}, 2^{-k+1}) \times (2^k, 2^{k+1})$ for every $k = 1, 2, \ldots$. Let

$$g(t,x) = \begin{cases} 2, & \text{if } (t,x) \in P_k\ (k = 1,2,\ldots); \\ -1, & \text{if } (t,x) \in N_k\ (k = 1,2,\ldots); \\ 0 & \text{otherwise.} \end{cases}$$

We show that $G(t) = \int_0^\infty g(t,x)\,dx = 0$ for every t. This is obvious if t is not in any of the intervals $(2^{-k}, 2^{-k+1})$, since in that case, $g(t,x) = 0$ for every x. If, however, $t \in (2^{-k}, 2^{-k+1})$, then the section function g_t is 2 on the interval $(2^{k-1}, 2^k)$, -1 on the interval $(2^k, 2^{k+1})$, and 0 everywhere else, and thus its integral is indeed zero. Thus, $G(t) = 0$ for every t, which implies $\int_0^1 G(t)\,dt = 0$.

We now show that $\int_0^\infty \left(\int_0^1 g(t,x)\,dt \right) dx \neq 0$. Let $I(x) = \int_0^1 g(t,x)\,dt$. If $x \in (0,1)$, then $g(t,x) = 0$ for every $t \in [0,1]$, giving $I(x) = 0$. If $x \in (1,2)$, then $I(x) = \int_{1/2}^1 2\,dt = 1$. If, however, $x \in (2^k, 2^{k+1})$ for some positive integer k, then

$$I(x) = \int_{2^{-k}}^{2^{-k+1}} (-1)\,dt + \int_{2^{-k-1}}^{2^{-k}} 2\,dt = 0.$$

This implies

$$\int_0^\infty \left(\int_0^1 g(t,x)\,dt \right) dx = \int_0^\infty I(x)\,dx = 1.$$

In the previous example the function g is not continuous, but it can be made continuous by a simple alteration of the construction. Let

$$f(t,x) = \begin{cases} 2 \cdot \sin(2^k \pi t) \cdot \sin(2^{-k+1} \pi x), & \text{if } (t,x) \in P_k\ (k = 1,2,\ldots); \\ -\sin(2^k \pi t) \cdot \sin(2^{-k} \pi x), & \text{if } (t,x) \in N_k\ (k = 1,2,\ldots); \\ 0, & \text{otherwise.} \end{cases}$$

It is easy to check that f is continuous and bounded on the set $[0,1] \times [0,\infty)$, $F(t) = \int_0^\infty f(t,x)\,dx = 0$ for every t, but $\int_0^\infty \left(\int_0^1 f(t,x)\,dt \right) dx > 0$.

Example 8.29. Let $h(t,x) = \int_0^t f(u,x)\,du$ for every $(t,x) \in \mathbb{R}^2$, where f is the function from the previous example. It is easy to check that h is continuous everywhere, and the improper integral $H(t) = \int_0^\infty h(t,x)\,dx$ is convergent for every $t \in [0,1]$. This follows from the fact that for every $t \in [0,1]$, we have $h(t,x) = 0$ if x is large enough.

Now, $D_1 h = f$ is continuous and bounded on $[0,1] \times [0,\infty)$. However, $H'(t) = \int_0^\infty D_1 h(t,x)\,dx = \int_0^\infty f(t,x)\,dx = 0$ cannot hold for every $t \in [0,1]$, since that would imply that H is constant. However, $H(0) = 0$ and

$$H(1) = \int_0^\infty h(1,x)\,dx = \int_0^\infty \left(\int_0^1 f(t,x)\,dt \right) dx > 0.$$

This shows that Theorem 8.24 does not hold for improper integrals, not even when $D_1 f$ is continuous and bounded on the set $[a,b] \times [c,\infty)$.

The questions about the continuity, differentiability, and integrability of parametric improper integrals are analogous to similar questions for series of functions. In fact, the connection is more than simple analogy. Consider the series of functions $\sum_{n=1}^\infty f_n(t)$, convergent on the interval $[a,b]$. Let f be the sum of the series on $[a,b]$. Let $f(t,x) = f_n(t)$ for every $t \in [a,b]$, $n \le x < n+1$, and $n = 1,2,\ldots$. It is easy to see that

$$\int_1^\infty f(t,x)\,dx = \sum_{n=1}^\infty f_n(t) = f(t)$$

for every $t \in [a,b]$. Thus, if we establish sufficient conditions that ensure that $\int_1^\infty \frac{\partial f}{\partial t}(t,x)\,dx = f'(t)$, then we also have sufficient conditions for the term-by-term differentiability of the function series $\sum_{n=1}^\infty f_n(t)$. The same holds for the continuity of the integral or its integrability.

Since uniform continuity played a pivotal role in the corresponding theorems (Theorems 7.40 and 7.42) for function series, it is not at all surprising that we need a similar notion for the case of parametric improper integrals.

Definition 8.30. Let f be defined on the set $[a,b] \times [c,\gamma)$, with $a,b,c \in \mathbb{R}$, $a < b$, and $c < \gamma \le \infty$. Suppose that the Riemann integral $\int_c^d f(t,x)\,dx$ exists for every $t \in [a,b]$ and $c < d < \gamma$.

We say that the *parametric improper integral $\int_c^\gamma f(t,x)\,dx$ is uniformly convergent on* $[a,b]$ if the improper integral $\int_c^\gamma f(t,x)\,dx$ is convergent for every $t \in [a,b]$, and for every $\varepsilon > 0$ there exists $d \in (c,\gamma)$ such that $\left| \int_\omega^\gamma f(t,x)dx \right| < \varepsilon$ for every $d \le \omega < \gamma$ and $t \in [a,b]$.

Remark 8.31. Let $\varepsilon > 0$ be fixed. For a fixed t, the convergence of the improper integral $\int_c^\gamma f(t,x)\,dx$ implies that for every $\varepsilon > 0$ there exists $d \in (c,\gamma)$ such that $\left|\int_\omega^\gamma f(t,x)\,dx\right| < \varepsilon$ for every $d \le \omega < \gamma$. However, the d "threshold" depends not only on ε, but also on t. Thus, it is possible that for a fixed ε the supremum of the d threshold values, corresponding to different parameters $t \in [a,b]$, is γ, and there is no $d < \gamma$ that works for *every* parameter t. The uniform convergence of the integral $\int_c^\gamma f(t,x)\,dx$ on $[a,b]$ means exactly the existence of such a common threshold, independent of t, for every $\varepsilon > 0$.

In each of the following three theorems, we assume that f is defined on the set $[a,b] \times [c,\gamma)$, that for every $t \in [a,b]$ the Riemann integral $\int_c^d f(t,x)\,dx$ exists for each $c < d < \gamma$, and furthermore, that the improper integral $\int_c^\gamma f(t,x)\,dx$ is convergent and its value is $F(t)$.

Theorem 8.32. *Let f be continuous on the set $[a,b] \times [c,\gamma)$. If the improper integral $\int_c^\gamma f(t,x)\,dx$ is uniformly convergent on $[a,b]$, then F is continuous on $[a,b]$.*

Proof. Choose a sequence of numbers $c < d_n < \gamma$ such that $\lim_{n\to\infty} d_n = \gamma$. Let $F_n(t) = \int_c^{d_n} f(t,x)\,dx$ for every $t \in [a,b]$ and $n = 1,2,\ldots$. By Theorem 8.22, F_n is continuous for every n.

If we can prove that the sequence of functions F_n converges uniformly to F on the interval $[a,b]$, then Theorem 7.12 will imply the continuity of F on $[a,b]$.

Let $\varepsilon > 0$ be fixed. Since the integral $\int_c^\gamma f(t,x)\,dx$ is uniformly convergent on $[a,b]$, there exists $c < d < \gamma$ such that $\left|\int_\omega^\gamma f(t,x)\,dx\right| < \varepsilon$ for every $d \le \omega < \gamma$ and $t \in [a,b]$. However, $d_n \to \gamma$, and thus there exists n_0 such that $d < d_n < \gamma$ for every $n \ge n_0$. Then for every $n \ge n_0$ and $t \in [a,b]$, we have

$$|F(t) - F_n(t)| = \left|\int_c^\gamma f(t,x)\,dx - \int_c^{d_n} f(t,x)\,dx\right| = \left|\int_{d_n}^\gamma f(t,x)\,dx\right| < \varepsilon.$$

This proves that F_n converges uniformly to F on the interval $[a,b]$. □

Theorem 8.33. *Let f be integrable on the box $[a,b] \times [c,d]$ for every $c < d < \gamma$. If the improper integral $\int_c^\gamma f(t,x)\,dx$ is uniformly convergent on $[a,b]$, then F is integrable on $[a,b]$, and*

$$\int_a^b F(t)\,dt = \int_c^\gamma \left(\overline{\int_a^b} f(t,x)\,dt\right) dx, \tag{8.46}$$

meaning also that the improper integral on the right-hand side exists.

Proof. Again, choose a sequence of numbers $c < d_n < \gamma$ such that $\lim_{n\to\infty} d_n = \gamma$, and let $F_n(t) = \int_c^{d_n} f(t,x)\,dx$ for every $t \in [a,b]$ and $n = 1,2,\ldots$. By Theorem 8.23, F_n is integrable on $[a,b]$, and

$$\int_a^b F_n(t)\,dt = \int_c^{d_n} \left(\overline{\int_a^b} f(t,x)\,dt \right) dx \tag{8.47}$$

for every n. In the proof of Theorem 8.32 we showed that the sequence of functions F_n converges uniformly to F on the interval $[a, b]$. Thus by Theorem 7.16, F is integrable on $[a, b]$, and $\lim_{n\to\infty} \int_a^b F_n(t)\,dt = I$, where $I = \int_a^b F\,dt$. This means that the sequence of integrals on the right-hand side of (8.47) converges to I as $n \to \infty$. This holds for every sequence $d_n < \gamma$, $d_n \to \gamma$, and then, by the transference principle, we find that the limit

$$\lim_{\omega\to\gamma-0} \int_a^\omega \left(\overline{\int_a^b} f(t,x)\,dt \right) dx$$

exists, and its value is I. (If $\gamma = \infty$, then by $\omega \to \gamma - 0$ we mean $\omega \to \infty$.) In other words, the improper integral on the right-hand side of (8.46) is convergent, and its value is I. □

Theorem 8.34. *Suppose that the partial derivative $D_1 f$ exists and is continuous on the set $[a, b] \times [c, \gamma)$, and that the improper integral $\int_c^\gamma D_1 f(t, x)\,dx$ is uniformly convergent on $[a, b]$. Then F is differentiable on $[a, b]$, and $F'(t) = \int_c^\gamma D_1 f(t, x)\,dx$ for every $t \in [a, b]$.*

Proof. Let $I(t) = \int_c^\gamma D_1 f(t, x)\,dx$ for every $t \in [a, b]$. By Theorem 8.32, the function f is continuous on $[a, b]$.

Let $u \in [a, b]$ be arbitrary. Applying Theorem 8.33 to the interval $[a, u]$ and to the function $D_1 f$, we get that

$$\int_a^u I(t)\,dt = \int_c^\gamma \left(\int_a^u D_1 f\,dt \right) dx = \int_c^\gamma (f(u,x) - f(a,x))\,dx = F(u) - F(a).$$

Therefore, the function $F - F(a)$ is the integral function of the function I. Since I is continuous, it follows that F is differentiable, and $F'(t) = I(t)$, for every $t \in [a, b]$. (See [7, Theorem 15.5].) □

Remark 8.35. One can show that in Theorem 8.34, the condition on the continuity of the partial derivative $D_1 f$ can be replaced by the condition that $D_1 f$ is bounded on the box $[a, b] \times [c, d]$ for every $c < d < \gamma$.

In order to apply Theorems 8.32–8.34, we need conditions that guarantee the uniform convergence of parametric improper integrals and are easy to check. The following theorem formalizes one such condition.

Theorem 8.36. *Let the Riemann integral $\int_c^d f(t, x)\,dx$ exist for every $c < d < \gamma$ and $t \in [a, b]$. If there is a function $M\colon [c, \gamma) \to \mathbb{R}$ such that $|f(t, x)| \le M(x)$ holds for every $t \in [a, b]$ and $x \in [c, \gamma)$, and if furthermore, the improper integral*

$\int_c^\gamma M(x)\,dx$ is convergent, then the parametric improper integral $\int_c^\gamma f(t,x)\,dx$ is uniformly convergent on $[a,b]$.

Proof. By the majorization principle for improper integrals, the improper integral $\int_c^\gamma f(t,x)\,dx$ is convergent for every $t \in [a,b]$. (See [7, Theorem 19.18].) Let $\varepsilon > 0$ be fixed. Since $\int_c^\gamma M(x)\,dx$ is convergent, there exists $c < d < \gamma$ such that $\int_d^\gamma M(x)\,dx < \varepsilon$. Now, $|f(t,x)| \le M(x)$ holds for every $(t,x) \in [a,b] \times [c,\gamma)$, and thus

$$\left| \int_\omega^\gamma f(t,x)\,dx \right| \le \int_\omega^\gamma |f(t,x)|\,dx \le \int_\omega^\gamma M(x)\,dx < \varepsilon,$$

for every $t \in [a,b]$ and $d \le \omega < \gamma$. □

Example 8.37. Let $f(t,x) = (e^{-x} - e^{-tx})/x$ for every positive t, x. Then $D_1 f(t,x) = e^{-tx}$ is continuous, and the parametric improper integral $\int_0^\infty e^{-tx}\,dx$ is uniformly convergent on $[a,b]$ for every $0 < a < b$. Indeed, $|e^{-tx}| \le e^{-ax}$ holds for every $t \in [a,b]$ and $x > 0$. Since the improper integral $\int_0^\infty e^{-ax}dx$ is convergent, we can apply Theorem 8.36.

Thus, it follows from Theorem 8.34 that our calculations in (8.44) was justified for every positive t, and thus the value of the integral $\int_0^\infty \frac{e^{-x}-e^{-tx}}{x}\,dx$ is indeed $\log t$ for every $t > 0$.

A more general result can be found in Exercise 8.10 (which does not use Theorems 8.36 and 8.34).

Example 8.38. Let $f(t,x) = (x^t - 1)/\log x$ for every t and $x > 0$. The function $D_1 f(t,x) = x^t$ is continuous, and the parametric improper integral $\int_0^1 x^t\,dx$ is uniformly convergent on $[a,b]$ for every $-1 < a < b$. Indeed, $|x^t| \le x^a$ holds for every $t \in [a,b]$ and $0 < x \le 1$, and the improper integral $\int_0^1 x^a\,dx$ is convergent. Thus, we can apply Theorem 8.36. (Of course, we apply the theorem to a $(\gamma, c]$-type interval, instead of a $[c,\gamma)$-type interval, but it should be obvious that this is irrelevant.)

Thus, it follows from Theorem 8.34 that our calculations in (8.41) were justified for every $t > -1$. We get, in fact, that $\int_0^1 \frac{x^t-1}{\log x}\,dx = \log(t+1)$ holds for every $t > -1$.

As an application of the theorems above, we will investigate the Γ function, defined by a parametric improper integral as follows.

We show that the integral $\int_0^\infty x^{t-1} \cdot e^{-x}dx$ is convergent if $t > 0$. Let us inspect the integrals $\int_0^1$ and $\int_1^\infty$ separately. The integral $\int_1^\infty x^{t-1}\,e^{-x}dx$ is convergent for all t. We can prove this using the majorization principle: For every t, we have $x^{t-1}\,e^{-x} < x^{-2}$ if x is large enough. Since $\int_1^\infty x^{-2}\,dx$ is convergent, so is $\int_1^\infty x^{t-1} \cdot e^{-x}dx$.

If $t \ge 1$, then $\int_0^1 x^{t-1} \cdot e^{-x}dx$ is an ordinary Riemann integral. If $0 < t < 1$, then we can apply the majorization principle: since if $x \in (0,1]$, then $|x^{t-1} \cdot e^{-x}| \le x^{t-1}$ and the integral $\int_0^1 x^{t-1}\,dx$ is convergent, we know that the integral $\int_0^1 x^{t-1} \cdot e^{-x}dx$ is also convergent.

Definition 8.39. The value of the integral $\int_0^\infty x^{t-1} \cdot e^{-x} dx$ is denoted by $\Gamma(t)$ for every $t > 0$.

Thus the Γ function is defined on $(0, \infty)$. The Γ function appears in many applications, and it is the most important special function among the nonelementary functions. First we show that

$$\Gamma(t+1) = t \cdot \Gamma(t) \tag{8.48}$$

for all $t > 0$. Using integration by parts we obtain

$$\int_a^b x^t \cdot e^{-x} dx = -\int_a^b x^t \cdot (e^{-x})' dx = -\left[x^t e^{-x}\right]_a^b + t \cdot \int_a^b x^{t-1} \cdot e^{-x} dx$$

for every $0 < a < b$. Letting $a \to 0$ and $b \to \infty$, we obtain, for every $t > 0$,

$$\int_0^\infty x^t \cdot e^{-x} dx = t \cdot \int_0^\infty x^{t-1} \cdot e^{-x} dx.$$

That is, we get (8.48).

Next we show that $\Gamma(n) = (n-1)!$ for every positive integer n. Since $\Gamma(1) = \int_0^\infty e^{-x} dx = 1 = 0!$, this is true if $n = 1$. If $\Gamma(n) = (n-1)!$, then (8.48) gives $\Gamma(n+1) = n \cdot \Gamma(n) = n \cdot (n-1)! = n!$, and so the statement follows by induction.

Now we turn to the investigation of the Γ function in detail.

Theorem 8.40. *The function* Γ *is infinitely differentiable on the half-line* $(0, \infty)$, *and*

$$\Gamma^{(k)}(t) = \int_0^\infty x^{t-1} \cdot (\log x)^k \cdot e^{-x}\, dx \tag{8.49}$$

for every $t > 0$ *and every nonnegative integer* k.

Proof. Let $0 < a < b$ and $k \in \mathbb{N}$ be fixed. By applying Theorem 8.36, we prove that the integrals $\int_1^\infty x^{t-1} \cdot (\log x)^k \cdot e^{-x}\, dx$ and $\int_0^1 x^{t-1} \cdot (\log x)^k \cdot e^{-x}\, dx$ are uniformly convergent on $[a, b]$. It is enough to find functions $f\colon [1, \infty) \to \mathbb{R}$ and $g\colon (0, 1] \to \mathbb{R}$ such that the improper integrals $\int_1^\infty f\, dx$ and $\int_0^1 g\, dx$ are convergent, and $x^{t-1} \cdot (\log x)^k \cdot e^{-x} \le f(x)$ and $x^{t-1} \cdot |\log x|^k \cdot e^{-x} \le g(x)$ hold for every $x \ge 1$ and $x \in (0, 1]$. If $t \in [a, b]$ and $x \ge 1$, then

$$x^{t-1} \cdot (\log x)^k \cdot e^{-x} \le x^{b-1} \cdot (\log x)^k \cdot e^{-x}, \tag{8.50}$$

and thus we can choose f to be the function on the right-hand side of (8.50). The convergence of the improper integral $\int_1^\infty f\, dx$ follows from the fact that $f(x) \le 1/x^2$ for x large enough. If $t \in [a, b]$ and $0 < x \le 1$, then

$$x^{t-1} \cdot |\log x|^k \cdot e^{-x} \le x^{a-1} \cdot |\log x|^k. \tag{8.51}$$

Therefore, we can choose g to be the function on the right-hand side of (8.51). The convergence of the improper integral $\int_1^\infty g\,dx$ follows from the fact that $g(x) \le x^{(a/2)-1}$ holds for x small enough and the integral $\int_0^1 x^{(a/2)-1}\,dx$ is convergent.

Let $\Gamma_1(x) = \int_1^\infty x^{t-1}e^{-x}\,dx$. Applying Theorem 8.34 repeatedly, we find that the function Γ_1 is infinitely differentiable on the interval $[a,b]$, and $\Gamma_1^{(k)}(x) = \int_1^\infty x^{t-1}\cdot(\log x)^k\cdot e^{-x}\,dx$ for every $t \in [a,b]$ and every positive integer k. Similarly, if $\Gamma_0(x) = \int_0^1 x^{t-1}e^{-x}\,dx$, then Theorem 8.34 implies that the function Γ_0 is infinitely differentiable on the interval $[a,b]$, and $\Gamma_0^{(k)}(x) = \int_0^1 x^{t-1}\cdot(\log x)^k\cdot e^{-x}\,dx$ for every $t \in [a,b]$ and every positive integer k. Since $\Gamma = \Gamma_0 + \Gamma_1$ and $0 < a < b$ was arbitrary, it follows that Γ is differentiable on $(0,\infty)$, and (8.49) holds. □

It follows from the previous theorem that $\Gamma''(t) = \int_0^\infty x^{t-1}(\log x)^2\cdot e^{-x}\,dx$ for every $t > 0$. Thus, Γ'' is positive, i.e., Γ is strictly convex on the half-line $(0,\infty)$. Now we show that

$$\lim_{x\to 0+0}\Gamma(x) = \lim_{x\to\infty}\Gamma(x) = \infty. \tag{8.52}$$

Indeed,

$$\Gamma(t) > \int_0^1 x^{t-1}\cdot e^{-x}\,dx > \int_0^1 x^{t-1}\cdot e^{-1}\,dx = \frac{1}{et}$$

for every $t > 0$, and

$$\Gamma(t) > \int_2^\infty x^{t-1}\cdot e^{-x}\,dx > \int_2^\infty 2^{t-1}\cdot e^{-x}\,dx = \frac{2^t}{2e^2}$$

for every $t > 1$, from which (8.52) is obvious. Now, (8.52) implies that Γ cannot be monotone on the whole half-line $(0,\infty)$; hence Γ' has a root. Furthermore, since Γ' is strictly increasing on $(0,\infty)$, Γ' has exactly one root. Let this root be t_0. Then Γ is strictly decreasing on $(0,t_0)$ and strictly increasing on (t_0,∞). The equalities $\Gamma(1) = \Gamma(2) = 1$ imply $1 < t_0 < 2$.

It is simple to check that if $t, s > 0$, then the integral $\int_0^1 x^{t-1}(1-x)^{s-1}\,dx$ is convergent. If t or s is a positive integer, then the value of the integral can be computed easily (see [7, Exercise 19.42]). Our next aim is to compute the integral for every $t, s > 0$.

Notation 8.41. We denote the value of the integral $\int_0^1 x^{t-1}(1-x)^{s-1}\,dx$ by $B(t,s)$ for every $t, s > 0$.

Theorem 8.42. *For every $t, s > 0$ we have*

$$B(t,s) = \frac{\Gamma(t)\cdot\Gamma(s)}{\Gamma(t+s)}. \tag{8.53}$$

Proof. Let t, s be fixed positive numbers. The substitution $x = \cos^2 \varphi$ transforms the integral $\int_0^1 x^{t-1}(1-x)^{s-1}\,dx$ into

$$B(t,s) = 2 \cdot \int_0^{\pi/2} (\cos\varphi)^{2t-1}(\sin\varphi)^{2s-1}\,d\varphi. \tag{8.54}$$

The substitution $x = u^2$ transforms the integral defining $\Gamma(t)$ into $\Gamma(t) = 2 \cdot \int_0^\infty u^{2t-1}e^{-u^2}\,du$. Similarly, $\Gamma(s) = 2 \cdot \int_0^\infty v^{2s-1}e^{-v^2}\,du$.

Let A_R denote the intersection of the disk centered at the origin with radius R and the first plane quadrant. In other words, let A_R be the set of points given by the polar coordinates (r, φ) that satisfy $0 \le r \le R$ and $0 \le \varphi \le \pi/2$. Consider the two-variable function $f(u,v) = u^{2t-1}v^{2s-1}e^{-u^2-v^2}$ on the set A_R. Applying the theorem on substitution by polar coordinates (Theorem 4.25), we obtain

$$\begin{aligned}
\int_{A_R} f(u,v)\,du\,dv &= \int_{[0,R]\times[0,\pi/2]} f(r\cos\varphi, r\sin\varphi)\cdot r\cdot dr\,d\varphi = \\
&= \int_{[0,R]\times[0,\pi/2]} (\cos\varphi)^{2t-1}(\sin\varphi)^{2s-1}\cdot r^{2(t+s)-1}e^{-r^2}\,dr\,d\varphi = \\
&= \int_0^{\pi/2} (\cos\varphi)^{2t-1}(\sin\varphi)^{2s-1}\,d\varphi \cdot \int_0^R r^{2(t+s)-1}e^{-r^2}\,dr = \\
&= \frac{1}{2}\cdot B(t,s)\cdot \int_0^R r^{2(t+s)-1}e^{-r^2}\,dr.
\end{aligned} \tag{8.55}$$

If $R \to \infty$, then the last integral of (8.55) converges to $\Gamma(t+s)/2$, and thus

$$\lim_{R\to\infty} \int_{A_R} f(u,v)\,du\,dv = \frac{1}{4}\cdot B(t,s)\cdot\Gamma(t+s). \tag{8.56}$$

Since $[0, R/2]^2 \subset A_R \subset [0,R]^2$ and f is nonnegative, we have

$$\int_{[0,R/2]^2} f(u,v)\,du\,dv \le \int_{A_R} f(u,v)\,du\,dv \le \int_{[0,R]^2} f(u,v)\,du\,dv. \tag{8.57}$$

The value of the integral on the left-hand side is

$$\int_{[0,R/2]^2} u^{2t-1}v^{2s-1}e^{-u^2-v^2}\,du\,dv = \int_0^{R/2} u^{2t-1}e^{-u^2}\,du \cdot \int_0^{R/2} v^{2s-1}e^{-v^2}\,dv,$$

which converges to $\Gamma(t)\Gamma(s)/4$ as $R \to \infty$. Similarly, the integral on the right-hand side of (8.57) also converges to $\Gamma(t)\Gamma(s)/4$ as $R \to \infty$. The squeeze theorem claims that

$$\lim_{R\to\infty} \int_{A_R} f(u,v)\,du\,dv = \frac{1}{4}\cdot\Gamma(t)\cdot\Gamma(s).$$

Comparing this with (8.56) gives the statement of the theorem. □

Our next result presents an interesting connection between the sine function and the function Γ.

Theorem 8.43. *For every $x \in (0,1)$, we have*

$$\Gamma(x) \cdot \Gamma(1-x) = \frac{\pi}{\sin \pi x}. \tag{8.58}$$

Proof. Let $P_n(x) = x \cdot \prod_{n=1}^{n} \left(1 - \frac{x^2}{n^2}\right)$. Then by Theorem 7.97, we have $\lim_{n\to\infty} P_n(x) = (\sin \pi x)/\pi$ for every $x \in \mathbb{R}$. It is well known that

$$\Gamma(x) = \lim_{n\to\infty} \frac{n^x n!}{x(x+1)\cdots(x+n)} \tag{8.59}$$

(see Exercise 19.43 of [7]). If we apply this equality with $1-x$ in place of x and then take the product of the equality obtained and (8.59), we get

$$\Gamma(x) \cdot \Gamma(1-x) = \lim_{n\to\infty} \frac{1}{P_n(x)} \cdot \frac{n}{n+1-x} = \frac{\pi}{\sin \pi x} \cdot 1. \qquad \square$$

For more on the function Γ see Exercises 8.16–8.24. (See also exercises 19.40–19.46 in [7].)

It is easy to see that the improper integral $\int_1^\infty (\sin x/x)\, dx$ is convergent (see [7, Example 19.20.3]). Since $(\sin x)/x$ is continuous and bounded on $(0,1]$, it is integrable there. Thus, the improper integral $\int_0^\infty (\sin x/x)\, dx$ is also convergent. As another application of parametric integration we compute the exact value of this integral.

Theorem 8.44.

$$\int_0^\infty \frac{\sin x}{x}\, dx = \frac{\pi}{2}.$$

The proof uses the following theorem, which gives a sufficient condition for uniform convergence. Notice that this criterion is similar to the Abel criterion (Corollary 7.33), which gives a sufficient condition for the uniform convergence of series of functions. (For another criterion—the analogue to the Dirichlet criterion—see Exercise 8.12.)

Theorem 8.45. *Let the function $g(t,x)$ be defined and bounded on the set $[a,b] \times [c,\gamma)$, and let the section function g_t be monotone for every $t \in [a,b]$. If the improper integral $\int_c^\gamma h(x)\, dx$ is convergent, then the parametric improper integral $\int_c^\gamma g(t,x) \cdot h(x)\, dx$ is uniformly convergent on $[a,b]$.*

Proof. Let $|g(t,x)| \le K$ for every $(t,x) \in [a,b] \times [c,\gamma)$. Let $\varepsilon > 0$ be fixed, and choose $c < d < \gamma$ such that $\left|\int_u^v h(x)\,dx\right| < \varepsilon$ for every $d \le u < v < \gamma$. Let $t \in [a,b]$ and $d \le \omega < \gamma$ be given. By the second mean value theorem of integration [7, Theorem 15.8], for every $\omega < \Omega < \gamma$ there exists $\omega < u < \Omega$ such that

$$\int_\omega^\Omega g(t,x)\cdot h(x)\,dx = g(t,\omega)\cdot \int_\omega^u h(x)\,dx + g(t,\Omega)\cdot \int_u^\Omega h(x)\,dx,$$

and thus

$$\left|\int_\omega^\Omega g(t,x)\cdot h(x)\,dx\right| \le 2K\cdot\varepsilon. \tag{8.60}$$

This holds for every $d \le \omega < \Omega < \gamma$. Then by the Cauchy criterion for improper integrals [7, Theorem 19.15], the improper integral $\int_c^\gamma g(t,x)\cdot h(x)\,dx$ is convergent.

If Ω converges to γ from the left, then (8.60) gives

$$\left|\int_\omega^\gamma g(t,x)\cdot h(x)\,dx\right| \le 2K\cdot\varepsilon$$

for every $t \in [a,b]$ and $d \le \omega < \gamma$. This proves the uniform convergence of the improper integral in question. □

Proof of Theorem 8.44. Consider the function $e^{tx}\sin x$, where t is a nonzero constant. The indefinite integral of this function can be computed by partial integration. We get that

$$\int e^{tx}\sin x\,dx = \frac{e^{tx}}{1+t^2}\cdot(t\sin x - \cos x) + c, \tag{8.61}$$

which can also be double-checked directly. If $t < 0$, then $\lim_{x\to\infty} e^{tx} = 0$, and applying the Newton–Leibniz formula for improper integrals [7, Theorem 19.8] yields

$$\int_0^\infty e^{tx}\sin x\,dx = \frac{1}{1+t^2}. \tag{8.62}$$

Now consider the function

$$f(t,x) = e^{tx}\cdot\frac{\sin x}{x} \qquad (x \ne 0), \qquad f(t,0) = 1.$$

By Theorem 8.45, the parametric improper integral $\int_0^\infty f(t,x)\,dx$ is uniformly convergent on the interval $[-K,0]$ for every $K > 0$, since $(t,x) \in [-K,0]\times[0,\infty)$ implies $e^{tx} \le 1$, e^{tx} is monotone as a function of x for every t, and the improper integral $\int_0^\infty (\sin x/x)\,dx$ is convergent. Let

$$F(t) = \int_0^\infty e^{tx} \cdot \frac{\sin x}{x}\,dx$$

for every $t \le 0$. By Theorem 8.32, F is continuous on $[-K, 0]$ for every $K > 0$. Thus F is continuous on the half-line $(-\infty, 0]$.

Let $a < b < 0$. Then $D_1 f(t,x) = e^{tx} \sin x$ is continuous, and the improper integral $\int_0^\infty e^{tx} \sin x\,dx$ is uniformly convergent on $[a,b]$, since $|D_1 f(t,x)| \le e^{tx} \le e^{bx}$ there, and the improper integral $\int_0^\infty e^{bx}\,dx$ is convergent.

Thus, applying Theorem 8.34 and (8.62), we obtain that F is differentiable on the interval $[a,b]$, and $F'(t) = 1/(1+t^2)$ there. Sinçe this holds for every $a < b < 0$, we have $F'(t) = 1/(1+t^2)$ for every $t < 0$. Therefore, $F(t) = \operatorname{arc\,tg} t + c$ for every $t < 0$ with an appropriate constant c.

Now, $|F(t)| \le \int_0^\infty e^{tx}\,dx = 1/t$ for every $t < 0$ and $\lim_{t\to-\infty} F(t) = 0$. It follows that $c = \pi/2$ and $F(t) = \operatorname{arc\,tg} t + (\pi/2)$ for every $t < 0$. Since F is continuous from the left at 0, we get

$$\int_0^\infty \frac{\sin x}{x}\,dx = F(0) = \frac{\pi}{2},$$

which proves the theorem. □

Exercises

8.8. Show that the primitive function of $(x-1)/\log x$ is not an elementary function. (Use the following generalization of Liouville's theorem [7, Theorem 15.31]. Let $f_1, \ldots, f_n, g_1, \ldots, g_n$ be rational functions, and suppose that $f_i - f_j$ is not constant for every $1 \le i < j \le n$. If $\int \sum_{i=1}^n e^{f_i} g_i\,dx$ can be expressed in terms of elementary functions, then each of the indefinite integrals $\int e^{f_i} g_i\,dx$ can be expressed in terms of elementary functions.) (S)

8.9. Let $f(t,x) = t/(1+(tx)^2)$ for every $t, x \in \mathbb{R}$. Show that the improper integral $\int_0^\infty f(t,x)\,dx$ is convergent for every t, but its value (as a function of t) is not continuous at $t = 0$ from either side.

8.10. Let $f\colon [0,\infty) \to \mathbb{R}$ be continuous, and let the improper integral $\int_1^\infty \frac{f(x)}{x}\,dx$ be convergent. Show that

$$\int_0^\infty \frac{f(ax) - f(bx)}{x}\,dx = f(0) \cdot \log \frac{b}{a}$$

for every $a, b > 0$. (H)

8.11. Find the value of the following integrals for all $a, b > 0$.

(i) $\int_0^\infty \left(\frac{e^{-ax} - e^{-bx}}{x}\right)^2 dx$;

(ii) $\int_0^\infty \frac{e^{-ax} - e^{-bx}}{x} \cdot \sin cx\, dx$ $(c \in \mathbb{R})$;

(iii) $\int_0^\infty \frac{e^{-ax} - e^{-bx}}{x} \cdot \cos cx\, dx$ $(c \in \mathbb{R})$.

8.12. Suppose that

(i) the function $g(t, x)$ is defined on the set $[a, b] \times [c, \gamma)$,
(ii) the section function g_t is monotone for every $t \in [a, b]$,
(iii) the sequence of functions $g(t, x_n)$ $(n = 1, 2, \ldots)$ converges to 0 uniformly on $[a, b]$ for every $x_n < \gamma$, $x_n \to \gamma$,
(iv) $h\colon [c, \gamma) \to \mathbb{R}$ is integrable on $[c, d]$ for every $c < d < \gamma$, and
(v) there exists $K > 0$ such that $\left|\int_c^d h(x)\, dx\right| \le K$ for every $c < d < \gamma$.

Show that the parametric improper integral $\int_c^\gamma g(t, x) \cdot h(x)\, dx$ is uniformly convergent on $[a, b]$.

8.13. Let r, c be positive numbers. Show that the area of the domain whose boundary is the curve $|x|^c + |y|^c = r^c$ is $\frac{2}{c} \cdot \frac{\Gamma(1/c)^2}{\Gamma(2/c)} \cdot r^2$. Check that the case $c = 2$ gives the area of the disk.

8.14. Find the integral $F(t) = \int_0^{\pi/2} \log(1 + t \cdot \sin^2 x)\, dx$ for every $t > -1$. (S)

8.15. Show that the integral in the previous exercise is also convergent for $t = -1$ and its value is $-\pi \cdot (\log 2)$. (This gives a new solution for part (b) of [7, Exercise 19.20].) (H)

8.16. Show that $\int_0^\infty \frac{x^{t-1}}{(1+x)^{t+s}} = \frac{\Gamma(t) \cdot \Gamma(s)}{\Gamma(t+s)}$ for every $t, s > 0$. (H)

8.17. Show that

(i) $\int_0^\infty \frac{x^{-t}}{1+x}\, dx = \frac{\pi}{\sin \pi t}$ for every $0 < t < 1$, and

(ii) $\int_0^\infty \frac{dx}{1+x^t} = \frac{\pi/t}{\sin(\pi/t)}$ for every $t > 1$. (H)

8.18. Show that $\int_0^{\pi/2} (\sin\varphi)^{2x-1}\, d\varphi = 2^{2x-2} \cdot \frac{\Gamma(x)^2}{\Gamma(2x)}$ for every $x > 0$. Compare this with the value of $\int_0^{\pi/2} (\sin x)^n\, dx$ for integer values of n given by [7, Theorem 15.12]. (See the proof of Theorem 3.28.) (H)

8.19. Show that $\int_0^{\pi/2} (\operatorname{tg}\varphi)^{2x-1}\, d\varphi = \frac{1}{2} \cdot \frac{\pi}{\sin \pi x}$ for every $0 < x < 1$. (H)

8.20. The aim of the following exercises is to prove

$$\int_0^\infty \frac{\sin t}{t^s}\,dt = \frac{\pi}{2\Gamma(s)\sin\frac{s\pi}{2}} \tag{8.63}$$

for $0 < s < 2$.

(i) Show that $\Gamma(s) = t^s \cdot \int_0^\infty x^{s-1}e^{-tx}\,dx$ for every $t > 0$.

(ii) Show that for $0 < a < b$, we have

$$\int_a^b \frac{\sin t}{t^s}\,dt = \frac{1}{\Gamma(s)} \cdot \int_a^b \left(\int_0^\infty x^{s-1}e^{-tx}\sin t\,dx\right)dt.$$

(iii) Show that for $0 < a < b$, the improper integral $F(t) = \int_0^\infty x^{s-1}e^{-tx}$ $\sin t\,dx$ is uniformly convergent on $[a, b]$. Use this to prove

$$\int_a^b \frac{\sin x}{x^s}\,dx = \frac{1}{\Gamma(s)} \cdot \int_0^\infty x^{s-1} \cdot \left(\int_a^b e^{-tx}\sin t\,dt\right)dx.$$

(iv) Show that for $0 < a < b$, we have

$$\int_a^b \frac{\sin t}{t^s}\,dt = -\frac{1}{\Gamma(s)} \cdot \int_0^\infty \frac{x^{s-1}}{1+x^2} \cdot e^{-bx} \cdot (x \cdot \sin b + \cos b)\,dx + \\ + \frac{1}{\Gamma(s)} \cdot \int_0^\infty \frac{x^{s-1}}{1+x^2} \cdot e^{-ax} \cdot (x \cdot \sin a + \cos a)\,dx.$$

(v) Show that

$$\lim_{b\to\infty} \int_0^\infty \frac{x^{s-1}}{1+x^2} \cdot e^{-bx} \cdot (x \cdot \sin b + \cos b)\,dx = 0$$

and

$$\lim_{a\to 0} \int_0^\infty \frac{x^{s-1}}{1+x^2} \cdot e^{-ax} \cdot (x \cdot \sin a + \cos a)\,dx = \int_0^\infty \frac{x^{s-1}}{1+x^2}\,dx.$$

(vi) Show that

$$\int_0^\infty \frac{\sin x}{x^s}\,dx = \frac{1}{\Gamma(s)} \cdot \int_0^\infty \frac{x^{s-1}}{1+x^2}\,dx.$$

(vii) Show that

$$\int_0^\infty \frac{x^{s-1}}{1+x^2}\,dx = \frac{\pi}{2\sin\frac{s\pi}{2}},$$

using the results of Exercise 8.17.

8.21. Show that for $0 < s < 1$, we have

$$\int_0^\infty \frac{\cos t}{t^s}\, dt = \frac{\pi}{2\Gamma(s)\cos\frac{s\pi}{2}}.$$

8.22. Show that for $1 < s < 3$, we have

$$\int_0^\infty \frac{\cos t - 1}{t^s}\, dt = \frac{\pi}{2\Gamma(s)\cos\frac{s\pi}{2}}. \text{ (H)}$$

8.23. Show that for every $s > 1$, the improper integral $\int_0^\infty \sin x^s\, dx$ is convergent, and its value is $\frac{1}{s}\cdot\Gamma\left(\frac{1}{s}\right)\cdot\sin\frac{\pi}{2s}$. (H)

8.24. Show that for every $s > 1$, the improper integral $\int_0^\infty \cos x^s\, dx$ is convergent, and its value is $\frac{1}{s}\cdot\Gamma\left(\frac{1}{s}\right)\cdot\cos\frac{\pi}{2s}$.

8.4 Sets with Lebesgue Measure Zero and the Lebesgue Criterion for Integrability

As we saw before, if a bounded function is continuous apart from the points of a set of Jordan measure zero, then the function is integrable (see Theorem 4.14). However, this condition is not necessary for a function to be integrable. For example, every monotone function $f\colon [a,b] \to \mathbb{R}$ is integrable, but the set of points where a monotone function is not continuous can be dense in $[a,b]$, and such a set is not of measure zero. Another example is provided by the Riemann function, which is integrable on every interval but is discontinuous on $\mathbb{Q}$, an everywhere dense set.

Still, it is true that if a function f is bounded, then the integrability of f depends on how small the set of its points of discontinuity is. The precise form of this statement is given by Lebesgue's[7] theorem (Theorem 8.52 below), and our next aim is to prove this result. The monotone functions and the Riemann function have only countably many discontinuities, and thus their integrability will also follow immediately from Lebesgue's theorem. In fact, as we will see presently, every bounded function having only countably many points of discontinuity is necessarily integrable.

A set $A \subset \mathbb{R}^p$ has Jordan measure zero if and only if for every $\varepsilon > 0$ there exist finitely many boxes $B_1, \ldots, B_n$ whose union covers A and $\sum_{i=1}^n \mu(B_i) < \varepsilon$. Lebesgue realized that allowing countably many covering boxes turns the condition of having measure zero into a notion that suits our goals perfectly.

Definition 8.46. A set $A \subset \mathbb{R}^p$ is said to have *Lebesgue measure zero* if for every $\varepsilon > 0$ there exist countably many boxes such that the sum of their Jordan measures is smaller than ε, and their union covers A.

---

[7] Henri Lebesgue (1875–1941), French mathematician.

Remark 8.47. We can also require the covering boxes to be open in the definition of Lebesgue measure zero. Indeed, choose boxes $B_1, B_2, \dots$ such that they cover A and $\sum_{i=1}^{\infty} \mu(B_i) < \varepsilon/2$. Then replace every B_i with an open box R_i containing B_i and having $\mu(R_i) = 2 \cdot \mu(B_i)$.

Lemma 8.48. *If a set A has Jordan measure zero, then it also has Lebesgue measure zero.*

Proof. The statement is clear from the definitions. (Recall that every finite set is countable.) □

According to the following theorem, the converse of Lemma 8.48 is not true.

Theorem 8.49. *Every countable set has Lebesgue measure zero.*

Proof. Let $A \subset \mathbb{R}^p$ be countable, and let (a_k) be an enumeration of the elements of A. For a given $\varepsilon > 0$, cover a_k with a box B_k of Jordan measure $\varepsilon/2^k$ $(k = 1, 2, \dots)$. Then the boxes B_k cover A, and $\sum_{k=1}^{\infty} \mu(B_k) = \varepsilon$. Since ε was arbitrary, it follows that A has Lebesgue measure zero. □

We know that not every countable set has Jordan measure zero (e.g., if the set is not bounded or dense in a box); thus there exist sets with Lebesgue measure zero whose Jordan measure is not zero. However, we will show that the two definitions of having measure zero are equivalent to each other for an important class of sets.

Theorem 8.50. *If the set A is bounded, closed, and has Lebesgue measure zero, then it has Jordan measure zero.*

Proof. Let $\varepsilon > 0$ be fixed. By Remark 8.47, there exist open boxes $R_1, R_2, \dots$ that cover A and also satisfy $\sum_{k=1}^{\infty} \mu(R_k) < \varepsilon$. Applying Borel's theorem (Theorem 1.31), we get that $A \subset \bigcup_{k=1}^{N} R_k$ for N large enough, and thus $\overline{\mu}(A) \le \varepsilon$. Since ε was arbitrary, we have $\mu(A) = 0$. □

Before proving Lebesgue's theorem, we need to prove one more lemma.

Lemma 8.51. *If the set A_n has Lebesgue measure zero for every $n = 1, 2, \dots$, then the set $A = \bigcup_{n=1}^{\infty} A_n$ also has Lebesgue measure zero.*

Proof. Let $\varepsilon > 0$ be fixed. Since A_n has Lebesgue measure zero, there exist boxes $B_{n,1}, B_{n,2}, \dots$ that cover A_n and also satisfy $\sum_{k=1}^{\infty} \mu(B_{n,k}) < \varepsilon/2^n$. The boxes $B_{n,k}$ $(n, k = 1, 2, \dots)$ cover A.

We know that the boxes $B_{n,k}$ $(n, k = 1, 2, \dots)$ can be listed in a single sequence J_i $(i = 1, 2, \dots)$ (see [7, Theorem 8.5]). By Theorem 6.30,

$$\sum_{i=1}^{\infty} \mu(J_i) = \sum_{n=1}^{\infty} \sum_{k=1}^{\infty} \mu(B_{n,k}) \le \sum_{n=1}^{\infty} \frac{\varepsilon}{2^n} = \varepsilon,$$

which proves our statement. □

Theorem 8.52. (Lebesgue's theorem) *Let $A \subset \mathbb{R}^p$ be Jordan measurable. A function $f\colon A \to \mathbb{R}$ is integrable on A if and only if it is bounded and the set of its points of discontinuity has Lebesgue measure zero.*

Proof. Let $f\colon A \to \mathbb{R}$ be given, and let D denote the set of points where f is not continuous. If $x \in D$, then there is an $\varepsilon > 0$ such that every neighborhood of x has a point y with $|f(y) - f(x)| > \varepsilon$. Let us denote by D_n the set of points $x \in D$ that satisfy this condition for $\varepsilon = 1/n$. In other words, $x \in D_n$ if and only if every neighborhood of x has a point y such that $|f(y) - f(x)| > 1/n$. Obviously, $D = \bigcup_{n=1}^{\infty} D_n$.

First we show that if f is integrable, then D has Lebesgue measure zero. By Lemma 8.51, it is enough to prove that D_n has Lebesgue measure zero for every n. Let n and $\varepsilon > 0$ be fixed. Since f is integrable, we have $\Omega_F(f) < \varepsilon/n$ for a suitable partition $F = \{A_1, \ldots, A_k\}$.

Since $A_1, \ldots, A_k$ are Jordan measurable, it follows from Theorem 3.9 that $\mu(\partial A_i) = 0$ for every i. Let $E = \bigcup_{i=1}^{k} \partial A_i$. Then $\mu(E) = 0$.

Suppose that $A_i \cap (D_n \setminus E) \neq \emptyset$, and let $x \in A_i \cap (D_n \setminus E)$. Then $x \notin \partial A_i$ (since $x \notin E$), and thus $x \in \operatorname{int} A_i \cap D_n$. Therefore, by the definition of D_n, there exists a point $y \in \operatorname{int} A_i$ such that $|f(y) - f(x)| > 1/n$. This implies that $\omega(f; A_i) = M_i - m_i \geq 1/n$. If J denotes the set of indices i that satisfy $A_i \cap (D_n \setminus E) \neq \emptyset$, then

$$\frac{\varepsilon}{n} > \Omega_F(f) = \sum_{i=1}^{k} (M_i - m_i) \cdot \mu(A_i) \geq \sum_{i \in J} (M_i - m_i) \cdot \mu(A_i) \geq$$
$$\geq \sum_{i \in J} \frac{1}{n} \cdot \mu(A_i),$$

and thus $\sum_{i \in J} \mu(A_i) < \varepsilon$. We have proved that $D_n \setminus E$ can be covered by the union of finitely many Jordan measurable sets of total measure less than ε (namely, by the sets A_i $(i \in J)$). Since $D_n \subset (D_n \setminus E) \cup E$ and $\mu(E) = 0$, it follows that $\overline{\mu}(D_n) \leq \varepsilon$. Since ε was arbitrary, it follows that $\mu(D_n) = 0$. Therefore, the set D_n has Lebesgue measure zero by Lemma 8.48, and this is what we wanted to prove.

Now let f be bounded and let D have Lebesgue measure zero; we prove that f is integrable. It is enough to prove that f has arbitrarily small oscillatory sums.

Let $\varepsilon > 0$ be fixed. Since A is Jordan measurable, there are boxes $B_1, \ldots, B_N \subset A$ such that $\mu(A \setminus K) < \varepsilon$, where $K = \bigcup_{i=1}^{N} B_i$. Note that K is a bounded and closed set.

By Remark 8.47, there are open boxes $R_1, R_2, \ldots$ that cover D and satisfy $\sum_{k=1}^{\infty} \mu(R_k) < \varepsilon$. If $x \in K \setminus D$, then f is continuous at x, and thus there exists $\delta(x) > 0$ such that $|f(y) - f(x)| < \varepsilon$ for every $y \in B(x, \delta(x))$.

The union of the open boxes R_k $(k = 1, 2, \ldots)$ and open balls $B(x, \delta(x))$ $(x \in K \setminus D)$ cover K. By Borel's theorem (Theorem 1.31), there exist finitely many of these open sets that also cover K. Consider such a finite covering system $E_1, \ldots, E_m$. Let $A_0 = A \setminus K$ and $A_i = (E_i \cap K) \setminus \bigcup_{j<i} E_j$ for every $i =$

$1, \ldots, m$. Then $F = \{A_0, A_1, \ldots, A_m\}$ is a partition of A. Note that for every $1 \le i \le m$, A_i is a subset of one of the sets $E_1, \ldots, E_m$.

Let J denote the set of indices i such that $A_i \subset E_j$, where E_j is one of the open boxes R_k $(k = 1, 2, \ldots)$. Since $\sum_{k=1}^{\infty} \mu(R_k) < \varepsilon$, we have $\sum_{i \in J} \mu(A_i) < \varepsilon$.

If $i \notin J$ and $A_i \subset E_j$, then E_j is necessarily one of the balls $B(x, \delta(x))$. Now, $|f(y) - f(x)| < \varepsilon$ holds for every $y \in B(x, \delta(x))$, and thus $\omega(f; A_i) = M_i - m_i \le 2\varepsilon$.

Let $|f(x)| \le M$ for every $x \in A$. Then we have

$$\begin{aligned}
\Omega_F(f) &= \sum_{i=0}^{m} (M_i - m_i) \cdot \mu(A_i) = \\
&= (M_0 - m_0) \cdot \mu(A_0) + \sum_{i \in J} (M_i - m_i) \cdot \mu(A_i) + \sum_{i \notin J} (M_i - m_i) \cdot \mu(A_i) \le \\
&\le 2M \cdot \varepsilon + 2M \cdot \varepsilon + 2\varepsilon \cdot \mu(A) = \\
&= 2(2M + \mu(A)) \cdot \varepsilon.
\end{aligned}$$

Since ε was arbitrary, this implies that f is integrable. □

Corollary 8.53. *If $f \colon A \to \mathbb{R}$ is bounded and the set of points where f is not continuous is countable, then f is integrable on A.*

Proof. The statement follows from Theorems 8.49 and 8.52. □

Note that there exist integrable functions whose set of discontinuities is not countable, for example, the function f that is defined as $f(x) = 1$ if x is in the Cantor set C, and $f(x) = 0$ if $x \notin C$ (see Remark 4.15.1).

Remark 8.54. The difference between having Lebesgue measure zero and having Jordan measure zero lies in allowing coverings by countably many boxes instead of allowing coverings by finitely many boxes. As we saw, this seemingly small change significantly expands the family of sets of zero measure.

We can alter the definition of the outer measure in a similar manner. Let us cover the set $A \subset \mathbb{R}^p$ in every possible ways by countably many boxes, and take the sum $\sum_{n=1}^{\infty} \mu(B_n)$ of the Jordan measures of the covering boxes B_n. The infimum of these sums is called the **Lebesgue outer measure of the set** A, and it is denoted by $\overline{\lambda}(A)$.

Obviously, $\overline{\lambda}(A) \le \overline{\mu}(A)$ for every bounded set A. The equality does not always hold; e.g., the set $A = [0, 1] \cap \mathbb{Q}$ has $\overline{\lambda}(A) = 0$ (since A is countable), but $\overline{\mu}(A) = 1$.

We have multiple ways of defining the **Lebesgue inner measure**. We know that if B is Jordan measurable and $A \subset B$, then $\underline{\mu}(A) = \mu(B) - \overline{\mu}(B \setminus A)$ (see Exercise 3.11). Based on this observation, we can define the Lebesgue inner measure of the set A by the formula $\underline{\lambda}(A) = \mu(B) - \overline{\lambda}(B \setminus A)$, where B is an arbitrary box covering A. If $A \subset \mathbb{R}^p$ is not bounded, then let $\underline{\lambda}(A)$ be the supremum of the numbers $\underline{\lambda}(C)$, where C is an arbitrary subset of A.

Another possible definition of $\underline{\lambda}(A)$ is the supremum of the numbers $\overline{\lambda}(F)$, where F is an arbitrary *closed* subset of A. One can show that these two definitions give the same number $\underline{\lambda}(A)$.

Following the analogy with Jordan measurability, we say that a bounded set A is **Lebesgue measurable** if $\underline{\lambda}(A) = \overline{\lambda}(A)$. The number $\underline{\lambda}(A) = \overline{\lambda}(A)$ (or infinity) is called the **Lebesgue measure** of the set A, denoted by $\lambda(A)$.

There is a simpler method for finding the Lebesgue measurable sets. By Exercise 3.12, the bounded set A is Jordan measurable if and only if $\overline{\mu}(H) = \overline{\mu}(H \cap A) + \overline{\mu}(H \setminus A)$ for every bounded set H. Following the analogy, we can say that a set $A \subset \mathbb{R}^p$ (bounded or not) is Lebesgue measurable if $\overline{\lambda}(H) = \overline{\lambda}(H \cap A) + \overline{\lambda}(H \setminus A)$ for every set $H \subset \mathbb{R}^p$. One can prove that for bounded sets these two definitions of Lebesgue measurability are equivalent to each other.

It is easy to see that if A is bounded, then $\underline{\mu}(A) \le \underline{\lambda}(A) \le \overline{\lambda}(A) \le \overline{\mu}(A)$. It follows that *every Jordan measurable set is automatically Lebesgue measurable as well.* In fact, the family of Lebesgue measurable sets is much bigger than the family of Jordan measurable sets; e.g., every countable set is Lebesgue measurable, but there are countable sets that are not Jordan measurable, since countable unbounded sets and the countable sets that are dense in a ball are not Jordan measurable. Another useful property of Lebesgue measurability is the fact that *if the sets $A_1, A_2, \ldots$ are Lebesgue measurable, then their union is also Lebesgue measurable.* (The Jordan measure does not have this property, since the singletons are Jordan measurable, but $\mathbb{Q} \cap [0, 1]$ is not.) One can also prove that if the Lebesgue measurable sets $A_1, A_2, \ldots$ are mutually disjoint, then

$$\lambda\left(\bigcup_{n=1}^{\infty} A_n\right) = \sum_{n=1}^{\infty} \lambda(A_n).$$

These properties make Lebesgue measure extremely useful. The Lebesgue measure and the integral built on it (which is called the Lebesgue integral) form the basis of measure theory.

8.5 Two Applications of Lebesgue's Theorem

We know that if $g\colon A \to [c, d]$ is integrable on the box A and $f\colon [c, d] \to \mathbb{R}$ is continuous, then $f \circ g$ is also integrable. (See [7, Theorem 14.35] and page 128 of this volume.) Now with the help of Lebesgue's theorem, we prove that if we switch the order of the functions, then the statement does not remain true: plugging a continuous function into an integrable function does not always yield an integrable function.

Theorem 8.55. *There exist a continuous (moreover, differentiable) function $f\colon [0, 1] \to [0, 1]$ and an integrable function $g\colon [0, 1] \to [0, 1]$ such that $g \circ f$ is not integrable on $[0, 1]$.*

Proof. First we show that for every nonempty set $F \subset \mathbb{R}$, the function $\operatorname{dist}(\{x\}, F) = \inf\{|x-y| \colon y \in F\}$ is continuous and is furthermore Lipschitz on $\mathbb{R}$. Let $x, y \in [0,1]$ and let $z \in F$ be arbitrary. Then

$$\operatorname{dist}(\{y\}, F) \le |y-z| \le |y-x| + |x-z|.$$

Since this holds for every $z \in F$, we have $\operatorname{dist}(\{y\}, F) \le |y-x| + \operatorname{dist}(\{x\}, F)$. By switching the points x and y, we get $\operatorname{dist}(\{x\}, F) \le |y-x| + \operatorname{dist}(\{y\}, F)$, i.e., $|\operatorname{dist}(\{x\}, F) - \operatorname{dist}(\{y\}, F)| \le |x-y|$. Thus the function $\operatorname{dist}(\{x\}, F)$ is indeed Lipschitz.

Let $F \subset [0,1]$ be a fixed closed set that is not Jordan measurable. (Such a set exists; see Exercise 3.15.) Consider the function $f(x) = \int_0^x \operatorname{dist}(\{t\}, F)\, dt$ $(x \in [0,1])$. Obviously, f is monotonically increasing, $f(0) = 0$, and $f(1) \le 1$. Also, f is differentiable on $[0,1]$ by [7, Theorem 15.5]. We show that the set $f(F)$ has Jordan measure zero. Let $0 = x_0 < x_1 < \ldots < x_n = 1$ be a partition of the interval $[0,1]$ into n parts of equal length. If $F \cap [x_{i-1}, x_i] \ne \emptyset$, then $\operatorname{dist}(\{x\}, F) \le 1/n$ for every $x \in [x_{i-1}, x_i]$-re, and thus

$$f(x_i) - f(x_{i-1}) = \int\limits_{x_{i-1}}^{x_i} \operatorname{dist}(\{t\}, F)\, dt \le (x_i - x_{i-1}) \cdot \frac{1}{n} = \frac{1}{n^2}.$$

Since $f(F)$ is covered by the intervals $[f(x_{i-1}), f(x_i)]$ for which $F \cap [x_{i-1}, x_i] \ne \emptyset$, we find that $\overline{\mu}(f(F)) \le n/n^2 = 1/n$. This holds for every n, and thus $\overline{\mu}(f(F)) = 0$.

We need another property of the function f. We prove that if $x_0 \in [0,1] \setminus F$, then $f(x_0) \notin f(F)$. Indeed, if $x_0 \notin F$, then $(x_0 - \delta, x_0 + \delta) \cap F = \emptyset$ for an appropriate $\delta > 0$. Then the function $\operatorname{dist}(\{x\}, F)$ is positive, and thus f is strictly monotonically increasing on the interval $(x_0 - \delta, x_0 + \delta)$. Since f is monotonically increasing, we have $f(x) \le f(x_0 - \delta) < f(x_0)$ if $x \in F$, $x \le x_0$, and $f(x) \ge f(x_0 + \delta) > f(x_0)$ if $x \in F$, $x \ge x_0$.

Let $g(x) = 1$ if $x \in f(F)$, and $g(x) = 0$ if $x \notin f(F)$. By Theorem 2.7 the set $f(F)$ is closed, and thus the function g is continuous everywhere outside of the points of the set $f(F)$. Since $\overline{\mu}(f(F)) = 0$, it follows from Theorem 8.52 that g is integrable on $[0,1]$.

Now we prove that $g \circ f$ is not integrable. If $x_0 \in \partial F$, then $x_0 \in F$, $f(x_0) \in f(F)$ and $g(f(x_0)) = 1$. On the other hand, every neighborhood of x_0 has a point x such that $x \notin F$. As we saw, $f(x) \notin f(F)$ in this case; thus $g(f(x)) = 0$. This implies that $g \circ f$ is not continuous at the points of ∂F. Since F is not Jordan measurable, it follows that ∂F does not have Jordan measure zero, by Theorem 3.9. The set ∂F is bounded and closed, and thus by Theorem 8.50, ∂F cannot have Lebesgue measure zero. Then, by Theorem 8.52, $g \circ f$ is not integrable on $[0,1]$. □

We know that there exist unbounded derivatives (see [7, Example 13.46]). These derivatives have a primitive function, but they are not integrable. By applying Lebesgue's theorem we will now construct a function whose derivative is bounded but is not integrable. This gives an example of a *bounded* function that is not integrable but still has a primitive function.

The underlying idea of the construction is the following. Let

$$f(x) = \begin{cases} x^2 \sin(1/x), & \text{if } x \neq 0 \\ 0, & \text{if } x = 0 \end{cases}.$$

This function is everywhere differentiable, its derivative is bounded on $[-1, 1]$, but f' is not continuous at 0 (see [7, Example 13.43]). With the help of this function, we construct a function that behaves similarly at each point of a "big" set. For every $\alpha < \beta$, let

$$g_{\alpha,\beta}(x) = \frac{(x-\alpha)^2(\beta-x)^2}{\beta-\alpha} \cdot \sin \frac{1}{(x-\alpha)(\beta-x)}$$

for every $x \in (\alpha, \beta)$. Then $g_{\alpha,\beta}$ is differentiable on (α, β), and

$$g'_{\alpha,\beta}(x) = \frac{2}{\beta-\alpha} \cdot (x-\alpha)(\beta-x)((\alpha+\beta)-2x) \cdot \sin \frac{1}{(x-\alpha)(\beta-x)} + \\ + \frac{2x-(\beta+\alpha)}{\beta-\alpha} \cdot \cos \frac{1}{(x-\alpha)(\beta-x)}$$

for every $x \in (\alpha, \beta)$. It is easy to see that $|g'_{\alpha,\beta}(x)| \leq 2(\beta-\alpha)^2 + 1$ for every $x \in (\alpha, \beta)$.

If $(x-\alpha)(\beta-x) = 1/(k\pi)$ $(k \in \mathbb{N}^+)$, then

$$g'_{\alpha,\beta}(x) = \frac{2x-(\beta+\alpha)}{\beta-\alpha} \cdot (-1)^k.$$

It is clear that in every right-hand-side neighborhood of α and in every left-hand-side neighborhood of β the function $g'_{\alpha,\beta}$ takes values that are arbitrarily close to 1 and also takes values that are arbitrarily close to -1.

We need to prove the following facts for our construction.

Lemma 8.56. *An open set $G \subset \mathbb{R}$ is connected if and only if it is an open interval. Every open set $G \subset \mathbb{R}$ can be written as the union of disjoint open intervals.*

Proof. Any two points of an open interval can be connected by a segment; thus by part (i) of Theorem 1.22, it is connected. (Of course, this can also be proven directly.)

We now prove that if the open set $G \subset \mathbb{R}$ is connected, then G is an open interval. Let $\alpha = \inf G$ and $\beta = \sup G$. Then $(\alpha, \beta) \subset G$, since $x \in (\alpha, \beta) \setminus G$ would imply that $((\alpha, x) \cap G) \cup ((x, \beta) \cap G)$ is a partition of G into nonempty disjoint open sets, but such a partition does not exist, since G is connected. Since G is open, $G \subset (\alpha, \beta)$ is also true, and thus $G = (\alpha, \beta)$.

The second part of the lemma follows from the observation above and from Theorem 1.22. □

Theorem 8.57. *There exists a differentiable function* $f\colon [0,1] \to \mathbb{R}$ *such that* f' *is bounded but not integrable on* $[0,1]$.

Proof. Let F be a bounded, closed, and not Jordan measurable set. (By Exercise 3.15, such a set exists.) We may assume that $F \subset (0,1)$. Let $G = (0,1) \setminus F$; thus G is open. Let us denote by $\mathcal{G}$ the set of the components of G. By Lemma 8.56, every element of $\mathcal{G}$ is an open interval, and $G = \bigcup \mathcal{G}$.

Define the function f as follows. For $x \in F \cup \{0,1\}$, let $f(x) = 0$. If $x \in (0,1) \setminus F = G$, then x is in one of the components of G. If $x \in (\alpha,\beta) \in \mathcal{G}$, then let $f(x) = g_{\alpha,\beta}(x)$. Thus we have defined f at every $x \in [0,1]$.

We show that f is everywhere differentiable. If $x \in (0,1) \setminus F$, then x is in one of the components (α,β). Since $f(x) = g_{\alpha,\beta}(x)$ there, it is clear that f is differentiable at x. For every such x, we have $|f'(x)| = |g'_{\alpha,\beta}(x)| \le 2 \cdot 1^2 + 1 = 3$.

Next we prove that if $x \in F$, then f is differentiable at x, and $f'(x) = 0$. Let $y \in [0,1]$ be arbitrary. If $y \in F \cup \{0,1\}$, then $f(y) = 0$. If $y \in (0,1) \setminus F$ and $y \in (\alpha,\beta) \in \mathcal{G}$, then $|f(y)| \le (y-\alpha)^2 \le (y-x)^2$ for every $x \le \alpha$, and $|f(y)| \le (\beta-y)^2 \le (x-y)^2$ for every $x \ge \beta$. We have proved that $|f(y)| \le (x-y)^2$ for every $y \in [0,1]$. It follows that

$$\lim_{y \to x} \frac{f(y) - f(x)}{y - x} = 0,$$

i.e., $f'(x) = 0$. Thus, f is differentiable everywhere on $[0,1]$, with $|f'(x)| \le 3$ for every $x \in [0,1]$.

We prove that f' is discontinuous at every boundary point of F. If $x_0 \in \partial F$, then every neighborhood U of x intersects $(0,1) \setminus F$, and thus it intersects one of the intervals $(\alpha,\beta) \in \mathcal{G}$. It follows that U contains at least one of the points α and β. Therefore, f takes values arbitrarily close to both 1 and -1 on U. This holds for every neighborhood of x_0, and thus f' is not continuous at x_0.

Since F is not Jordan measurable, ∂F cannot have Jordan measure zero. Now ∂F is bounded and closed, and thus by Theorem 8.50, ∂f does not have Lebesgue measure zero. By applying Theorem 8.52 we get that f' is not integrable on $[0,1]$. □

8.6 Some Applications of Integration in Number Theory

Each of the applications we discuss in this section is based on the fact that the values of certain integrals are small.

1. A lower estimate for the number of primes not greater than n. Let $\pi(n)$ denote the number of primes not greater than n. The celebrated **prime number theorem** states that $\pi(n) \sim n/\log n$; i.e.,

$$\lim_{n \to \infty} \frac{\pi(n)}{n/\log n} = 1.$$

The theorem was proved independently by Hadamard and de la Vallée Poussin[8] in 1896. Since 1896 many other proofs of the theorem have been found, but every known proof is rather complicated. In this section we are concerned only with the weaker statement that $\pi(n)$ is at least a constant multiple of $n/\log n$.

Let $[1,\ldots,n]$ denote the least common multiple of the numbers $1,\ldots,n$.

Lemma 8.58. *For every positive integer n we have $[1,\ldots,n] > 2^{n-2}$.*

Proof. Since $x - x^2 > 0$ for every $x \in (0,1)$ and $x - x^2 < 1/4$ for every $x \in (0,1)$, $x \neq 1/2$, it follows that the value of the integral $\int_0^1 (x-x^2)^k\,dx$ is between 0 and $1/4^k$. On the other hand,

$$\int_0^1 (x-x^2)^k\,dx = \int_0^1 \sum_{i=0}^k (-1)^i \binom{k}{i} x^{k+i}\,dx = \sum_{i=0}^k (-1)^i \binom{k}{i} \cdot \frac{1}{k+i+1} =$$

$$= \frac{A}{[1,\ldots,(2k+1)]},$$

where A is an integer, since each of the numbers $k+i+1$ $(i=0,\ldots,k)$ is a divisor of $[1,\ldots,(2k+1)]$. We know that the integral is positive, and thus $A \geq 1$. Therefore,

$$\frac{1}{[1,\ldots,(2k+1)]} \leq \frac{A}{[1,\ldots,(2k+1)]} = \int_0^1 (x-x^2)^k\,dx < \frac{1}{4^k}$$

follows, i.e., $[1,\ldots,(2k+1)] > 4^k$. If n is odd and $n = 2k+1$, then $[1,\ldots,n] > 4^k = 2^{n-1}$. However, if n is even and $n = 2k$, then $[1,\ldots,n] \geq [1,\ldots,(2k-1)] > 4^{k-1} = 2^{n-2}$. □

With the help of this lemma, we can easily estimate the number of primes smaller than n. Obviously, if $m \leq n$, then every prime in the prime factorization of m is at most n. It follows that $[1,\ldots,n]$ equals the product of prime powers p^α, where $p \leq n$ and α is the largest exponent such that $p^\alpha \leq n$. Denote the number of primes not larger than n by $\pi(n)$. Then we have

$$[1,\ldots,n] = \prod_{p\leq n} p^\alpha \leq \prod_{p \leq n} n = n^{\pi(n)},$$

and thus

$$\pi(n) \geq \frac{\log([1,\ldots,n])}{\log n}. \tag{8.64}$$

[8] Charles Jean de la Vallée Poussin (1866–1962), Belgian mathematician.

Comparing this to Lemma 8.58 yields $\pi(n) > \log 2 \cdot (n-2)/\log n$. Since $\log 2 > 0.69$, we get the following lower estimate:

Theorem 8.59. *For every n large enough we have $\pi(n) > 0.69 \cdot \frac{n}{\log n}$.* □

The value of the integral $\int_0^1 (x-x^2)^k \, dx$ can be computed exactly: it is easy to prove, using integration by parts, that $\int\limits_0^1 x^m(1-x)^n \, dx = \frac{m! \cdot n!}{(m+n+1)!}$ for every $m, n \in \mathbb{N}$. (This also follows from Theorem 8.42.) Thus, $\int_0^1 x^k(1-x)^k \, dx = (k!)^2/(2k+1)!$. Applying Stirling's formula, we get

$$\int\limits_0^1 (x-x^2)^k \, dx \sim \sqrt{\frac{\pi}{4k}} \cdot \frac{1}{4^k}. \tag{8.65}$$

Then the argument of the proof of Lemma 8.58 gives $[1,\ldots,n]/2^n \to \infty$ if $n \to \infty$. However, the lower estimate of $\pi(n)$ cannot be improved by this observation; (8.65) does not give a constant better than $\log 2$. To improve the constant we need an estimate of the form $[1,\ldots,n] \ge c^n$, where $c > 2$.

The method above can be used to get such an estimate if in place of $x - x^2$, we choose an appropriate polynomial with integer coefficients. For example, let $f(x) = x(1-x)(2x-1)$. It is not hard to see that $|f(x)|$ takes its maximum at the points $\frac{1}{2} \pm \frac{1}{2\sqrt{3}}$, and the value of its maximum is $1/(6\sqrt{3})$. Thus, $0 < \int_0^1 f^{2k} \, dx < 1/(6\sqrt{3})^{2k}$. On the other hand, the argument used in the proof of Lemma 8.58 gives that the integral is of the form $A/[1,\ldots,(6k+1)]$, where A is a positive integer. Thus

$$1/[1,\ldots,(6k+1)] < 1/(6\sqrt{3})^{2k},$$

i.e., $[1,\ldots,(6k+1)] > (6\sqrt{3})^{2k}$ for every k. Now, for every $n \ge 7$ there exists $k \ge 1$ such that $6k+1 \le n < 6k+7$. Then $2k > (n-7)/3$, and thus

$$[1,\ldots,n] \ge [1,\ldots,(6k+1)] > (6\sqrt{3})^{2k} \ge c^{n-7},$$

where $c = (6\sqrt{3})^{1/3} = \sqrt[3]{2} \cdot \sqrt{3} = 2.1822\ldots$. It follows that $\pi(n) > 0.78 \cdot n/\log n$ for every n large enough.

To further improve our estimate, we need polynomials with integer coefficients that satisfy that $\rho(f) = (\max_{0 \le x \le 1} |f(x)|)^{1/d}$ is as small as possible, where d is the degree of the polynomial. (The value of ρ for the polynomial $x - x^2$ is $1/2$, and for the polynomial $x(1-x)(2x-1)$ is $1/2.1822\ldots$.) Every value $\rho(f)$ gives an approximation $[1,\ldots,n] > (1/\rho(f))^{n-a}$ with some constant a.

This method can be used to further improve on the constant $2.1822\ldots$, but not essentially. The truth is that for every $\varepsilon > 0$ we have

$$(e-\varepsilon)^n < [1,\ldots,n] < (e+\varepsilon)^n$$

for n large enough. This statement is basically equivalent to the prime number theorem in the sense that each statement follows easily from the other.

It is well known that the method described above does not give the approximation $[1,\ldots,n] > (e-\varepsilon)^n$ for every ε, since the infimum of the numbers $\rho(f)$ is strictly larger than $1/e$.

However, it is not clear whether we can get a better estimate using multivariable polynomials with integer coefficients. So far it has not been proved by this method that the estimate $[1,\ldots,n] > (e-\varepsilon)^n$ $(n > n_0(\varepsilon))$ holds for every $\varepsilon > 0$. For the details we refer to [10].

2. The irrationality of π and the transcendence of e. One can prove the irrationality of e by the following argument. It is easy to check that for every positive integer n, the value of the integral $I_n = \int_0^1 x^n \cdot e^x \, dx$ is a linear combination of the numbers 1 and e with integer coefficients. Therefore, if $e = p/q$, where p, q are positive integers, then $I_n = A_n/q$ for some positive integer A_n, and thus $I_n \geq 1/q$ for every n. On the other hand, $I_n \to 0$ if $n \to \infty$, which is impossible.

Now we give a very similar (but somewhat more complicated) proof of the irrationality of π. We consider the integral $J_n = \int_0^1 f(x) \sin rx \, dx$, where r is a rational number and $f(x) = \frac{1}{n!} \cdot x^n(1-x)^n$.

Lemma 8.60. *If g is a single-variable polynomial with integer coefficients and $h(x) = x^n g(x)/n!$, then $h^{(k)}(0)$ is an integer for every k.*

Proof. Let $h(x) = \sum_{i=n}^m \frac{c_i}{n!} \cdot x^i$, where $c_n, \ldots, c_m$ are integers. Then $h^{(k)}(0) = 0$ for every $k < n$ and $k > m$. On the other hand, if $n \leq k \leq m$, then $h^{(k)}(0) = (c_k/n!) \cdot k! = c_k(n+1) \cdot \ldots \cdot k$ is also an integer. □

Theorem 8.61. *If $0 < r \leq \pi$ is rational, then at least one of the numbers $\sin r$ and $\cos r$ is irrational.*

Proof. Let $f(x) = \frac{1}{n!} \cdot x^n(1-x)^n$. Applying Lemma 8.60 to the polynomial $g(x) = (1-x)^n$ yields that the numbers $f^{(k)}(0)$ are integers for every k. Since $f(1-x) = f(x)$, we have $f^{(k)}(1) = (-1)^k f^{(k)}(0)$, and thus the numbers $f^{(k)}(1)$ are also integers.

We compute the integral $J_n = \int_0^1 f(x) \sin rx \, dx$ by applying partial integration $2n$ times, consecutively. We get that

$$J_n = -\left[f(x)\frac{\cos rx}{r}\right]_0^1 + \frac{1}{r}\int_0^1 f'(x) \cos rx \, dx$$

$$= -\frac{1}{r}\left[f(1)\cos r - f(0)\right] + \frac{1}{r}\left[f'(x)\frac{\sin rx}{r}\right]_0^1 - \frac{1}{r^2}\int_0^1 f''(x) \sin rx \, dx = \ldots$$

$$= -\frac{1}{r}\left[f(1)\cos r - f(0)\right] + \frac{1}{r^2} f'(1) \sin r - \frac{1}{r^3}\left[f''(1)\cos r - f''(0)\right] - \ldots$$

$$+\frac{1}{r^{2n+1}}\left[f^{(2n)}(1)\cos r - f^{(2n)}(0)\right]. \quad (8.66)$$

Let $0 < r \le \pi$ and suppose that each of the numbers r, $\sin r$, $\cos r$ is rational. Let q be the common denominator of the numbers $1/r$, $\sin r$, and $\cos r$. Since $f^{(k)}(0)$ and $f^{(k)}(1)$ are integers for every k, it follows from (8.66) that $J_n = A_n/q^{2n+2}$, where A_n is also an integer. Now, $f(x) > 0$ and $\sin rx > 0$ for every $x \in (0, 1)$ (the latter follows from $0 < r \le \pi$), and thus $J_n > 0$. Thus we have $A_n \ge 1$ and $J_n \ge 1/q^{2n+2}$.

On the other hand, $f(x)\sin rx \le 1/n!$ for $x \in [0, 1]$, and thus $J_n \le 1/n!$. It follows that $1/q^{2n+2} \le J_n \le 1/n!$, which is impossible for n large enough, since $n!/q^{2n+2} \to \infty$. □

Corollary 8.62. *The number π is irrational.*

Proof. The numbers $\sin \pi$, $\cos \pi$ are rational; thus, by Theorem 8.61, π cannot be rational. □

The proof of the following theorem follows a similar argument. Recall that a number is said to be **algebraic** if it is a root of a nonzero polynomial having integer coefficients. A number is said to be **transcendental**, if it is not algebraic.

Theorem 8.63. *The number e is transcendental.*

Proof. Let us assume that e is an algebraic number. Then we would have

$$a_n e^n + a_{n-1}e^{n-1} + \ldots + a_0 = 0, \quad (8.67)$$

where $a_0, a_1, \ldots, a_n$ are integers, and $a_0 \ne 0$. Let f be an arbitrary polynomial. If the degree of f is m, then by applying integration by parts $m + 1$ times we get

$$\int_0^k f(x)e^{-x}dx = -\left(f(k) + f'(k) + \ldots + f^{(m)}(k)\right)e^{-k} +$$
$$+ \left(f(0) + f'(0) + \ldots + f^{(m)}(0)\right). \quad (8.68)$$

If we multiply (8.68) by $a_k e^k$ and add the equations that we get for $k = 0, 1, \ldots, n$, then applying (8.67), we obtain

$$\sum_{k=0}^{n} a_k e^k \int_0^k f(x)e^{-x}dx = -\sum_{k=0}^{n} a_k \left[f(k) + f'(k) + \ldots + f^{(m)}(k)\right]. \quad (8.69)$$

Next we will construct a polynomial f such that the left-hand side of (8.69) is small, while the right-hand side is a nonzero integer. The resulting contradiction will prove our theorem.

Let N be a positive integer. By Lemma 8.60, if g is a polynomial with integer coefficients and $h(x) = x^N g(x)/N!$, then $h^{(i)}(0)$ is an integer for every i. It follows that if g is a polynomial with integer coefficients, a is an integer, and $f(x) = (x-a)^N g(x)/(N-1)!$, then $f^{(i)}(a)$ is an integer and it is divisible by N for every i. Indeed, $f(x) = N \cdot h(x-a)$, where $h(x) = x^N g(x+a)/N!$. Since $g(x+a)$ is a polynomial with integer coefficients, $h^{(i)}(0)$ is an integer, and thus $f^{(i)}(a) = N \cdot h^{(i)}(0)$ is divisible by N for every i.

Let N be a prime satisfying $N > |a_0| \cdot n$. Let

$$f(x) = \frac{1}{(N-1)!} x^{N-1}(x-1)^N (x-2)^N \ldots (x-n)^N.$$

Then $f^{(i)}(k)$ is an integer and is divisible by N for every $i = 0, 1, \ldots$ and $k = 1, \ldots, n$. We prove that $f^{(i)}(0)$ is divisible by N for every i, except when $i = N-1$ (this is why we have x with the exponent $(N-1)$ instead of N).

Indeed, $f(x) = \sum_{i=N-1}^{M} c_i x^i/(N-1)!$, where $c_{N-1} = (\pm n!)^N$ and $c_N, \ldots, c_M$ are integers. It follows that

$$f^{(i)}(0) = \begin{cases} 0, & \text{if } i \le N-2, \\ (\pm n!)^N, & \text{if } i = N-1, \\ c_i \cdot N \cdot (N+1) \cdots i, & \text{if } i \ge N. \end{cases}$$

We can see that $N \mid f^{(i)}(0)$ for $i \neq N-1$. On the other hand, $f^{(N-1)}(0) = (\pm n!)^N$ is not divisible by N, since N is a prime and $n < N$.

Based on what we have proved above, we can see that every term on the right-hand side of (8.69) is divisible by N, except for the term $a_0 f^{(N-1)}(0)$. Thus, the right-hand side of (8.69) is a nonzero integer.

On the other hand, if $0 \le x \le n$, then

$$|f(x)| \le \frac{1}{(N-1)!} n^{(n+1)N} = \frac{A^N}{(N-1)!}, \text{ where } A = n^{n+1}.$$

Thus, the absolute value of the left-hand side of (8.69) is at most

$$(n+1) \max(|a_0|, |a_1|, \ldots, |a_n|) \, e^n \cdot n \cdot \frac{A^N}{(N-1)!} = C \cdot \frac{A^N}{(N-1)!},$$

where C and A are positive integers, independent of N. Since $A^N/(N-1)! \to 0$ as $N \to \infty$, it follows that for N large enough, the absolute value of the left-hand side of (8.69) is smaller than 1. However, the absolute value of the right-hand side is at least 1, since the right-hand side is a nonzero integer. This is a contradiction, proving the theorem. □

We remark that the number π is also transcendental. However, the proof of this fact is more complicated.

8.7 Brouwer's Fixed-Point Theorem

The fixed-point theorems are important in analysis and its applications, since they can often be used to prove the existence of certain numbers, vectors, or other mathematical objects. (We encountered an example in the proof of the open mapping theorem, which used Banach's fixed-point theorem (Theorem 2.36). The following result—one of the fundamental theorems of the topology of Euclidean spaces—is one of the most important fixed-point theorems.

Theorem 8.64. (Brouwer's[9] fixed-point theorem) *Every continuous mapping that maps a closed ball into itself has a fixed point.*

In the single-variable case the theorem states that if f is a continuous function mapping the interval $[a, b]$ into itself, then f has a fixed point. This follows from the Bolzano–Darboux theorem: if $f\colon [a, b] \to [a, b]$ and $g(x) = f(x) - x$ for every $x \in [a, b]$, then $g(a) \geq 0$ and $g(b) \leq 0$. Since g is continuous, it must have a root in $[a, b]$, which is a fixed point of f. The theorem is much harder to prove in higher dimensions. We give a proof that uses theorems on the differentiation of vector-valued functions.

From now on, let us denote by B_p the open unit ball $\{x \in \mathbb{R}^p\colon |x| < 1\}$, by $\overline{B}_p$ the closed unit ball $\{x \in \mathbb{R}^p\colon |x| \leq 1\}$, and by S_p its boundary, i.e., let $S_p = \{x \in \mathbb{R}^p\colon |x| = 1\}$. Clearly, it is enough to prove the theorem for $\overline{B}_p$.

Brouwer's fixed-point theorem is equivalent to the following statement.

Theorem 8.65. *There is no continuous mapping $f\colon \overline{B}_p \to S_p$ that leaves every point of S_p fixed.*

By equivalence we mean that these two theorems can be derived from each other. If the mapping $f\colon \overline{B}_p \to S_p$ is continuous and $f(x) = x$ for every $x \in S_p$, then the mapping $g(x) = -f(x)$ is continuous, it maps the ball $\overline{B}_p$ into itself, but it does not have a fixed point. Thus Theorem 8.65 follows from Theorem 8.64.

Now let $f\colon \overline{B}_p \to \overline{B}_p$ be continuous with no fixed points, i.e., let $f(x) \neq x$ for every $x \in \overline{B}_p$. Consider the half-line $\overrightarrow{f(x), x}$ starting from $f(x)$ and going through x. This intersects S_p at only one point other than $f(x)$; let this point be $g(x)$. It is not hard to see that the function g is continuous on $\overline{B}_p$. Thus, $g\colon \overline{B}_p \to S_p$ is continuous and $g(x) = x$ for every $x \in S_p$. However, this is impossible, based on Theorem 8.65. That is, Theorem 8.65 implies Theorem 8.64.

We begin the proof of Theorem 8.64 by proving Theorem 8.65 under stricter conditions.

Lemma 8.66. *There is no mapping f such that*

(i) *f is continuously differentiable on an open set containing $\overline{B}_p$,*
(ii) *$f(\overline{B}_p) = S_p$, and*

---

[9] Luitzen Egbertus Jan Brouwer (1881–1966), Dutch mathematician.

(iii) $f(x) = x$, *for every* $x \in S_p$.

Proof. Suppose that there exists such a mapping f. A sketch of the argument is as follows. Let

$$f_t(x) = (1-t) \cdot f(x) + t \cdot x$$

for every $t \in [0,1]$ and $x \in \overline{B}_p$. We show that if $t < 1$ and t is close enough to 1, then the mapping f_t maps the ball B_p onto itself bijectively, and furthermore, $\det f_t'(x) > 0$ for every $x \in B_p$. Then, using the integral transform formula, we get

$$\gamma_p = \int_{B_p} \det f_t'(x)\,dx, \tag{8.70}$$

where γ_p denotes the measure of B_p. We also prove—and this is a key component of the proof—that $\int_{B_p} \det f_t'(x)\,dx$ is a polynomial in the variable t. Since (8.70) holds on an interval $(1-\delta, 1)$, it has to hold everywhere. Nevertheless, for $t = 0$ we have $f_t = f$ and (as we will also show) $\det f'$ is zero everywhere, i.e., (8.70) does not hold. This contradiction will prove our lemma.

Let us go into the details. Let the components of f be $f_1, \ldots, f_p$. The partial derivatives $D_i f_j$ are continuous on $\overline{B}_p$; thus they are bounded there. Let $|D_i f_j(x)| \le K$ for every $x \in \overline{B}_p$ and $i, j = 1, \ldots, p$. By Lemma 2.33, $|f(y) - f(x)| \le Kp^2 \cdot |y - x|$ for every $x, y \in \overline{B}_p$. It is clear that $f_t \colon \overline{B}_p \to \overline{B}_p$ and $f_t(x) = x$ for every $x \in S_p$ and $t \in [0,1]$. If $x \ne y$, then

$$\begin{aligned} |f_t(y) - f_t(x)| &\ge t \cdot |y - x| - (1-t) \cdot |f(y) - f(x)| \ge \\ &\ge t \cdot |y - x| - (1-t) \cdot Kp^2 |y - x| = \\ &= (t - (1-t)Kp^2) \cdot |y - x| > 0, \end{aligned}$$

assuming that t is close enough to 1. On the other hand, $f_t'(x) = (1-t) \cdot f'(x) + t \cdot I$ (where I is the identity), and thus the matrix of $f_t'(x)$ is arbitrarily close to the identity matrix if t is close enough to 1. Therefore, there exists $\delta > 0$ such that if $1 - \delta < t \le 1$, then the mapping f_t is injective on B_p and its Jacobian determinant is positive for every $x \in B_p$.

Let $1 - \delta < t \le 1$ be fixed. We prove that $f_t(B_p) = B_p$. By the open mapping theorem, $f_t(B_p)$ is an open set, and thus $f_t(B_p) \subset B_p$. Suppose that $f_t(B_p) \ne B_p$, and choose a point $x \in B_p \setminus f_t(B_p)$. Let $y = f_t(0)$; then $y \in B_p$ and $y \in f_t(B_p)$. Thus, the segment $[x, y]$ intersects the boundary of $f_t(B_p)$; let $z \in [x, y] \cap \partial f_t(B_p)$. Then $z \in B_p$, and

$$z \in \partial f_t(B_p) \subset \operatorname{cl} f_t(B_p) \subset \operatorname{cl} f_t(\overline{B}_p) = f_t(\overline{B}_p),$$

since by Theorem 2.7, $f_t(\overline{B}_p)$ is a closed set. Thus there exists a point $u \in \overline{B}_p$ such that $f_t(u) = z$. Here $u \in S_p$ is impossible, since $u \in S_p$ would imply $f_t(u) = u \in$

S_p. Thus $u \in B_p$. However, in this case $z \in f_t(B_p)$ and $z \notin \partial f_t(B_p)$, since $f_t(B_p)$ is an open set. This is a contradiction, which proves that $f_t(B_p) = B_p$.

By the measure transform formula (Theorem 4.22, formula (4.13)), we have (8.70). Note that this is true for every $1 - \delta < t \le 1$.

Now we prove that $\int_{B_p} \det f_t'(x)\, dx$ is a polynomial in t. Indeed, in the matrix of the mapping $f_t'(x)$ the ith term of the jth row is $(1-t) \cdot D_i f_j(x)$ if $i \ne j$, and $(1-t) \cdot D_j f_i(x) + t$ if $i = j$. Thus, the determinant $\det f_t'(x)$ is a sum of the form $\sum_{i=1}^N g_i(t) \cdot h_i(x)$, where g_i is a polynomial and h_i is a continuous function for every i. Integrating this with respect to x on the ball B_p yields the polynomial $P(t) = \sum_{i=1}^N c_i \cdot g_i(t)$, where $c_i = \int_{B_p} h_i(x)\, dx$.

We know that $P(t) = \gamma_p$, for every $1 - \delta < t \le 1$. This is possible only if P is the constant function equal to γ_p. In particular, for $t = 0$, we have

$$\gamma_p = P(0) = \int_{B_p} \det f_0'(x)\, dx = \int_{B_p} \det f'(x)\, dx.$$

However, by condition (ii) we have $f(B_p) \subset S_p$; thus the interior of $f(B_p)$ is empty. By the open mapping theorem it follows that $f'(x)$ cannot be injective at any point $x \in B_p$. That is, $\det f'(x) = 0$ for every $x \in B_p$, and thus $\int_{B_p} \det f'(x)\, dx = 0$. This is a contradiction, which proves that there is no function satisfying conditions (i), (ii), and (iii). □

Proof of Theorem 8.64. Suppose that the mapping $f\colon \overline{B}_p \to \overline{B}_p$ is continuous but has no fixed points. First we prove that there is a polynomial with the same properties. (From now on, by a polynomial we mean a mapping every component of which is a polynomial.)

The function $|f(x) - x|$ is continuous on $\overline{B}_p$, and thus by Weierstrass's theorem (Theorem 1.51), it has a smallest value. Since, by assumption, $f(x) \ne x$ for every $x \in \overline{B}_p$, this smallest value is positive; i.e., there exists $\delta > 0$ such that $|f(x) - x| > \delta$ for every $x \in \overline{B}_p$.

We extend f to the whole space in the following way: let $f(x) = f(x/|x|)$ for $|x| > 1$. It is easy to see (using the fact that the mapping $x \mapsto x/|x|$ is continuous on the set $\mathbb{R}^p \setminus \{0\}$ and f is continuous on $\overline{B}_p$) that the extended function is continuous everywhere.

Let $\varepsilon > 0$ be fixed, and apply Weierstrass's approximation theorem (Theorem 1.54) to a box containing $\overline{B}_p$. We get that there exists a polynomial g such that $|f(x) - g(x)| < \varepsilon$ for every $x \in \overline{B}_p$. Consider the polynomial $h = (1-\varepsilon) \cdot g$. If $|x| \le 1$, then $|h(x)| \le (1-\varepsilon) \cdot (1+\varepsilon) = 1 - \varepsilon^2 < 1$, i.e., h maps the closed ball $\overline{B}_p$ into itself. On the other hand, for every $|x| \le 1$, we have

$$|h(x) - x| \ge |f(x) - x| - |f(x) - g(x)| - |g(x) - h(x)| > \delta - \varepsilon - \varepsilon(1+\varepsilon) > 0$$

for ε small enough. We have constructed a polynomial h such that h maps $\overline{B}_p$ into itself and h has no fixed point in $\overline{B}_p$.

Since h is uniformly continuous on the closed sphere $\overline{B}(0,2)$ and $|h(x)| \le 1 - \varepsilon^2$ for every $|x| \le 1$, there exists $\eta > 0$ such that $|h(x)| < 1$ for every $|x| < 1 + \eta$. The polynomial h does not have a fixed point in the ball $G = B(0, 1+\eta)$, since $1 < |x| < 1 + \eta$ implies $|h(x)| < 1$, and thus $h(x) \neq x$.

For every $|x| < 1 + \eta$, consider the half-line $\overrightarrow{h(x), x}$ with endpoint $h(x)$ passing through x. This half-line intersects S_p at a single point; let this point be $s_0(x)$. The function s_0 is defined in the ball G, it maps the ball $\overline{B}_p$ into S_p, and $s_0(x) = x$ for every $x \in S_p$. If we can prove that s_0 is continuously differentiable on G, then we will obtain a contradiction to Lemma 8.66, and our proof will be complete.

Since the mapping h is continuously differentiable, the mapping s_0 also has to be continuously differentiable, intuitively. The precise proof goes as follows.

If $x, y \in \mathbb{R}^p$ and $x \neq y$, then the half-line $\overrightarrow{y, x}$ consists of the points $y + t(x - y)$, where $t \ge 0$. The point $y + t(x - y)$ is in S_p exactly if $|y + t(x - y)| = 1$. Let

$$\begin{aligned} q(x, y, t) &= |y + t(x - y)|^2 = \langle y + t(x - y), y + t(x - y) \rangle = \\ &= |y|^2 + 2\langle y, x - y \rangle \cdot t + |x - y|^2 \cdot t^2. \end{aligned} \tag{8.71}$$

We know that $y + t(x - y) \in S_p$ if and only if $q(x, y, t) = 1$. For a fixed x, y, q is a second-degree polynomial of the variable t, with a positive leading coefficient. If $|y| < 1$, then $q(x, y, 0) = |y|^2 < 1$, and thus there exists exactly one $t \ge 0$ such that $q(t) = 1$. We have proved that for every $|y| < 1$ and $x \neq y$ the half-line $\overrightarrow{y, x}$ intersects S_p at a single point. If this point of intersection is $s(x, y)$, then $s(x, y) = y + t(x - y)$, where $q(x, y, t) = 1$.

The function $s(x, y)$ is defined on the open set

$$U = \{(x, y) \colon x \in \mathbb{R}^p,\ |y| < 1,\ x \neq y\} \subset \mathbb{R}^{2p}.$$

If $(x, y) \in U$, then by (8.71), $q(x, y, t) = 1$ is a quadratic equation in t, which has exactly one nonnegative root. By applying the formula for the roots of a quadratic equation, we get $t = R_1 + \sqrt{R_2}$, where R_1 and R_2 are rational functions of the variables x, y (defined everywhere on U). Thus, s is continuously differentiable on U.

Let us return to the polynomial h and the mapping s_0. Since $|x| < 1 + \eta$ implies $|h(x)| < 1$ and $h(x) \neq x$, we have $s_0(x) = s(x, h(x))$. Since both $s(x, y)$ and h are continuously differentiable, s_0 is also continuously differentiable. □

As we saw, Theorem 8.65 is an easy consequence of Brouwer's fixed-point theorem.

With the help of the method used in the previous proof we now prove an interesting topological theorem, which can be formulated as follows. The surface of the three-dimensional ball cannot be "combed" without having a "cowlick." Let us say that the surface of the sphere, i.e., the set S_3, is covered by hair. Combing this hair means that for every $x \in S_3$ the mop of hair at the point x leans on S_3 in some direction $v(x)$, where the unit vector $v(x)$ is a continuous function of the point x. In other words, combing the sphere requires the existence of a continuous mapping $v\colon S_3 \to S_3$ such that $v(x)$ is the unit vector perpendicular to x for every $x \in S_p$.

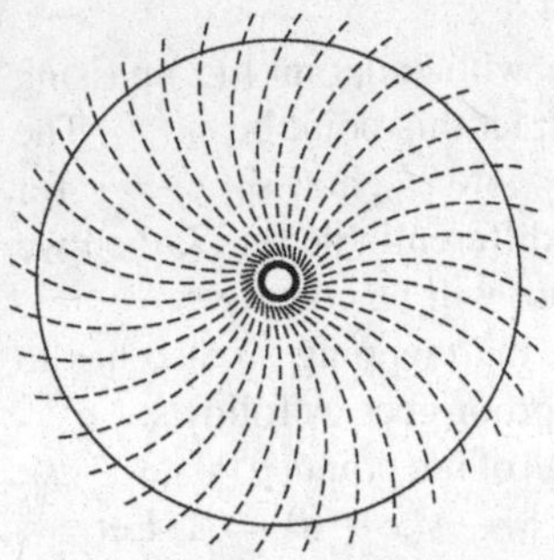

$p = 3$, simple "cowlick"

8.2. Figure

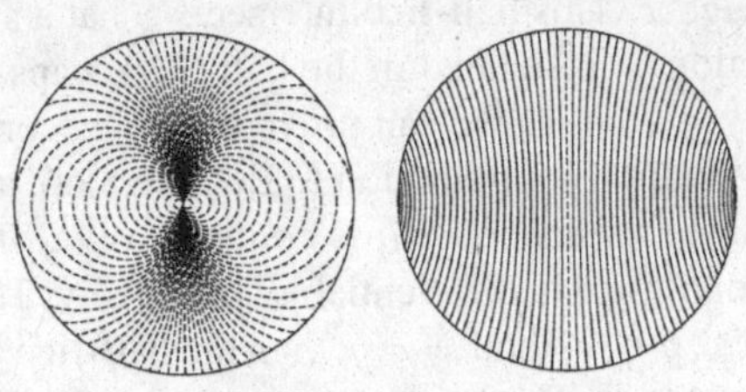

$p = 3$, double "cowlick"
the upper semi-sphere and the connected lower semi-sphere

8.3. Figure

Theorem 8.67. *If p is odd, then there is no continuous mapping $v\colon S_p \to S_p$ such that $\langle v(x), x\rangle = 0$ for every $x \in S_p$.*

We give only an outline of the proof. The missing parts are similar to the proof of Brouwer's fixed-point theorem. Let us assume that there exists a continuous mapping $v\colon S_p \to S_p$ such that $\langle v(x), x\rangle = 0$ for every $x \in S_p$. First we prove that there also exists a continuously differentiable function with these properties.

Extend v to the whole space, using the formula $v(rx) = rv(x)$ ($|x| = 1$, $r \geq 0$). The extended mapping v is everywhere continuous. Applying Weierstrass's approximation theorem (Theorem 1.54) to the coordinate functions of v on a cube containing S_p, we get a polynomial $f\colon \mathbb{R}^p \to \mathbb{R}^p$ satisfying $|v(x) - f(x)| < 1/2$ for every $x \in S_p$. Let

$$g(x) = f(x) - \langle f(x), x\rangle \cdot x$$

for every x. Then the coordinate functions of g are also polynomials, $\langle g(x), x\rangle = 0$ for every $x \in S_p$, and $g \neq 0$ on the set S_p. The latter statement follows from the fact that $|x| = 1$ implies $\langle v(x), x\rangle = 0$, and thus

$$|\langle f(x), x\rangle| = |\langle f(x) - v(x), x\rangle| < \frac{1}{2}.$$

Therefore,

$$|g| \geq |f| - \frac{1}{2} \geq |v| - |v - f| - \frac{1}{2} > 1 - \frac{1}{2} - \frac{1}{2} = 0$$

on the set S_p. Since f is continuous, $g \neq 0$ in an appropriate open set G containing the set S_p. The mapping $w = g/|g|$ is continuously differentiable on G, and $\langle w(x), x\rangle = 0$ for every $x \in S_p$.

We have proved that we can choose v to be continuously differentiable. We use the notation v instead of w, and let $v(rx) = rv(x)$ for every $|x| = 1$ and $r \geq 0$. The expanded function v is continuously differentiable on the set $\mathbb{R}^p \setminus \{0\}$, and $|v(x)| = |x|$ for every x.

We now prove that for $t > 0$ small enough,

$$\{x + tv(x)\colon |x| = r\} = \{x\colon |x| = r \cdot \sqrt{1+t^2}\} \tag{8.72}$$

for every $1 \leq r \leq 2$. If $|x| = r$, then $|x + tv(x)| = \sqrt{r^2 + t^2 r^2} = r \cdot \sqrt{1+t^2}$, since $v(x)$ is perpendicular to x and $|v(x)| = r$. We have proved that the left-hand side of (8.72) is a subset of the right-hand side. To prove the other direction, let $|b| = r \cdot \sqrt{1+t^2}$, where $1 \leq r \leq 2$. For $t > 0$ small enough, the mapping $h(x) = b - tv(x)$ maps the bounded and closed set $F = \{x\colon 1/2 \leq |x| \leq 3\}$ into itself. Since v is continuously differentiable on a neighborhood of the set F, it follows that v is Lipschitz on F (this is easy to prove with the help of Lemma 2.33, using the fact that F is bounded and closed). It follows that for t small enough, h is a contraction on F. Thus, by Banach's fixed-point theorem, there is $x \in F$ such that $h(x) = x$, i.e., $x + tv(x) = b$. Since x and $v(x)$ are perpendicular to each other and $|v(x)| = |x|$, we have $|x| = r$, and we have proved (8.72).

8.4. Figure

It also follows from v being Lipschitz that for t small enough, the mapping $V_t(x) = x + tv(x)$ is injective on F. Now by (8.72), the mapping V_t maps the set $K = \{x\colon 1 \leq |x| \leq 2\}$ into the set $\{x\colon \sqrt{1+t^2} \leq |x| \leq 2\sqrt{1+t^2}\}$, whose measure is $(1+t^2)^{p/2} \cdot (2^p - 1) \cdot \gamma_p$. According to the measure transform formula,

$$\int_K \det V_t'\, dx = (1+t^2)^{p/2} \cdot (2^p - 1) \cdot \gamma_p \tag{8.73}$$

for every t small enough. (We use the fact that for t small enough, $\det V_t'(x)$ is positive.) However, the left-hand side of (8.73) is a polynomial in t. If p is odd, then the right-hand side is not a polynomial in t for any interval $(1, 1+\delta)$, which is a contradiction. □

Remark 8.68. **1.** If p is even, then S_p can be combed, i.e., there exists a continuous mapping $v\colon S_p \to S_p$ such that $\langle v(x), x\rangle = 0$ for every $x \in S_p$. For the $p = 2$ case, see Figure 8.4. In general, if $p = 2n$, then the mapping

$$v(x_1, \ldots, x_{2n}) =$$
$$= (x_2, -x_1, x_4, -x_3, \ldots, x_{2n}, -x_{2n-1})$$

works.

2. It is not hard to reduce Brouwer's fixed-point theorem to Theorem 8.67; see [8].

8.8 The Peano Curve

In 1890, Giuseppe Peano[10] realized that *there exists a continuous curve in the plane that covers a whole square of the plane, i.e., whose range is an entire square of the plane.* A curve with this property is called a **Peano curve**. The existence of such curves means that some continuous curves behave differently from what our intuition would suggest. Therefore, if we want to ensure that a curve is as we imagine curves in general, we need stronger conditions than continuity (e.g., differentiability or rectifiability).

Below, we give two constructions of Peano curves.

3	2
0	1

15	12	11	10
14	13	8	9
1	2	7	6
0	3	4	5

8.5. Figure

I. Let $Q = [0,1] \times [0,1]$. The lines defined by the equations $x = k/2^n$ and $y = k/2^n$ (where $k = 1, \ldots, 2^n - 1$) divide the square Q into 4^n squares of side length 2^{-n}. In the first step of our construction we enumerate these squares as $Q_0^n, \ldots, Q_{4^n-1}^n$ for every n, with the following properties.

(i) For every n and $0 < i \le 4^n - 1$ the squares Q_{i-1}^n and Q_i^n are adjacent; i.e., they have a common side.

(ii) For every n and $0 < i \le 4^n - 1$, we have $Q_i^n = Q_{4i}^{n+1} \cup Q_{4i+1}^{n+1} \cup Q_{4i+2}^{n+1} \cup Q_{4i+3}^{n+1}$.

Let $Q_0^0 = Q$. Let $n \ge 0$, and suppose we have an enumeration $Q_0^n, \ldots, Q_{4^n-1}^n$ satisfying (i). Divide Q_0^n into four nonoverlapping congruent squares. We

[10] Giuseppe Peano (1858–1932), Italian mathematician.

can enumerate these squares as $Q_0^{n+1}, \ldots, Q_3^{n+1}$ such that for every $i = 1, 2, 3$, the squares Q_{i-1}^{n+1} and Q_i^{n+1} are adjacent. Furthermore, one of the sides of Q_3^{n+1} lies on the common side of the squares Q_0^n and Q_1^n. Now divide Q_1^n into four nonoverlapping congruent squares and enumerate these smaller squares as $Q_4^{n+1}, \ldots, Q_7^{n+1}$ such that for each of $i = 4, 5, 6, 7$, the squares Q_{i-1}^{n+1} and Q_i^{n+1} are adjacent. Furthermore one of the sides of Q_7^{n+1} lies on the common side of the squares Q_1^n and Q_2^n. (It is easy to see that this is always possible.) Continuing the process yields the enumerations Q_i^{n+1} $(i = 0, \ldots, 4^{n+1} - 1)$, which satisfy both (i) and (ii). Figure 8.5 gives a possible enumeration for $n = 1$ and $n = 2$.

Let $t \in [0, 1]$, and let $0.a_1a_2\ldots$ denote the representation of t in the number system of base 4. By property (ii), the squares $Q_{a_1}^1$, $Q_{4a_1+a_2}^2$, $Q_{4^2a_1+4a_2+a_3}^3$, $\ldots$ are nested in each other (i.e., each contains the next one), and thus they have a common point by Cantor's theorem (Theorem 1.25). Let $\gamma(t)$ denote the (single) common point of these squares. Then γ maps the interval $[0, 1]$ into the square Q.

For every point $x \in Q$, there exist nested squares $Q_{i_1}^1 \supset Q_{i_2}^2 \supset \ldots$ such that $\bigcap_{n=1}^{\infty} Q_{i_n}^n = \{x\}$. Property (ii) implies that for suitable digits $a_n = 0, 1, 2, 3$ we have $i_n = 4^{n-1}a_1 + 4^{n-2}a_2 + \ldots + a_n$. Therefore, by the definition of γ, we have $\gamma(0, a_1a_2\ldots) = x$. Since $x \in Q$ was arbitrary, this proves $\gamma([0, 1]) = Q$.

We now prove that γ is continuous. We show that if $|t_2 - t_1| < 1/4^n$, then

$$|\gamma(t_2) - \gamma(t_1)| \leq 2\sqrt{2}/2^n. \tag{8.74}$$

Let $t_1 = 0.a_1a_2\ldots$ and $t_2 = 0.b_1b_2\ldots$. The condition $|t_2 - t_1| < 1/4^n$ does not necessarily imply that the first n digits of t_1 and t_2 are the same, but it is true that if $i = 4^{n-1}a_1 + 4^{n-2}a_2 + \ldots + a_n$ and $j = 4^{n-1}b_1 + 4^{n-2}b_2 + \ldots + b_n$, then $|i - j| \leq 1$. Thus, Q_i^n and Q_j^n are either coincident or adjacent. Since $\gamma(t_1) \in Q_i^n$ and $\gamma(t_2) \in Q_j^n$, and furthermore, the diameter of Q_i^n and Q_j^n is $\sqrt{2}/2^n$, (8.74) follows.

Let $\varepsilon > 0$ be arbitrary. If n is so large that $2\sqrt{2}/2^n < \varepsilon$ holds, then (8.74) implies $|\gamma(t_2) - \gamma(t_1)| < \varepsilon$ whenever $|t_2 - t_1| < 1/4^n$. Let $\gamma = (f, g)$. Then $|f(t_2) - f(t_1)| < \varepsilon$ and $|g(t_2) - g(t_1)| < \varepsilon$ for every pair of numbers $t_1, t_2 \in [0, 1]$ that satisfies $|t_2 - t_1| < 1/4^n$. Therefore, f, g and γ are continuous.

II. First we construct a continuous map from the Cantor set C onto $C \times C$. Let $x \in C$, and let the representation of x in the base-3 number system be $x = 0.a_1a_2\ldots$, where $a_i \in \{0, 2\}$ for every i. Let $\varphi(x) = 0.a_1a_3a_5\ldots$ and $\psi(x) = 0.a_2a_4a_6\ldots$. Then φ and ψ map the set C into itself, and φ and ψ are continuous, since $x, y \in C$ and $|x - y| < 1/3^{2n}$ imply $|\varphi(x) - \varphi(y)| \leq 1/3^n$ and $|\psi(x) - \psi(y)| \leq 1/3^n$. It can be easily verified that the map $x \mapsto (\varphi(x), \psi(x))$ $(x \in C)$ maps C onto $C \times C$.

Let $f\colon [0, 1] \to [0, 1]$ be the Cantor function (see Exercise 3.28), let $g_1(x) = f(\varphi(x))$, and let $g_2(x) = f(\psi(x))$ for every $x \in C$. Then the map $x \mapsto (g_1(x), g_2(x))$ maps the set C onto $[0, 1] \times [0, 1]$.

Finally, extend g_1 and g_2 to the interval $[0, 1]$ such that whenever (α, β) is an interval contiguous to the set C (i.e., a component of the open set $(0, 1) \setminus C$), then the extension is linear on the closed interval $[\alpha, \beta]$. It is easy to show that the resulting functions are continuous on $[0,1]$. Then the mapping $x \mapsto (g_1(x), g_2(x))$ $(x \in [0,1])$ defines a Peano curve.

By slightly altering the second construction we can map the interval $[0,1]$ continuously onto the unit cube of $\mathbb{R}^3$, or more generally, onto the unit cube of $\mathbb{R}^n$ as well. Moreover, $[0,1]$ can be mapped continuously onto the "infinite-dimensional" unit cube in the following sense.

Theorem 8.69. *There exists an infinite sequence of continuous functions* $g_1, g_2, \ldots$ *mapping the interval* $[0,1]$ *into itself with the following property: for every sequence* $x_i \in [0,1]$ *there exists a number* $t \in [0,1]$ *such that* $g_i(t) = x_i$ *for every* $i = 1, 2, \ldots$.

Proof. Let the representation of $x \in C$ in the base-3 number system be $0.a_1a_2\ldots$. Let us define the numbers $\varphi_1(x)$, $\varphi_2(x), \ldots$ as follows. Partition the positive integers into infinitely many disjoint infinite subsets (e.g., let $A_i = \{2^i \cdot k - 2^{i-1} : k \in \mathbb{N}^+\}$ $(i = 1, 2, \ldots)$). If the ith set of the partition is $\{n_1, n_2, \ldots\}$, then let $\varphi_i(x) = 0, a_{n_1}a_{n_2}\ldots$.

Let $g_i(x) = f(\varphi_i(x))$ for $x \in C$, and extend g_i to $[0,1]$ such that the extension is linear in the closure of each interval contiguous to C. It is easy to see that the functions g_i satisfy the conditions of the theorem. □

Chapter 9
Hints, Solutions

Hints

Chapter 1

1.27 Cover every isolated point of A with a ball B such that $A \cap B$ has a single element, and apply Lindelöf's theorem.

1.31 For every star A, choose an open ball $B(a, r)$ such that $B(a, r)$ contains the center of the star A, its boundary intersects all three segments of A, and r and the coordinates of a are rational numbers.

The set $B \setminus A$ has three components. Choose points p_1, p_2, p_3 from each component having rational coordinates. Assign the quadruple (B, p_1, p_2, p_3) to A. Show that if A_1 and A_2 are disjoint, then their corresponding quadruples are different from each other.

1.35 The answer to the second question is negative. Apply Borel's theorem.

1.40 Let $\mathcal{H}(A)$ denote the sets we get from A in the way described in the exercise. Using (1.8), show that we can get every element of $\mathcal{H}(A)$ by applying the operations at most four times, and only the sets ext ∂ ext int A and ext ∂ ext ext A require four operations.

1.45 Show that the set $\{x \in A\colon f(x) < n\}$ is countable for every n, using Exercise 1.27.

1.49 Notice that if $f(x, y) = 0$ for every point with $y \le x^2$ or $y \ge 3x^2$, then the function f restricted to any line that goes through the origin is continuous at the origin. Construct such a function that is continuous everywhere outside of the origin, but not continuous at the origin itself.

1.52 There exists such a polynomial. Try a polynomial of the form $p^2 + q^2$, where p and q are polynomials, at every point at most one of them is zero, but they can take small values simultaneously.

1.54 The statement is not true. Find a counterexample in which f is not bounded.

M. Laczkovich and V.T. Sós, *Real Analysis*, Undergraduate Texts in Mathematics, https://doi.org/10.1007/978-1-4939-7369-9_9

1.69 First show that f is constant on every segment contained in G and parallel to one of the axes. Using this, show that f is constant on every ball contained in G. Finally, show that f is constant on G, using the condition that G is connected.

1.71 Try a polynomial of the form $f(x) + g(y)$, with single-variable polynomials f and g.

1.80 (h) The answer is yes. Show that $v - u \leq \sqrt[3]{4(v^3 - u^3)}$ for every $u < v$; then use this inequality to prove

$$\lim_{(x,y)\to(0,0)} \frac{\sqrt[3]{x^3 + y^4} - x}{\sqrt{x^2 + y^2}} = 0.$$

1.86 Show that $(y - x^2)(y - 2x^2)$ is a counterexample.

1.90 Apply the results of Exercise 1.69.

1.96 These are the functions $f(x, y) = g(x) + h(y)$, where $g, h\colon \mathbb{R} \to \mathbb{R}$, and g is differentiable.

Chapter 2

2.13 The statement is false. Find a counterexample for the case $p = 1,\ q = 2$.

2.14 We need to prove that if the (finite or infinite) derivative of the single-variable continuous function $f\colon [a, b] \to \mathbb{R}$ exists everywhere and $f'(x) \neq 0$ for every $x \in [a, b]$, then f is strictly monotone on $[a, b]$.

Chapter 3

3.28 Part (a) is trivial. For proving part (b), notice that every monotone function $f\colon [a, b] \to \mathbb{R}$ is continuous if its range is an interval.

3.30 Show that the lines going through the sides of the polygon cut it into nonoverlapping convex polygons and that every convex polygon can be cut into nonoverlapping triangles using the diagonals starting from a vertex.

3.33 First show that if H has measure zero, then A also has measure zero. For the proof use the fact that the statement of the exercise is true for convex sets by Theorem 3.26. In the general case show that the boundary of A has measure zero.

Chapter 4

4.6 (g) Transform the set A into a ball using an appropriate substitution; then substitute with polar coordinates.

4.9 The difference of the functions $x \mapsto \overline{\int_c^d} f_x\, dy$ and $x \mapsto \underline{\int_c^d} f_x\, dy$ is nonnegative, with integral zero on $[a, b]$. This reduces the statement to the following: if g is nonnegative and integrable on $[a, b]$ and $\int_a^b g\, dx = 0$, then the set $\{x \in [a, b] : g(x) = o\}$ is dense in $[a, b]$. See [7, Exercise 14.12].

Chapter 5

5.2 Let F and G be primitive functions. Apply the result of Exercise 1.90 to $F - G$.

5.6 Start from a point x and walk along the closed polygonal line T until you arrive at a vertex y that you have already reached before. Then, starting from y, obtain a closed polygonal line T_1 that does not intersect itself. Delete T_1 from T, and repeat this process until T is exhausted.

5.17 Let δ be small enough and choose a contained polygon with minimal number of vertices and diameter finer than δ. Show that the sides of this polygon do not cross each other for δ small enough.

5.21 Apply the differentiation rule for inverse functions and the formula of the integral transform.

5.23 The planes going through the sides of the polyhedron decompose it into convex polyhedra. Connecting an interior point of a convex polyhedron P with the vertices of P gives a partition of P into pyramids.

Chapter 6

6.2 (a) Give a closed form for the partial sums using the identity

$$\frac{1}{n^2+2n} = \frac{1}{2n} - \frac{1}{2(n+2)}.$$

A similar method can be used for the series (b), (c), and (d).

6.3 The left-hand side of the inequality is less than $1 - \frac{1}{2^c} + \frac{1}{3^c} - \ldots - \frac{1}{(2n)^c}$. Deduce from the inequality that if $c > 1$, then the sequence of the partial sums of the series $\sum_{n=1}^{\infty} 1/n^c$ is bounded.

6.4 Give the upper bound $N/10^{k-1}$ to the sum $\sum_{10^{k-1} \le a_n < 10^k} 1/a_n$, where N denotes how many numbers there are with k digits that do not contain the digit 7.

6.8 Apply the Cauchy–Schwarz–Bunyakovsky inequality ([7, Theorem 11.19]).

6.9 We can assume that (a_n) is monotonically decreasing. Show that $a_n \ge 0$ for every n, and apply $\lim_{n\to\infty} \sum_{i=n}^{2n} a_i = 0$ (from the Cauchy criterion).

6.10 Let $c > 1$. Put $I_k = \{n\colon 2^k \le s_n < 2^{k+1}\}$. Suppose that $I_k \ne \emptyset$, and let $I_k = \{a, a+1, \ldots, b\}$. Then

$$\sum_{n\in I_k} \frac{a_n}{(s_n)^c} \le \frac{1}{2^{ck}} \cdot \sum_{n=a}^{b} a_n = \frac{s_b - s_{a-1}}{2^{ck}} \le \frac{s_b}{2^{ck}} \le \frac{2^{k+1}}{2^{ck}}.$$

Deduce from this that $\sum a_n/(s_n)^c$ is convergent.

To prove the divergence of the series $\sum a_n/s_n$, estimate the sum $\sum_{n\in I_k} a_n/s_n$ from below, assuming that $a_n/s_n < 1/3$ for every n large enough. If $a_n \ge s_n/3$ for infinitely many n, then it is clear that $\sum a_n/s_n$ is divergent.

6.11 Put $J_k = \{n\colon 2^{-k} \le r_n < 2^{-k+1}\}$. For $c < 1$ estimate the sums $\sum_{n \in J_k} a_n/(r_n)^c$ from above, and estimate the sum $\sum_{n \in J_k} a_n/r_n$ from below.

6.13 Let $[1,2] \times [0,1/2]$ be the first rectangle. At every step, choose the largest possible rectangle in H that does not overlap any of the previous rectangles and such that the sequence of the upper right-hand points of these rectangles is

$$\left(\frac{3}{2}, \frac{2}{3}\right), \left(\frac{5}{4}, \frac{4}{5}\right), \left(\frac{7}{4}, \frac{4}{7}\right), \left(\frac{9}{8}, \frac{8}{9}\right), \left(\frac{11}{8}, \frac{8}{11}\right), \left(\frac{13}{8}, \frac{8}{13}\right), \left(\frac{15}{8}, \frac{8}{15}\right), \ldots$$

in this order.

6.14 The series is convergent if and only if $k = m$. Show that for integers $0 < a < b$, we have

$$\frac{1}{b} \cdot \log n \le \sum_{i=0}^{n} \frac{1}{ib + a} \le \frac{1}{a} + \frac{1}{b} + \frac{1}{b} \cdot \log n.$$

6.18 Show that if the series is conditionally convergent, then it has a reordering with arbitrarily large and arbitrarily small partial sums.

6.36 The series is divergent. By splitting the sum $s_n = n + \sqrt{n} + \sqrt[3]{n} + \ldots + \sqrt[n]{n}$ appropriately and estimating both parts, show that $s_n < n \cdot \log n$ for every n large enough.

6.37 Let $b_n = (a_n)^{1-1/\log n}$. Show that if $a_n \le 1/n^2$, then $b_n \le c/n^2$, and if $a_n \ge 1/n^2$, then $b_n \le c \cdot a_n$. Then use the majorization principle.

6.40 (a) Show that the series has a majorant of the form $\sum c/n^b$, where $b > 1$.

(b) Show that $a_n \ge c/n$ for every n large enough, with a constant $c > 0$.

6.59 Apply the formula $\sin x \sin y = \frac{1}{2}\left(\cos(x-y) - \cos(x+y)\right)$.

6.60 Let s_n be the nth partial sum of the series. By assumption, the sequence $(s_1 + \ldots + s_n)/n$ is convergent. Changing the first element of the series appropriately, we can ensure that $(s_1 + \ldots + s_n)/n$ converges to zero. Let $n \cdot a_n$ be bounded from below. Multiplying the series by a suitable positive number, we may assume that $a_n \ge -1/n$ for every n.

We need to show that if $a_n \ge -1/n$ for every n and $(s_1 + \ldots + s_n)/n \to 0$, then $s_n \to 0$. Let $\varepsilon > 0$ be fixed, and suppose that $s_n \ge 2\varepsilon$ for infinitely many n. Show that $s_n, s_{n+1}, \ldots, s_{n+k} > \varepsilon$ for every such n and $k < \varepsilon n$, which contradicts $(s_1 + \ldots + s_n)/n \to 0$. Similarly, if $s_n \le -2\varepsilon$, then $s_n, s_{n-1}, \ldots, s_{n-k} < \varepsilon$ for every $k < \varepsilon n$, which leads to another contradiction.

6.61 Show that $\cos nx$ does not tend to zero for any x. Show that if $x \ne k\pi$, then $\sin nx$ and $\sin n^2 x$ do not tend to zero. For the proof use the addition formula for the sine function and the formula $\cos 2x = 2\cos^2 x - 1$.

Chapter 7

7.3 Show that

$$|x| - p_{n+1}(x) = (|x| - p_n(x)) \cdot \left(1 - \frac{|x| + p_n(x)}{2}\right)$$

and

$$|x| - p_n(x) \le |x| \cdot \left(1 - \frac{|x|}{2}\right)^n$$

for every n and for every $|x| \le 1$, respectively.

7.7 Let (r_n) be an enumeration of $\mathbb{Q} \cap [0,1]$. For every n, choose disjoint intervals $I_{n,1}, \ldots, I_{n,n}$ such that $I_{n,k} \subset (r_k, r_k + (1/n))$ for every $k = 1, \ldots, n$. Let f_n be zero everywhere outside of the intervals $I_{n,k}$, and let the maximum of f_n be $1/k$ on $I_{n,k}$ for every $k = 1, \ldots, n$.

7.10 Show that the sequence of functions $\sin nx$ works.

7.11 Show that for every countable set $H \subset [a,b]$, (f_n) has a subsequence that is convergent at every point of H. Apply this result to the set $\mathbb{Q} \cap [a,b]$. Let $f_{n_k} \to f$ on the set $\mathbb{Q} \cap [a,b]$. Show that the function f is monotone on $\mathbb{Q} \cap [a,b]$. Extend f to $[a,b]$ as a monotone function, and let H be the set of points where f is discontinuous. Show that if a subsequence of (f_{n_k}) is convergent at the points of H, then it is convergent on the whole of $[a,b]$.

7.12 Let the limit function be f, and let $\varepsilon > 0$ be fixed. Choose $\delta > 0$ according to the uniform continuity of f. Show using the monotonicity of f_n that if $a = x_0 < \ldots < x_k = b$ is a partition of mesh $< \delta$ and $|f_n(x_i) - f(x_i)| < \varepsilon$ for every $i = 1, \ldots, k$, then $|f_n - f| < 2\varepsilon$ on $[a,b]$.

7.16 Construct x as the intersection of a sequence of nested closed intervals. Let $a_1 < b_1$, $b_1 - a_1 < 1$ be arbitrary. Choose the index n_1 such that the difference quotient of the function f_n between the points a_1, b_1 is within distance 1 of the difference quotient d_1 of the function f between these points, for every $n \ge n_1$. Choose $c_1 \in (a_1, b_1)$ such that $|f'_{n_1}(c_1) - d_1| < 1$, and let $a_1 < a_2 < c_1 < b_2 < b_1$, $b_2 - a_2 < 1/2$ be points with $|f'_{n_1}(x) - d_1| < 1$ for every $x \in [a_2, b_2]$. Choose the index $n_2 > n_1$ such that the difference quotient of the functions f_n between the points a_2, b_2 is within distance 1/2 of the difference quotient d_2 of the function f between these points, for every $n \ge n_2$. Choose $c_2 \in (a_1, b_1)$ such that $|f'_{n_2}(c_2) - d_2| < 1/2$, and let $a_2 < a_3 < c_2 < b_3 < b_2$, $b_3 - a_3 < 1/3$ be points with $|f'_{n_2}(x) - d_2| < 1$ for every $x \in [a_3, b_3]$. Show that by repeating this process we obtain a subsequence (f_{n_k}) and a point $x \in \bigcap_{k=1}^{\infty} [a_k, b_k]$ that satisfy the conditions.

7.19 Let $\lim_{n\to\infty} f_n = f$, and let f_n be uniformly equicontinuous. Let $\varepsilon > 0$ be fixed, and choose a $\delta > 0$ according to the uniform equicontinuity. Show that if f_n

is closer to f than ε at the base points of a partition of mesh $< \delta$, then $|f_n - f| < 2\varepsilon$ on $[a, b]$. Show that this implies the uniform convergence of the sequence.

If the sequence is uniformly convergent, then f is continuous. Let $\varepsilon > 0$ be fixed, and choose n_0 such that $|f_n(x) - f(x)| < \varepsilon$ holds for every $n > n_0$ and $x \in [a, b]$. Choose $\delta > 0$ according to the uniform continuity of $f_1, \ldots, f_{n_0}, f$. Show that if $x, y \in [a, b]$ and $|x - y| < \delta$, then $|f_n(x) - f_n(y)| < 3\varepsilon$ for every n and $x, y \in [a, b]$.

7.22 The statement is false even for continuous functions. Find an example in which f_n is zero outside of the interval I_n, where the intervals I_n are mutually disjoint, and the maximum of f_n on I_n is $1/n$.

7.27 Show that the value of the sum $\sum_{k=n}^{2n} (\sin kx)/k$ at the point $x = \pi/(4n)$ is larger than a positive number independent of n.

7.33 Show that $(f(y_n) - f(x_n))/(y_n - x_n)$ falls between the numbers

$$\min\left(\frac{f(x_n) - f(a)}{x_n - a}, \frac{f(y_n) - f(a)}{y_n - a}\right) \quad \text{and} \quad \max\left(\frac{f(x_n) - f(a)}{x_n - a}, \frac{f(y_n) - f(a)}{y_n - a}\right).$$

7.35 Let x be fixed. For every $n \in \mathbb{N}^+$ and $h \neq 0$ we have

$$\begin{aligned} \frac{f(x+h) - f(x)}{h} &= \sum_{i=0}^{n-1} b^i \frac{\cos(a^i(x+h)) - \cos(a^i x)}{h} + \\ &+ b^n \frac{\cos(a^n(x+h)) - \cos(a^n x)}{h} + \\ &+ \sum_{i=n+1}^{\infty} b^i \frac{\cos(a^i(x+h)) - \cos(a^i x)}{h} \stackrel{\text{def}}{=} \\ &\stackrel{\text{def}}{=} A_n(h) + B_n(h) + C_n(h). \end{aligned}$$

Construct a sequence $h_n \to 0$ such that $(f(x + h_n) - f(x))/h_n \to \infty$ using the following argument. Choose h_n such that the middle term $B_n(h_n)$ is large, and every term of $C_n(h_n)$ is nonnegative. The first term $A_n(h)$ will not cause problems, since

$$|A_n(h)| \leq \sum_{i=0}^{n-1} b^i \cdot \frac{|a^i h|}{|h|} = \frac{(ab)^n - 1}{ab - 1} < \frac{(ab)^n}{ab - 1}$$

for every $h \neq 0$, and we can choose $B_n(h)$ to be larger than this.

7.36 Show that if $n = 1$, then the function $g(x) = (a/s) \cdot \sin sx$ works, assuming that s is large enough. For $n > 1$, define the functions f_n recursively.

7.37 Let f be of the form $\sum_{n=0}^{\infty} f_n$, where each f_n is infinitely differentiable, and the series $\sum_{n=0}^{\infty} f_n^{(k)}$ of the kth derivatives is uniformly convergent on every bounded interval and for every k. Show that the functions f_n can be chosen in such a way that f satisfies the conditions of the exercise.

7.43 Write $(1+(1/n))^{n+b}$ in the form $e^{(n+b)\cdot\log(1+(1/n))}$, and then use the power series of $\log(1+x)$ around 0 and the power series of e^x around 1.

7.45 Put $F(x) = \sum_{n=0}^{\infty}\left(\frac{x^{3n+1}}{3n+1} - \frac{x^{3n+2}}{3n+2}\right)$. Clearly, F can be obtained by integrating the power series $\sum_{n=0}^{\infty}(x^{3n} - x^{3n+1})$ term by term. Find the sum of this power series, and use this to find $F(x)$; then prove that the formula of F holds for $x = 1$, and substitute $x = 1$.

7.47 Show that if $c > -1$, then the terms of the series have alternating signs from some index on and the absolute values of the terms is monotone decreasing, converging to 0. Thus the series is convergent. Then apply Abel's continuity theorem to calculate the sum.

Show that for $c \leq -1$ the terms of the series do not converge to zero.

7.48 Show that for $c > 0$ the nth partial sum of the series is $(-1)^n\binom{c-1}{n}$, and this converges to zero as $n \to \infty$.

Show that the sum of the series is infinity for $c < 0$.

7.49 Use Stirling's formula to prove the convergence of the series.

7.52 The statement is not true. Show that if $\sum_{n=0}^{\infty} a_n$ is a convergent series of nonnegative terms and if $\sum_{n=0}^{\infty} n \cdot a_n$ is divergent, then the left-hand derivative of the function $f(x) = \sum_{n=0}^{\infty} a_n \cdot x^n$ ($|x| \leq 1$) at 1 is infinity.

7.53 The statement is false.

7.58 Let $x_0 \in (-r, r)$ and $x \in (x_0 - \delta, x_0 + \delta)$ be fixed. Then we have $|x_0| + |x - x_0| < r$. By the binomial theorem we have

$$x^n = \sum_{i=0}^{n}\binom{n}{i} x_0^{n-i} \cdot (x - x_0)^i. \tag{9.1}$$

Replace x^n in the power series $\sum_{n=0}^{\infty} a_n x^n$ by the right-hand side of (9.1), then reorder the series according to the exponents of $x - x_0$. Show that all series appearing in the argument above are absolutely convergent, and thus the sum of the original series does not change during these operations.

7.60 (a) Since $a_0 = f(a) = b$, it follows that

$$g(f(x)) = \sum_{n=0}^{\infty} b_n\left(\sum_{k=1}^{\infty} a_k \cdot (x - a)^k\right)^n \tag{9.2}$$

on $(a - \delta, a + \delta)$. Let $|x - a| < \delta$. Perform the exponentiation of the nth term on the right-hand side of (9.2) (i.e., multiply the appropriate infinite series by itself n

times), then reorder the resulting series according to the exponents of $x - a$. Prove that all series appearing in the argument above are absolutely convergent, and thus the sum of the original series does not change during these operations.

7.64 Let J be a bounded closed interval in I. Show that if $x_0 \in J$ and the radius of convergence of the series $\sum_{n=0}^{\infty} \frac{f^{(n)}(x_0)}{n!}(x-x_0)^n$ is $> \delta$, then $|f^{(n)}(x_0)| \le (c \cdot n)^n$ for every n with some c depending only on δ. Then apply Theorem 7.48.

7.65 Let A_k denote the set of points $x_0 \in J$ for which there exists $\delta > 0$ such that $|f^{(n)}(x)| \le (k \cdot n)^n$ for every $x \in (x_0 - \delta, x_0 + \delta)$ and $n > 0$. By Theorem 7.59, $J = \bigcup_{k=1}^{\infty} A_k$. It is sufficient to prove that $J = A_k$ for an appropriate k. Assume that this is false, and let $x_k \in J \setminus A_k$ for every k. Choose a convergent subsequence of (x_k). Show that if this subsequence converges to x_0, then $x_0 \in \bigcup_{k=1}^{\infty} A_k$ leads to a contradiction.

7.66 Let $M_1 < M_2 < \ldots$ be positive even integers with $M_k > \max\{|f(x)|: |x| \le 2k+2\}$ $(k = 1, 2, \ldots)$. Show that the power series $\sum_{k=1}^{\infty} (x/k)^{M_k}$ is convergent everywhere. Show that if its sum is $g(x)$, then $g(x) > f(x)$ for every $|x| \ge 1$.

7.71 (iv) Apply the binomial theorem (7.17) to find the power series of $\sqrt{1-4x}$.

7.76 Let $s_n^{(0)} = s_n = a_1 + \ldots + a_n$, $s_n^{(1)} = \frac{s_1 + \ldots + s_n}{n}$, $s_n^{(2)} = \frac{s_1^{(1)} + \ldots + s_n^{(1)}}{n}$, etc. Suppose that $s_n^{(k)}$ is convergent. Changing a_1 if necessary, we may assume that $s_n^{(k)} \to 0$. Prove by induction on i that $\lim_{n\to\infty} s_n^{(k-i)}/n^i = 0$ for every $i = 1, \ldots, k$.

7.78 Suppose that $|a_n/n^k| \le C$ for every $n = 1, 2, \ldots$. Then $e^{1/(1+x)} \le e + C \cdot \sum_{n=0}^{\infty} n^k |x|^n$ for every $|x| < 1$. Show that if $|x| < 1$, then $\sum_{n=0}^{\infty} n^k x^k = p(x)/(1-x)^{k+1}$, where p is a polynomial. Thus $(1-|x|)^{k+1} \cdot e^{1/(1+x)}$ is bounded on $(-1, 1)$, which is impossible.

7.80 Use the fact (not proved in this book) that if the 2π-periodic function f is continuous everywhere and is monotone on an open interval I, then its Fourier series represents f at every point of I.

7.85 Show that

$$\sum_{j=0}^{N} \sin\left(\frac{2\pi n}{N+1} \cdot j\right) = \sum_{j=0}^{N} \cos\left(\frac{2\pi n}{N+1} \cdot j\right) = 0$$

for every integer $1 \le n \le N$. Deduce from this that $f(0) \le N+1$ for the function f given in the exercise; then apply this result to the function $f(x+c)$.

7.89 Define the function as $\sum_{k=1}^{\infty} a_k \cos kx$, with an absolutely convergent $\sum_{k=1}^{\infty} a_k$.

7.90 Show that if $\sum_{n=1}^{\infty} |a_n \cos nx| < \infty$ at a point $x = p\pi/q$, where p, q are integers and q is odd, then $\sum_{n=1}^{\infty} |a_n| < \infty$.

7.91 Show that for every n there exist c_n and α_n such that $a_n = c_n \cdot \cos\alpha_n$ and $b_n = c_n \cdot \sin\alpha_n$. Then $a_n \cos nx + b_n \sin nx = c_n \cdot \cos(nx - \alpha_n)$. Show that if $\sum_{n=1}^{\infty} |a_n \cos nx + b_n \sin nx| \le 1$ on the interval I, then $\sum_{n=1}^{\infty} \int_I |c_n| \cdot \cos^2(nx - \alpha_n)\, dx \le |I|$. Show that $\int_I \cos(2nx - 2\alpha_n)\, dx \to 0$, and then use this to prove $\sum_{n=1}^{\infty} |c_n| < \infty$. Finally, show that $|a_n| + |b_n| \le 2|c_n|$ for every n.

7.92 Let $\left|\sum_{n=1}^{N} a_n \sin nx\right| \le K$ for every x and N. Then $\left|\sum_{n=N}^{2N} a_n \sin nx\right| \le 2K$. Apply this to the point $x = \pi/(4N)$ to get that $(n \cdot a_n)$ is bounded.

For the converse, see [2, 7.2.2, p. 114].

7.93 If the series is uniformly convergent, then

$$\lim_{N\to\infty} \left| \sum_{n=N}^{2N} a_n \sin nx \right| = 0.$$

The argument applied in the previous exercise gives $n \cdot a_n \to 0$.

For the converse, see [2, 7.2.2, p. 114].

7.96 Let $\varepsilon > 0$ be fixed and let $F\colon 0 = x_0 < x_1 < \ldots < x_n = 2\pi$ be a partition with $\Omega_F(f) < \varepsilon$. Let g be a step function that takes the value m_i on (x_{i-1}, x_i) $(i = 1, \ldots, n)$. Show that the absolute value of every Fourier coefficient of $f - g$ is at most $2\pi \cdot \varepsilon$, then apply the result of the previous exercise.

7.105 In the proof of Theorem 7.93 we showed that the right-hand side of (7.46) is 2-replicative. Use the same ideas here.

7.107 (i) (b): Multiply both sides of (7.60) by $n!$. Notice that every term, except for the middle one, is present twice on the right-hand side when n is even. Notice also that $\binom{2k}{k}$ is an even number, since $\binom{2k}{k} = 2 \cdot \binom{2k-1}{k-1}$.

7.117 (i) Show that the decimal representation of $\sqrt{2}$ has infinitely many nonzero digits. Use this to prove that the value of the lim sup is 1. (ii) The value of the lim inf depends on whether the decimal representation of $\sqrt{2}$ has infinitely many zero digits. If it has infinitely many zero digits, then the value of the lim inf is zero; otherwise, its value is 1. Unfortunately, it is not known whether there are infinitely many zero digits in the decimal representation of $\sqrt{2}$ (this is an old open problem in number theory).

7.119 Show that the sequence a_n/n converges to $\liminf_{n\to\infty}(a_n/n)$.

Chapter 8

8.4 If f is linear, then the statement is obtained by applying integration by parts. Next prove the statement when f is piecewise linear. In the general case replace f by a piecewise linear function that is equal to f at the base points of a fine partition of $[a, b]$, and estimate the difference.

8.10 First show that $\int_\delta^\infty (f(ax)-f(bx))/x\,dx = \int_{a\delta}^{b\delta}(f(t)/t)\,dt$ for every $\delta > 0$. Find the limit of this integral, using that $f(0)-\varepsilon < f(t) < f(0)+\varepsilon$ for every t small enough.

8.15 Show that the improper integral $\int_0^{\pi/2} \log(t\cdot\sin^2 x+1)\,dx$ is uniformly convergent on the interval $[-1,0]$.

8.16 Use the substitution $x = y/(1+y)$ in the integral defining $B(t,s)$.

8.17 (i) Apply the result of the previous exercise with $s = 1-t$; then apply Theorem 8.43. (ii) Use the substitution $x^t = y$.

8.18 Use (8.54) with $t = 1/2$ and $s = x$. If $2x-1$ is an even integer, then use the formula $2\sqrt{\pi}\cdot\Gamma(2x) = 4^x\cdot\Gamma(x)\Gamma(x+1/2)$ (see [7, exercise 19.46]).

8.19 Use (8.54) with $t = 1-x$ and $s = x$; then apply Theorems 8.42 and 8.43.

8.22 Reduce the statement to (8.63) using integration by parts.

8.23 Reduce the statement to (8.63) using the substitution $x^s = y$.

Solutions

Chapter 1

1.9 There is no such set, since $x \in \text{int}\, A$ implies that $B(x, r) \subset \text{int}\, A$, with an appropriate $r > 0$.

1.10 (i) Let $x \notin \partial A \cup \partial B$; we need to prove that $x \notin \partial(A \cup B)$. If $B(x, r) \subset A$ or $B(x, r) \subset B$ for some $r > 0$, then $B(x, r) \subset A \cup B$, and thus $x \in \text{int}(A \cup B)$, and $x \notin \partial(A \cup B)$. Suppose that $B(x, r) \not\subset A$ and $B(x, r) \not\subset B$ for every $r > 0$. The condition $x \notin \partial A \cup \partial B$ yields $B(x, r) \cap A = \emptyset$ and $B(x, r) \cap B = \emptyset$ for an appropriate $r > 0$. Then $B(x, r) \cap (A \cup B) = \emptyset$, whence $x \notin \partial(A \cup B)$.

(ii) Let $H^c = \mathbb{R}^p \setminus H$ for every set $H \subset \mathbb{R}^p$. By the definition of a boundary point, $\partial H^c = \partial H$ for every $H \subset \mathbb{R}^p$. Applying (i) to the sets A^c, B^c gives

$$\partial(A \cap B) = \partial((A \cap B)^c) = \partial(A^c \cup B^c) \subset \partial A^c \cup \partial B^c = \partial A \cup \partial B.$$

1.22 By Theorem 1.17 it is enough to prove that if $x_n \in A'$ and $x_n \to x$, then $x \in A'$, i.e., x is a limit point of the set A. For every $\varepsilon > 0$ we have $x_n \in B(x, \varepsilon/2)$ if n is large enough. For such an n the set $B(x_n, \varepsilon/2) \cap A$ is infinite, since $x_n \in A'$. Since $B(x_n, \varepsilon/2) \subset B(x, \varepsilon)$, it follows that $B(x, \varepsilon) \cap A$ is infinite. This holds for every $\varepsilon > 0$, and thus $x \in A'$.

1.36 (a) Let $A \subset \mathbb{R}^2$ be the graph of $1/x$, and let B be the x-axis.

(b) Let $A = \{n + (1/n) : n = 2, 3, \ldots\}$, and let $B = \mathbb{N}$.

1.52 The polynomial $x^2 + (xy - 1)^2$ works.

1.59 Since h is continuous on A, h is bounded there by Weierstrass's theorem. Suppose that $|h(x)| \le M$ for every $x \in A$. By Weierstrass's approximation theorem for single-variable continuous functions, there exists a single-variable polynomial s such that $||t| - s(t)| < \varepsilon/M$ for every $|t| \le 1$. (This also follows from Exercise 7.3.) The function $g(x) = M \cdot s(h(x)/M)$ is a polynomial, and

$$||h(x)| - g(x)| = M \cdot \left| \left| \frac{h(x)}{M} \right| - s\left(\frac{h(x)}{M} \right) \right| < M \cdot \frac{\varepsilon}{M} = \varepsilon$$

for every $x \in A$.

1.60 First, let $n = 2$. By the previous exercise, there is a polynomial g such that $||h_1(x) - h_2(x)| - g(x)| < \varepsilon$ for every $x \in A$. Then the polynomials $g_1 = (h_1 + h_2 + g)/2$ and $g_2 = (h_1 + h_2 - g)/2$ satisfy the conditions, since

$$\max(h_1(x), h_2(x)) = ((h_1(x) + h_2(x)) + |h_1(x) - h_2(x)|)/2$$

and

$$\min(h_1(x), h_2(x)) = ((h_1(x) + h_2(x)) - |h_1(x) - h_2(x)|)/2$$

for every x. In the general case the statement of the exercise follows by induction on n.

1.61 If $a = b$, then let $g_{a,b}$ be the constant function equal to $f(a)$. If $a = (a_1, \ldots, a_p) \neq b = (b_1, \ldots, b_p)$, then there is an index $1 \le i \le p$ such that $a_i \neq b_i$. It is easy to see that $g_{a,b}(x) = c \cdot x_i + d$ satisfies the conditions, for suitable real numbers c, d.

1.62 For every $b \in A$ let $g_{a,b}$ be as in the previous exercise. Since $g_{a,b}$ is continuous and $g_{a,b}(b) = f(b)$, there exists a neighborhood $U(b)$ of b such that $g_{a,b}(x) > f(x) - (\varepsilon/2)$ for every $x \in U(b)$. The open sets $U(b)$ $(b \in A)$ cover A, and thus by Borel's theorem, finitely many of these also cover A. Let $b_1, \ldots, b_n$ be points of A such that $A \subset \bigcup_{i=1}^n U(b_i)$. Let $G_a = \max(g_{a,b_1}, \ldots, g_{a,b_n})$. Then $G_a(a) = f(a)$, and $G_a(x) > f(x) - (\varepsilon/2)$ for every $x \in A$. By the statement of Exercise 1.60, there exists a polynomial g_a such that $|G_a(x) - g_a(x)| < \varepsilon/2$ for every $x \in A$. Clearly, g_a satisfies the conditions.

1.63 Let g_a be as in the previous exercise. Since g_a is continuous and $g_a(a) < f(a) + \varepsilon$, there exists a neighborhood $V(a)$ of a such that $g_a(x) < f(x) + \varepsilon$ for every $x \in V(a)$. The open sets $V(a)$ $(a \in A)$ cover A, and thus by Borel's theorem, there exist finitely many of them that still cover A. Let $a_1, \ldots, a_k$ be points of A such that $A \subset \bigcup_{i=1}^k V(a_i)$. Let $G = \min(g_{a_1}, \ldots, g_{a_k})$. Then $f(x) - \varepsilon < G(x) < f(x) + \varepsilon$ for every $x \in A$. By the statement of Exercise 1.60, there is a polynomial g such that $|G(x) - g(x)| < \varepsilon$ for every $x \in A$. Then we have $|f(x) - g(x)| < 2\varepsilon$ for every $x \in A$.

1.71 Let f be a single-variable polynomial that takes its local maxima at a and b $(a \neq b)$ (e.g., the polynomial $2x^2 - x^4$ takes its strict and absolute maximum at the points ± 1). The two-variable polynomial $p(x, y) = f(x) - y^2$ has local maxima at $(a, 0)$ and $(b, 0)$. But p does not have a local minimum, since the section function $p_c(y) = f(c) - y^2$ does not have a local minimum at $y = d$ for any (c, d).

1.80 (h) The answer is yes. First we show that $v - u \le \sqrt[3]{4(v^3 - u^3)}$ for every $u < v$. It is easy to see that $(v-u)^2 \le 4(u^2 + uv + v^2)$. Multiplying this by $(v - u)$ and taking the cube root of both sides yields the desired inequality.

Let $f(x, y) = \sqrt[3]{x^3 + y^4}$. We have $f'_x(0,0) = 1$ and $f'_y(0,0) = 0$. Thus, in order to prove that f is differentiable at the origin we have to show that

$$\lim_{(x,y)\to(0,0)} \frac{\sqrt[3]{x^3 + y^4} - x}{\sqrt{x^2 + y^2}} = 0. \tag{9.3}$$

Applying the previous inequality to $u = x$ and $v = \sqrt[3]{x^3 + y^4}$ gives

$$0 \le \frac{\sqrt[3]{x^3 + y^4} - x}{\sqrt{x^2 + y^2}} \le \frac{\sqrt[3]{4 \cdot ((x^3 + y^4) - x^3)}}{\sqrt{x^2 + y^2}} = \frac{\sqrt[3]{4} \cdot y^{4/3}}{\sqrt{x^2 + y^2}} \to 0.$$

Then, by the squeeze theorem, we get (9.3).

1.82 If f is differentiable at a, then $f(x) = \alpha \cdot (x-a) + f(a) + \varepsilon(x)(x-a)$, where $\alpha = f'(a)$ and $\lim_{x \to a} \varepsilon(x) = 0$. For every $b \in \mathbb{R}$ and $(x, y) \in \mathbb{R}^2$, we have

$$g(x,y) = \alpha \cdot (x-a) + 0 \cdot (y-b) + \eta(x,y),$$

with $\eta(x,y) = \varepsilon(x) \cdot (x-a)$. Since

$$\frac{|\eta(x,y)|}{|(x,y)-(a,b)|} \le \frac{|\varepsilon(x)| \cdot |x-a|}{|x-a|} = |\varepsilon(x)|,$$

we have

$$\lim_{(x,y)\to(a,b)} \frac{\eta(x,y)}{|(x,y)-(a,b)|} = 0,$$

and thus g is differentiable at (a, b).

Chapter 2

2.7 These are the functions $g(x+y)$ where g is differentiable. Indeed, for c fixed, the derivative of $h_c(x) = f(x, c-x)$ is $D_1 f(x, c-x) + D_2 f(x, c-x) \cdot (-1) = 0$ by $D_1 f = D_2 f$. Thus $h_c' = 0$, and h_c is constant. If $h_c = g(c)$, then $f(x, c-x) = g(c)$ for every c and x, and the substitution $c = x + y$ gives $f(x,y) = g(x+y)$.

2.13 The statement is false. Let $f(t) = (\cos t, \sin t)$ for every $t \in \mathbb{R}$. Now $f(0) = f(2\pi)$, but there is no $c \in \mathbb{R}$ such that $f'(c)$ is the vector $(0, 0)$, since $|f'(c)| = 1$ for every c.

2.17 (a) The determinant of the Jacobian matrix at the point (x, y) is $e^{2x} \neq 0$.
(b) Suppose that $f(x,y) = f(u,v)$. Then $e^x = |f(x,y)| = |f(u,v)| = e^u$, and thus $x = u$. From $e^x \cos y = e^u \cos v$ and $e^x \sin y = e^u \sin v$ we get $\cos y = \cos v$ and $\sin y = \sin v$, which implies $v = y + 2k\pi$ with an integer k. If (x, y) and (u, v) are points of an open disk of radius π, then $|v - y| < 2\pi$, and $y = v$. (In fact, this argument shows that f is injective in every horizontal open strip of width 2π.)
(c) The function $\varphi(x,y) = \left(\log\sqrt{x^2+y^2}, \operatorname{arc\,tg}(y/x)\right)$ works.

Chapter 3

3.5 There are boxes $R_1, \ldots, R_n$ such that $\sum_{i=1}^n \mu(R_i) < \overline{\mu}(A) + \varepsilon/2$. Let $R_i = [a_{i,1}, b_{i,1}] \times \ldots \times [a_{i,p}, b_{i,p}]$ for every i, and put

$$R_{i,\delta} = [a_{i,1} - \delta, b_{i,1} + \delta] \times \ldots \times [a_{i,p} - \delta, b_{i,p} + \delta]$$

$(i = 1, \ldots, n)$. It is clear that if δ is small enough, then $\mu(R_{i,\delta}) < \mu(R_i) + \varepsilon/(2n)$ for every i. Then $G = \bigcup_{i=1}^n \operatorname{int} R_{i,\delta}$ is open,

$$A \subset \bigcup_{i=1}^n R_i \subset \bigcup_{i=1}^n \operatorname{int} R_{i,\delta} = G,$$

and

$$\overline{\mu}(G) \leq \sum_{i=1}^{n} \mu(R_{i,\delta}) < \sum_{i=1}^{n} \mu(R_i) + (\varepsilon/2) < \overline{\mu} + \varepsilon.$$

3.13 Let $A = \bigcup_{n=1}^{\infty}(a_n, b_n)$. Suppose that $I_1, \ldots, I_k \subset A$ are nonoverlapping closed intervals. By Borel's theorem, there exists N such that $\bigcup_{i=1}^{k} I_i \subset \bigcup_{n=1}^{N}(a_n, b_n)$. Then, by Theorem 3.6,

$$\sum_{i=1}^{k} |I_i| \leq \sum_{n=1}^{N}(b_n - a_n) < \sum_{n=1}^{\infty}(b_n - a_n).$$

Since this holds for every finite system of nonoverlapping closed intervals $I_1, \ldots, I_k$ of A, we have $\underline{\mu}(A) \leq \sum_{n=1}^{\infty}(b_n - a_n)$ by the definition of the inner measure.

3.14 Let (r_n) be an enumeration of $\mathbb{Q} \cap [0,1]$. If J_n is an open interval containing r_n and shorter than $\varepsilon/2^n$, then the set $G = \bigcup_{n=1}^{\infty} J_n$ works. Indeed, by the previous exercise we have $\underline{\mu}(G) < \varepsilon$, and $\mathbb{Q} \cap [0,1] \subset G$ implies $\overline{\mu}(G) \geq 1$.

3.15 Let G be the set defined in the previous exercise, with $\varepsilon = 1/2$. Then $F = [0,1] \setminus G$ is closed. Since $\mathbb{Q} \cap [0,1] \subset G$, the interior of F is empty, and thus $\underline{\mu}(F) = 0$. Now, $\overline{\mu}(F) = 1 - \underline{\mu}([0,1] \cap G) \geq 1 - (1/2)$ (see Exercise 3.10). Modifying the construction, we can have $\overline{\mu}(F) > 1$.

3.17 (a) The constant 1 set function is nonadditive but is translation-invariant, normalized, and nonnegative.

(b) Let $H = \bigcup_{n \in \mathbb{Z}}[n, n + (1/2)]$, and let $m(A) = 2 \cdot \mu(A \cap H)$ for every Jordan measurable set $A \subset \mathbb{R}$. The function m is not translation-invariant, but is additive, normalized, and nonnegative. We can construct a similar example in $\mathbb{R}^p$ as well.

(c) The set function $m(A) = 2 \cdot \mu(A)$ is not normalized, but it is additive, translation-invariant, and nonnegative.

(d) Constructing an additive, translation-invariant, normalized function that is not nonnegative is more difficult. We need to use the fact that for every irrational $\alpha > 0$ there exists a function $f\colon \mathbb{R} \to \mathbb{R}$ such that f is additive (i.e., $f(x + y) = f(x) + f(y)$ for every $x, y \in \mathbb{R}$), $f(1) = 1$, and $f(\alpha) = -1$. (Such a function is necessarily discontinuous at every point and not bounded on any interval; see [7, Exercise 10.94]. Let $m(A) = f(\mu(A))$ for every Jordan measurable set $A \subset \mathbb{R}$. The function m is translation-invariant, additive, and normalized. However, $m(A) = -1$ for every A with $\mu(A) = \alpha$.

3.26 The numbers 1/4, 3/4, 1/10, 3/10, 7/10, 9/10 are in C and have finite decimal representations. Find more!

3.37 If Λ is not invertible, then its range is a proper linear subspace of $\mathbb{R}^p$ that does not have interior points. (It is easy to see that the linear space generated by a ball of $\mathbb{R}^p$ is $\mathbb{R}^p$ itself.) In this case the sets $\Lambda(A)$ and $\Lambda(B)$ have no interior points either; thus they are nonoverlapping.

Let Λ be invertible, and suppose that $\Lambda(A)$ and $\Lambda(B)$ have a common interior point $y \in \operatorname{int}\Lambda(A) \cap \operatorname{int}\Lambda(B)$. Then $B(y,\varepsilon) \subset \Lambda(A) \cap \Lambda(B)$ for a suitable $\varepsilon > 0$. Let $\Lambda^{-1}(y) = x$. Since Λ is continuous at x, there exists $\delta > 0$ such that $\Lambda(B(x,\delta)) \subset B(y,\varepsilon) \subset \Lambda(A) \cap \Lambda(B)$. Since Λ is invertible, it follows that $B(x,\delta) \subset A \cap B$. This is a contradiction, since we assumed that A and B were nonoverlapping.

Chapter 5

5.17 Let $g\colon [a,b] \to \mathbb{R}^2$ be a simple closed curve, let $\Gamma = g([a,b])$, and let a number $0 < \delta < \operatorname{diam}\Gamma$ be fixed. There exists $\delta_1 > 0$ such that if $x, y \in \Gamma$ and $|x-y| < \delta_1$, then one of the arcs connecting x and y in Γ has diameter less than δ. Let $P = (p_0, p_1, \ldots, p_{n-1}, p_n = p_0)$ be a polygon finer than δ_1 (i.e., $|p_{i-1} - p_i| < \delta_1$ for every $i = 1, \ldots, n$) with minimal number of vertices. We show that P does not intersect itself. Suppose this is not true, that is, that there are segments $[p_{i-1}, p_i]$ and $[p_{j-1}, p_j]$ intersecting each other, where $1 \le i$ and $i+1 < j \le n$. Then

$$|p_{i-1} - p_{j-1}| + |p_i - p_j| \le |p_{i-1} - p_i| + |p_{j-1} - p_j| < 2\delta_1,$$

and thus $\min(|p_{i-1} - p_{j-1}|, |p_i - p_j|) < \delta_1$. If $|p_{i-1} - p_{j-1}| < \delta_1$, then one of the two arcs of the curve Γ between p_{i-1} and p_{j-1} has diameter less than δ. Delete the part of the polygon P that lies on this arc, and replace it by the segment $[p_{i-1}, p_{j-1}]$. We have obtained a polygon in Γ finer than δ_1 and having a smaller number of vertices than P, which is impossible. We handle the case $|p_i - p_j| < \delta_1$ similarly.

Chapter 6

6.2 (a) Since $1/(n^2+2n) = (1/2)\cdot(1/n - 1/(n+2))$, we have

$$\sum_{n=1}^{N} \frac{1}{n^2+2n} = \frac{1}{2}\cdot\sum_{n=1}^{N}\left(\frac{1}{n} - \frac{1}{n+2}\right) = \frac{1}{2}\cdot\left(1 + \frac{1}{2} - \frac{1}{N+1} - \frac{1}{N+2}\right).$$

Thus the partial sums of the series tend to 3/4, so the series is convergent with sum 3/4.

(b) If we leave out the first term in the series in (a), then we get the series in (b). Thus the partial sums of this new series tend to $(3/4) - (1/3) = 5/12$, so it is convergent with sum 5/12.

(c) Since $1/(n^3 - n) = (1/2)\cdot(1/(n-1) - 2/n + 1/(n+1))$, we have

$$\sum_{n=2}^{N} \frac{1}{n^3-n} = \frac{1}{2}\cdot\sum_{n=2}^{N}\left(\frac{1}{n-1} - \frac{2}{n} + \frac{1}{n+1}\right) = \frac{1}{2}\cdot\left(1 - \frac{1}{2} - \frac{1}{N} + \frac{1}{N+1}\right).$$

Thus the partial sums of the series tend to 1/4, so the series is convergent with sum 1/4.

6.5 Let s_n be the nth partial sum of the series. Then

$$\begin{aligned}a_1 + 2a_2 + \ldots + na_n &= (a_1 + \ldots + a_n) + (a_2 + \ldots + a_n) + \ldots + (a_n) = \\ &= s_n + (s_n - s_1) + \ldots + (s_n - s_{n-1}) = \\ &= (n+1)s_n - (s_1 + \ldots + s_n)\end{aligned}$$

and

$$\frac{a_1 + 2a_2 + \ldots + na_n}{n} = \frac{n+1}{n} s_n - \frac{s_1 + \ldots + s_n}{n}. \tag{9.4}$$

If $\lim_{n\to\infty} s_n = A$, then $(s_1 + \ldots + s_n)/n \to A$, and so the right-hand side of (9.4) converges to zero.

6.17 Let (r_n) be an enumeration of the rational numbers, with $r_0 = 0$. We show that the series $\sum_{n=1}^{\infty}(r_n - r_{n-1})$ satisfies the condition. Indeed, for every real number A there exists a sequence r_{n_k} of distinct rational numbers that converges to A. By reordering the sequence we may assume that $0 = n_0 < n_1 < n_2 < \ldots$. The kth partial sum of $\sum_{k=1}^{\infty}\left(\sum_{n=n_{k-1}+1}^{n_k}(r_n - r_{n-1})\right)$ is r_{n_k}. Since $r_{n_k} \to A$ as $k \to \infty$, the bracketed series is convergent, and its sum is A.

6.56

$$\frac{a_n}{n} = \frac{s_1 + \ldots + s_n}{n} - \frac{s_1 + \ldots + s_{n-1}}{n} \to A - A = 0.$$

6.60 Let s_n be the nth partial sum of the series. By assumption, $(s_1 + \ldots + s_n)/n$ is convergent. Changing the first element of the series appropriately, we may assume that $(s_1 + \ldots + s_n)/n$ converges to zero. We may assume that $n \cdot a_n$ is bounded from below; otherwise, we switch to the series $\sum_{n=1}^{\infty}(-a_n)$. Multiplying the series by a suitable positive number, we may also assume that $a_n \ge -1/n$ for every n.

We need to prove that if $a_n \ge -1/n$ for every n and $(s_1 + \ldots + s_n)/n \to 0$, then $s_n \to 0$. Let $0 < \varepsilon < 1/2$ be given, and suppose that $s_n \ge 2\varepsilon$ for infinitely many n. If $s_n \ge 2\varepsilon$ and $k \le \varepsilon n$, then

$$s_{n+k} = s_n + a_{n+1} + \ldots + a_{n+k} \ge 2\varepsilon - k \cdot \frac{1}{n+1} > \varepsilon.$$

This implies

$$(s_1 + \ldots + s_{n+[\varepsilon n]}) - (s_1 + \ldots + s_n) > [\varepsilon n] \cdot \varepsilon > \varepsilon^2 n - \varepsilon,$$

which is impossible for n large enough, since $(s_1 + \ldots + s_{n+[\varepsilon n]})/n \to 0$, and $(s_1 + \ldots + s_n)/n \to 0$.

Now let $s_n \le -2\varepsilon$ for infinitely many n. If $s_n \le -2\varepsilon$ and $k < \varepsilon n/2$, then

$$s_{n-k} = s_n - a_n - \ldots - a_{n-k+1} \le -2\varepsilon + k \cdot \frac{1}{n-k} < -\varepsilon.$$

Thus

$$(s_1 + \ldots + s_n) - (s_1 + \ldots + s_{n-[\varepsilon n/2]}) < -[\varepsilon n/2] \cdot \varepsilon < -\frac{\varepsilon^2 n}{2} + \varepsilon,$$

which is impossible for n large enough, since $(s_1 + \ldots + s_{n-[\varepsilon n]})/n \to 0$, and $(s_1 + \ldots + s_n)/n \to 0$. We have shown that $|s_n| < 2\varepsilon$ for every n large enough; thus $s_n \to 0$.

6.61 We prove that $\cos nx$ does not converge to zero for any x. Indeed, if $\cos nx \to 0$, then $\cos 2nx = 2\cos^2 nx - 1 \to -1$, which is a contradiction.

Next we show that $\sin nx$ and $\sin n^2 x$ do not converge to zero for any $x \neq k\pi$. Indeed, if $\sin nx \to 0$, then

$$\cos nx \cdot \sin x = \sin(n+1)x - \sin nx \cdot \cos x \to 0.$$

Since $\cos nx$ does not converge to zero, it follows that $\sin x = 0$ and $x = k\pi$. Finally, if $\sin n^2 x \to 0$, then

$$\cos n^2 x \cdot \sin(2n+1)x = \sin(n+1)^2 x - \sin n^2 x \cdot \cos(2n+1)x \to 0.$$

Since $|\cos n^2 x| = \sqrt{1 - \sin^2 n^2 x} \to 1$, we have $\sin(2n+1)x \to 0$. Thus,

$$2\sin 2nx \cdot \cos x = \sin(2n+1)x + \sin(2n-1)x \to 0,$$

and either $\sin 2nx \to 0$ or $\cos x = 0$. In both cases $x = k\pi/2$. If k is odd, then $\sin n^2 x = \pm 1$ for every n odd, which is impossible. Hence k is even, and $x = (k/2) \cdot \pi$.

Chapter 7

7.17 Let $\varepsilon > 0$ be fixed. By the Cauchy criterion (Theorem 7.9), there exists N such that $|f_n'(x) - f_m'(x)| < \varepsilon$ for every $x \in I$ and $n, m \geq N$. Since $(f_n(x_0))$ is convergent, there exists an index M such that $|f_n(x_0) - f_m(x_0)| < \varepsilon$ for every $n, m \geq M$. Let $n, m \geq \max(N, M)$ and $x \in I$. Denote the function $f_n - f_m$ by $h_{n,m}$. By our choices of N and M we have $|h_{n,m}'(x)| < \varepsilon$ for every $x \in I$, and $|h_{n,m}(x_0)| < \varepsilon$. By the mean value theorem, for every $x \in I$ there is $c \in [x_0, x]$ such that $h_{n,m}(x) - h_{n,m}(x_0) = h_{n,m}'(c) \cdot (x - x_0)$. Thus

$$|h_{n,m}(x)| \leq \varepsilon \cdot |x - x_0| + |h_{n,m}(x_0)| < \varepsilon \cdot |I| + \varepsilon,$$

where $|I|$ is the length of interval I. We have proved that $|f_n(x) - f_m(x)| < (|I| + 1) \cdot \varepsilon$ holds for every $x \in I$ and $n, m \geq N_1$, where $N_1 = \max(N, M)$. Since ε was arbitrary, it follows from Theorem 7.9 that the sequence of functions (f_n) converges uniformly to some function f on I.

We have to prove $f'(x) = g(x)$ for every $x \in I$. Let $\varepsilon > 0$ be fixed, and let N be as above. For $n, m \geq N$, we have $|h'_{n,m}(x)| < \varepsilon$ for every $x \in I$ and then, by the mean value theorem,

$$|h_{n,m}(y) - h_{n,m}(x)| \leq \varepsilon \cdot |y - x|$$

for every $x, y \in I$. Since $\lim_{m\to\infty} h_{n,m}(z) = f_n(z) - f(z)$ for every $z \in I$,

$$|f_n(y) - f(y) - f_n(x) + f(x)| \leq \varepsilon \cdot |y - x|$$

follows, and thus

$$\left| \frac{f_n(y) - f_n(x)}{y - x} - \frac{f(y) - f(x)}{y - x} \right| \leq \varepsilon \tag{9.5}$$

for every $x, y \in I$, $y \neq x$, and $n \geq N$. Fix x. Since $f'_n(x) \to g(x)$, we can choose an index $n \geq N$ such that $|f'_n(x) - g(x)| < \varepsilon$ holds. According to the definition of the derivative, there exists $\delta > 0$ such that

$$\left| \frac{f_n(y) - f_n(x)}{y - x} - f'_n(x) \right| < \varepsilon$$

for every $y \in (x - \delta, x + \delta) \setminus \{x\}$. Comparing this with (9.5), we get

$$\left| \frac{f(y) - f(x)}{y - x} - g(x) \right| \leq \varepsilon + \left| \frac{f_n(y) - f_n(x)}{y - x} - f'_n(x) \right| + |f'_n(x) - g(x)| < 3\varepsilon$$

for $y \in (x - \delta, x + \delta)$, $y \neq x$. Since ε was arbitrary, f is differentiable at x, and $f'(x) = g(x)$.

7.33 Let

$$m_n = \min\left(\frac{f(x_n) - f(a)}{x_n - a}, \frac{f(y_n) - f(a)}{y_n - a} \right)$$

and

$$M_n = \max\left(\frac{f(x_n) - f(a)}{x_n - a}, \frac{f(y_n) - f(a)}{y_n - a} \right)$$

for all n. It is clear that $m_n \to f'(a)$ and $M_n \to f'(a)$ if $n \to \infty$. Let $p_n = (a - x_n)/(y_n - x_n)$ and $q_n = (y_n - a)/(y_n - x_n)$. Then $p_n, q_n > 0$ and $p_n + q_n = 1$. Since

$$\frac{f(y_n) - f(x_n)}{y_n - x_n} = p_n \cdot \frac{f(a) - f(x_n)}{a - x_n} + q_n \cdot \frac{f(y_n) - f(a)}{y_n - a},$$

we have $m_n \leq (f(y_n) - f(x_n))/(y_n - x_n) \leq M_n$. Thus the statement follows by the squeeze theorem.

7.35 We proved the continuity of f in Example 7.37. Let x be fixed. For every $n \in \mathbb{N}^+$ and $h \neq 0$ we have

$$\begin{aligned}\frac{f(x+h)-f(x)}{h} &= \sum_{i=0}^{n-1} b^i \frac{\cos(a^i(x+h)) - \cos(a^i x)}{h} + \\ &+ b^n \frac{\cos(a^n(x+h)) - \cos(a^n x)}{h} + \\ &+ \sum_{i=n+1}^{\infty} b^i \frac{\cos(a^i(x+h)) - \cos(a^i x)}{h} \stackrel{\text{def}}{=} \\ &\stackrel{\text{def}}{=} A_n(h) + B_n(h) + C_n(h).\end{aligned}$$

We construct a sequence $h_n \to 0$ with $(f(x+h_n) - f(x))/h_n \to \infty$. The idea is to choose h_n such that $B_n(h_n)$ is large and every term of $C_n(h_n)$ is nonnegative. The first sum $A_n(h_n)$ will not cause any problems. Indeed, by Theorem 10.24 the cosine function has the Lipschitz property, and

$$|A_n(h_n)| \leq \sum_{i=0}^{n-1} b^i \cdot \frac{|a^i h_n|}{|h_n|} = \frac{(ab)^n - 1}{ab-1} < \frac{(ab)^n}{ab-1}$$

for every $h_n \neq 0$. The quantity $B_n(h_n)$ can be chosen to be larger than that.

We distinguish two cases. If $\cos(a^n x) \leq 0$, then let h_n be the smallest positive number such that $a^n(x+h_n) = 2k\pi$, where k is an integer. In other words, let k be the integer such that $2(k-1)\pi \leq a^n x < 2k\pi$, and let $h_n = (2k\pi - a^n x)/a^n$. Then $0 < h_n < 2\pi/a^n$. Since the numerator of the fraction in the definition of $B_n(h_n)$ is at least 1, we must have

$$B_n(h_n) \geq b^n \cdot \frac{1}{h_n} > b^n \cdot \frac{a^n}{2\pi} = \frac{(ab)^n}{2\pi}.$$

If $i > n$, then $a^i(x+h_n)$ also has the form $2m\pi$ with an integer m, since a is an integer. Thus the numerators of the terms of the series defining $C_n(h_n)$ are nonnegative, and then the sum itself is nonnegative. Therefore, we have

$$\frac{f(x+h_n) - f(x)}{h_n} > -\frac{(ab)^n}{ab-1} + \frac{(ab)^n}{2\pi} + 0 = (ab)^n \left(\frac{1}{2\pi} - \frac{1}{ab-1} \right). \quad (9.6)$$

Next suppose that $\cos(a^n x) > 0$. Let h_n be the largest negative number such that $a^n(x+h_n) = (2k-1)\pi$, where k is an integer. In other words, let k be the integer such that $(2k-1)\pi < a^n x \leq (2k+1)\pi$, and let $h_n = ((2k-1)\pi - a^n x)/a^n$. Then $0 > h_n > -2\pi/a^n$. Since the numerator of the fraction in the definition of $B_n(h_n)$ is at most -1, we have

$$|B_n(h_n)| \ge b^n \cdot \frac{1}{|h_n|} > b^n \cdot \frac{a^n}{2\pi} = \frac{(ab)^n}{2\pi}.$$

If $i > n$, then $a^i(x + h_n)$ also has the form $(2m-1)\pi$ with an integer m, since a was an odd integer. Thus the numerators of the terms of the series defining $C_n(h_n)$ are nonpositive, and then the sum itself is again nonnegative, since $h_n < 0$. Thus (9.6) holds in this case as well. We have defined h_n for every n. By assumption, $ab - 1 > 2\pi$, and thus (9.6) implies $(f(x+h_n) - f(x))/h_n \to \infty$ as $n \to \infty$. Clearly, this implies that f is not differentiable at x. Since x was arbitrary, f is nowhere differentiable.

7.36 If $n = 1$, then $g(x) = (a/s) \cdot \sin sx$ is a function satisfying the conditions, assuming that $s > (|a|+1)/\varepsilon$.

Let $n > 1$, and suppose that g is infinitely differentiable, $g^{(n-1)}(0) = a$, $g^{(i)}(0) = 0$ for every $0 \le i < n-1$, and $|g^{(i)}(x)| < \varepsilon/K$ for every $0 \le i < n-1$ and $|x| < K$. Obviously, the function $G(x) = \int_0^x g(t)\,dt$ is infinitely differentiable, $G^{(n)}(0) = a$, $G^{(i)}(0) = 0$ for every $0 \le i < n$, and $|G^{(i)}(x)| < \varepsilon$ for every $0 \le i < n$ and $|x| < K$.

7.37 We will give f in the form $\sum_{n=0}^{\infty} f_n$, where f_n is infinitely differentiable for every n, and the series $\sum_{n=0}^{\infty} f_n^{(k)}$ of the kth derivatives is uniformly convergent on every bounded interval and for every k. By Theorem 7.42 it follows that f is infinitely differentiable, and

$$f^{(k)}(x) = \sum_{n=0}^{\infty} f_n^{(k)}(x) \tag{9.7}$$

for every x. Let the sequence (a_n) be given, and let f_0 be the constant function equal to a_0. If $n > 0$ and the functions $f_0, \ldots, f_{n-1}$ are already defined, then choose a function f_n that is infinitely differentiable and such that

$$f_n^{(n)}(0) = a_n - \sum_{i=0}^{n-1} f_i^{(n)}(0),$$

$f_n^{(i)}(0) = 0$ for every $0 \le i < n$, and $|f_n^{(i)}(x)| < 1/2^n$ for every $0 \le i < n$ and $|x| < n$. (By the previous exercise, such a function exists.) In this way we have defined functions f_n for every n.

The series on the right-hand side of (9.7) is uniformly convergent on the interval $(-K, K)$ for every $K > 0$. Indeed, if $n > \max(K, k)$, then $|f_n^{(k)}(x)| < 1/2^n$ follows from our construction, for every $|x| < K$. Thus we can apply the Weierstrass criterion. Then it follows from Theorem 7.42 that f is infinitely differentiable. By (9.7) we have $f^{(k)}(0) = a_k$, since the right-hand side of (9.7) at $x = 0$ is

$$\sum_{n=0}^{k-1} f_n^{(k)}(0) + f_k^{(k)}(0) + \sum_{n=k+1}^{\infty} 0 = a_k.$$

7.102 The denominator of B_{2n} is the product of the primes p for which $p-1 \mid 2n$. This follows from the fact that $B_{2n} - \sum_{p-1|2n} 1/p$ is an integer. For a proof of this statement see [5, Theorem 118, p. 91].

Chapter 8

8.8 Substituting $x = e^t$ gives

$$\int \frac{x-1}{\log x}\, dx = \int \frac{e^{2t} - e^t}{t}\, dt.$$

Therefore, it is enough to prove that the right-hand side is not elementary. Using the statement given in the exercise, this follows from the fact that the integral $\int (e^t/t)\, dt$ is not elementary. This is proved in [7, Examples 15.32.1].

8.14 The two-variable function $\log(1 + t \cdot \sin^2 x)$ is continuous on the set $A = (-1, \infty) \times [0, \pi/2]$, and thus the integral exists for every $t > -1$. Since the partial derivative

$$\frac{\partial}{\partial t} \log(t \cdot \sin^2 x + 1) = \frac{\sin^2 x}{t \cdot \sin^2 x + 1}$$

is also continuous on the set A, it follows from Theorem 8.24 that F is differentiable, and

$$F'(t) = \int_0^{\pi/2} \frac{\sin^2 x}{t \cdot \sin^2 x + 1}\, dx$$

for every $t > -1$. We can calculate the integral on the right-hand side by the substitution $\operatorname{tg} x = u$. For $t > -1$ and $t \neq 0$, we get

$$F'(t) = \int_0^\infty \frac{u^2/(1+u^2)}{(tu^2/(1+u^2)) + 1} \cdot \frac{1}{1+u^2}\, du = \int_0^\infty \frac{u^2}{((t+1)u^2+1)(u^2+1)}\, dx =$$
$$= \frac{1}{t} \cdot \int_0^\infty \left[\frac{1}{u^2+1} - \frac{1}{(t+1)u^2+1}\right] du = \frac{1}{t} \cdot \left[\operatorname{arc\,tg} u - \frac{\operatorname{arc\,tg}(\sqrt{t+1} \cdot u)}{\sqrt{t+1}}\right]_0^\infty =$$
$$= \frac{\pi}{2}\left(\frac{1}{t} - \frac{1}{t \cdot \sqrt{t+1}}\right) = \frac{\pi}{2} \cdot \frac{1}{(1+\sqrt{t+1}) \cdot \sqrt{t+1}}.$$

Since $F(0) = 0$, we have

$$F(t) = \frac{\pi}{2} \cdot \int_0^t \frac{dt}{(1+\sqrt{t+1}) \cdot \sqrt{t+1}}.$$

We can calculate this integral by the substitution $v = \sqrt{t+1}$. We get that

$$F(t) = \pi \cdot \left[\log(1 + \sqrt{t+1}) - \log 2\right].$$

Notation

M. Laczkovich and V.T. Sós, *Real Analysis*, Undergraduate Texts in Mathematics, https://doi.org/10.1007/978-1-4939-7369-9

References

1. Arnold, V.I.: Mathematical Methods of Classical Mechanics. Graduate Texts in Mathematics, vol. 60, 2nd edn. Springer, New York (1989)
2. Edwards, R.E.: Fourier Series. Graduate Texts in Mathematics, vol. 64, 2nd edn. Springer, Berlin (1979)
3. Erdős, P., Surányi, J.: Topics in the Theory of Numbers. Springer, New York (2003)
4. Euler, : Foundations of Differential Calculus. Springer, New York (2000)
5. Hardy, G.H., Wright, E.M.: An Introduction to the Theory of Numbers, 4th edn. Clarendon Press, Oxford (1975)
6. Horn, R.A., Roger, A.: Matrix Analysis, 2nd edn. Cambridge University Press, Cambridge (2013)
7. Laczkovich, M., Sós, V.T.: Real Analysis. Foundations and Functions of One Variable. Undergraduate Texts in Mathematics, vol. 47. Springer, New York (2015)
8. Milnor, J.: Analytic proofs of the "hairy ball theorem" and the Brouwer fixed point theorem. Am. Math. Mon. **85**, 521–524 (1978)
9. Moise, E.E.: Geometric Topology in Dimensions 2 and 3. Graduate Texts in Mathematics, vol. 47. Springer, New York (1977)
10. Pritsker, I.E.: The Gelfond-Schnirelman method in prime number theory. Canad. J. Math. **57**, 1080–1101 (2005)
11. Riesz, F.: Sur le théorème de Jordan. Acta Sci. Math. **9**, 154–162 (1939). Reprinted in: Frédéric Riesz: *Oeuvres Complètes*, vol. I, pp. 173–181. Akadémiai Kiadó, Budapest (1960)
12. Rudin, W.: Principles of Mathematical Analysis. International Series in Pure and Applied Mathematics, 3rd edn. McGraw-Hill Book Co., New York (1976)
13. Thomassen, C.: The Jordan-Schönflies theorem and the classification of surfaces. Am. Math. Mon. **99**(2), 116–130 (1992)
14. Thorpe, J.A.: Elementary Topics in Differential Geometry. Undergraduate Texts in Mathematics. Springer, New York (1994). Corrected reprint of the 1979 original
15. Tolstov, G.P.: Fourier Series. Dover Publications, New York (1976)
16. Zygmund, A.: Trigonometric Series. Cambridge University Press, Cambridge (1993)
17. http://www.uam.es/personal_pdi/ciencias/cillerue/Curso/zeta2.pdf
18. http://mathworld.wolfram.com/RiemannZetaFunctionZeta2.html
19. http://claymath.org/millennium-problems/riemann-hypothesis

M. Laczkovich and V.T. Sós, *Real Analysis*, Undergraduate Texts in Mathematics, https://doi.org/10.1007/978-1-4939-7369-9

Index

M. Laczkovich and V.T. Sós, *Real Analysis*, Undergraduate Texts in Mathematics, https://doi.org/10.1007/978-1-4939-7369-9

Printed in the United States
By Bookmasters